WETLAND SOILS

SOILS

Genesis, Hydrology, Landscapes,
and Classification

WETLAND SOILS

Genesis, Hydrology, Landscapes, and Classification

Edited by

J. L. Richardson
M. J. Vepraskas

LEWIS PUBLISHERS

Boca Raton London New York Washington, D.C.

Library of Congress Cataloging-in-Publication Data

Catalog record is available from the Library of Congress

Visit the CRC Press Web site at www.crcpress.com

© 2001 by CRC Press LLC
Lewis Publishers is an imprint of CRC Press LLC

No claim to original U.S. Government works
International Standard Book Number 1-56670-484-7
1-56670-484-7Printed in the United States of America   3 4 5 6 7 8 9 0
Printed on acid-free paper

Preface

Anyone dealing with wetlands needs to understand the properties and functions of the soils found in and around wdetlands. The ability to identify wetland soils is at the core of wetland delineation. Wetland restoration revolves around techniques that are designed to restore the chemical reactions that occur in these soils. These chemical processes cause the soil to become anaerobic, and this condition requires special adaptations in plants if they are to survive in a wetland environment. "Wetland soils" is a general term for any soil found in a wetland. The term "hydric soil" was introduced by Cowardin et al. (1979) for wetland soils. Hydric soil has been redefined for jurisdictional purposes by the USDA's National Technical Committee for Hydric Soils as:

"... soils that are formed under conditions of saturation, flooding, or ponding long enough during the growing season to develop anaerobic conditions in the upper part" (Hurt et al., 1998).

Hydric soils are the principal subject of this text.

This book fills a large gap in the wetlands literature. No previous book has been devoted solely to the subject of hydric soils and their landscapes, hydrology, morphology, and classification. Several publications focus on a portion of the topics covered in this book, notably Mausbach and Richardson (1994); Richardson et al. (1994); Vepraskas (1994); Hurt et al. (1998); and Vepraskas and Sprecher (1997). The problem with each of these is that they are too focused and specialized to be used as texts for college level courses.

We assembled a team of scientists to develop a comprehensive book on hydric soils that could be used as a text in college courses and as a reference for practicing professionals. The text is intended for individuals who have, or are working toward, a B.S. degree in an area other than soil science. It is intended to prepare individuals to work with real wetlands outdoors, and all chapters have been written by individuals with extensive field experience.

The authors of this text describe a diverse range of soils that occur in and around wetlands throughout North America. These wetlands are widely recognized as consisting of three main components: hydric soils, hydrophytic vegetation, and wetland hydrology. We believe that the hydric soils are the most important component of the three. While most wetlands could be identified and their functions understood if the site's hydrology were known, an individual wetland's hydrology is far too dynamic for field workers to fully understand it without long-term monitoring studies. Some morphological aspects of hydric soils, however, can be used to evaluate a site's hydrology. As noted by Cowardin et al. (1979), soils are long-term indicators of wetland conditions. Soils can be readily observed in the field, and unlike hydrology, their characteristics remain fairly constant throughout the course of a year. They are not as readily altered as plants, which can be removed by plowing for example. The publications of Vepraskas (1992) on redoximorphic features and Hurt et al. (1998) on hydric soil field indicators have placed in the hands of field workers essential tools for delineation of soils into hydric and nonhydric categories. This book explains how soil morphology can be used as a field tool to evaluate soil hydrology and soil biogeochemical processes. A recurring theme in this text is that hydric soils are components of a landscape whose soils have been altered by hydrologic and biogeochemical processes.

We have organized the book into three parts. Part I examines the basic concepts, processes, and properties of aspects of hydric soils that pertain to virtually any hydric soil. We recognize that most users of this text will not be soil scientists, so the first chapter is a general overview that introduces important terms and concepts. The second chapter explains the historic development of the concept of a hydric soil, while the following chapters examine soil hydrology, chemistry, biology, soil organic matter, and the development and use of the hydric soil field indicators.

Part II of the text is devoted to the soils in specific kinds of wetlands. We have chosen to classify wetlands following Brinson's (1993) hydrogeomorphic model (HGM). This model considers hydrology and landscape as two dominant factors that create differences among wetlands and cause individual wetlands to vary in the types of functions they perform. Water is so dynamic that it is difficult to assess its role in wetlands unless long-term observations are made at various places in and around the wetland. Part III of the text is devoted to special wetland conditions that we feel need more emphasis, such as the wetland soils composed of sands, organic soils in northern North America, prairie wetlands in the midwestern U.S., wetlands in saline, dry climates, and wetlands with modified hydrology.

The terminology used throughout the text is that developed for the field of soil science. The soils discussed are described and classified according to the conventions of the USDA's Natural Resources Conservation Service (Soil Survey Staff, 1998). Common wetland terms, such as *fen, peatland,* or *pocosin,* are used only to illustrate a particular concept. We believe that most soil science terms are rigidly defined and are used consistently throughout the U.S. and much of the world. On the other hand, some of the common wetland terms (e.g., fen, bog) are defined differently across the U.S., while the exact meanings of others (e.g., peatland, pocosin) are not clear. While the terminology of the hydric soil field indicators (Hurt et al., 1998) may be new to many readers, each indicator is rigidly defined, field tested, and can be used to define a line on a landscape that separates hydric and upland soils.

J. L. Richardson
M. J. Vepraskas

REFERENCES

Brinson, M. M. 1993. *A Hydrogeomorphic Classification for Wetlands.* Tech. Rept. WRP-DE-4, U.S. Army Engineer Waterways Experiment Station, Vicksburg, MS.

Cowardin, L. M., V. Carter, F. C. Golet, and E. T. LaRoe. 1979. *Classification of Wetlands and Deepwater Habitats of the United States.* U.S. Fish and Wildlife Service, U.S. Government Printing Office, Washington, DC.

Hurt, G. W., P. M. Whited, and R. F. Pringle (Eds.). 1998. *Field Indicators of Hydric Soils in the United States.* USDA Natural Resources Conservation Service. Fort Worth, TX.

Mausbach, M. J. and J. L. Richardson. 1994. Biogeochemical processes in hydric soils. *Current Topics in Wetland Biogeochemistry* 1:68–127. Wetlands Biogeochemistry Institute, Louisiana State University, Baton Rouge, LA.

Richardson, J. L., J. L. Arndt, and J. Freeland. 1994. Wetland soils of the prairie potholes. *Adv. Agron.* 52:121–171.

Soil Survey Staff. 1998. *Keys to Soil Taxonomy.* 8th ed. USDA, Natural Resources Conservation Service, U.S. Government Printing Office, Washington, DC.

Vepraskas, M. J. 1992. *Redoximorphic Features for Identifying Aquic Conditions.* Tech. Bull. 301. North Carolina Agr. Res. Serv. Tech. Bull. 301, North Carolina State Univ., Raleigh, NC.

Vepraskas, M. J. and S. W. Sprecher (Eds.). 1997. *Aquic Conditions and Hydric Soils: The Problem Soils.* SSSA Spec. Publ. No. 50, Soil Science Society of America, Madison, WI.

About the Editors

J. L. Richardson is professor of soil science at North Dakota State University in Fargo and is a frequent consultant for wetland soil/water problems for government and industry.

Dr. Richardson received his Ph.D. from Iowa State University in soil genesis, morphology, and classification. He is a member of the American Society of Groundwater Scientists and Engineers, the National Water Well Association, the North Dakota Professional Soil Classifiers, the Society of Wetland Scientists, the Soil Science Society of America, and the National Technical Committee for Hydric Soils. He is author of over 80 peer-reviewed or edited articles related to wetlands, wet soils, or water movement in landscapes.

M. J. Vepraskas is professor of soil science at North Carolina State University in Raleigh where he conducts research on hydric soil processes and formation. He currently works with consultants and government agencies on solving unique hydric soil problems throughout the U.S.

Dr. Vepraskas received his Ph.D. from Texas A & M University. He is a member of the American Association for the Advancement of Science, American Society of Agronomy, International Society of Soil Science, North Carolina Water Resources Association, Soil Science Society of North Carolina, Society of Wetland Scientists, and the National Technical Committee for Hydric Soils. He is a Fellow of the Soil Science Society of America. In 1992, he authored the technical paper, "Redoximorphic Features for Identifying Aquic Conditions," which has become the basis for identifying hydric soils in the U.S.

Contributors

J. L. Arndt
Petersen Environmental, Inc.
1355 Mendota Heights Rd.
Mendota Heights, MN

Jay C. Bell
Department of Soil, Water, and Climate
University of Minnesota
St. Paul, MN

Janis L. Boettinger
Department of Plants, Soils, and
  Biometeorology
Utah State University
Logan, UT

Scott D. Bridgham
Department of Biological Sciences
University of Notre Dame
Notre Dame, IN

Mark M. Brinson
Biology Department
East Carolina University
Greenville, NC

V. W. Carlisle
Professor Emeritus
Soil and Water Science Department
University of Florida
Gainesville, FL

Mary E. Collins
Soil and Water Science Department
University of Florida
Gainesville, FL

Christopher B. Craft
School of Public and Environmental Affairs
Indiana University
Bloomington, IN

R. A. Dahlgren
Soils and Biogeochemistry
Department of Land, Air, and Water Resources
University of California
Davis, CA

C. V. Evans
Department of Geology
University of Wisconsin-Parkside
Kenosha, WI

S. P. Faulkner
Wetland Biogeochemistry Institute
Louisiana State University
Baton Rouge, LA

J. A. Freeland
Northern Ecological Services, Inc.
Reed City, MI

Willie Harris
Soil and Water Science Department
University of Florida
Gainesville, FL

W. A. Hobson
Urban Forester
City of Lodi
Lodi, CA

G. W. Hurt
National Leader for Hydric Soils
USDA, NRCS
Soil and Water Science Department
University of Florida
Gainesville, FL

Carol A. Johnston
Natural Resources Research Institute
University of Minnesota
Duluth, MN

R. J. Kuehl
Soil and Water Science Department
University of Florida
Gainesville, FL

David L. Lindbo
Soil Science Department
North Carolina State University
Plymouth, NC

Maurice J. Mausbach
Soil Survey and Resource Assessment
USDA Natural Resources Conservation
 Service
Washington, DC

J. A. Montgomery
Environmental Science Program
DePaul University
Chicago, IL

W. Blake Parker
Hydric Soils
Verona, MS

Chein-Lu Ping
University of Alaska — Fairbanks
Agriculture and Forestry Experiment
 Station
Palmer Research Center
Palmer, AK

M. C. Rabenhorst
Department of Natural Resource Sciences
University of Maryland
College Park, MD

J. L. (Jimmie) Richardson
Department of Soil Science
North Dakota State University
Fargo, ND

S. W. Sprecher
U.S. Army Corps of Engineers
South Bend, IN

J. P. Tandarich
Hey & Associates
Chicago, IL

James A. Thompson
Department of Agronomy
University of Kentucky
Lexington, KY

Karen Updegraff
Natural Resources Research Institute
Duluth, MN

M. J. Vepraskas
North Carolina State University
Department of Soil Science
Raleigh, NC

Frank C. Watts
USDA, Natural Resources Conservation
 Service
Baldwin, FL

P. M. Whited
Natural Resources Conservation Service
Wetland Science Institute
Hadley, MA

We dedicate this book to the following Unsung Heroes

The development of the concept of hydric soils, as well as the procedures used to identify them, were developed over a period of at least 40 years with contributions coming from many people as part of the national soil survey program. Early work on hydric soils began with soil scientists working for the USDA's Soil Conservation Service, which is now the Natural Resources Conservation Service. These field scientists evaluated soils in wetlands as part of the national program to map soils in the U.S. However, a few people, and a few people only, brought the idea of hydric soils and their value forward nationally. We recognize below three individuals who were instrumental in developing and improving how hydric soils are identified in the U.S.

Dr. Warren C. Lynn was among the first to study the landscape processes that form hydric soils. Dr. Lynn is a Research Soil Scientist for the National Soil Survey Laboratory in Lincoln, NE. He received his B.S. and M.S. degrees from Kansas State University, and his Ph.D. in Soil Science from the University of California at Davis. Dr. Lynn's research has been focused in the areas of pedology that support the National Cooperative Soil Survey. Specifically he has worked on Histosols, Vertisols, and on improving methods to evaluate the minerals in soils. His contributions to wetland soils center on his development of the USDA's Wet Soils Monitoring Project. In cooperation with universities throughout all portions of the U.S., Dr. Lynn began scientific studies to monitor landscapes that are documenting the morphology, water table fluctuations, and oxidation–reduction dynamics of key hydric soils in a landscape setting. The network of monitoring stations has been expanded over the years to cover soils in eight states across the U.S. These data represent the quantitative science backbone for development of the hydric soil field indicators that are now used to identify hydric soils in the U.S. Dr. Lynn quietly altered our thinking from profile hydrology to landscape hydrology.

W. Blake Parker formulated the concept of hydric soils and developed the field criteria for their identification. Blake is a graduate of Auburn University. From 1977 to 1984 Blake, as an employee of the USDA Natural Resources Conservation Service (then Soil Conservation Service), worked with the U.S. Army Corps of Engineers, U.S. Environmental Protection Agency, and the U.S. Fish and Wildlife Service to develop the methodology needed for delineation of wetlands based on hydric soils and hydrophytic vegetation. He then worked with the National Wetlands Inventory project as a soil scientist for 4 years. He developed the first definition of hydric soils and the first National List of Hydric Soils. Later he was assigned to the U.S. Army Corps of Engineers Waterways Experiment Station and advised their research programs on wetland soils and hydrology.

He served as a long-time member of the National Technical Committee for Hydric Soils, which is the body responsible for defining and identifying the hydric soils in the U.S.

DeWayne Williams will be remembered as a teacher who trained many of the USDA's soil scientists in how to use field indicators to mark hydric soil boundaries. His training forced soil mappers to recognize that hydric soil identification had to use different procedures that were more precise than those used to prepare soil maps for the national soil survey program. DeWayne worked as a soil scientist for the USDA's Natural Resources Conservation Service for more than 40 years. He earned a B.S. degree in Soil Science from Texas A&M University. DeWayne's contributions include surveying soils in the U.S., India, Russia, Mexico, Canada, China, North Korea, and Puerto Rico. He has contributed to hydric soils in the U.S. by developing rigid standards for describing the soil morphology and landscapes of hydric soils. He recognized early that hydric soils could be identified by key characteristics that occurred at specific depths in the soil. He was also a major early worker in the development of regional hydric soil indicators. He was a charter member of the National Technical Committee for Hydric Soils, and served on the Committee for 10 years.

From 1991 until his retirement in 1996, DeWayne worked almost full time training USDA and Corps of Engineers wetland delineators in hydric soil identification. DeWayne now spends considerable time trying to increase food production in North Korea.

Both editors salute these scientists as the pathfinders who started us on the trail that led to this book. We owe them more than we can say in words for their personal and professional contributions. … Thanks ever so much!

Contents

Part III. Wetland Soils with Special Conditions

PART I

Basic Principles of Hydric Soils

CHAPTER 1

Basic Concepts of Soil Science

S. W. Sprecher

INTRODUCTORY OVERVIEW OF SOIL

This chapter provides an introduction to soil description in the field, soil classification, and soil survey. The terminology and approach used are those of the Soil Survey Staff of the U.S. Department of Agriculture Natural Resources Conservation Service (USDA–NRCS), the federal agency with primary responsibilities for defining and cataloging hydric soils in the U.S. Topics covered include the information necessary to complete the soils portion of wetland delineation forms and some common soil science terminology that experience has shown may be misunderstood by wetland scientists who have had no formal training in soil science.

The various disciplines that study soils define "soil" according to how they use it. Civil engineers emphasize physical properties; geologists emphasize degree of weathering; and agriculturalists focus on the properties of soil as a growth medium. "Pedology" is the branch of soil science that studies the components and formation of soils, assigning them taxonomic status, and mapping and explaining soil distributions across the landscape. It provides the perspective from which the USDA Soil Survey Program regards soils and is also the perspective of this book. A pedologic definition of soil is:

> The unconsolidated mineral or organic matter on the surface of the earth that has been subjected to and shows the effects of genetic and environmental factors of: climate (including water and temperature effects), and macro- and microorganisms, conditioned by relief, acting on parent material over a period of time. The product-soil differs from the material from which it is derived in many physical, chemical, biological, and morphological properties and characteristics. (Soil Science Society of America, 1997.)

Here soil is seen to have natural organization and to be biologically active. This inherent organization results from climatic and biological forces altering the properties of the materials of the earth's surface. Because these soil-forming forces exert progressively less influence with depth, they result in more or less horizontal layers that are termed "soil horizons" (Figure 1.1). Individual kinds of soil are distinguished by their specific sequence of horizons, or "soil profile." The characteristics and vertical sequences of these soil horizons vary in natural patterns across the landscape.

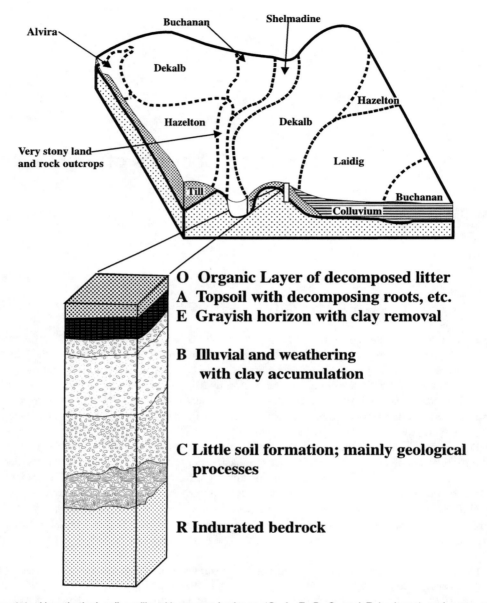

Figure 1.1   Hypothetical soil profile with master horizons (O, A, E, B, C, and R horizons) and surrounding landscape, including other mapped soils on the landscape (dashed lines). (Adapted from Lipscomb, G. H. 1992. *Soil Survey of Monroe County, Pennsylvania*. USDA–SCS in cooperation with the Penn. State Univ. and Penn. Dept. Envir. Resources, U.S. Govt. Printing Office, Washington, DC.)

Organic Soils and Mineral Soils

There are two major categories of soils, organic soils and mineral soils, which differ because they form from different kinds of materials. Organic soil forms from plant debris. These soils are found in wetlands because plant debris decomposes less rapidly in very wet settings. Organic soils are very black, porous, and light in weight, and are often referred to as "peats" or "mucks."

Mineral soils, on the other hand, form from rocks or material transported by wind, water, landslide, or ice. Consequently, mineral soil materials consist of different amounts of sand, silt,

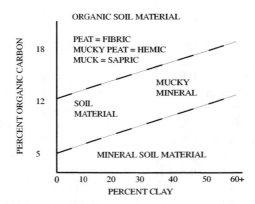

Figure 1.2 Levels of clay and organic carbon that define distinctions between organic and mineral soil materials (bold line). An uncommon but important subset of mineral materials is "mucky mineral" soil materials (carbon and clay contents between the dashed and bold lines). (USDA–NRCS. 1998. Field indicators of hydric soils in the United States, version 4.0. G.W. Hurt, P.M. Whited, and R.F. Pringle (Eds.) USDA–NRCS, Fort Worth, TX.)

and clay, and constitute the majority of the soils in the world. They occur both within and outside of wetlands.

Distinguishing between organic and mineral soils is important, because the two categories are described and classified differently. In practice, mineral and organic soils are separated on the basis of organic carbon levels. The threshold carbon contents separating organic and mineral soils are shown in Figure 1.2. Organic matter concentrations above these levels dominate the physical and chemical properties of the soil. It is extremely difficult to estimate organic carbon content in the field unless you train yourself using samples of known carbon concentration. In general, if the soil feels gritty or sticky, or resists compression, it is mineral material; if the soil material feels slippery or greasy when rubbed, has almost no internal strength, and stains the fingers, it may be organic. Highly decomposed organic material is almost always black; brownish horizons without discernible organic fibers are almost always mineral.

The USDA–NRCS currently recognizes three classes of organic matter for field description of soil horizons: sapric, hemic, and fibric materials. Differentiating criteria are based on the percent of visible plant fibers observable with a hand lens (i) in an unrubbed state and (ii) after rubbing between thumb and fingers 10 times (Table 1.1). "Sapric," "hemic," and "fibric" roughly correspond to the older terms "muck," "mucky peat," and "peat," respectively. Complete details on identifying sapric, hemic, and fibric materials are given in Chapter 6.

Table 1.1 Percent Volume Fiber Content of Sapric, Hemic, and Fibric Organic Soil Horizons

Horizon Descriptor	Horizon Symbol	Proportion of Fibers Visible with a Hand Lens	
		Unrubbed	Rubbed
Sapric	Oa	< 1/3	< 1/6
Hemic	Oe	1/3–2/3	1/6–2/5
Fibric	Oi	> 2/3	> 2/5

From Soil Survey Staff. 1975. *Soil Taxonomy: A Basic System of Soil Classification for Making and Interpreting Soil Surveys.* USDA–SCS Agric. Handbook 436, U.S. Govt. Printing Office, Washington, DC.

Soil Horizons

As previously noted, soils are separated largely on the basis of the types of horizons they have and the horizons' properties. Horizons, in turn, are differentiated from each other by differences in organic carbon content, morphology (color, texture, etc.), mineralogy, and chemistry (pH, Fe redox status, etc.). Most people are aware that mineral soils have a dark, friable topsoil and lighter colored, firmer subsoil. Below the subsoil is geologic material that has not yet weathered into soil; this may be alluvium, decomposed rock, unweathered bedrock, or other materials. In very general terms, pedologists call the topsoil in mineral soils the "A horizon," the subsoil the "B horizon," the underlying parent material the "C horizon," and unweathered rock, the "R horizon" (Figure 1.1). Pedologists also recognize a light-colored "E horizon" that may be present between the A and B horizons. Organic soils contain organic horizons ("O horizons"). Each kind of master horizon (A, B, C, E, and O horizon) is usually subdivided into different subhorizons. The approximately 20,000 named soils in the United States are differentiated from each other on the basis of the presence and sequence of these different subhorizons, as well as external factors such as climate, hydrologic regime, and parent material. Pedologists study the earth's surface to a depth of about 2 meters; parent material differences at greater depths usually are not considered.

SOIL DESCRIPTIONS FOR WETLAND DELINEATION FORMS

When describing soils, wetland delineators need to include the following features in their soil descriptions: horizon depths, color, redoximorphic features (formerly called mottles), and an estimate of texture. These important soil characteristics change with depth and help differentiate horizons within the soil profile. Other features, too, should be described if pertinent to the study at hand. Formal procedures for describing soils can be found in the *Soil Survey Manual* (Soil Survey Division Staff, 1993) and the *Field Book for Describing and Sampling Soils* (Schoeneberger et al. 1998).

The soil surface is frequently covered by loose leaves and other debris. This is not considered to be part of the soil and is scraped off. Below this layer the soil may contain organic or mineral soil material. If organic material is present, the soil surface begins at the point where the organic material is partially decomposed.

The depth of the top and bottom of each horizon is recorded when describing soils; the top of the first horizon is the soil surface. Subsequent horizons are distinguished from those above by change in soil color, texture, or structure, or by changes in presence or absence of redoximorphic features.

Soil Colors

The most obvious feature of a soil body or profile is its color. Because the description of color can be subjective, a system to standardize color descriptions has been adopted. The discipline of soil science in the United States uses the Munsell color system to quantify color in a standard, reproducible manner. The *Munsell® Soil Color Charts* (GretagMacbeth, Munsell Corporation, 1998) will be used here to explain soil color determination in the field because most U.S. soil scientists are more familiar with the traditional format than with more recent, alternative formats.

The *Munsell Soil Color Charts* are contained in a 15×20-cm 6-ring binder of 11 pages, or charts. Each chart consists of 29 to 42 color chips. The Munsell system notes three aspects of color, in the sequence "Hue Value/Chroma," for example, 10YR 4/2 (Plate 1). All the chips on an individual chart have the same hue (spectral color). Within a particular hue — that is, on any one color chart — values are arrayed in rows and chromas in columns. Hue can be thought of as the quality of pigmentation, value the lightness or darkness, and chroma the richness of pigmentation (pale to bright).

Specifically, hue describes how much red (R), yellow (Y), green (G), blue (B), or purple (P) is in a color. Degree of redness or yellowness, etc., is quantified with a number preceding the letter, e.g., 2.5Y. Most soil hues are combinations of red and yellow, which we perceive as shades of brown. These differences in hue are organized in the Munsell color charts from reddest (10R) to yellowest (5Y), with the chips of each hue occupying one page of the charts. The sequence of charts, from reddest to yellowest, is as follows (also, see Plate 2):

10R	2.5YR	5YR	7.5YR	10YR	2.5Y	5Y

Reddest	red-yellow mixes	Yellowest

When determining soil hue from the Munsell charts, it is helpful to ask yourself if the soil sample is as red or redder than the colors on a particular page of the charts. Most soils in the United States have 10YR hues, so start with that chart unless your local soil survey report describes most soils as having a different hue.

Subsoils containing minerals with reduced iron (Fe(II)) may be yellower or greener than hue 5Y. Such colors are represented on the color charts for gley, or the "gley pages" (Plate 2). These have neutral hue (N) or hues of yellow (Y), green (G), blue (B), or purple (P). Soil horizons with colors found on the gley charts are generally saturated with water for very long periods of time and may be found in wetlands (Environmental Laboratory 1987; USDA–NRCS 1998).

Value denotes darkness and lightness, or simply the amount of light reflected by the soil or a color chip. For instance, the seven chips in column 2 of the 10YR chart (Plate 1) each have different values, but all have chroma of 2 and hue of 10YR. A-horizon colors usually have low value (very dark to black) because of staining by organic matter. Colors of hydric soil field indicators (Chapter 8 of this book) frequently need to be determined below the zone of organic staining where values are higher than 3 or 4; the exceptions are when a hydric soil feature is made up of organic matter, or when organic staining continues down the soil profile for several decimeters (Chapter 8).

Chroma quantifies the richness of pigmentation or concentration of hue. High-chroma colors are richly pigmented; low-chroma colors have little pigmentation and are dull and grayish. Chromas are columns on the color charts (Plate 1). Note how the colors on the left seem to be more dull and washed out than those on the right of the color chart. B horizons (subsoils) that are waterlogged and chemically reduced much of the year have much of their pigment "washed out" of them; like the low-chroma color chips, they too are grayish.

Soil colors seldom match any Munsell color chip perfectly. Standard NRCS procedures require that Munsell colors be read to the nearest chip and not be interpolated between chips. Recent NRCS guidance for hydric soil determination, however, requires that colors be noted as equal to, greater than, or less than critical color chips (USDA–NRCS 1998; see also Chapter 8). Colors should not be extrapolated beyond the range of chips in the color book.

Because soil colors vary with differences in light quality, moisture content, and sample condition, samples should be read under standard conditions. Color charts are designed to be read in full, mid-day sunlight, because soils appear redder late in the day than they do at mid-day. The sun should be at your back so the sunlight strikes the soil sample and color chips at a right (90 degree) angle. Sunglasses should not be worn when reading soil colors because their lenses remove parts of the color spectrum from the light reaching the eye.

Wetland delineators should describe soils on the basis of moist colors. To bring a soil specimen to the moist state, slowly spray water onto the sample until it no longer changes color. The soil is too wet if it glistens and should be allowed to dry until its surface is dull. The soil specimen should be gently broken open, and the color read off the otherwise undisturbed, open face. Both the inside and outsides of natural soil aggregates can be read this way.

Matrix and Special Features

The predominant color of a soil horizon is known as its matrix color, that is, the color that occupies more than half the volume of the horizon. If a horizon has several colors and none occupies 50% of the volume, the investigator should describe the various colors and report percent volume for each. Often soil aggregates have different colors outside and inside; these, too, should be noted separately.

Mottles are small areas that differ from the soil matrix in color. Mottles that result from waterlogging and chemical reduction are now called "redoximorphic features" (Soil Survey Staff 1992). These features are listed as part of the field indicators for hydric soils and should be described carefully when filling out wetland data sheets (Chapter 8; Vepraskas 1996). Chemical reduction is not the only source of color differences within the soil. Other causes of color differences within a horizon include recently sloughed root material (often reddish), root decomposition (very dark grey to black), decomposition of pebbles or rocks (usually an abrupt, strong contrast with the surrounding matrix), and carbonate accumulation (white).

The USDA–NRCS soil sampling protocols require a description of mottle color, abundance, size, contrast, and location (Soil Survey Division Staff 1993, pp. 146–157; Vepraskas 1996). Colors of redoximorphic features should be described with standard Munsell notation. Classes of abundance, size, and contrast are found in Table 1.2. Abundance is the percent of a horizon that is occupied by a particular feature. Abundance should be determined using diagrams for estimating proportions of mottles; these usually accompany commercial soil color books and can also be found in the USDA–NRCS literature (Soil Survey Division Staff 1993). Most people overestimate the abundance of mottles without the use of some aid.

Color contrast is how much the mottle colors differ from the matrix color. The appropriate terms are "faint" (difficult to see), "distinct" (easily seen), or "prominent" (striking, obvious). Quantitative definitions of these terms are presented in Table 1.2 and Figure 1.3. It is also useful to note if redoximorphic features are oriented in some specific way, such as along root channels, on faces of fracture planes, etc. (see Chapter 7 for further details).

Table 1.2   Abundance, Size, and Contrast of Mottles

Mottle Abundance[1]		Mottle Size[1]	
Few	<2%	Fine	<5 mm
Common	2–20%	Medium	5–15 mm
Many	>20%	Coarse	>15 mm

Mottle Contrast[2] (see also Figure 1.3)			
	Hues on Same Chart (e.g., both colors 10YR)	**Hue Difference on Chart (e.g., 10YR vs. 7.5YR)**	**Hue Difference Two Charts or More (e.g., 10YR vs 5YR)**
Faint	≤2 units of value, and ≤1 unit of chroma	≤1 unit of value and ≤1 unit of chroma	Hue differences of 2 or more charts are distinct or prominent
Distinct	Between faint and prominent	Between faint and prominent	0 to <2 units of chroma and/or value
Prominent	At least 4 units in value and/or chroma	At least 3 units in value and/or chroma	At least 2 units in value and/or chroma

[1] From Soil Survey Division Staff. 1993. *Soil Survey Manual.* USDA–SCS Agric. Handbook 18. U.S. Govt. Printing Office, Washington, DC.
[2] USDA–NRCS. 1998. Field indicators of hydric soils in the United States, Version 4.0. G. W. Hurt, P. M. Whited, and R. F. Pringle (Eds.) USDA–NRCS, Forth Worth, TX.

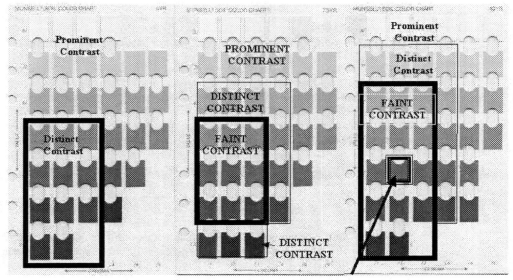

Matrix Color

Figure 1.3   Contrast for redoximorphic features, with respect to a matrix color of 10YR 4/2. The left-hand figure depicts ranges of distinct and prominent color contrasts when the redoximorphic feature is two or more color charts redder than the matrix. (For example, soil matrix is 10YR 4/2, and the redoximorphic feature has hue of 5YR.) The middle figure depicts ranges of faint, distinct, or prominent contrast for features that are one hue page different from the matrix. (For example, soil matrix is 10YR 4/2, and the redoximorphic feature has hue of 7.5YR.) The right-hand figure depicts ranges of faint, distinct, or prominent color contrast for features that differ in value and/or chroma but share a common hue. (For example, matrix is 10YR 4/2, and the redoximorphic feature is also 10YR but differs in value and/or chroma.) (Adapted from Sprecher, S. W. 1999. Using the NRCS hydric soil indicators with soils with thick A horizons. WRP Tech. Note SG-DE-4.1. U.S. Army Engineer Waterways Experiment Station, Vicksburg, MS.)

Soil Texture

Wetland delineation protocols of the USDA–NRCS (1998) and the U.S. Army Corps of Engineers (Environmental Laboratory 1987) both require that wetland scientists distinguish between sandy and non-sandy soils. It is further recommended that more precise estimates of relative sand, silt, and clay contents be recorded on soil data forms because these affect so many other properties of soils. The relative percentage of sand, silt, and clay is referred to as "soil texture." Soil texture is usually recorded for each soil horizon. Unfortunately, the terminology of soil texture is confusing because some of the same terms are used to describe both (1) individual soil particles and (2) mixtures of particles (e.g., sand, silt, and clay).

Individual earthen particles range in size from boulders to microscopic clay crystals. Soil textures are determined on the basis of particles having diameters of 2 mm and smaller. The USDA soil texture system identifies three classes of particles: sand, silt, and clay (Table 1.3). A fourth class, coarse fragments, is also recognized (i.e., gravels >2 mm, rocks, etc.), but coarse fragments are disregarded when determining the USDA texture of a soil.

Sand particles feel at least slightly gritty when rubbed between the fingers. Silt materials feel like flour when rubbed. Most clays feel sticky when rubbed. Sand and silt particles tend to be roughly spheroidal, with either smooth or rough edges. Clay particles are mostly flat and plate-like; they have a large surface area that influences soil chemical characteristics. Notice that there is no such thing as a "loam" particle. "Loam" is the name for a mixture of particles of different sizes.

Table 1.3 Sizes of Soil Particle Classes

Class	Size
Sand	0.05–2 mm
Silt	0.002–0.05 mm
Clay	<0.002 mm (<2 microns)
Coarse fragments (not considered for soil texture analysis)	>2 mm

From Soil Survey Division Staff. 1993. *Soil Survey Manual.* USDA–SCS Agric. Handbook 18. U.S. Govt. Printing Office, Washington, DC.

Only rarely do natural soils consist entirely of one size class of particles; most of the time sand, silt, and clay are present in a mixture in soil horizons. The USDA defines twelve different combinations, called textural classes, for describing and classifying soils by texture (Figure 1.4). All percentages are on a dry weight basis.

Notice in Figure 1.4 that "sand," "silt," and "clay" are names of both individual particles and soil textures. If a soil sample is >90% sand- or silt-sized particles, the texture of the sample is named "sand" or "silt," respectively, after the dominant size fraction. However, less than half of the mass of a soil can be clay-sized particles and the material may still be called "clay"; this is because of the dominant influence of clay particles on overall soil properties.

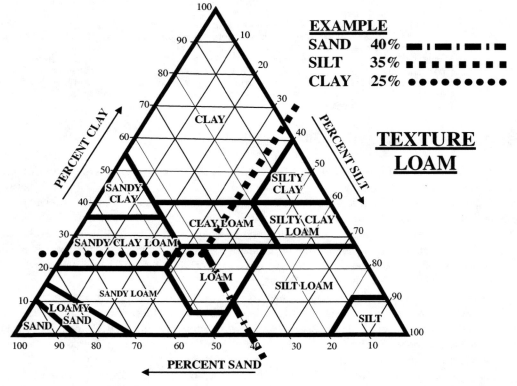

Figure 1.4 Soil texture triangle with example of a loam soil sample. Read 40% sand-sized particles along the bottom axis from right to left and follow the 40% line upward at 60 degrees to the left; "25% clay-sized particles" is read off the clay axis on the left side of the triangle, and "35% silt-sized particles" is read off the right axis. These three lines intersect in the "loam" area of the triangle, so the sample has a loam textural classification. (Adapted from Sprecher, S. W. 1991. Introduction to hydric soils, instructional slide set. U.S. Army Engineer Waterways Experiment Station, Vicksburg, MS.)

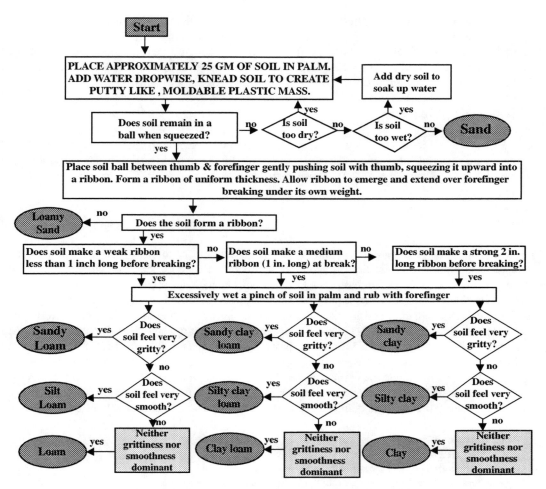

Figure 1.5 Flow chart for estimating soil texture by feel. To estimate soil texture, first wet the soil in the palm of your hand to its state of greatest malleability. It may take a couple of minutes of manipulation to wet the smaller clay aggregates. If the soil gets too wet and puddles, just add more dry soil and rework it to optimum malleability. After the soil is adequately moistened, follow the flow chart by trying to make a ball and then a ribbon of the soil. A soil's ability to hold a ribbon shape reflects its clay content. Grittiness or smoothness of the ribboned soil indicates high content of sand or silt, respectively. Note that no provision has been made for the texture "silt." This omission is not serious because pure silt is uncommon and the difference between silt and silt loam is inconsequential in most routine wetlands work. (Adapted from Thien, S. J. 1979. A flow diagram for teaching texture-by-feel analysis. *J. Agron. Ed.* 8:54–55.)

With training and practice, soil scientists can learn to estimate soil textures in the field by rubbing a moistened soil sample between their fingers and testing for properties such as ductility, grittiness, smoothness, stickiness, resistance to pressure, and cohesiveness. This art is locally specific because of regional variations in organic matter and clay mineralogy. Wetland investigations for regulatory purposes, however, usually do not require the field accuracy necessary for soil mapping. Routine wetland investigations should record whether a soil horizon is generally sandy, silty, clayey, or loamy, even if the notes are accompanied by a disclaimer about accuracy. This level of accuracy can be achieved by using a widely accepted flow chart for estimating soil textures (Figure 1.5).

The boundary between sandy and loamy hydric soil indicators (USDA–NRCS 1998; Chapter 8) is the boundary between loamy fine sand (sandy soil indicators) and loamy very fine sand (loamy and clayey soil indicators). A rule of thumb for determining whether the indicators for "sandy" or "loamy or clayey" soils should be used is to take a moist soil sample and roll it into a 1-in. ball. Drop the ball into the palm of your hand from a height of about 25 cm (10 in.). If no ball can be formed or if the ball falls apart when dropped, then use the indicators for sandy soils. If the ball stays intact after dropping, use the indicators for "loamy or clayey" soils.

Mucky Mineral Textures

When the organic matter content of a mineral soil horizon is intermediate between organic and mineral soil materials, it is said to have a "mucky modified mineral texture," such as mucky sand or mucky sandy loam (Figure 1.2). These textures can only be learned by practicing with soil samples of known contents of clay and organic carbon.

Other Features

Formal soil descriptions include numerous distinctions in addition to horizon color, texture, and redoximorphic features. Soil structure is a property that describes the aggregation of soil particles, and the presence of large cracks and root channels. It affects root growth and water movement through the soil. Terminology for soil structure is based on the concept of the natural soil aggregate (soil "ped") and its size, shape, and strength of expression. The details are beyond the scope of this chapter but are available in standard soils texts and NRCS publications (Soil Survey Staff 1975, pp. 474 to 476; Soil Survey Division Staff 1993, pp. 157 to 163). Wetland delineation data sheets seldom require that soil structure be noted, but often redoximorphic features are found at horizon boundaries where water temporarily perches, such as at the contact where a horizon with well-developed structure overlies a horizon with minimal structure.

Other features that wetland delineators should be aware of when describing hydric soils include:

- Density of roots (especially abrupt changes in root density)
- Compacted or cemented layers (such as plow or traffic pans)
- Different kinds of iron or manganese segregations
- External factors such as geomorphic position, water table depth, etc.

The details of these features are described in professional soils publications (Soil Survey Staff, 1975, especially Appendix I; Soil Survey Division Staff 1993, pp. 59-196; and Vepraskas 1996).

KINDS OF SOIL HORIZONS

Soil scientists use the features discussed above to characterize individual soil horizons down through the soil profile; the major layers ("master horizons") recognized by U.S. pedologists (soil scientists) are O, A, E, B, C, and R horizons (Figure 1.1). Pedology distinguishes several varieties of each of the master horizons; the most significant of these subordinate horizons for the purposes of wetland science are listed in Table 1.4. Few soils have all of the master horizons, and probably no soil has all of the subhorizons listed in Table 1.4. The wetland delineator usually inspects soil to 50 cm and, therefore, usually sees only the O and A horizons and the top of the E or B horizons, if present. A trained soil scientist, on the other hand, generally wants to investigate lower horizons as well, in order to understand the relation of the surface horizons to the landscape and its hydrology.

Organic horizons (O horizons) are most prominent in organic soils, or "Histosols." In large closed depressions (for example, bogs or pocosins), organic horizons may form a bed of peat or muck 1 m or more thick; 40 cm is the minimum thickness of peat or muck required for a soil to

Table 1.4   Subordinate Horizons of Greatest Significance to Wetland Science

Horizon	Significance
Oi	Fibric organic matter (little decomposition)
Oe	Hemic organic matter (intermediate decomposition)
Oa	Sapric organic matter (high decomposition)
Ap	Plowed A horizon
Bw	Weathering, weakly developed B horizon
Bt	Increase in illuvial clay in B horizon
Bg	Gleying significant
Btg	Increase in illuvial clay and significant gleying
Bh	Humus rich subsoil, spodic horizon

be classified as a Histosol. Subordinate distinctions among O horizons are the Oi horizon (fibric, or, little decomposed), Oa horizon (sapric, or highly decomposed), and Oe horizon (hemic, of intermediate decomposition) (Table 1.1). Although organic horizons are not present in most mineral soils, when present as 1- to 2-cm thick Oa horizons, they can be important hydric soil field indicators (Chapter 8). When they do occur in mineral soils, it is usually at the soil surface, unless the soil is buried by mineral matter washed in from flooding or upslope erosion.

In most soils, the uppermost mineral layer, or topsoil, is referred to as the A horizon. It is important to recognize the A horizon because hydric soils are usually identified from features immediately below it. The A horizon is usually the darkest layer in the soil (moist value/chroma of 3.5/2 or darker in most hydric soil situations). It usually has more roots and is more friable or crumbly than lower horizons. These distinctions result from biological activity and organic matter accumulation in the A horizon. Most natural A horizons vary in thickness from approximately 5 to 30 cm, but some are thicker. Plowing may obscure A horizon features because of mixing with subsoil materials. Plowed soil surfaces are referred to as "Ap" horizons. Ap horizons can be identified by the abrupt, sharp lower boundary at the depth of a plow blade — generally 15 to 25 cm, depending on local agricultural practices.

The E horizon, when present, is a layer from which clay and iron oxides have been leached ("eluviated"). The E horizon is typically lighter in color than the rest of the soil above and below, usually gray to white. It is important to recognize E horizons because their low chroma and high value can be mistaken for evidence of wetness and Fe reduction. Many, but not all, E horizons have a texture similar to that of the A horizon. E horizons of hydric soils typically contain redox concentrations (i.e., reddish mottles). E horizons are underlain by a layer having a higher content of clay (Bt horizon) or transported organic material (Bh horizon).

The B horizon is the layer of most obvious mineral weathering. The B horizon is also the layer into which material translocates from the overlying E and A horizons. The B horizon has soil peds (coherent aggregates) unless the soil is nearly pure sand. Upland B horizons have the colors (generally browns) of the iron minerals that weather out of the original parent material. Wetland B horizons are grayer due to reduction and removal of iron pigmenting minerals (see Bg horizons below and Chapter 7).

Most hydric soils have a subsoil horizon that is seasonally anaerobic due to high water tables and chemical reduction. This is termed a Bg horizon; the "g" indicates processes of "gleying," that is, chemical reduction of iron or manganese. Matrix colors of Bg horizons are usually gray, with chromas of 2 or less and values of 4 or more, usually with redox concentrations (reddish mottles); Bg colors are not restricted to the Munsell gley charts. Not all Bg horizons are indicative of hydric soils; for example, deeper water tables may create Bg horizons below a depth of 30 cm, which is not shallow enough for the soil to be hydric.

Bt horizons are zones where clay accumulates from above ("illuviates"), often from an E horizon. The increased clay content is significant to hydric soils because water can perch in and on top of Bt horizons and cause redoximorphic features to form. Such perched water, however, may not be present long enough or frequently enough to cause the formation of gray matrices or hydric soils.

Bh horizons are the dark subsoil horizons often found in sandy soils under coniferous vegetation, especially in the Southeast Coastal Plain and in glacial outwash plains. Their morphology is distinctive: they almost always underlie a white E horizon; they are black to dark reddish brown; and their boundaries with horizons above and below are usually very sharp. The *Corps of Engineers Wetlands Delineation Manual* (Environmental Laboratory 1987) refers to these as "organic pans" and considers them to be a hydric soil indicator; after a decade of investigations, it has been learned that Bh horizons are not necessarily diagnostic for hydric soils (USDA–NRCS 1998).

The geologic material in which soils form is termed the C horizon, if unconsolidated, or R horizon, if bedrock. Many soils in fluvial settings have only an A and a C horizon, entirely lacking O, E, and B horizons. C horizons retain the structure and color of the original parent material. In the fluvial setting, the C horizon would retain evidence of sedimentary stratification, whereas B horizons in the same setting would have developed enough structure that the boundaries between depositional strata are obliterated. Few wetlands have an R horizon because most depressional areas are deeper than 2 m to bedrock.

SOIL TAXONOMY

Soil Taxonomy (Soil Survey Staff 1975, 1998) is the most comprehensive classification system used to catalog soils in the United States. Wetland scientists need to be familiar with the highest level of the system and with a handful of subordinate distinctions in order to understand concepts and terminology in the hydric soils literature.

Soil Taxonomy is a hierarchical taxonomy with six levels (Order, Suborder, Great Group, Subgroup, family, and series; see Table 1.5). The highest level is comprised of twelve soil Orders (Table 1.6); soil Orders are based on fundamental differences in soil genesis. The second level, the Suborder, often indicates the hydrologic regime of the soil or its annual precipitation inputs. Sometimes the third level (Great Group) and often the fourth level (Subgroup) carry information about soil hydrology. All four levels are communicated in the taxonomic name. The fifth level (family) provides information about soil texture and mineralogy, among other things. The sixth

Table 1.5 Hierarchy of Soil Taxonomy and Example Using Wakeley Soil Series

Level	Distinctions	Example	Significance
Order	Major soil forming processes	Entisol	Minimal soil development
Suborder	Moisture regime, parent material, secondary processes	Aquent	Aquic moisture regime
Great group	Diagnostic layers, base status, horizon expression, water perching	Epiaquent	Perched water table (WT)
Subgroup	Moisture regime refinements	Aeric epiaquent	WT is dominantly below 25 cm
Family	Texture, mineralogy, temperature, acidity	Sandy over clayey, mixed, nonacid, frigid aeric epiaquents	Outwash over lacustrine, no mineral dominates, cold
Series	Comparable to species in plant taxonomy	Wakeley	
Phase	Slope, flooding, surface texture, etc.	Wakeley mucky sand	Surface horizon is mucky sand

Note: The complete family name of the Wakeley series is "sandy over clayey, mixed, nonacid, frigid Aeric Epiaquents."

Table 1.6   Soil Orders (highest level of *Soil Taxonomy*)

Order	Suffix in Taxonomic Name	Significance	Typical Location
Alfisols	-alf	Significant clay illuviation and high base status	Cool, humid forests
Andisols	-and	Significant presence of volcanic glass	Areas of volcanic ash deposition
Aridisols	-id	Desert climate	Deserts
Entisols	-ent	Minimal soil development	Sands; recent deposits
Gelisols	-el	Permafrost	High latitude & elevation
Histosols	-ist	Formed in deposits of organic material	Wet closed depressions
Inceptisols	-ept	Young soil with incipient development	Active landforms nationwide
Mollisols	-oll	Thick, dark A horizons	Prairies
Oxisols	-ox	High content of iron oxides	Tropics
Spodosols	-od	Subsoil horizon of humus and Al / Fe sesquioxides	Sandy glacial outwash or SE coastal plain
Ultisols	-ult	Significant clay illuviation and low base status	Warm, humid forests
Vertisols	-ert	Shrink/swell activity due to clays	Clay beds, esp. south-central US

level is the soil series name, e.g., "Sharkey" or "Myakka" or "Wakeley." These series names are usually taken from a town or geographic feature associated with the soil and can be thought of as comparable to the binomial species name in the Linnaean classification systems of plants and animals. As of 1997 approximately 20,000 soil series (i.e., different types of soil) were recognized in the United States. Most soil maps in the United States include distinctions between soil phases, which are subsets of the series, much as varieties are subsets of plant or species.

Soil taxonomic names at the Subgroup level (for example, "Aeric Epiaquept") are artificial constructs consisting of (i) a modifier, (ii) a prefix, (iii) an infix, and (iv) a suffix; each part uses terms that identify the higher levels of the taxonomy. Take for example the Subgroup "Aeric Epiaquent" (Table 1.5). The suffix "ent" indicates that the soil is an Entisol, which is a soil with little profile development (Tables 1.5 and 1.6). The infix "Aqu" indicates that the soil is seasonally saturated and anaerobic; its Suborder is Aquent. The prefix "Epi" indicates that the water table is perched or ponded, making it an Epiaquent. The modifier "Aeric" indicates that the wetness problems in this Subgroup are moderate but not extreme; "aeric" connotes "aerated."

The connotative translation code for the constituent parts of soil names is found in *Soil Taxonomy* (Soil Survey Staff 1975) and in many soils textbooks. Most of the distinctions in *Soil Taxonomy* are not significant to hydric soils work; the most pertinent are listed in Table 1.7.

Table 1.7   Words and Phrases from Soil Taxonomy That Have Particular Significance to Wetland Science

Word or Phrase	Meaning
Aqu-	An aquic (or seasonally reducing) moisture regime (e.g., Aqualf). Soils with a different syllable in the Suborder (second) position have drier moisture regimes (e.g., Udalf).
Epi- vs. endo-	A perched water table (e.g., Epiaqualf) in contrast with a water table rising from the bottom of the soil (Endoaqualf).
Aeric	Somewhat ameliorated wetness limitations (e.g., Aeric Epiaqualf). The water table is within 75 cm of the soil surface.
Histic	High organic matter content in the soil surface and usually formed under extreme wetness (e.g., Histic Edoaquoll).
Mollic taxa and Mollisol order (suffix is "oll")	Thick, dark A horizons, which make hydric soil identification difficult because redoximorphic features tend to be masked by organic matter to considerable depth (e.g., Mollic Natrustalf; and Typic Endoaquoll).
Fluv-	Alluvial deposition; possible flooding hazard (e.g., Fluvaquent).
Vertic taxa and Vertisol order (suffix is "ert")	High clay contents with high shrink–swell capacity; hydrologic inputs are usually surficial rather than from below (e.g., Vertic Epiaquept; and Aeric Epiaquert).

Table 1.8 Natural Drainage Classes

Drainage Class	Water Levels
Very Poorly Drained	At or near surface much of growing season.
Poorly Drained	At shallow depths periodically during growing season.
Somewhat Poorly Drained	Wet at shallow depths for significant periods of growing season.
Moderately Well Drained	In rooting zone for short periods during growing season.
Well Drained	Too deep to hinder normal plant growth but shallow enough for optimal crop growth.
Somewhat Excessively Drained	Very deep due to sandiness or rare due to shallow, sloped impermeable layer.
Excessively Drained	Very deep and rare. Water holding worse than for somewhat excessively drained soils.

From Soil Survey Division Staff. 1993. *Soil Survey Manual.* USDA–SCS Agric. Handbook 18. U.S. Govt. Printing Office, Washington, DC.

The formal criteria for hydric soils (Chapter 2) utilize *Soil Taxonomy* as part of the computerized data filter from which the national hydric soils list is developed. Hydric soils are limited to certain taxa. As *Soil Taxonomy* is updated the hydric soil criteria are updated also (Chapter 2).

NATURAL DRAINAGE CLASSES

The system of natural drainage classes (Table 1.8) was developed to group soils into seven classes with similar limitations regarding availability of soil water for crop production, ranging from too little water (for example, high on sand dunes) to too much water (for example, in depressions) (Soil Survey Staff 1951). These range from excessively drained soils, which need irrigation for profitable crop production, to very poorly drained soils, which are so wet that artificial drainage is not economical except for high-value crops.

Drainage classes add information not available from texture alone. A common misconception is that all sandy soils are well drained. Poor drainage is controlled by shallowness of water tables and minimal runoff as well as by hydraulic conductivity, so sandy soils in depressions can be very poorly drained (Chapter 3).

It is useful to think of drainage classes as falling across the continuum of the hydrologic gradient (Figure 1.6). Generally, the lower in the landscape, the more poorly drained a soil. There are numerous exceptions to this generalization, however, such as divergent flow on low slopes (see Chapter 3), seepage slopes, natural levees on stream banks (see Chapter 12), and poorly drained flats on drainage divides (see Chapter 14). Most hydric soils are either very poorly drained or poorly drained, unless they have been hydrologically modified or unless formal hydric soil criteria for flooding or ponding are met.

SOIL SURVEY

The National Cooperative Soil Survey is the United States' program to map the soils of the nation, their distribution, properties, and potentials and limitations for land use. The fundamental concept of the United States soil survey is the soil map unit. The map unit is an abstract concept describing the kinds of soils generally mapped together. In this regard, soil mapping is analogous to vegetation mapping. The legend of a vegetation map may include a map unit of "Red Oak Forest." Not all plants within areas so mapped are red oaks; similarly, for example, not all soils within areas mapped as "Sharkey clay" are Sharkey soils. In both cases, the dominant plant species or soil series within the map unit is the one after which the unit is named, but there can be numerous inclusions of other plants or soils.

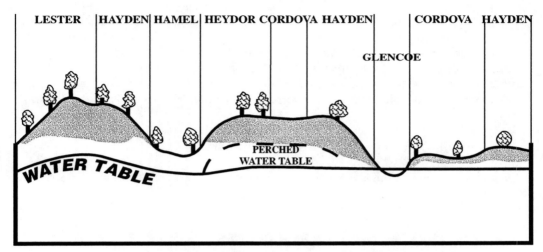

Figure 1.6 Schematic of landscape positions for different natural drainage classes. This example was taken from Hennepin County, Minnesota. (Adapted from Lueth, R. A. 1974. *Soil Survey of Hennepin County, Minnesota.* USDA–SCS in cooperation with the Minn. Agric. Exp. Station, U.S. Govt. Printing Office, Washington, DC.)

Most soil maps in the nation in areas with a history of agriculture were made by second-order surveys (scales of 1:12,000 to 1:30,000). The minimum size delineation on a second-order map is 0.6 to 4 hectares (depending on scale), and most map delineations are considerably larger because of constraints on map legibility. First-order soil maps cover a smaller land area and are more detailed; and third- and fourth-order maps cover larger land areas and are less detailed. It is not recommended to make site-specific hydric soil determinations from the office using second-, third-, or fourth-order soil survey information alone because of natural soil variability and the presence of inclusions within soil mapping units. Onsite investigations are required. Note also that most soil maps were not made with hydric soils in mind. The concept of a hydric soil was developed after the majority of the nation's land had already been mapped.

Second-order soil surveys map soils at the level of the soil phase, which is a subset of the soil series. Typical distinctions made at the level of the soil phase are slope, flooding frequency, and surface texture. Many soil series have both hydric and non-hydric phases, even within the same county. Take for example two neighboring soils in Levy County, Florida, both of them dominated by the Myakka soil series (Kriz 1995). Map unit 37 is the phase "Myakka mucky sand, occasionally flooded" and is dominated by hydric soils; map unit 38 is the phase "Myakka sand" and is dominated by non-hydric soils. A soil in the field is not necessarily hydric just because it is named on a hydric soils list at the level of the series.

Soil survey reports can provide wetland scientists with detailed descriptions of soils and their properties, inventories of soils of possible interest and their geographic distribution, and lists of potentials and limitations for use and management. They are written from several years of experience in the county of the survey and are correlated with information about similar soils in other counties in the region. The soil survey information will become increasingly useful for geographic information systems, too, as the NRCS digitizes and computerizes more of the nations's soil maps.

SUMMARY

Wetland soil investigations utilize the same protocols used for standard soil survey projects. If the study is limited to hydric soils determinations, it usually suffices to describe horizon depths, color, redoximorphic features, and textural class. In mineral soils many hydric soil determinations are made below the A horizon, usually in the B horizon; however, be aware of alternative horizons and features that may be present at these shallow depths. Some hydric soils must be determined from features composed of soil organic matter. Prior to an onsite investigation for any purpose it is useful to consult local soil survey reports. Appropriate use of soil survey reports, however, requires familiarity with soil mapping conventions, including map scale, the concept of natural drainage classes, map unit inclusions, and terms in *Soil Taxonomy* that apply to soil wetness.

ACKNOWLEDGMENTS

Permission to publish this chapter has been granted by the Chief of Engineers, U.S. Army Corps of Engineers.

REFERENCES

Environmental Laboratory. 1987. *Corps of Engineers Wetlands Delineation Manual.* Technical Report Y-87-1. U.S. Army Engineer Waterways Experiment Station, Vicksburg, MS.

GretagMacbeth, Munsell®, Corporation. 1998. Munsell® soil color charts. GretagMacbeth Corporation, New Windsor, NY.

Kriz, D. M. 1995. List of hydric soils by county. pp. 155–409. In V. C. Carlisle (Ed.) *Hydric Soils of Florida Handbook,* 2nd ed. Florida Association of Environmental Soil Scientists, Gainesville, FL.

Lipscomb, G. H. 1992. *Soil Survey of Monroe County, Pennsylvania.* USDA–SCS in cooperation with the Penn. State Univ. and Penn. Dept. Envir. Resources, U.S. Govt. Printing Office, Washington, DC.

Lueth, R. A. 1974. *Soil Survey of Hennepin County, Minnesota.* UDA–SCS in cooperation with the Minn. Agric. Exp. Station, U.S. Govt. Printing Office, Washington, DC.

Schoeneberger, P. J., Wysocki, D. A., Benham, E. C., and Broderson, W. D. 1998. Field book for describing and sampling soils. Natural Resources Conservation Service, USDA, National Soil Survey Center, Lincoln, NE.

Soil Science Society of America. 1997. *Glossary of Soil Science Terms.* Soil Science Soc. of Am., Madison, WI.

Soil Survey Division Staff. 1993. *Soil Survey Manual.* USDA–SCS Agric. Handbook 18. U.S. Govt. Printing Office, Washington, DC.

Soil Survey Staff. 1951. *Soil Survey Manual.* USDA–SCS Agric. Handbook 18, U.S. Govt. Printing Office, Washington, DC.

Soil Survey Staff. 1975. *Soil Taxonomy: A Basic System of Soil Classification for Making and Interpreting Soil Surveys.* USDA–SCS Agric. Handbook 436, U.S. Govt. Printing Office, Washington, DC.

Soil Survey Staff. 1992. National soil taxonomy handbook, Issue 16. USDA–SCS, Washington, DC.

Soil Survey Staff. 1998. *Keys to Soil Taxonomy,* 8th ed. USDA–NRCS, Washington, DC.

Sprecher, S. W. 1991. Introduction to hydric soils, instructional slide set. U.S. Army Engineer Waterways Experiment Station, Vicksburg, MS.

Sprecher, S. W. 1999. Using the NRCS hydric soil indicators with soils with thick A horizons. WRP Tech. Note SG-DE-4.1. U.S. Army Engineer Waterways Experiment Station, Vicksburg, MS.

Thien, S. J. 1979. A flow diagram for teaching texture-by-feel analysis. *J. Agron. Ed.* 8:54–55.

USDA–NRCS. 1998. Field indicators of hydric soils in the United States, Version 4.0. G. W. Hurt, P. M. Whited, and R. F. Pringle (Eds.) USDA–NRCS, Fort Worth, TX.

Vepraskas, M. J. 1996. Redoximorphic features for identifying aquic conditions. North Carolina Agr. Res. Serv. Tech. Bull. 301. North Carolina State Univ., Raleigh, NC.

CHAPTER 2

Background and History of the Concept of Hydric Soils

Maurice J. Mausbach and W. Blake Parker

INTRODUCTION

This chapter describes how the concept of hydric soils was developed over about a 10-year period and how it continues to develop as we learn more about the processes that lead to the formation of wetland soils. The discussion has been broken into three major sections: (1) the background of the development of hydric soils, (2) the evolving nature of the definition, and (3) the concept of hydric soil field indicators. In the background section, we cover the initial activities of the development of hydric soils through the publication of the first national list of hydric soils. In the section on the evolving nature of hydric soils, we discuss the impact of legislation on the development of hydric soil criteria and definition. And finally, we discuss the concept and use of hydric soil indicators in the identification of hydric soils in the field.

Material addressed in this section is based on the notes and correspondence in the official files of the Natural Resources Conservation Service (formerly the Soil Conservation Service), minutes from all the meetings of the National Technical Committee for Hydric Soils, publications of this committee, and *Federal Register* notices.

A hydric soil is one that is normally associated with wetlands and hydrophytic vegetation. In other words, "hydric soil" was defined by observing the connection between the vegetation in classic wetlands and the soils that help support it. Some of the characteristics that support hydric soils are wetness or saturation during the growing season of plants, and anaerobic conditions in the root zone of plants. Key soil properties related to hydrophytic vegetation are the length of time that a soil must be saturated, the location of the water table relative to the soil surface and plant roots, and the period of the year that represents the growing season of the plants common to the area.

As noted in Chapter 1, soils are identified in the field by their morphologic characteristics. Wetland soils are mostly identified by soil colors that are related to the duration of saturation and reducing conditions in the soil. Hydrophytic vegetation, on the other hand, is related to anaerobic conditions in the soil, as opposed to reducing conditions, and the length of time a soil remains saturated during the plants' active growing period. One of the main questions that remains to be resolved for hydric soils is the length of time they must be saturated to support hydrophytic vegetation.

1-56670-484-7/01/$0.00+$.50

This period of saturation depends on temperature, organic matter and moisture content, and the microbial activity in the soil. Thus, in some periods of the year when the soil is warm and has a fresh supply of labile or readily decomposable organic matter, the soil becomes anaerobic in a matter of a few days and supports conditions that favor growth of hydrophytic vegetation. Conversely, in cool periods or when plants are relatively dormant, the development of anaerobic conditions due to saturation may take a few weeks or more. For these reasons, the length of time required for saturation in a hydric soil is still given in general terms and is tied to the development of anaerobic conditions. Field morphological indicators are used as physical evidence that saturation is of sufficient duration to favor the growth of hydrophytic vegetation.

Anaerobic or anoxic conditions are important because hydrophytes are adapted to grow in oxygen-limiting soil conditions by transmitting oxygen from the atmosphere to the root. This must occur during the growing season. Thus, defining a growing season becomes another issue. The growing season for native plant species is often different from that of commonly grown crops. Since wetlands developed under native vegetation, the growing season for hydrology is defined as that of the native species for an area. This becomes confusing since the general public sometimes associates growing season with common crops in an area. To further confuse the issue, there is a second growing season that is used to generate a list of hydric soils. Although it is related to the growing season for hydrology, the growing season for hydric soils is very general and is tied to soil temperature regimes in *Soil Taxonomy* (Soil Survey Staff 1975). It is only used in a computer program that generates a list of hydric soils from the national soil survey database.

HISTORY OF THE DEVELOPMENT OF THE HYDRIC CLASS OF SOILS

Introduction

In the early part of the 1970s, the U.S. Department of the Interior's Fish and Wildlife Service (USFWS) proposed completing an inventory of wetlands in the United States. The inventory was to be made largely using remote sensing techniques such as aerial photography, satellite imagery, and other available information related to wetlands. Part of the "other" information included the national inventory of soils conducted by the National Cooperative Soil Survey (NCSS) in the United States. The soil survey information was to be the foundation for the National Wetlands Inventory (NWI), since wetlands could be identified on soil maps that were already available for more than 75% of the cropland area. However, for the soil inventory maps to be useful in wetland inventories, it was necessary to identify soils normally associated with wetlands as described by Cowardin et al. (1979). The USFWS asked the Natural Resources Conservation Service (NRCS), formerly the Soil Conservation Service (SCS) of the U.S. Department of Agriculture (USDA) to develop the criteria and protocols for identifying soils found in wetlands.

The NRCS is the lead agency for the NCSS, which is responsible for mapping soils throughout the U.S. Individual soils shown on the map are called soil series and are normally named after local geographic sites such as towns and cities where they are first identified. Map units may consist of one dominant soil or, where they occur in patterns too intricate to map separately, can contain two or three dominant soils (Figure 2.1). A landscape diagram of the soil map shown in Figure 2.1 is given in Figure 2.2. Note that the wetland soils are on flat, wide interfluves and in wet areas at the head of drainage ways. For use in wetland inventory activities, the USFWS wanted a list of the names of wetland soils, so they could identify map units associated with wetlands, or map units that included wetlands if the map unit contained more than one kind of soil.

Cowardin, et al. (1979) coined the term hydric soil in their publication, "Classification of Wetlands and Deepwater Habitats of the United States." Though hydric soils were not specifically defined in this publication, it was understood they were a necessary part of the definition of wetlands. The authors defined wetland as having one or more of the following three attributes:

- The land periodically supports hydrophytes predominantly
- The substrate is predominantly undrained hydric soil
- The substrate is non-soil and is saturated with water or is covered by water some time during the growing season of each year.

In cooperation with the effort that the USFWS was beginning with the NWI, the SCS agreed to lead the development of a definition (classification) of hydric soil and to provide a list of hydric soils for use in the NWI. Work began on developing a class of hydric soils concurrent with the development of the Cowardin et al. (1979) publication. The main objective of the hydric soil definition or classification was to define a class of soils that correlated closely with hydrophytic vegetation and to produce a list of hydric soils that could be used with the soil survey maps to assist in developing the NWI maps.

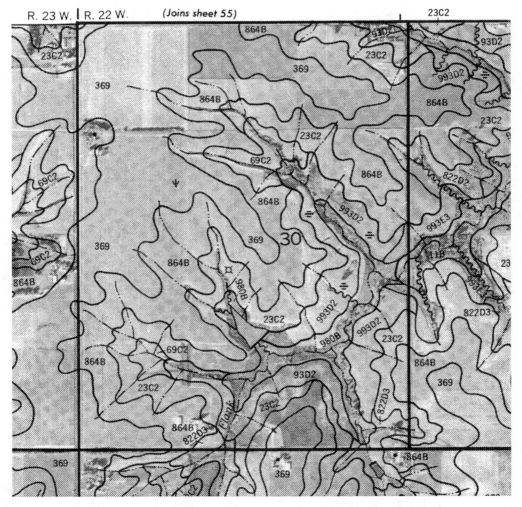

Figure 2.1 Example of a soil map from Warren County, Iowa. Map sheet 62. Soil map unit 369 is Winterset silty clay loam, 0 to 2% slopes; 864B is Grundy silt clay loam, 2 to 5% slopes; 23C2 is Arispe silty clay loam, 5 to 9% slopes, moderately eroded; 93D2 is Adair–Shelby clay loams, 9 to 14% slopes, moderately eroded; 993D2Armstrong–Gara loams, 9 to 14% slopes moderately eroded; and 69C2 Clearfield silty clay loam, 5 to 9% slopes, moderately eroded. (From Bryant, Arthur A. and John R. Worster. 1978. *Soil Survey of Warren County, Iowa.* USDA, Soil Conservation Service, Washington, DC.)

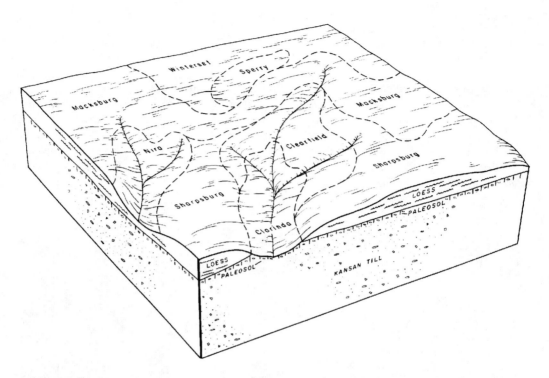

Figure 2.2   Relationship of slope and parent material to soil of the Macksburg–Sharpsburg–Winterset associ-
ation, Warren County, Iowa. The Winterset, Clearfield, Sperry, and Clarinda soils are hydric. (From
Bryant, Arthur A. and John R. Worster. 1978. *Soil Survey of Warren County, Iowa.* USDA, Soil
Conservation Service, Washington, DC.)

A core group of soil scientists from the SCS and biologists from the USFWS and SCS was
formed to investigate the classification and morphology of soils that occur in wetlands. Keith
Young and W. Blake Parker, both soil scientists, and Carl Thomas, National Biologist of the SCS,
led the group.

The initial strategy was to define the concept and criteria for identifying hydric soils, then to
make field visits to areas that were considered good examples of wetlands, and finally to determine
soil properties and classifications of soils that were associated with these wetlands. The working
hypothesis was that soils classified in Aquic Suborders would always correlate with wetlands and
that these classes of *Soil Taxonomy* (Soil Survey Staff, 1975) could be used to develop a list of
hydric soils. In addition, it was hypothesized that a common morphology could be associated with
soils commonly found in wetlands.

Early Concepts of Hydric Soils — 1977 to 1983

As the core group began to make field studies, some initial questions were: (1) how long does
it take hydric soils to form, and (2) how long does a soil have to be saturated in order to support
growth of hydrophytes? Ironically, these same questions have yet to be answered completely,
although scientists have developed research studies to help address them.

Working Definition of Hydric Soil

The initial working definition of a hydric soil was:

Hydric soils are soils with water at or near the surface for most of the growing season or the soil is saturated long enough to support plants that grow well in a wet environment.

It was thought that all soils with aquic moisture regimes would meet this definition. The definition of the aquic moisture regime:

"… implies a reducing regime that is free of dissolved oxygen because the soil is saturated by ground water or by water of the capillary fringe" (Soil Survey Staff, 1975).

Implications of this definition are that, at the highest categories in *Soil Taxonomy* such as Typic Subgroups of Aquic Suborders, the whole soil is saturated. Conversely, in Aquic Subgroups of Suborders other than Aquic, only part of the soil is saturated, and most likely only the lower parts. *Soil Taxonomy* does not specify a duration of saturation but suggests that it is at least a few days. Saturation does have to continue long enough to cause formation of the morphological properties used as taxonomic criteria of aquic moisture regimes. Many of these taxonomic indicators involve the colors produced by iron reduction, which implies the soil profile is saturated long enough to become anaerobic (free of dissolved oxygen) and reducing (see Chapter 7 for further discussion).

Morphological Approach

Early concepts of hydric soils suggested a group of soils that are normally associated with wetlands and hydrophytic vegetation. Thus, early field studies concentrated on correlating "hydric soils" to hydrophytic vegetation. Correlating hydric soils to vegetation was generally a process of describing the morphology of the soils and summarizing common morphological features, mainly soil color in areas dominated by hydrophytic vegetation. Since *Soil Taxonomy* is based on morphological criteria, the morphological approach seemed logical. As a result of these field studies, the team observed that most hydric soils:

- Have dominant colors in the matrix as follows: (1) if there is mottling (redox concentrations), the chroma is 2 or less, and (2) if there is no mottling, chroma is 1 or less.
- Have three wetness conditions: (1) Typic or similar Subgroups that meet the wetness requirements of Typic; (2) Aeric or similar Subgroups that do not meet the wetness requirements of Typic; and (3) other Subgroups with or without wetness requirements of Typic.
- Histosols, except Folists, were also considered hydric.

These observations are very close to the criteria used to distinguish Aquic Suborders and Subgroups in *Soil Taxonomy*. At these initial stages in the development of hydric soil criteria, the team was using the scientific basis of *Soil Taxonomy* in the hydric soil classification. The dark or gray colors represent accumulation of organic matter and reduction of iron oxides in the soil. Mottled color patterns of gray and red represent alternating zones of reduced and oxidized iron, indicative of conditions associated with the top of the water table (the boundary between saturated and unsaturated soil). Eventually, the team modified the initial definition of hydric soils to align them more closely with the definition of aquic moisture regimes. In 1980, a list and definition of hydric soils was distributed to the State SCS staffs for testing and review. The definition was:

Hydric soils are soils that for a significant period of the growing season have reducing conditions (soil is virtually free of oxygen) in the major part of the root zone and are saturated (a soil is considered saturated at the depth at which water stands in an unlined bore hole or when all pores are filled with water. Soil temporarily saturated as a result of controlled flooding or irrigation are excluded from hydric soils) within 25 cm of the surface. Most hydric soils have properties that reflect dominant colors in the matrix as follows: (1) if there is mottling (redox concentrations), the chroma is 2 or less, and (2) if there is no mottling, the chroma is 1 or less.

Important parts of the definition are: (1) the concept that hydric soils are saturated at the surface during the vegetative growing season; and (2) that gray soil colors and mottles (redox concentrations) are associated with wetland soils.

Comments received on this definition of hydric soils and the accompanying list of hydric soils suggested:

- Soil water does not have to be virtually free of oxygen because soil microorganisms will quickly deplete available oxygen
- A class of obligate and facultative hydric soils is needed
- Aeric Subgroups may not be hydric in the southern U.S.
- *Soil Taxonomy* should not be used in the hydric soil criteria because not all aquic moisture regimes are presently reducing or saturated but are related to the presence of morphology associated with wetness
- Designation of hydric status must be at the series level
- Drained soils should not be listed on the hydric soil list

The main concern with use of the aquic moisture regime and subsequent classification in *Soil Taxonomy* is that in the keys to Suborders the key reads:

"... have an aquic moisture regime or are artificially drained and have characteristics associated with wetness" (Soil Survey Staff, 1975).

The phrase "or artificially drained" includes soils in the aquic moisture regime that may not presently have the saturation required for hydric soils. Also, the use of soil characteristics associated with wetness, such as redox concentrations and redox depletions (Chapter 7), may be related to relict conditions of the soil and may not always indicate present hydrology or wetness. The relict conditions may reflect a previously wetter regime which is no longer present due to incision of streams (down cutting), resulting in water tables at greater depths. Relict conditions could also result from climate change or from human activities of drainage of wetlands or protection of areas from floods, such as through construction of levees or dikes.

As a result of these comments, the committee revised the definition and criteria slightly and asked for comments from SCS State staffs and asked the State staffs to prepare lists of hydric soils using the definition and criteria. The criteria still relied on the use of the aquic moisture regime and presence of morphological indicators of wetness within 25 cm of the soil surface.

National Technical Committee for Hydric Soils

The group developing the hydric soils classification was formalized into the National Technical Committee for Hydric Soils (NTCHS) in 1981. Its mission was to "finalize the hydric soil definition and to prepare an approved list of hydric soils." The original team included soil scientists from SCS, the SCS National Biologist, a USFWS biologist, and two university experts in wetland soils. The initial instructions to the committee stressed the intent of the hydric soil definition to identify soils that:

- Favor the production and regeneration of hydrophytic vegetation
- Have a high degree of correlation between hydrophytic plant communities and hydric soils
- Eliminate areas protected from flooding or that are drained
- Eliminate artificially wet areas created by human influences
- Include wet areas from natural factors such as beaver ponds

These charges reflected concerns at the time that areas of human-induced wetness, such as seepage along irrigation canals and ponds, would be interpreted or regulated as wetlands. Another concern was that areas that had been drained or protected from flooding should not be considered hydric or wetland.

In 1983 the NTCHS was expanded to include members from the U.S. Army Corps of Engineers (COE) and Environmental Protection Agency (EPA). The COE and EPA use hydric soils in determining wetlands as part of the Clean Water Act (Environmental Laboratory 1987). The list of functions for the NTCHS was modified to:

- Develop the definition and criteria for hydric soils
- Develop procedures for reassessing the criteria and the list of hydric soils
- Develop an operational list of hydric soils and distribute it to SCS state offices and cooperators
- Coordinate activities with the National Wetland Plant List Review Panel
- Provide continuing technical leadership in the formulation, evaluation, and application of criteria for hydric soils

The NTCHS has been expanded over the years and now consists of six university members representing different areas of the country, and representatives from the U.S. Forest Service and Bureau of Land Management. The functions of the committee remain the same, except that it has taken on responsibility for coordinating the development of hydric soil field indicators and other techniques that may be used to identify hydric soils.

Standardization of Criteria 1983–1985

Movement Away from Morphology

The NTCHS summarized feedback from the 1983 definition of hydric soils and accompanying state-developed lists of hydric soils and noticed an alarming inconsistency among state lists such that a soil series considered hydric in one state was listed as non-hydric in the neighboring state. Because of this inconsistency of lists among states, the NTCHS concentrated on developing a standardized procedure to generate the list of hydric soils. This procedure required a common data source. The NCSS has such a database and, at the time, it was called the Soil Interpretations Record (SIR). The SIR is a national database that contains soil property records for all soil series recognized in the National Cooperative Soil Survey in the United States (Mausbach, et al. 1989). The database contains Soil Taxonomic Class and properties related to wetland soils, including natural drainage class, water table depths, flooding and ponding frequency and duration, and the time of year for which the data are representative. The use of this database limited the NTCHS to the properties already recorded in the database. Other properties, such as oxygen content, or reduction/oxidation potential, would have been useful, but were not in the database. It is important to note that these criteria were meant for use in a computer program and were not intended for the field identification of hydric soils.

Taxonomy–Water Tables–Drainage Class

The natural drainage class of soils (Chapter 1) refers to the frequency and duration of wetness similar to the conditions under which the soil developed (Soil Survey Staff 1993). The seven drainage classes are defined in Table 1.8. The definition of the drainage class is closely tied to growth of mesophytic crops. Soils in poorly and very poorly drained classes must be artificially drained for successful growth of mesophytic crops and thus are often associated with wetlands. Soils that are somewhat poorly drained have wetness characteristics that markedly restrict the growth of mesophytic crops. Wetness in somewhat poorly drained soils can be either of short duration and close to the surface or of longer duration and deeper in the root zone.

The SIR also contains the Land Capability Classification (LCC) (SCS Staff 1961) of the soils at the class and subclass levels. The classes and subclasses of the LCC are given in Tables 2.1 and

Table 2.1   Description of Land Capability Classes

Class	Description
	Land Suited to Cultivation and Other Uses
Class I	Soils have few limitations that restrict their use. Subclasses are not used within this class.
Class II	Soils have some limitations that reduce the choice of plants or require moderate conservation practices.
Class III	Soils have severe limitations that reduce the choice of plants or require special conservation practices or both.
Class IV	Soils have very severe limitations that restrict the choice of plants, require very careful management, or both.
	Land Limited in Use — Generally Not Suited to Cultivation
Class V	Soils have little or no erosion hazard but have other limitations impractical to remove that limit their use largely to pasture, range, woodland, or wildlife food and cover. These soils are mostly wet, stony or have climatic limitations.
Class VI	Soils have severe limitations that make them generally unsuited to cultivation and limit their use largely to pasture or range, woodland, or wildlife food and cover.
Class VII	Soils have very severe limitations that make them unsuited to cultivation and that restrict their use largely to grazing, woodland, or wildlife.
Class VIII	Soil and landforms have limitations that preclude their use for commercial plant production and restrict their use to recreation, wildlife, or water supply, or to esthetic purposes.

From SCS Staff. 1961.

2.2. Land in classes I to IV is considered suitable for crop production, with land in class IV having the greatest number of limitations. The "w" subclass of the LCC indicates that excess water is a limitation to crop production and is used as a modifier of the class. Thus, land in subclass IVw has more severe wetness limitations for crop production than land in subclass IIw. Land in Class VIIIw has the most severe wetness limitations.

The NTCHS designed criteria that utilized the natural drainage class, water table, flooding and ponding, and Land Capability Classification as well as the aquic moisture regime of *Soil Taxonomy*. They arbitrarily created growing season periods based on soil temperature regimes in *Soil Taxonomy* (National Technical Committee for Hydric Soils 1985). The growing seasons are very general and roughly correspond to initiation of plant growth in the spring. The criteria were used in a computer program to generate a list of hydric soils for review by the State SCS staffs. The resulting criteria were:

1. All Histosols except Folists, or
2. Aquic or Alboll Suborders, or Salorthid Great Groups that have water tables less than 1.5 ft during the growing season and which are either
   a. poorly drained and have a land capability classification of IIw–VIIIw, or
   b. are somewhat poorly drained and have a land capability classification of IVw–VIIIw; or
3. Soils with frequent flooding or ponding of long duration or very long duration that occurs during the growing season and that have a land capability classification of IVw–VIIIw.

Comments received on this list of hydric soils suggested that capability classification could not be used because subclasses were based on a hierarchy. In this hierarchy, the wetness factor is second to erosion and ahead of soil and climatic factors. Thus, a soil with both climatic limitation and wetness problems would have a "w" subclass, but the class may be determined by the severity of the climatic factor rather than the wetness factor. For example, a wet soil without a climatic limitation may be classed as IIw, but a similar soil with a climatic limitation may be classed as IIIw.

Others commented that: (1) *Soil Taxonomy* criteria do not identify all hydric soils; (2) a number of SIRs are missing drainage class information; (3) a number of soils with aquic moisture regimes do not have water tables close to the surface; (4) the criteria do not support the definition; and (5)

Table 2.2   Description of Land Capability Subclasses

Subclass Notation	Description
(e) Erosion	Erosion subclass is made up of soils in which the susceptibility to erosion is the dominant problem or hazard in their use.
(w) Excess water	Excess water subclass is made up of soils in which excess water is the dominant hazard or limitation in their use.
(s) Soil limitations	Soil limitation within the rooting zone subclass includes shallowness or rooting depth, stones, low moisture-holding capacity, low fertility that is difficult to correct, and salinity or presence of sodium.
(c) Climatic limitations	Climatic limitation subclass is made up of soils where the climate (temperature or lack of moisture) is the only major hazard or limitation.

From SCS Staff. 1961.

the flooding and ponding criteria include well and excessively drained soils. Some of these comments related to the SIR database and the need to complete the necessary information in it, but others related to deficiencies in the criteria, such as having well or excessively drained hydric soils on the computer-generated list.

First Published Hydric Soil Definition and Criteria — 1985

The NTCHS redrafted the hydric soil definition and criteria and replaced Land Capability Subclass with drainage class information and water table depths. The final definition and criteria were (National Technical Committee for Hydric Soils 1985):

Definition — A hydric soil is a soil that in its undrained condition is saturated, flooded, or ponded long enough during the growing season to develop anaerobic conditions that favor the growth and regeneration of hydrophytic vegetation.

The use of the phrase "in its undrained condition" is a direct tie to *Soil Taxonomy* and relates to the phrase "unless artificially drained." *Soil Taxonomy* is a system designed to eliminate the temporary effects of human beings on soil properties used in the classification criteria. For example, if artificial drainage systems are not maintained, the soil will quickly revert to its wet condition. The phrase, "in its undrained condition," means that even if a soil has been drained or otherwise protected from flooding, the soil is still considered hydric. The NTCHS wanted to use information from the SIR, and that information is mostly tied to the natural conditions of the soil. Because hydric soils are classified using conditions under which they formed, additional information is needed to make wetland determinations.

1985 Criteria (NTCHS 1985)

1. All Histosols except Folists, or
2. Soils in Aquic Suborders, Aquic Subgroups, Albolls Suborder, Salorthids Great Group, or Pell Great Groups of Vertisols that are:
   a. somewhat poorly drained and have water table less than 0.5 ft from the surface at some time during the growing season, or
   b. poorly drained or very poorly drained and have either:
      (1) water table at less than 1.0 ft from the surface at some time during the growing season if permeability is equal to or greater than 6.0 in/hr in all layers within 20 inches, or
      (2) water table at less than 1.5 ft from the surface at some time during the growing season if permeability is less than 6.0 in/hr in any layer within 20 inches, or
3. Soils that are ponded during any part of the growing season, or
4. Soils that are frequently flooded for long duration or very long duration during the growing season.

The criteria for water table depths are tied to the definition of drainage classes (Soil Survey Staff 1993). As previously discussed, the most persistent and highest water tables are associated with the very poorly drained soils, while lower and shorter-duration water tables are associated with the somewhat poorly drained soils. Because the SIR database does not include duration at which a water table stays at or above a certain depth, drainage class reflects the duration of the water table in the hydric soil criteria. Since somewhat poorly drained soils are interpreted as having relatively short-duration water tables, the depth requirement for water tables in these soils is less than 0.5 ft (at the soil surface). The use of permeability class relates to ease of drainage of excess water from the soil and, to some extent, to the capillary fringe above the free water table. It is important to notice that water table depths and permeability are used in connection with the presence of an aquic moisture regime. Flooding and ponding are separate criteria and are not subject to the requirement for an aquic moisture regime. Because flooding and ponding stand alone in the criteria, some hydric soils that are periodically inundated are well or excessively drained. This has been and remains an issue with some users of the hydric soil list.

The first National List of Hydric Soils of the United States was published in 1985 (NTCHS 1985). It was used by the USFWS in the NWI, by EPA and COE in wetland determinations, and by the NRCS in wetland determinations for Swampbuster provisions of the 1985 Food Security Act.

EVOLVING NATURE OF THE HYDRIC SOIL DEFINITION AND CRITERIA

Impact of Government Regulations

Food Security Act of 1985 and 1989 Federal Wetlands Manual

The passage of the Food Security Act (FSA) of 1985 played a significant role in the use of hydric soil definition, criteria, and lists. It passed into law the definition of a wetland as an area meeting three criteria: hydrophytic vegetation, hydric soils, and hydrology. The FSA greatly expanded the use of the list of hydric soils from its original purpose in the mapping of wetlands in the NWI to regulatory uses. Rules and regulations developed by the Department of Agriculture for the FSA allowed the use of only two criteria, hydric soils and vegetation, in areas where hydrology had not been modified. These changes in the use of the hydric soil list, definition, and criteria placed increased pressure on the hydric soil definition and criteria with respect to length of time needed for a soil to become anaerobic. Increasingly, groups were citing the hydric soils criteria as indicating 7 days of saturation, flooding, or ponding as the length of time required for a soil to become anaerobic.

Section 404 of the Clean Water Act

In 1972 the U.S. Congress enacted the Federal Water Pollution Control Act (Public Law 92-500,33 U.S.C. 1251), currently known as the Clean Water Act, to address the rapidly degrading quality of the nation's waters. Its objective is to maintain and restore the chemical, physical, and biological integrity of the waters of the United States. Section 404 of this Act authorizes the Secretary of the Army, acting through the Chief of Engineers, to issue permits for the discharge of dredge or fill material into the waters of the United States, including wetlands.

After the Clean Water Act was passed and Section 404 was designated as the responsibility of the COE, the COE proceeded to define the term *wetlands* and, in turn, developed a wetland delineation manual. The COE Wetlands Delineation Manual was developed for identifying and delineating wetlands regulated by Section 404 of the Clean Water Act (Environmental Laboratory 1987).

The definition used for determining wetlands in Section 404 is as follows:

Those areas that are inundated or saturated by surface or ground water at a frequency and duration sufficient to support, and that under normal circumstances do support, a prevalence of vegetation typically adapted for life in saturated soil conditions. Wetlands generally include swamps, marshes, bogs, and similar areas.*

Explicit in the definition is the consideration of three environmental parameters: hydrology, soil, and vegetation. Positive wetland indicators of all three parameters are normally present in wetlands. The definition of wetlands contains the phrase "under normal circumstances," which was included because there are instances in which the vegetation or soils in a wetland have been inadvertently or purposely removed or altered as a result of recent natural events or human activities.

Wetlands Delineation Manuals

In 1989 the Federal Interagency Committee for Wetland Delineation (Federal Interagency Committee for Wetland Delineation 1989) was formed as a result of directives from the Bush administration, to develop a unified manual for the identification of wetlands. At that time the COE had its 1987 wetlands manual (Environmental Laboratory 1987), the EPA had its version of a wetlands manual (Sipple 1988a and 1988b), the USFWS used the Cowardin (1979) manual, and the NRCS had a wetlands identification manual (SCS Staff 1986). Each addressed the specific agency's responsibility toward identifying and protecting wetlands, and, as the public perceived, delineation of wetland areas varied according to the manual used. The 1989 *Federal Wetlands Manual* was developed for use by all agencies involved in the delineation of jurisdictional wetlands. The manual was in use from late 1989 to 1991, when agencies were instructed to use the 1987 COE manual for wetland determinations. The major reason for discontinuing use of the 1989 manual was the perception by the public that it increased the area of wetlands being regulated by bringing the wetland boundary "too far up the hill." The 1989 wetlands manual used, verbatim, the hydric soil criteria, which were developed for a database search, as the hydrology criterion, which had to be applied in the field. Moreover, users then misinterpreted the hydric soil criteria by stating that a water table could be as much as 1.5 ft from the surface and still meet wetland hydrology requirements. This misinterpretation was due to the convention of listing water tables in the SIR by 0.5 ft increments. The NTCHS could have easily used less than or equal to 1.0 ft. in place of less than 1.5 ft in the criteria. The intention of the NTCHS was to require that water tables be within 1.0 ft of the soil surface for a soil to be hydric. The NTCHS has since adjusted the criteria for hydric soils to use the "less than or equal to" phrase for the specific water table depths in the second criterion.

In addition to these developments, the NTCHS had received comments criticizing the implied 7 days' duration of saturation for anaerobic conditions to develop as interpreted from language in the National Soils Handbook (Soil Survey Staff 1983). In response to these comments, they reviewed the recent literature and research on wetland soils with respect to anaerobic conditions in the upper part of the soil as related to sandy soils, duration of wetness, and depth of wetness. A duration for saturation of 1 week was added to the criteria in a 1987 revision (National Technical Committee for Hydric Soils 1987). In 1990, the NTCHS made a significant change to the criteria by increasing the period for saturation from 1 week to 2 weeks or more during the growing season based on recent research. This change did not affect the list of hydric soils because the Soil Interpretation Record distinguishes high water table duration as a "few weeks." This change was later deleted from the criteria (see Present Definition and Criteria of Hydric Soils later in this chapter). All of these changes were published in the *Federal Register.***

---

* The definition is recorded in the *Federal Register,* 1982, Vol. 42, page 37128, for the Corps of Engineers.
** Third addition of Hydric Soil of the U.S. *Federal Register,* October 11, 1991, Vol. 56, No. 198, page 51371. Changes in definition and soils in 1993, *Federal Register,* October 6, 1993, Vol. 58, No. 192, page 52078. Changes to the definition in 1994, *Federal Register,* July 13, 1994, Vol. 59, No. 133, page 35680.

The SCS and NTCHS also conducted field tests in the southeastern coastal plain and added a special criterion for sandy soils based in part on the potential capillary rise in these very sandy soils. This criterion requires the water table to be at the surface for these soils.

Concepts of Obligate, Facultative–Wet, Facultative, Facultative–Upland Hydric Soils

Hydrophytic vegetation is grouped into obligate, facultative–wet, facultative, facultative–upland, and upland classes depending on the probability that the plant is associated with wetlands. As early as 1981 William H. Patrick, Jr., in a letter to the NTCHS, suggested that hydric soils be grouped into subclasses such as obligate and facultative hydric soils. He suggested that these subdivisions would help eliminate the idea that all hydric soils are associated with wetlands. The concept of subdivisions of hydric soils was rejected at the time because it would be difficult to consistently separate the groups, especially since the NTCHS was having trouble standardizing the list among states at the time. The NTCHS, however, had continued to be concerned that some soils on the list were considered well drained or even excessively drained. These soils are mostly flooded soils.

The concept of subdividing the list of hydric soils surfaced again in 1992, and the NTCHS, under the leadership of Maurice Mausbach and DeWayne Williams, developed criteria for separating hydric soils of different degrees of wetness to correspond to the classes of hydrophytic vegetation. These criteria were as follows:

Group 1. Wettest hydric soils (similar to obligate plant group)
  Hydric soils classified in the Histosols Order, Histic Subgroups, Humic Subgroups, Humaque-
  pts, Umbraquults, Suflaquents, Hydraquents and Umbraqualf Great Groups, and thapto-histic
  Subgroups, or soils in the very poorly drained drainage class.

Group 2. Hydric soil (similar to facultative–wet plants)
  Hydric soils classified as Salorthids, or poorly drained soils, or a combination of poorly drained
  and very poorly drained soil with water table less than or equal to 1.0 ft from the surface.

Group 3. Hydric soils (similar to facultative plants)
  Hydric soils left over from other groups.

Group 4. Hydric soils that are rarely associated with wetlands (similar to facultative–upland plants)
  Hydric soils that do not have an Aquic moisture regime or that have a drainage class of
  moderately well, well, somewhat excessively, and excessively drained.

While it is of academic interest, the concept of subdividing the list of hydric soils based on these criteria has never been implemented. It is described here so it is not forgotten, because it may be required to solve problems that arise in the future.

Present Definition and Criteria of Hydric Soils

The rules and regulations for the Swampbuster portion of the FSA of 1985 specify that changes in the definition and criteria of hydric soils be published as a notice in the *Federal Register.** The most recent hydric soil definition is:

A hydric soil is a soil that formed under conditions of saturation, flooding, or ponding long enough during the growing season to develop anaerobic conditions in the upper part.

* *Federal Register,* February 24, 1995, Vol. 60, No. 37, page 10349.

The wording of the definition has evolved since the 1985 version mostly in an attempt to clarify misconceptions. The phrase "in its undrained condition" has been removed and replaced with "formed under conditions" of saturation, flooding, and ponding. The phrases mean the same and require only that a hydric soil form under conditions of wetness. Therefore, soils that have been altered by humans, such as by use of surface or tile drainage, are considered hydric soils because they formed under anaerobic conditions. The phrase of the 1985 definition, "that favor the growth and regeneration of hydrophytic vegetation," has been replaced by the phrase "in the upper part." Again the meaning of the definition has not changed. The NTCHS removed the reference to hydrophytic vegetation because the definition of wetlands includes hydric soils and hydrophytic vegetation. The committee wanted to avoid the appearance of a circular argument in the definition of wetlands and thus wanted a definition of hydric soils without reference to hydrophytic vegetation. The term, "in the upper part" of the present definition relates to the root zone. It is generally recognized that hydrophytic vegetation is adapted for growth in soils with anaerobic conditions in the active root zone.

The most current criteria for hydric soils are:

1. All Histosols except Folists, or
2. Soil in Aquic Suborders, Great Groups, or Subgroups, Albolls Suborder, and Aquisalids, Pachic Subgroups, or Cumulic Subgroups that are:
   a. Somewhat poorly drained with a water table equal to 0.0 feet from the surface during the growing season, or,
   b. Poorly drained or very poorly drained soil that have either:
      (1) water table equal to 0.0 feet from the surface during the growing season if textures are coarse sand, sand, or fine sand in all layers within 20 inches, or for other soils
      (2) water table at less than or equal to 0.5 feet from the surface during the growing season if permeability is equal to or greater than 6.0 inches/hour in all layers within 20 inches, or
      (3) water table at less than or equal to 1.0 feet from the surface during the growing season if permeability is less than 6.0 inches/hour in all layers within 20 inches, or
3. Soils that are frequently ponded for long duration or very long duration during the growing season, or
4. Soils that are frequently flooded for long duration or very long duration during the growing season.

Criteria 2 and 3 have been modified since 1985. The second criterion has been updated to reflect changes in *Soil Taxonomy* (Soil Survey Staff 1996). The Pell Great Group has been replaced by Aquic Suborders and Subgroups, and the Salorthid Great Group has been replaced by the Great Group Aquisalids. The NTCHS added the Pachic and Cumulic Subgroups because they supersede the Aquic Subgroup in some cases. The NTCHS clarified the depth to water tables by changing the wording from "less than" in the 1985 criteria to "less than or equal to." This change was only for clarification and did not change water table depth requirements or the list of hydric soils. In the present criteria, the NTCHS has added a "sandy soil" part to subsection "b" of the second criterion to require saturation to the surface in very sandy soils. This is directly related to the height of the capillary fringe, which is small in these sandy soils.

Criterion 3 has been changed to match the wording of Criterion 4 by adding the phrase "for long or very long duration." Again, this change was made to clarify the original intention of the committee that flooding and ponding are of equal importance in determining a soil to be hydric.

CONCEPTS OF HYDRIC SOIL CRITERIA AND FIELD INDICATORS

As previously discussed, the definition and criteria for hydric soils were developed to aid in the mapping of wetlands for the NWI. The criteria were designed to generate a list of hydric soils using available information from the national soil database. The list was meant to be used with soil survey maps to locate probable areas of hydric soils that are usually associated with hydrophytic vegetation and wetlands. The NWI also used topographic maps and other geospatial information to help confirm the presence of wetlands. However, as the use of the criteria and list of hydric soils expanded to include the identification and delineation of jurisdictional wetlands, the criteria were being used outside of their original concept. The duration of saturation and depths of water tables listed in the criteria became the definition and criteria for jurisdictional wetlands and field criteria for delineating these wetlands.

Following are some characteristics of the hydric soil criteria that make it difficult to use them for field identification of hydric soils:

1. Depths to water tables are listed as ranges in the SIR database and are meant to define the full range in depths to the water table in a soil over the time period listed for the occurrence of the water table. Thus, the range includes both the wettest and driest times during the specified period. The range also includes differences in the water regime among soils of a specific series. The convention used in developing the list was to use the shallowest water table depth of the range because the NTCHS wanted to capture all potential hydric soils for use with **additional** information in the NWI to identify possible areas of wetlands.
2. Water tables are difficult to observe in the field unless the field visit coincides with a wet period that is representative of the normal climatic conditions in the area. Most delineators are unable to wait for these conditions and must make a determination within a limited time frame.
3. Most delineators do not have access to all the information needed to correctly identify a soil series in the field, and many soil series have properties that span the water table depths and duration of hydric soils.
4. It is time consuming to classify soils in the field especially in delineating boundaries of hydric soils.
5. Drainage classes are not interpreted and applied uniformly across the U.S.

The present concept for the use of hydric soil criteria continues to be that of generating a list for use with soil survey information in mapping wetlands in the NWI. The criteria are not meant for use as field tools in identifying and delineating wetlands. As described in Chapter 3, actual measurement of water tables in the field requires installation of piezometers and collection of data over a number of years to obtain an average water table depth during the growing season. Most wetland delineators must make their determinations within a limited amount of time, sometimes in less than 2 hrs.

Relationship of Criteria to Indicators

Morphological indicators of hydric soils were proposed to allow wetland delineators to both identify and delineate hydric soils in the field (Chapter 8). The indicators were developed to reflect conditions specified in the hydric soil criteria that reflect the definition of hydric soils, specifically anaerobic conditions in the upper part. As noted previously, the criteria are used to generate a list of hydric soils that potentially meet the hydric soil definition using the NCSS database. The hydric soil field indicators are morphological properties of the soils that are used in the field to verify the presence of hydric soils (hydric soil criteria). They, too, reflect the definition of hydric soils.

The hydric soil indicators complement the criteria and are not to be considered new or different criteria for hydric soils. It is interesting to note that the initial activities of the group that defined hydric soils concentrated on the use of morphological indicators. The initial morphological criteria were eventually dropped because of a lack of uniformity among states. In retrospect, it became necessary to first define the water table conditions in the present hydric soil criteria, and then develop morphological indicators. The hydric soil indicators are discussed fully in Chapter 8.

Relationship of Hydric Soil Definition to Morphologic or Field Indicators

The hydric soil field indicators are designed to reflect the saturation and anaerobic requirements of the hydric soil definition (Hurt et al., 1998). The intent of the indicators is that they be used in conjunction with lists of hydric soils, soil maps, and landscape position when making a wetland determination. In areas that are hydrologically modified, the indicators, if present, reflect soil conditions prior to modification. While a site that has been hydrologically modified may have a hydric soil, it may not meet requirements for wetland hydrology and will not be a jurisdictional wetland. Wetland hydrology is a separate criterion for wetlands and thus must be verified using additional information. It is the concept of the developers of the hydric soil field indicators that they only be used in areas that are mapped as hydric soils or that have inclusions of hydric soils. The indicators are also used in areas of the landscape that are typically associated with wetland soils, in the case where a soil survey is not available.

REFERENCES

Bryant, Arthur A. and John R. Worster. 1978. *Soil Survey of Warren County, Iowa.* USDA, Soil Conservation Service, Washington, DC.

Cowardin, Lewis M., Virginia Carter, Francis C. Golet, and Edward T. LaRoe. 1979. *Classification of Wetlands and Deepwater Habitats of the United States.* U.S. Fish and Wildlife Service, U.S. Dept. of Interior, Washington, DC.

Environmental Laboratory. 1987. *Corps of Engineers Wetlands Delineation Manual.* Technical Report #-87-1, U.S. Army Engineer Waterways Experiment Station, Vicksburg, MS.

Federal Interagency Committee for Wetland Delineation. 1989. *Federal Manual for Identifying and Delineating Jurisdictional Wetlands.* U.S. Army Corps of Engineers, U.S. Environmental Protection Agency, U.S. Fish and Wildlife Service, and USDA Soil Conservation Service, Washington, DC.

Hurt, G. W., P. M. Whited, and R. F. Pringle (Eds.) 1998. *Field Indicators of Hydric Soils of the United States.* Version 4.0. USDA, NRCS, Fort Worth, TX.

Mausbach, Maurice J., David L. Anderson, and Richard W. Arnold. 1989. Soil survey databases and their uses. *Proceedings of the 1989 Summer Computer Simulation Conference.* Clema, J. K. (Ed.). July 24–27, 1989, Austin TX. Sponsored by the Soc. for Computer Simulation.

National Technical Committee for Hydric Soils (NTCHS). 1985. *Hydric Soils of the United States.* USDA Soil Conservation Service, Washington, DC.

National Technical Committee for Hydric Soils (NTCHS). 1987. *Hydric Soils of the United States.* USDA Soil Conservation Service, Washington, DC.

Sipple, W. S. 1988a. *Wetland Identification and Delineation Manual. Volume I. Rationale, Wetland Parameters, and Overview of Jurisdictional Approach.* U.S. Environmental Protection Agency. Revised Interim Final, April 1988.

Sipple, W. S. 1988b. *Wetlands Identification and Delineation Manual. Volume II. Field Methodology.* U.S. Environmental Protection Agency. Revised Interim Final, April 1988.

Soil Conservation Service (SCS) Staff. 1961. *Land-Capability Classification.* USDA Soil Conservation Service, Agriculture Handbook No. 210. Washington, DC.

Soil Conservation Service Staff (SCS). 1986. *Food Security Act Manual.* USDA, SCS, Washington, DC.

Soil Survey Staff. 1975. *Soil Taxonomy: A Basic System of Soil Classification for Making and Interpreting Soil Survey.* Agric. Handbook No. 436, U.S. Govt. Printing Office, Washington, DC.

Soil Survey Staff. 1993. *Soil Survey Manual.* USDA, Soil Conservation Service. Agric. Handbook No. 18, U.S. Govt. Printing Office, Washington, DC.

Soil Survey Staff. 1996. *Keys to Soil Taxonomy,* Seventh Edition. USDA, NRCS, Washington, DC.

CHAPTER 3

Hydrology of Wetland and Related Soils

J.L. Richardson, J.L. Arndt, and J.A. Montgomery

INTRODUCTION

Wetland hydrology involves the spatial and temporal distribution, circulation, and physicochemical characteristics of surface and subsurface water in the wetland and its catchment over time and space. Soils record the long-term spatial and temporal distribution and circulation of water because actions of water on soil parent material result in the formation of distinctive soil morphological characteristics. Soil morphology, as used here, is the field observable characteristics possessed by a soil such as soil texture, soil color, and soil structure, and the types of soil horizons present. These soil morphological characteristics, a subset of which is known as "hydric soil indicators" (Hurt et al. 1996), are directly related to a specific set of hydrologic parameters. Soil horizons, for instance, are layer-like soil morphological features that often develop in response to water movement. The study of wetland soils is, therefore, intimately linked to the study of hydrology because hydrology influences soil genesis and morphology.

Soil and Water

Soil, an admittedly complex material, results from the influence of five soil-forming factors (Jenny 1941): (1) organisms, (2) topography, (3) climate, (4) parent material, and (5) time. These factors affect and are affected by water. For example, the biota growing on and in soils are strongly influenced by water's presence, both directly because organisms require water to live, and indirectly because the amount of soil water influences oxygen availability in the soil matrix. Topography frequently directs and controls the flow of both surface and subsurface water to and from a wetland. Climate influences the amount and timing of water availability. Parent material affects the flow of water because it forms the matrix through which surface water infiltrates and through which groundwater flows. The weathering of parent material is directly influenced by water availability. Lastly, time is required for soil development to happen.

Soil also results from the action of four general soil-forming processes: (1) additions, (2) deletions, (3) transformations, and (4) translocations (Simonson 1959; Figure 3.1). Soil is the perfect

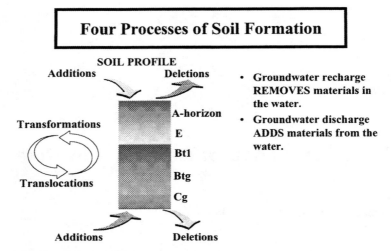

Figure 3.1 All four soil-forming processes involve water in some way.

media in which to study wetland hydrology because all four processes involve water in some way. Water adds material through deposition of eroded sediment and precipitation of dissolved minerals. It transforms soil material through weathering reactions. Water moves (translocates) both solids and dissolved material in mass flow within the soil itself. Water can entirely remove soil material that is dissolved by weathering reactions (transformations), or through erosion of the soil surface.

The study of water and its effects on soil is a unifying principle in soil investigations. The application of hydrologic principles can explain many aspects of hydric soil genesis and morphology that are discussed in detail in other chapters of this book. Similarly, with knowledge of hydrologic principles as a base, the study of hydric soil morphology and genesis relate important information about the nature of wetland hydrology.

Chapter Overview

The study of wetland hydrology requires an introduction to a few basic hydrologic principles. Specifically, *hydrodynamics* refers to the movement of groundwater and surface water to, through, and from a given wetland. Our use of the term *hydrodynamics* is exclusive of precipitation and evapotranspiration; however, we are not implying that precipitation, evapotranspiration, and other processes in the hydrologic cycle are irrelevant to an understanding of wetland hydrology. Indeed, the role of the hydrologic cycle in wetland hydrology is discussed further in the next section of this chapter. Most wetlands also exhibit temporal fluctuations in water levels, defined herein as *hydroperiod*. The water balance of an individual wetland is a fundamental, unique, and distinctive property in which gains equal losses, plus or minus changes in water storage. Water balance is discussed in detail in a later section. Hydrodynamics affects hydroperiod through controls on the water balance of a wetland. The focus of this chapter will be on hydrodynamics, with a brief discussion of hydroperiod. This discussion is followed by an examination of surface and subsurface water movement. Subsurface water movement is not easily observed and thus requires an introduction to the basic principles of shallow groundwater movement and the influence of both hillslope position and geometry on water movement. Other selected physical aspects of wetland hydrology will be discussed next, followed by a discussion of unsaturated flow and the importance of hydrodynamics at the edges of wetlands. Finally, we will describe the relationship between a hydrology–climatic sequence and soil morphology.

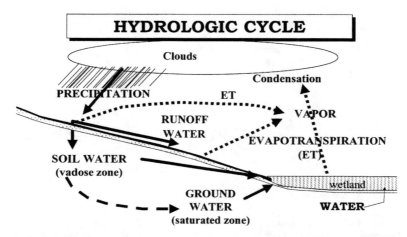

Figure 3.2 The global hydrologic cycle depicts various stages of water circulation through the environment. Precipitation strikes the earth where it can be intercepted and evaporated to the atmosphere, infiltrated into the soil, or run off as overland flow.

REVIEW OF BASIC HYDROLOGIC PRINCIPLES

The Hydrologic Cycle

The endless circulation of water between solid, liquid, and gaseous forms is called the hydrologic cycle. In order to place hydric soil morphology and genesis in the proper context, it is important to recognize that the hydrologic cycle and its associated processes occur at a multitude of spatial and temporal scales. In the broadest scale, water cycles from the oceans to the atmosphere to the land, then back to the oceans (Figure 3.2).

The oceans are the ultimate source and sink for water at the global scale. Evaporation and condensation are the processes by which water changes state from liquid to gas and gas to liquid. The energy that produces these transformations comes ultimately from the sun; however, the processes can operate at any scale from microscopic to global. Atmospheric convection, surface and subsurface flow serve as transport mechanisms. The atmosphere, rivers, lakes, wetlands, groundwater, glaciers, and adsorption to surfaces (interception) serve as temporary storage components of the cycle.

Because transport and change of state processes operate at any scale in the hydrologic cycle, water can cycle many times during its journey to and from the ocean. For example, water vapor in a freezing soil might condense on the surface of a growing ice crystal. When the resulting ice lens eventually melts, the liquid water could move downward into the water table, or it might be taken up by a plant root to be evaporated and released to the atmosphere. In the atmosphere it could condense in a thundercloud and fall as rain onto the surface of a lake, to be stored for days or months prior to evaporation, or it could be released to a stream, with eventual transport to the ocean. In all of its forms, water has a very high capacity to do work. Physical and chemical weathering processes depend on the presence of water.

Basic Water Chemistry, Structure, and Physics

While water is one of the most ubiquitous compounds found in nature, it is also arguably the most unique. A basic review of selected physical properties of water helps in evaluating weathering processes in soils and assessing water movement in saturated and unsaturated soils.

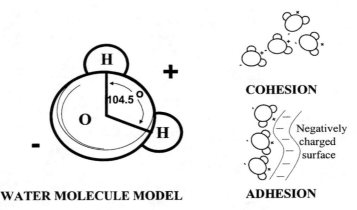

Figure 3.3   Structure of the water molecule. Note that the bond angle produces a dipole with opposing positive and negative regions. It is because of the charged dipole that water is attracted to itself (cohesion) and to other charged surfaces (adhesion).

Water consists of two atoms of monovalent, positively charged hydrogen (atomic symbol H) bound to one atom of divalent, negatively charged oxygen (atomic symbol O) (Figure 3.3). The bonds joining the atoms are strongly covalent; thus very large amounts of energy are required to break the bonds holding the water molecule together. The decomposition of water into its constituent atoms rarely occurs, and water molecules are very persistent in nature.

Water is also unusual in that it is found in solid, liquid, and gas states within a narrow temperature range that is characteristic of the earth's surface. These characteristics are the direct result of the configuration of the water molecule. The bond formed between the two hydrogen atoms and the oxygen atom is sharply angled at approximately 104.5°, which results in distinct positively and negatively charged regions around the water molecules (Pauling 1970; Figure 3.3). Chemists refer to molecules with distinct positive and negative regions as dipoles. Because water is strongly dipolar, it is strongly attracted to itself (cohesion) and to other charged surfaces (adhesion). An understanding of the cohesive and adhesive properties of water aids in the understanding of the physical state of water in the soil, water movement under saturated and unsaturated conditions, and water's ability to dissolve practically anything.

Water the "Universal Solvent"

On a simple level, chemists identify molecules by bond type. Covalent bonds involve electron sharing and are very strong. Ionic bonding involves electron transfers that result in much weaker bonds. Most minerals exhibit mixed bond types that are partly covalent and partly ionic. Molecules with purely ionic bonds are very soluble in dipolar liquids (solvents) such as water because the charged solvent molecules compete with the other atoms in a mineral solid for the bond. Once an atom or a charged portion of the ionic solid is removed from the mineral, the charged molecules of the solvent surround the ion and prevent it from bonding with a solid. Thus, common table salt, a mineral dominantly ionic in character, is much more soluble in water, a dipolar solvent, than in alcohol, which is not as strongly dipolar. Because of its ubiquitous presence and strongly dipolar nature, water is known as a "universal solvent" and is implicated in most, if not all, chemical weathering processes involving geologic and soil materials (Carroll 1970).

Gas Relationships: Aerobic and Anaerobic Conditions

The soil air component of an aerated soil consists of the same N_2, O_2, CO_2, and trace gasses, as the atmosphere. The proportions of oxygen and carbon dioxide change, however, in the soil air.

The change is in response to soil biota respiration, which consumes oxygen and releases carbon dioxide. Oxygen is replenished, and diffusion processes that are sufficiently rapid remove carbon dioxide such that soil microbial and plant root respiration is not inhibited. Enough carbon dioxide diffuses out of the soil so that excessive levels do not accumulate. Diffusion of gasses through water, however, is approximately 10,000 times slower than diffusion through air (Greenwood 1961). When water saturates an aerated soil, oxygen diffusion through the water is insufficient to maintain aerobic respiration, and aerobes die or become dormant (Gambrel and Patrick 1978). In order to survive under saturated conditions in the soil; organisms evolved adaptive processes to circumvent the lack of oxygen (anaerobic processes). The intensity and duration of these processes are controlled by the amount and persistence of water saturation in the soil, along with other factors.

Basic Hydrologic Principles Describing Groundwater Flow

In a very elementary way, the magnitude and persistence of groundwater saturation defines a hydric soil. However, groundwater is not a static entity. The dynamic nature of groundwater flow strongly influences the intensity and rate of soil chemical and physical processes that leave numerous morphologic indicators in soil. Thus, in addition to the presence or absence of a high water table in a soil, knowledge of the direction, magnitude, and rate of groundwater flow is necessary to place the morphological characteristics of hydric soils in the context of a wetland and its landscape. The direction, magnitude, and rate of groundwater flow are functions of the nature of the porous matrix through which the groundwater flows and the energy status of soil water.

Adhesion, Cohesion, and Capillarity

Soils are porous media containing varying proportions of living and dead organic matter; mineral particles of sand, silt, and clay; water and its dissolved constituents; and gasses. Liquid water interacts with soil solids by adsorption processes; these interactions are beyond the scope of this chapter. For our purposes, it is sufficient to say that *hydrophilic* surfaces attract and are wetted by water, and *hydrophobic* surfaces repel water and are not wetted by it, at least initially.

The interactions of adhesive and cohesive forces at solid/liquid interfaces can be described by a simple equation that represents equilibrium between these forces. For example, when a drop of water meets a solid surface, a contact angle (γ) is formed that represents equilibrium between the solid/liquid (σ_{sl}), liquid/gas (σ_{lg}), and solid/gas (σ_{sg}) interfacial tensions (Figure 3.4). At equilibrium, the magnitude of γ defines three classes of substances: (1) those that are not wet (*hydrophobic*, $\gamma \geq 90°$); (2) those that are partially wet (partially *hydrophilic*, $0 \leq \gamma \leq 90°$); and (3) those that are completely wet (*hydrophilic*, $\gamma = 0°$).

The preceding discussion of adhesive and cohesive forces can be extended to describe the phenomenon of capillary rise, which is defined as the height to which water in a capillary tube will rise relative to the free water or water table surface (Figure 3.5). At equilibrium, the adhesive and cohesive forces involved with the surface tension (σ) of water exactly balance the weight of the water in the capillary tube. The relationship is described by Equation 1 where Hc is the height of rise in the capillary tube; σ is the surface tension of water; γ is the contact angle between the solid and liquid as defined in Figure 3.5; r is the radius of the capillary tube; ρ is the density of water; and g is acceleration due to gravity.

$$Hc = 2\sigma(\cos \gamma)/r\rho g \qquad \text{(Equation 1)}$$

Equation 1 approximates the height of rise (Hc) in capillary tubes and, within limits, can be used to approximate the thickness of the capillary fringe that exists above the water table in soils with low organic matter. When considering a soil profile with a water table at some depth, we can

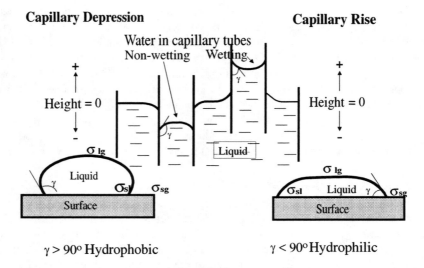

Figure 3.4   Contact angle (γ) between a solid and liquid interface determines two classes of substances. Those substances that have γ > 90 degrees are not wetted by the liquid and are hydrophobic. Those substances that have γ < 90 are wetted by the liquid and are hydrophilic. The upward movement of water ("capillary rise") in capillary pores characterizes hydrophilic solids. Hydrophobic solids exhibit capillary depression. Soils are usually thought of as hydrophilic for water; however, organic matter coatings on soil particles can render them partly to wholly hydrophobic. See the text for the explanation of the σ's. (Adapted from Kutilek, M. and D.R. Nielsen. 1994. *Soil Hydrology*. Catena Verlag, Cremlingen-Destedt, Germany.)

separate the profile into three distinct regions (vadose, capillary fringe, and saturated zones) defined by the physical state of water relative to the soil matrix (Figure 3.6a,b). The water table is defined as the equilibrium level of water in an unlined borehole of sufficient diameter so that capillary rise is negligible. Most of the water found in pore spaces below the water table is "free" water. Free water implies that it is not adsorbed to soil particles.

Capillary Rise

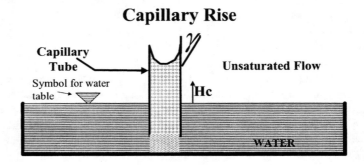

$$Hc = 2\sigma(\cos\gamma)/r\rho g$$

Figure 3.5   Height of capillary rise (Hc) relates to the surface tension (σ) of water and air at 20°C. This tension is about 72 dynes/cm; g is the resistance of gravity; and ρ is the weight or density of water. The capillary rise depends on the wetting of soil particles by water and air and the "effective" size of the pores (r) in the soil. Angle γ is the wetting angle between water and the substance. Angle γ is 0° in a fully wetted condition and approaches 90° or a more in repellent condition when no capillary rise occurs (see Figure 3.4).

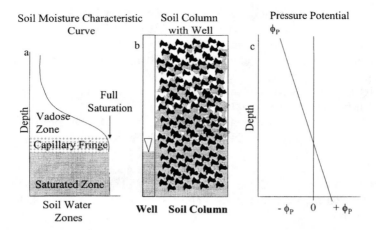

Figure 3.6 We can separate the water in a soil profile into three distinct regions: (1) the saturated zone, (2) the capillary fringe, and (3) water in the unsaturated or vadose zone. The pressure potential is positive below the water table and negative above the water table. The capillary fringe is characterized by near saturation with water under negative pressure. The capillary fringe is only a few centimeters thick in most surface soils.

A capillary fringe of varying thickness exists above the water table (Figure 3.6b). While this zone is nearly water-saturated, the water is adsorbed to soil particles to a greater degree than water below the water table.

The soil above the water table including the capillary fringe is in the unsaturated or *vadose* zone. This zone contains various amounts of water depending upon the pore size and the height in the soil above the water table. Water in this zone is strongly adsorbed to the soil particles, and many of the air-filled pores are contiguous to the soil surface and are connected to the atmosphere. The variation of the volumetric water content in the unsaturated zone depends upon the connectivity and size of the interconnected pores. Contiguous, very fine pores will be water filled to a considerable height above the water table. Pores that are large enough to drain more easily by gravity will be water filled to a lower height (U.S. Army COE, 1987).

Implications of the Physical States of Water for Jurisdictional Wetland Determinations

The impact of capillary fringe thickness on the wetland-hydrology parameter for wetland delineation is not specifically mentioned in the U.S. Army COE (1987) *Wetlands Delineation Manual*. With regard to a depth requirement for soil saturation in jurisdictional wetlands, the 1987 *Manual* only states that the wetland hydrology factor is met under conditions where:

"[t]he soil is saturated to the surface at some time during the growing season of the prevalent vegetation." (Paragraph 26.b.3), and

"[T]he depth to saturated soils will always be nearer to the surface due to the capillary fringe." (Paragraph 49.b.2)

Several scientists have utilized Equation 1 to calculate the height of capillary rise in soils by assuming constant values for σ, ρ, g, and γ. In pure quartz γ is 0°. Using these constants and expressing length units in centimeters, Equation 1 is simplified as:

$$Hc = 0.15/r \qquad \text{(Equation 2)}$$

If we assume that the average effective pore size diameter in medium sands is 0.01 cm, Hc corresponds to 15 cm (6 in). If we further assume that loams have an average pore size half that of medium sand (0.005 cm), Hc becomes 30 cm (12 in). Thus, a sandy soil, relatively uncoated with organic matter, with an average effective porosity diameter of 0.01 cm should have a saturated zone extending approximately 15 cm (6 in) above the free water surface. A loamy soil with an average porosity of 0.005 cm should have a saturated zone extending at least 30 cm (12 in) above the free water surface (Mausbach 1992).

Various U.S. Army COE district offices (e.g., St. Paul, MN District Office) have provided guidance on the saturation-depth requirement that includes the capillary fringe using Equation 2 to compute the height of rise (h). In general, it is assumed that a water table at 6 in will produce soil saturation to the surface in sandy soils (loamy sands and coarser), and a water table at 12 in will result in saturation to the surface in loamy, silty, and clayey soils (sandy loam and finer).

An assumption on the thickness of the capillary fringe that is based exclusively on texture, however, is frequently incorrect because the organic matter present in natural soils increases the contact angle (cf. Equation 1) and thus reduces the height of capillary rise (Schwartzendruber et al. 1954; Richardson and Hole 1978). Wetland soils in general, and Histosols or organic soils in particular, have thin capillary fringes due to the presence of large amounts of organic matter that can result in hydrophobic behavior, and strong soil structure that results in a large macropore volume. In many cases water repellency and the corresponding absence of a capillary fringe are observed in soils high in organic matter if the soils are sufficiently dry (Richardson and Hole 1978). Soils with even 2% organic matter can have strong structure with large macropores created from fine textured soils. The aggregates between the pores lack the continuous connection needed for capillarity. The presence of organic matter combined with the confounding effects of soil structure modifying the pore size distribution has been experimentally shown to result in a capillary fringe that is much thinner for the surface layers of most natural soils (Skaggs et al. 1994). Capillarity is normally less than if calculated using only texture because of lower wetting, larger pore size because of soil structure and plant roots, and abundant air circulation. Many researchers involved in quantification of the soil saturation requirement in jurisdictional wetlands now recommend that the capillary fringe be ignored when evaluating depth to saturation for the surface layers of most natural soils (Skaggs et al. 1994, 1995).

Energy Potentials and Water Movement.

A fundamental principle of fluid mechanics is that liquids flow from areas of high to low potential energy. The total potential energy (Φt) of a "particle" of water is the sum of various potential energies (potentials), including an osmotic potential (Φo), gravitational potential (Φg), and pressure potential (Φp).

Osmotic potential involves the potential energy arising from interactions between the dipolar water molecule and dissolved solids. While Φo is important for water flow in plants, it can usually be neglected in soil water flow except in saline soils. Gravitational potential is the potential energy of position, and can be described by the position of a particle of water above or below some reference datum. Similarly, pressure potential is the potential energy arising from both the pressure of the column of water above the water "particle" and the potential energy associated with adsorptive (adhesive) forces between the water molecule and soil solids. These two components of Φp oppose each other, where the pressure exerted on the particle by the overlying water column is considered a "positive potential," and the pressure due to adsorptive forces is considered a "negative potential."

Under saturated conditions (i.e., below the water table), the vast majority of water molecules are far enough removed from solid surfaces that adsorptive forces can be neglected. Φp, therefore, is simply due to the pressure of the column of water above the "particle" in question. Under these conditions, the "pressure potential" is positive. Above the water table, however, there is no column of free water above the zero pressure point except immediately after a rain. After a heavy rain, the larger pores in the soil fill with infiltrating and downward-moving gravitational water. Adsorptive forces usually dominate at other times, and the pressure potential is negative. Negative pressure potentials (tension) are commonly determined by soil tensiometers. When one considers a cylinder of soil with a water table at some depth, Φp is 0 at the water table, negative above the water table, and positive below the water table (cf. Figure 3.6c).

Darcy's Law

The first quantitative description of groundwater movement was developed as a result of Henry Darcy's 1856 studies to quantify water flow through sand filters used to treat the water supply for the city of Dijon, France. Darcy's experiment used manometers to determine the water pressure at varying locations in a cylinder filled with sand, into and out of which there was a constant discharge (Q). The height of water in the manometers relative to a reference level was the "hydraulic head" (H), and the difference in head (dH) between points in the sand divided by the length of the flow path between the points (dL) was the "hydraulic gradient" (Figure 3.7b). Darcy then compared Q for different sand textures and hydraulic gradients. He found that the rate of flow was directly and quantitatively related to (1) the hydraulic gradient (dH/dL), and (2) a factor called the "hydraulic conductivity" (K) that was a function of texture and porosity (Figure 3.7a).

Soils and geologic sediments usually form a more heterogeneous matrix for water flow than the sand filters investigated by Darcy. In most situations, the hydraulic conductivity of soils is a function of both soil structure and texture and can be further modified by the presence of large macropores along fractures and root channels. Texture is the relative proportion of sand-, silt-, and clay-size particles. Soil structure is the combination of primary soil particles into secondary units called *peds* (Brady and Weil 1998).

The complex spatial distribution of structure and texture combined with the presence of fractures and macropores in natural sediments can confound a Darcian interpretation of groundwater flow unless the characteristics of the flow matrix are taken into account. Laboratory-derived values of hydraulic conductivity are often quite different from field-derived hydraulic conductivity (K) values for the same material. Measurements of hydraulic conductivity are scale dependent. The influence of the nature of the flow matrix on groundwater movement is discussed in detail in a following section (*Soil Hydrologic Cycle and Hydrodynamics*).

Assumptions for Darcy's Law

Darcy's law was empirical in nature and was based on experimental observation. Subsequent research has shown that Darcy's law is not valid under conditions where the flow matrix is so fine textured that adsorptive forces become significant (cf. previous section on *Adhesion, Cohesion, and Capillarity*), or under conditions where hydraulic gradients are so steep that turbulent flow dominates. However, conditions where Darcy's law does not apply are rarely encountered, and it has become a fundamental tool for quantifying groundwater flow under saturated conditions. Darcy's observations have been validated under most conditions of groundwater flow when the variation of pore size distribution that affects hydraulic characteristics of the flow matrix is accounted for.

It should be emphasized that Darcy's manometers provided quantitative information regarding the total potential of water at the point of interest. In a theoretical exercise, Hubbert (1940) applied physics equations relating energy and work to prove that the elevations in Darcy's manometers (e.g., hydraulic head) were exactly equal to the total potential energy divided by the acceleration due to gravity. In other words, the elevations in manometers, which are simply monitoring wells, provide quantitative information on energy potentials and energy gradients that can be used in conjunction with information on hydraulic conductivity and flow path geometry to quantify all aspects of groundwater flow at the macroscopic scale.

Groundwater Flow (Q) = <u>K</u> (dH/dL)

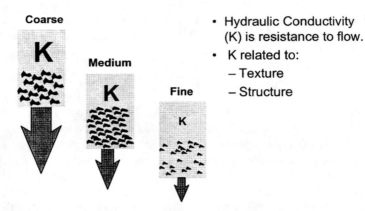

Figure 3.7a Saturated flow below the water table relates to Darcy's Law. A. The amount of flow is due to the saturated hydraulic conductivity (K), which is usually related to both structure and texture in soils.

Groundwater Flow (Q) = K (dH/dL)

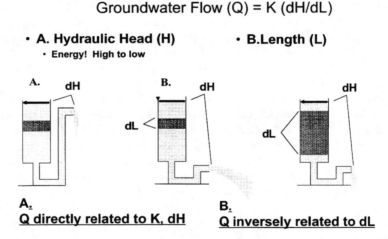

Figure 3.7b Saturated flow below the water table relates to Darcy's Law. B. The amount of flow is due to the hydraulic gradient (dH/dL) that creates the flow and is related to the amount of soil that the water flows through (dL) and the head difference (dH).

Methods of Determining the Nature of Groundwater Flow

The concepts of water flow developed above are routinely used to describe groundwater movement in and around wetlands. At a landform or landscape scale, however, it is important to understand how theory interacts with practice for better interpretations of results from groundwater studies.

Piezometers and Water Table Wells

The direction of groundwater flow is determined through the use of monitoring wells installed at various locations on the landscape; however, a distinction must be made between the two types of monitoring wells commonly used: water table wells and piezometers. Monitoring wells commonly consist of a plastic pipe slotted along a portion of its length and placed in boreholes dug below the water table.

Piezometers are monitoring "wells" that consist of a section of unslotted pipe that is open at both ends or a pipe slotted only at the bottom. The portion of the pipe that is slotted, or the open bottom, is screened with a "well fabric" to keep soil and sand out of the tube and let water in. A limited sand pack is used only in the zone being monitored or screened in the soil profile. Above this sand pack, the remaining area between the pipe casing and the borehole wall is filled with an impermeable material such as bentonite. When compared to an established reference elevation, the water level in the piezometer represents the hydraulic head at the slotted and screened interval. It should be emphasized that under conditions of active groundwater flow, the water level in a piezometer does not usually reflect the elevation of the water table surface.

Water table wells, on the other hand, are designed to identify the elevation (e.g., hydraulic head) of the water table surface at a given point in time. Water table wells most commonly consist of plastic pipe that is slotted to the surface or wells slotted at the bottom that have the annular space between the pipe casing and the sides of the borehole filled with coarse sand. The slots and the sand pack act to "short circuit" the piezometric effect or average out the pressure effect. In wetlands, the need to determine the standing water in the upper 15 or 30 cm (sand and other textures, respectively) requires the use of a shallow water table well or several shallow piezometers at a single location.

Hydraulic heads from at least two piezometers or a water table well are necessary to quantify the direction of groundwater flow. Water level elevations from water table wells placed at various points on the landscape can produce a topographic map of the water table surface that quantitatively illustrates the direction of groundwater flow: water will flow from groundwater mounds (i.e., high head) to groundwater depressions (i.e., low head) along this surface.

Furthermore, when water table wells are installed at the same location as one or more piezometers (a piezometer nest), the vertical direction of groundwater flow can be determined by comparing the elevations of water levels in the nested wells. When no difference in water elevations is observed, stagnant or no flow conditions are indicated (Figure 3.8A). If the elevation in the piezometer is lower than the elevation in the water table well, water flow is downward, indicating groundwater recharge (Figure 3.8B). If the reverse is true, then upward flow (groundwater discharge) is indicated (Figure 3.8C).

Darcy's law and its mathematical extensions give us the quantitative tools necessary to evaluate groundwater movement in near-surface aquifers. Water table elevations obtained from wells and piezometers indicate local hydraulic heads (H). Local pressure head is the distance between the water table and the screened interval of the piezometer. The distances between wells (L) and water elevations give us the hydraulic gradient in two or three dimensions. Stratigraphy obtained from well logs and actual samples, as well as single-well or multiple-well hydraulic tests, gives us an estimate of hydraulic conductivity within strata. The well and piezometer landscape positions and the magnitude of the water levels reflected in them can be used to relate groundwater recharge and discharge as components of the wetland water balance for a landscape. With these data, hydrology can be identified and hydric soil morphology can be placed in the context of groundwater flow on landscapes (Figure 3.9).

Cone of Depression

An analysis of pumping from a well installed below the water table uses the hydrology concepts developed above to demonstrate simply the interaction between saturated flow, the water table, and hydraulic gradient (Figure 3.10). When water is pumped from a well, the water table near the well is depressed as water is removed from the saturated zone and is pumped away. With further pumping, the water table depression progressively moves away from the well, with the water table surface forming the shape of an inverted cone. The shape of the water table depression in the vicinity of the well is appropriately called a *cone of depression*. The rate of water movement at the water table surface increases with increasing steepness of the water table surface, which represents the hydraulic

Wells and Piezometers

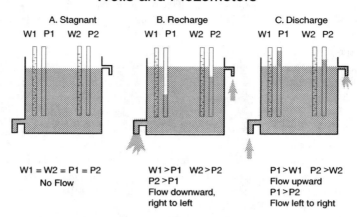

Figure 3.8   (A) Stagnant (no flow) conditions illustrated with two sets of wells (W1 and W2) and piezometers (P1 and P2). Piezometers measure the pressure or head of the water at the bottom of the piezometer tube. If the water level of the piezometer is equal to the water level in the well, the hydraulic gradient is 0 and there is no water flow. (B) Recharge conditions illustrated with two sets of wells (W1 and W2) and piezometers (P1 and P2). Piezometers measure the pressure or head of the water at the bottom of the piezometer tube. If the water level of the piezometer is lower than the water level in the well, the hydraulic gradient and water flow are downward. (C) Discharge conditions illustrated with two sets of wells (W1 and W2) and piezometers (P1 and P2). Piezometers measure the pressure or head of the water at the bottom of the piezometer tube. If the water level of the piezometer is higher than the water level in the well, the hydraulic gradient and water flow are upward.

gradient (dh/dl). As illustrated in Figure 3.10, water will flow faster along the sloping surface of the cone of depression than along the flat surface of the water table away from the cone of depression.

Plants withdrawing water by evapotranspiration produce a drawdown of the water table in a similar fashion, with the effects being more evident at the edge of the wetland where the soil surface is not ponded. Meyboom (1967) showed that phreatophytes (plants capable of transpiring and removing large amounts of water from saturated soil) at the edge of a wetland can change the direction and magnitude of water flow in and around wetlands.

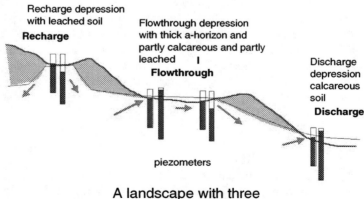

A landscape with three
Soil Types & Hydrology Conditions

Figure 3.9   The magnitude and position of groundwater recharge and discharge as components of the wetland water balance can be identified, and hydric soil morphology can be placed in the context of groundwater flow through the use of Darcy's law combined with well, piezometer, and hydraulic characteristics of the flow matrix.

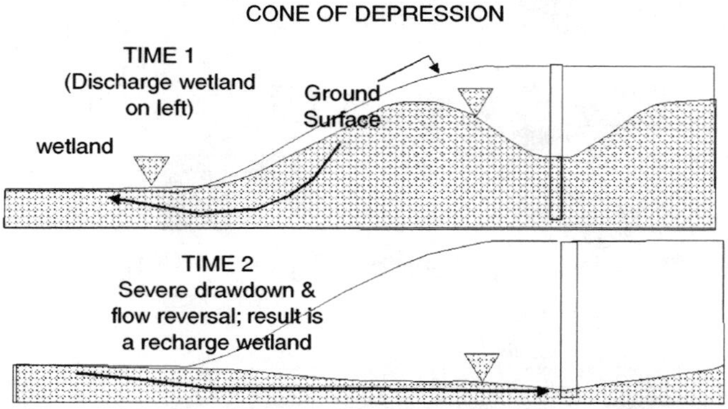

CONE OF DEPRESSION

TIME 1
(Discharge wetland
on left)

Ground
Surface

wetland

TIME 2
Severe drawdown &
flow reversal; result is
a recharge wetland

Figure 3.10 Schot (1991) observed that domestic water appropriation from a well field in Holland lowered water tables sufficiently to create a groundwater flow reversal in a nearby wetland. (Adapted from Schot, Paul. 1991. Solute transport by groundwater flow to wetland ecosystems. Ph.D. thesis, University of Utrecht, Geografisch Instituut Rijksuniversiteirt. 134p.)

As a broader application, Schot (1991) provided an example of the adverse effects of large-scale domestic groundwater appropriations on adjacent wetlands; these effects may become universal with increasing urbanization. Schot examined the progressive effects of well withdrawals on an adjacent wetland in Holland (a very simplified version is given in Figure 3.10). Prior to and immediately after the initiation of pumping, the wetland received discharge water from the upland. This type of wetland is known as a *discharge wetland* and would be considered a valuable rich-fen by the Europeans. However, drawdown of the water table by continuous pumping has resulted in a reversal of groundwater flow, such that the wetland now recharges the groundwater (*recharge wetland).* If pumping were discontinued, the wetland would revert to its natural state as a discharge wetland. If pumping continues, however, the wetland will continue to recharge the groundwater with potentially significant adverse effects to both the water supply and the integrity of the wetland itself. If the wetland water is contaminated, the suitability of the well water may be compromised as the wetland water mixes with the groundwater prior to withdrawal from the well. The wetland's hydrologic regime has changed, and the wetland now loses water to the groundwater instead of gaining water from it. The wetland will certainly get smaller. Depending on the water source, it might dry up altogether. Changes in the water chemistry could also occur because of the removal of the groundwater component to the wetland's water balance. Dissolved solids discharged to the wetland in the groundwater under natural conditions are now removed, and runoff and precipitation low in dissolved solids feed the wetland. The effects of this change dramatically alter the nutrient and plant community dynamics in the wetland, even if it does not desiccate entirely.

Anthropogenic alterations to the groundwater component of wetland hydrology have ramifications for wetland preservation and ecosystem functions and quality. Regional and local studies relating to the indirect effects of anthropogenic alterations to groundwater hydrology on wetland ecosystem function are, however, in their infancy.

Climate and Weather

The hydrologic cycle and climate are inextricably intertwined. *Climate* is the collective state of the earth's atmosphere for a given place within a specified, usually long, interval of time. Weather, on the other hand, is defined as the individual state of the atmosphere for a given place over a short time period. The distinction between weather and climate is important to the study of hydric soils. Hydric soils are assumed to reflect equilibrium between climate and landscape. The transient effects of wet and dry weather will usually not be reflected in hydric soil morphology because the effects

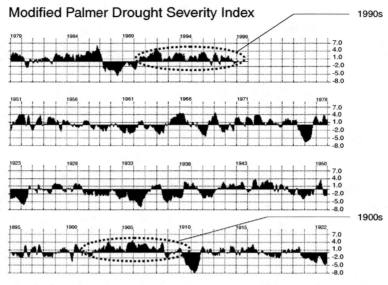

Modified Palmer Drought Severity Index

1990s

1900s

Minnesota - Division 06: 1895-1999 (Monthly Averages)

Figure 3.11 The Palmer Drought Severity Index (PDI) for Region 6, Minnesota. The data indicate that the period from 1990 through 1998 has been wetter than normal and is the wettest continuous period since 1905. These data are available on the Internet.

of weather occur over too short a period. Weather is reflected in piezometer, well observations, and other observations. The distinction between climate and weather, however, is blurred somewhat during long-term drought and pluvial periods. Climatic interpretations can have serious problems with regard to regulatory and scientific evaluation. Wetland hydrology during a long-term drought or pluvial period that lasts longer than a decade becomes the "norm" in the minds of people, especially in the case of seasonal wetlands or in wetlands of hydrologically altered areas. Often, relict soil morphology is suspected when it is the morphology that reflects the current local conditions best. The principal difficulty is one of context: is the period in question characteristic of normal conditions or not?

The Palmer Drought Severity Index, developed and used by the National Weather Service, indicates the severity of a given wet or dry period. This index is based on the principles of balance between moisture supply and demand, and it integrates the effects of precipitation and temperature over time. The index generally ranges from –6.0 to +6.0, but as illustrated in Figure 3.11, the index may even reach 8 in some extremes, with negative values denoting dry spells and positive values indicating wet spells. Values from 3 to –3 indicate normal conditions that do not include "severe" conditions. Break points at –0.5, –1.0, –2.0, –3.0, and –4.0 indicate transitions to incipient, mild, moderate, severe, and extreme drought conditions, respectively. The same adjectives are attached to the corresponding positive values to indicate wetter than normal conditions. An example of the Palmer Drought Severity Index applied to the period beginning 1895 and ending 1998 for the Minneapolis, Minnesota, area is shown in Figure 3.11.

Hydrogeomorphology

Geomorphology is the study of the classification, description, nature, origin, and development of landforms on the earth's surface. *Hydrogeomorphology* is the study of the interrelationships between landforms and processes involving water. Water erosion and deposition influence the genesis and characteristics of landforms. Conversely, characteristics of the landform influence surface and subsurface water movement in the landscape.

INPUTS = OUTPUTS + / - STORAGE

$$P + Ho + GWd = Gwr + So + ET + \Delta Storage$$

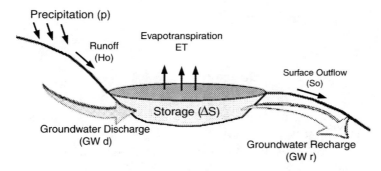

Figure 3.12 The hydrologic balance allows for a budget analysis of the water in the environment. By measuring the inputs and outputs along with changes in storage (ΔS), unknown parts of the cycle can be calculated. Various landscapes can be contrasted by knowing a few parameters.

Water Balance and Hydroperiod

The water balance equation describes the water balance in wetlands on the landscape (Figure 3.12). It is deceptively simple, stating that the sum of precipitation, runoff, and groundwater discharge (inputs) are equal in magnitude to the sum of evapotranspiration, surface outflow, and groundwater recharge (outputs), plus or minus a change in groundwater and surface water storage. The process (transpiration) by which plants uptake water and then evaporate some of it through their stomata to the atmosphere, and the process (evaporation) by which water is evaporated directly from the soil or plant surface directly to the atmosphere are combined and called evapotranspiration (ET). Water that infiltrates 30 cm or deeper below the ground surface is usually lost only through transpiration, with minimal evaporation. Some plants (phreatophytes) draw water directly from the water table. These plants consume large quantities of groundwater and can depress or lower the water table.

When averaged over time, the long-term water balance of an area dictates whether or not a wetland is present. Short-term variations in the water balance of a given wetland produce short-term fluctuations in the water table, defined herein as a wetland's *hydroperiod*. If inputs exceed outputs, balance is maintained by an increase in storage (i.e., water levels in the wetland rise). If outputs exceed inputs, balance is maintained by a decrease in storage (i.e., water levels in the wetland fall).

Slope Morphology and Landscape Elements

One of the strongest controls on the water balance of a wetland is topography. Runoff in particular is strongly controlled by topographic factors, including slope gradient, which influences the kinetic energy of runoff, and slope length, which influences the amount of water present at points on the landscape. These points are discussed in basic soil textbooks such as Brady and Weil's (1998) text. Most important for hydric soil genesis is the way in which slopes direct runoff to specific points on the landscape. Wetlands frequently occur at topographic positions on a hillslope that accumulate runoff water.

Landforms consist of slopes having distinctive morphologic elements with widely differing hydraulic characteristics (Figure 3.13). Subsurface water content progressively increases downslope as runoff from upslope positions is added to that of downslope positions. A low slope gradient and relatively low soil water content generally characterize the highest (summit) position. Slope gradients increase in the shoulder positions, generally reach a maximum in the backslope positions,

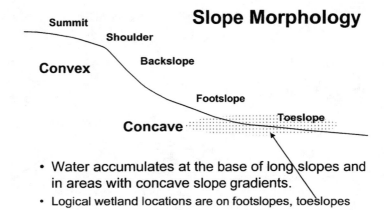

- Water accumulates at the base of long slopes and in areas with concave slope gradients.
- Logical wetland locations are on footslopes, toeslopes

Figure 3.13　Hillslope profile position. Wetlands are favored at hillslope profile positions where water volumes are maximized and slope gradients are low. (From Schoeneberger, P.J., D.A. Wysocki, E.C. Benham, and W.D. Broderson. 1998. *Field Book for Describing and Sampling Soils*. National Soil Survey Center, Natural Resources Conservation Service, USDA, Lincoln, NE.)

and then decrease in the footslope and toeslope (lowest) positions. Footslope and toeslope positions are characterized by maximum water content and minimum gradient. Based on runoff characteristics alone, footslopes and toeslopes in concave positions are logical locations for wetlands because they occur in areas of maximum water accumulation and infiltration.

Slopes exist in more than two dimensions. In three dimensions most slopes can be thought of as variations of divergent and convergent types (Figure 3.14). Divergent slopes (dome-like) disperse runoff across the slope, whereas runoff is collected on convergent (bowl-like) slopes. Plan-view maps of each slope type are shown in Figure 3.14. The presence of convergent and divergent slopes

Hillslope Geometry

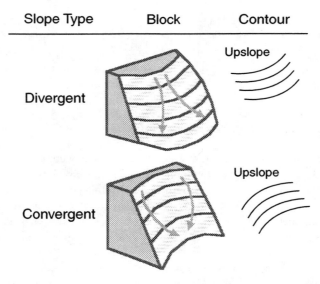

Figure 3.14　Hillslope geometry in three dimensions and two directions. Slopes can be thought of as convergent, divergent, and linear (not shown). (From Schoeneberger, P.J., D.A. Wysocki, E.C. Benham, and W.D. Broderson. 1998. *Field Book for Describing and Sampling Soils*. National Soil Survey Center, Natural Resources Conservation Service, USDA, Lincoln, NE.)

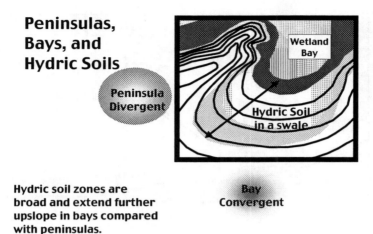

Peninsulas, Bays, and Hydric Soils

Peninsula Divergent

Wetland Bay

Hydric Soil in a swale

Bay Convergent

Hydric soil zones are broad and extend further upslope in bays compared with peninsulas.

Figure 3.15 Swales adjacent to wetland bays are convergent landforms that accumulate water. Divergent water-shedding slopes characterize peninsulas. Hydric soil zones tend to be broad and extend further upslope in bays compared with peninsulas.

on topographic maps indicates where runoff is focused and recharge is maximized. Convergent and divergent areas appear on topographic maps as depressions and knolls in uplands, and bays and peninsulas around wetlands, respectively.

Swales (low depression-like areas) located adjacent to bays in wetlands are in convergent locations, hence, they are characterized by low slope gradients, and they accumulate water. Infiltration and groundwater recharge are maximized, resulting in high water tables. Conversely, peninsulas are divergent landforms often characterized by steeper, water-shedding slopes. The steeper slopes result in both lower infiltration rates and slower groundwater recharge; hence, more precipitation runs off directly to the wetland. Hydric soil zones thus tend to be broad and extend further upslope in bays compared with peninsulas (Figure 3.15). The authors have consistently observed this relationship in the Prairie Pothole Region (PPR) and have frequently used these features for preliminary offsite assessments of wetlands in the region. They can be easily identified on topographic maps and on stereo pair aerial photographs.

The topographic controls on the surface runoff component of the water balance of a given wetland are usually easily understood and directly observable. Topography is also a significant control on the subsurface water-balance components of groundwater recharge and discharge. The relationship, however, is not necessarily direct. Soils and geologic sediments are of equal or greater importance and create situations in which the topographic condition is deceiving because the flow is actually hidden from view in an underground aquifer.

SOILS, WATER, AND WETLANDS

The Soil Hydrologic Cycle and Hydrodynamics

The term "wetland" implies wetness (involving hydrology) and land (involving soils and landscapes). Therefore, it is reasonable that an understanding of soil hydrology and soil–landscape relationships is necessary to understand wetland hydrodynamics. The soil hydrologic cycle (after Chorley 1978; Figure 3.16) is a portion of the global hydrologic cycle that includes progressively more detailed examination of water movement on and in the landscape.

Precipitation that falls on the landscape is the ultimate source of water in the soil hydrologic cycle (Figure 3.16). Precipitation water, which has infiltrated, percolates along positive hydraulic

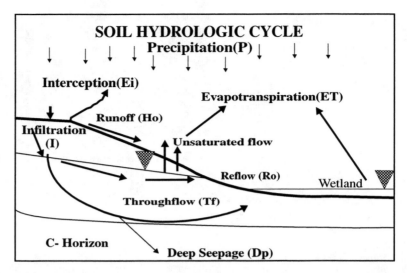

Figure 3.16 Soil hydrology includes precipitation, infiltration, surface vegetation interception and evapotranspiration, overland flow, throughflow, deep-water percolation and groundwater flow. One form of overland flow from a saturated soil is called the reflow. (Adapted from Chorley, R.J. 1978. The hillslope hydrological cycle, pp. 1–42, *in* M.J. Kirkby (Ed.) *Hillslope Hydrology.* John Wiley & Sons, New York.)

gradients until either the gradient decreases to zero, whereupon movement stops and then reverses via unsaturated flow, as water is removed by evapotranspiration, or water movement continues until the wetting front merges with the water table. At this point, groundwater recharge occurs and the water moves by saturated flow in the subsurface. This subsurface, saturated flow usually flows laterally and is called throughflow (Tf in Figure 3.16). Groundwater moving by throughflow may discharge at the soil surface and flow as overland flow. This process is termed *reflow* (Ro in Figure 3.16) and is often referred to as a seepage face. Along the way, some reflow can be lost by evapotranspiration if it comes near enough to the soil surface. Deep-water penetration is the water lost from the local flow system to fracture flow or deeper groundwater that is below the rooting zone of most plants. The amount of water moving as deep penetration is usually less than the amount moving as throughflow.

Landscape-scale or catchment-scale water budget approaches are appropriate for the analysis of wetland hydrodynamics and hydroperiod. The water budget can be expressed by the following budget equation, which is presented graphically in Figure 3.17.

$$P = Ei + Ho + I + \Delta S \qquad \text{(Equation 3)}$$

In Equation 3, P = precipitation input, Ei = amount of precipitation intercepted and evaporated, Ho = amount of Hortonian overland flow (traditional runoff), I = amount of infiltration, ΔS = change in surface storage. Plants are important in increasing infiltration and decreasing runoff and erosion (Bailey and Copeland, 1961). Once intercepted by the plant canopy, precipitation may evaporate to the atmosphere or continue flowing to the ground surface as canopy drip or stemflow. Precipitation that is intercepted by the plant canopy loses much of its kinetic energy when it falls or flows to the ground. The reduced kinetic energy results in less detachment and erosion of soil particles at the surface of the soil and less sealing of the pores necessary for water to infiltrate the soil surface.

Water that infiltrates into the soil begins to move downward as a wetting front when the soil surface becomes saturated. Large soil pores, called macropores, transfer water downward via gravity flow. Water that moves through highly conductive macropores can rapidly move past the wetting front (called *bypass flow;* Bouma 1990). Wetting fronts are frequently associated with the macro-

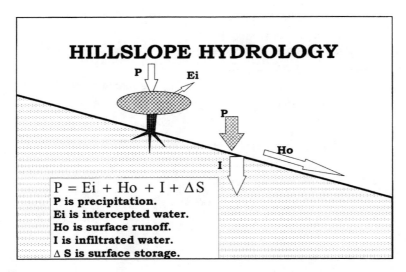

Figure 3.17   The surface of a soil separates the water into essentially three parts and two streams. The intercepted water (Ei) is sent back to the atmosphere. The water that reaches the surface is split into two flow paths: (1) overland flow (Ho) occurs rapidly to the nearby depression, and (2) the infiltrated water (I) (groundwater) moves much more slowly along complex paths. Though not readily seen, groundwater can be a very important component of the water balance of many wetlands.

pores as well; thus, the actual progression of the wetting front in a soil during and immediately after a precipitation event can be very complex.

Soil structure, texture, and biotic activity influence the size and number of macropores, which are most abundant near the soil surface and decrease in abundance with depth. This large number of macropores results in a concomitant progressive decrease in vertical saturated hydraulic conductivity (K_{vs}) with depth in the soil. Horizontal saturated hydraulic conductivity (K_{hs}), however, may remain high across landscapes, reflecting the higher concentrations of macropores in the surface soil horizons.

Transient groundwater flow systems associated with significant precipitation events can impact the hydroperiod of isolated, closed basins, depending on the relative amounts of surface run-on and groundwater flow that are discharged to the pond. The impacts of overland flow on hydroperiod are observed as a rapid rise in pond stage or water table of a given wetland due to the rapid overland flow from the catchment to the pond. The impacts of transient groundwater discharge on pond hydroperiod, however, are not as observable as the impacts of overland flow. The effects can occur over periods of days to weeks depending on the timing, magnitude, and intensity of the precipitation events and catchment geometry.

Shallow but extensive transient, saturated groundwater-flow systems can form in sloping upland soils in the wetland's catchment because of the influence of a permeable surface combined with the presence of a slowly permeable subsoil. Slowly permeable horizons in the soil profile, such as: argillic horizons, which have accumulated extra clay; fragipans, which are brittle horizons with low permeability; duripans, which are cemented horizons; and frozen soil layers that restrict downward groundwater flow. Lateral groundwater flow through the more permeable surface soil, however, is relatively unrestricted and is driven by a hydraulic gradient produced by the sloping ground surface within the wetland's catchment. The groundwater in this transient groundwater system flows slowly downslope. A portion of groundwater in these transient, shallow flow systems may be discharged to the soil surface upslope of the wetland as reflow, a component of runoff. Another portion is discharged to the wetland through seepage at the wetland's edge. A third portion remains as stored moisture when saturated flow ceases. The influence of groundwater discharge on a wetland's hydroperiod (producing a visible water level change) is not immediate because ground-

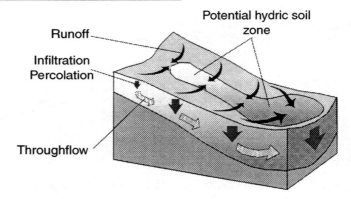

Figure 3.18 Illustration of soil hydrology on landscapes with multidirectional concave hillslopes. Water flow converges from the sides as well as from headslope areas. During precipitation events the saturated zone expands upslope to contribute to increased reflow.

water flow in soil–landscapes is slow relative to surface flow. Significant amounts of water, however, can be discharged to the pond over a period of days or weeks that can maintain the more rapid stage increases produced by surface flow.

The importance of hillslope geometry is illustrated in Figure 3.18. Concave hillslopes, particularly those that are concave in more than one direction, tend to concentrate overland flow, thus maximizing throughflow, interflow, and reflow. During precipitation events, the saturated zone that contributes to reflow increases in area upslope. These saturated areas are potential sites for the genesis of hydric soils.

Water flowing on soil–landscapes can occur as Hortonian overland flow (Ho) spawned by precipitation or snow-melt, or it may occur as reflow (Ro). Overland flow moves rapidly compared to groundwater. Overland flow contains little dissolved load but carries most of the sediment and usually leaves the sediment on wetland edges or the riparian zone (area along a stream bank) adjacent to stream channels. The magnitude of Hortonian overland flow is inversely proportional to the amount and type of ground cover. Ground cover, moreover, is related to land use.

The water budget for infiltrated water can be expressed by the following equation (after Chorley 1978), which is graphically presented in Figure 3.19:

$$I = T_f + D_p + ET + \Delta SW \qquad \text{(Equation 4)}$$

where I = infiltration, T_f = throughflow (also called lateral flow or interflow), D_p = deep water penetration, ET = evapotranspiration, and ΔSW = change in soil water. The units are usually inches or centimeters of water.

Effects of Erosion, Sedimentation, and Hydroperiod on Wetlands

Land-use changes in a wetland's catchment can alter the wetland's hydrodynamics. Tillage in prairie wetlands, for instance, results in increased runoff and discharge into the wetlands. One of our colleagues working on soils of prairie wetlands relates the story of how his parents had a pair of cinnamon teal nesting in their semipermanent pond in the pasture of their dairy operation. The parents switched from dairy to cropland and plowed the pasture that was the catchment for the

HILLSLOPE HYDROLOGY

$$I = Tf + Ro + Dp + ET + / - \Delta SW$$

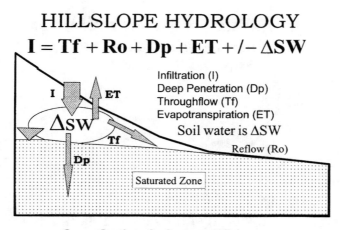

Infiltration (I)
Deep Penetration (Dp)
Throughflow (Tf)
Evapotranspiration (ET)
Soil water is ΔSW
Reflow (Ro)

Saturated Zone

Cross-Section of a Concave Hillslope

Figure 3.19   Water that infiltrates can (1) be used by plants or evaporated, (2) flow downslope in large pores, (3) flow away from the soil surface as deep water penetration, or (4) be added to or removed from the stored soil water. The downslope movement of groundwater (throughflow) discharges at pond edges. Much of the groundwater flows in transient, surficial groundwater flow systems formed in response to significant precipitation events.

pond. The pond became inundated more quickly in the spring; however, it also dried out much sooner and the nesting habitat was lost. The cinnamon teal became a fond memory!

High intensity rains on bare, tilled ground result in high levels of runoff and considerable erosion of the soil that fills depressions with sediment. Runoff and eroded sediments are transported downslope until they are deposited in low-relief areas, including wetlands, and fill the depressions to a degree that they no longer function as wetlands. Conversely, on well-vegetated landscapes more infiltration results in less sediment production. Freeland (1996) and Freeland et al. (1999) observed large amounts of recently deposited sediments as light-colored surface alluvium overlying buried A-horizons in wetlands surrounded by tilled land. No sediments, however, were observed on the soils in wetlands with catchments with native vegetation. Small depressions, in particular, are functionally impacted by even small amounts of sediment. The functions relating to storage of water are particularly disturbed by sediment.

Tischendorf (1968) noted that in 14 months of observation in the southeastern U.S., 55 rainstorms did not produce overland flow in the upper reaches of their forested watershed in Georgia, although 19 storms had enough intensity to produce runoff hydrographs. Flood peaks were related to saturated areas near streams. These areas enlarged during the storm event due to throughflow (interflow), and the associated reflow contributed to overland flow. Kirkham (1947) observed that with intense precipitation, the hilltops had vertical downward flow (recharge), the middle slopes were characterized by throughflow, and the base of slopes had upward flow or artesian discharge flow. Richardson et al. (1994) observed such flows after heavy rains around wetlands in the Prairie Pothole Region (Richardson et al. 1994). Runoff, however, is not common on the ground surface of forests or grasslands with good vegetation cover, primarily because of the associated high infiltration rates (Kirkby and Chorley 1967, Hewlett and Nutter 1970, Chorley 1978, Kramer et al. 1992, Gilley et al. 1996). The rate of overland flow can be as much as 3 km/hr (Hewlett and Nutter 1970). Groundwater flow is orders of magnitude slower than surface flow. For instance, groundwater flowing through coarse-textured sediments at 1 m/day is considered rapid (Chorley 1978), yet this flow rate is only 1/72,000 times that seen in typical surface runoff.

Urbanization also decreases infiltration and increases runoff. Retention ponds constructed to store stormwater runoff effectively behave as recharge ponds that hopefully help to recharge groundwater and wetlands. Obviously, wetland depressions have an important function in terms of

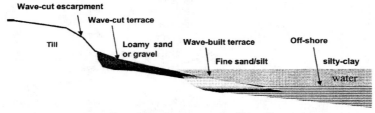

**Generalized Stratigraphy for
Semipermanent Prairie Wetlands
Fringing Wetland Edge**

- Lower gradient enhances infiltration.
- Coarse textures enhance infiltration and groundwater movement.

Figure 3.20   Fringing wetland edge with an escarpment created by wave erosion that expands the basin width, a wave-cut terrace that is covered with a veneer of gravel, and a wave-built terrace with fine sand and silt. Offshore sediments composed of silts and clays fill the basin and reduce water capacity.

sediment entrapment and runoff abatement if retention ponds are being engineered for use in urban settings, although some action will be needed periodically to remove the sediment from retention ponds and place it back on the landscape.

Fringing Wetlands and Wave Activity

Fringing wetlands of the Hydrogeomorphic Model Classification system are wetlands that border lakes, bays, and other large bodies of open water. They have an upland side and a side that yields to the open water, and are thus transitional from upland to open water conditions. During pluvial cycles, high water may rise over the emergent vegetation in fringing wetlands. Waves striking the shoreline during these times erode the shore and result in the subsequent formation of a distinctive landscape (Figure 3.20) that consists of (i) a wave-cut escarpment, (ii) a wave-cut terrace, and (iii) a wave-built terrace. These geomorphic features all have distinct soil textures and other physicochemical properties. The waves undercut the headlands in steeper areas creating a scarp (an erosional feature). The platform where the waves actually strike is a gently sloping, erosional landform called the wave-cut terrace. While the wave action enlarges the area of the basin, the attendant erosion of the uplands and deposition of the eroded material within the pond decreases overall basin depth and produces a depositional landform called a wave-built terrace that lies pondward of the wave-cut platform.

Although these geomorphic features are not formal indicators of the presence of wetland hydrology in jurisdictional wetlands, wetland scientists performing wetland delineations frequently use these features as secondary indicators of hydrology. These secondary features are incorporated into the "water marks," "drainage patterns," and "sediment deposits" commonly referred to in land ownership disputes around lakes and ponds. We are not referring to "wetland delineation" here but to legal ownership of the land, and such disputes have a far longer history than wetland delineation. Wave created water-marks around lakes are used to determine public vs. private ownership and access rights of the public around lakes in the Dakotas and Minnesota.

Effects of Saturated and Unsaturated Groundwater Flow on Wetlands

The preceding wave-cut and wave-built landscape is an example of how hydrology and landform interact to produce a distinctive hydrologic pattern in fringing-depressional wetlands. After intense runoff-producing precipitation events, the relatively level sand and gravels on the wave-cut terraces enhance infiltration of the runoff water. Beach sediments act as an aquifer, and the underlying sediments act as an aquitard, resulting in lateral groundwater flow. Once infiltrated, the water rapidly moves laterally along a hydraulic gradient through the coarse-textured beach sediments until it reaches the finer-textured silts and clays characteristic of the wave-built terrace. The silts and clays on the wave-built terrace are lower in hydraulic conductivity. Thus they transmit less water. This results in the development of a transient groundwater mound landward of the interface between the coarse-textured beach sediments and the fine-textured, near-shore depositional sediments deposited pondward from the wave-built terrace (Figure 3.20). This specific type of groundwater/surface water interaction with sediment and landform has been shown to have implications for groundwater discharge, salinization processes, and plant community distribution around Northern Prairie wetlands (Richardson and Bigler, 1984; Arndt and Richardson, 1989; 1993). These processes may be important hydrologic controls for wetlands outside the Northern Prairie region.

Flownet and Examples of Flownet Applications

Flownets

Darcy's law and its mathematical extensions have been employed in groundwater flow modeling since the mid-1800s. However, the presence of complex stratigraphy and topography, coupled with the need for numerous wells and piezometers necessary to characterize water conditions at a complex landscape scale, have limited the use of the Darcian relationships to small-scale studies or studies that deal with very homogeneous materials.

The influence of stratigraphy and topography on groundwater flow systems was not fully appreciated until the advent of numerical methods and computer programs that accurately model groundwater flow in two and three dimensions. One such method produces a flownet, which consists of a mesh of contoured *equipotential lines* and flow *streamlines*. Equipotential lines connect areas of equal hydraulic head along which no flow occurs. Streamlines indicate the path of groundwater flow and are orthogonal to equipotential lines.

A detailed description of numerical methods and procedures used to develop complex flownets is beyond the scope of this chapter. Detailed descriptions of the methods are in most basic groundwater hydrology texts and papers (e.g., Cedargren 1967, Freeze and Cherry 1979, Mills and Zwarich 1986, Richardson et al. 1992). However, simply put, numerical methods place a two- or three-dimensional rectangular network of grid points over the flow system, and Darcy's equation is applied to develop finite-difference expressions for the flow at each node. Boundary conditions and assumptions, coupled with actual and estimated values of hydrologic parameters at specific nodes, are used to interpolate values for these parameters at the remaining nodes. Seminal research encompassing landscape-scale groundwater modeling that was initiated in the 1960s (Toth 1963; Freeze and Witherspoon 1966, 1967, 1968) has expanded into an explosion of research into virtually all facets of groundwater flow and has resulted in the development of numerous groundwater models.

Figure 3.21 provides the salient characteristics of a flownet simulation using Version 5.2 of the program FLOWNET (Elburg et al. 1990). The figure represents the simple situation of groundwater flows in isotropic, homogeneous media with a water table that linearly declines in elevation from left to right. The height of the bars above the cross-section represents the hydraulic head and is equivalent to the water table elevation. Equipotential lines are dashed, streamlines are dotted, and the large arrow indicates the direction of groundwater movement. By convention, adjacent streamlines form stream tubes through which equal volumes of water flow. Fast groundwater flow is indicated in regions where streamlines are closely spaced. Conversely, slow flow is indicated by widely spaced streamlines.

"Flownets"

$$Q=K \frac{dH}{dL}$$

Darcy's Law

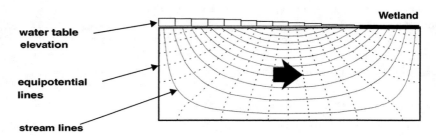

Figure 3.21   Arraying equipotential lines (lines of equal hydraulic head) perpendicular to groundwater stream-lines creates flownets.

Effects of Topography (1): Closed Basins, Glaciated Topography

The examples that follow use FLOWNET simulations to illustrate the impacts of topography and stratigraphy on wetland hydrology. Real-world examples from recent soil research are provided to reinforce the concepts present in the simulations.

FLOWNET computer modeling accurately simulates or depicts the effect of water table topography on the development of groundwater flow systems as examined in Toth (1963). We assume that the water table topography is a subdued reflection of the surface topography in areas with humid climates. The flownet simulation in Figure 3.22, therefore, illustrates that the presence of a long, regional slope of the water table will result in the development of a simple groundwater flow system. This flow system is characterized by (1) distinct upland recharge zone (upper left portion of the simulation), (2) a distinct zone of *throughflow* where groundwater is moving approximately horizontally in the middle of the simulation, and (3) a distinct zone of groundwater discharge into a wetland, lake, or river.

The simple flow system described above is in direct contrast to that produced when water table relief is high and complex (Toth 1963). In our FLOWNET simulation, short, choppy slopes that would be characteristic of hummocky glacial topography produce highly complex flow systems consisting of small, locally developed flow systems contained within progressively larger flow systems. The large, bold arrows in Figure 3.22, the second diagram, indicate both localized flow systems that are isolated from each other and the regional flow system. Groundwater flow within these local flow systems is driven by internal recharge and discharge characteristics. Flow can be with or counter to the regional flow as indicated by the bold arrows. If the water table configuration in Figure 3.22 is persistent, however, there is and will be no hydrologic groundwater connection between adjacent systems.

The presence of these complex flow systems has a significant impact on the regional hydrogeology. Soluble constituents released by weathering processes that occur during recharge will be transported to groundwater discharge areas. The soluble materials persist within the local discharge system unless removed by some surface transport mechanism, such as wind erosion during drought times or removal in a surface drain in pluvial times. In the Prairie Pothole Region (PPR), where surface drainage is limited or absent, the presence of numerous, hydrologically isolated local

**Hummocky topography results in many
local groundwater-flow systems.**

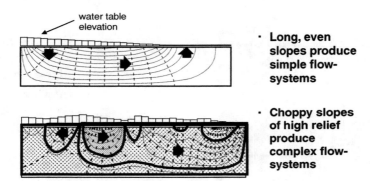

- Long, even slopes produce simple flow-systems

- Choppy slopes of high relief produce complex flow-systems

Figure 3.22 The upper diagram is a smooth topography with a simple flow pattern. The second indicates the presence of hummocky topography and poorly integrated surface drainage. This creates local flows within larger regional systems (Adapted from Toth, J. 1963. A theoretical analysis of ground-water flow in small drainage basins. *J. Geophys. Res.* 68:4197–4213.)

groundwater flow systems partly explain why one wetland may be fresh while a neighboring pond is extremely saline.

Effects of Topography (2): Breaks in Slope

Pfannkuch and Winter (1984) observed that breaks in slope, or areas where the slope gradient changes from steep to gentle or flat, were often points of groundwater discharge and were frequently occupied by seeps and sloping wetlands. Assuming that the water table is a subdued replica of the land surface, Figure 3.23A shows that their observations are confirmed by a flownet simulation. Water movement within broad, level flats between sloping areas is slow and limited by low hydraulic gradients. Groundwater discharge is focused at the foot of slopes where these hydraulic gradients decrease the greatest amount.

**Discharge is enhanced where there is
a break in the slope of the water table.**

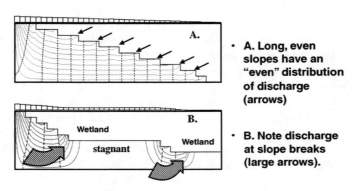

- A. Long, even slopes have an "even" distribution of discharge (arrows)

- B. Note discharge at slope breaks (large arrows).

Figure 3.23 FLOWNET simulation shows that breaks in slope are frequently groundwater discharge areas occupied by seeps and sloping wetlands.

**Intensity of edge-focused discharge is related
to aquifer wetland size.**

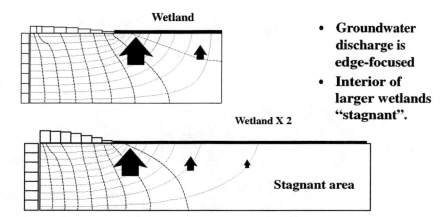

- Groundwater discharge is edge-focused
- Interior of larger wetlands "stagnant".

Figure 3.24   A FLOWNET illustration of the effect of wetland size and aquifer thickness on groundwater movement. As a wetland increases in size, the tendency is for groundwater to discharge at the wetland edge.

Effects of Topography (3): Wetland Size and Aquifer Thickness

Pfannkuch and Winter (1984) also noted that the intensity of edge-focused groundwater discharge is related to aquifer thickness and wetland size. Because hydraulic head is relatively constant across the ponded wetland surface, the hydraulic gradient decreases rapidly away from the edge. As can be seen in the simulations (Figure 3.24), the effect is magnified when the aquifer is thin and/or the wetland is large. The hydrologic implications are that groundwater discharge is always edge-focused in large ponded wetlands, and that the interior of such large wetlands can be considered to be relatively "stagnant" (or lacking flow) as far as groundwater flow is concerned. This effect is only enhanced when the wetland edge is also characterized by a break in slope (cf. Figure 3.23B for a simulation). The figure again illustrates the presence of edge-focused discharge and its resulting salinization characteristics.

Effects of Stratigraphy (1): The Effects of Layering

Sediment layering and sediment isotropy/anisotropy are extremely important hydraulic characteristics when considering groundwater flow into and out of wetlands. The FLOWNET simulations discussed above assume topography as the only variable. The flow matrix for these simulations is assumed to be homogeneous, with an isotropic hydraulic conductivity. A sediment layer is isotropic if the hydraulic conductivity within the layer is the same in all directions, and is anisotropic if the hydraulic conductivity differs with direction within the layer. Sediment homogeneity and isotropy are rarely encountered in soil–landscapes. Layering of sediment strata of differing hydraulic conductivity is the usual condition and is caused by the differential action of erosive and depositional processes over time. Most sediments are anisotropic due to depositional and packing processes that favor the lateral orientation of flat, nonspherical particles, and the fact that roots are concentrated near the surface and decrease in abundance with depth. In addition, soil-forming processes create structure and horizons in soils that strongly influence hydraulic conductivity of soils.

In general, lateral groundwater flow is favored over vertical groundwater flow especially in the soil zone, because of (1) the presence of soil horizons and sediment layers of varying hydraulic

Effects of Sediment Layering

Flow Cross-sections

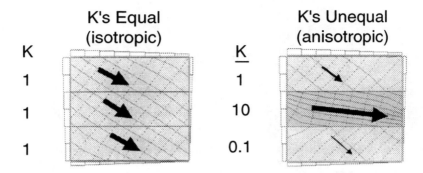

Layering favors lateral flow

Figure 3.25   The effect of layering by soil texture, density or structure creates an increase in lateral flow potential (right-side diagram) when contrasted to the isotropic flow potential (left side) of homogeneous strata.

conductivity, and (2) the presence of anisotropy that favors lateral flow within a given layer (i.e., higher hydraulic conductivity in the horizontal direction). FLOWNET simulations (Figure 3.25) show that layering, in any order, strongly favors lateral flow because of the high flow velocities that are characteristic of the more conductive layer. Given the same hydraulic gradient, flow is much slower in the less conductive layers and is directed primarily downward. The result is that the majority of the flow occurs laterally in the conductive layers. The layer with the lowest hydraulic conductivity limits the speed of downward groundwater flow, and the layer with the highest hydraulic conductivity limits the speed of lateral groundwater flow.

A technique developed by hydrogeologists, determines the composite horizontal and vertical hydraulic conductivity (Kh and Kv, respectively) for a given stratigraphic section composed of layers of varying hydraulic conductivity (Maasland and Haskew 1957; Freeze and Cherry 1979, p. 32–34). This compositing technique reinforces the significance of the layering impact on groundwater flow. Figure 3.26 provides a situation near a solid waste landfill facility, where the near surface stratigraphy consists of interbedded Pleistocene lacustrine strand and near-shore sediments that vary in texture from clay loam to fine sandy loam. The compositing technique applied to this situation yielded a Kh/Kv ratio of 8000. In other words, for the entire section, groundwater flow was 8000 times faster in the horizontal direction when compared to the vertical direction. In this situation, which contains rather typical sediment layers and hydraulic conductivities, it is obvious that groundwater flow would occur almost entirely within the coarse textured layers and would be lateral in nature. In the field, it is not uncommon for layered heterogeneity to lead to regional composite Kh/Kv values on the order of 100:1 to 1000:1 (Freeze and Cherry 1979).

The impacts of layering are particularly important for transient saturated flow in soils because soils are layered entities that consist of horizons that vary in structure, texture, and hydraulic conductivity. Consider an Alfisol on a slope above a wetland with a well-granulated loamy A horizon, a silty, platy E horizon, and a clay-textured Bt horizon. After a significant precipitation event, water would infiltrate the soil surface and percolate downward; however, the Bt horizon that is low in hydraulic conductivity would limit vertical flow. Throughflow would occur preferentially in the granulated A horizon and the platy E horizon. Groundwater flow would be directed laterally downslope and would resurface as edge-focused discharge at the periphery of the wetland. If rainfall events were frequent enough and of sufficient magnitude, groundwater transferred laterally and

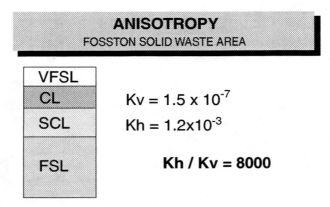

Figure 3.26 The concept of anisotropy is that differences between lateral flow and downward flow exist in soils (or rocks). The most restrictive layer (slowest Kv) governs downward movement, and the least restrictive layer (fastest Kh) governs lateral flow.

downward through soil surface horizons would accumulate on the soil surface at discharge locations and could maintain saturation for a long enough period for hydric soils to develop. This mechanism explains the presence of hydric soils in and adjacent to the bottoms of swales with no evidence of surface inundation, and it also explains the presence of a hydric soil ring above the ponded portions of wetlands.

Effects of Stratigraphy (2): Fine and Coarse Textured Lenses

The presence of soil horizons and sediments with contrasting hydraulic conductivity can have a great impact on both groundwater flow and the resulting presence and hydrologic characteristics of wetlands on the landscape. We can compare groundwater flow in an idealized landscape with a homogeneous flow matrix (cf. Figure 3.21) to a similar landscape containing a sand lens embedded in the homogeneous materials (Figure 3.27). Hydraulic gradients are the same in both illustrations.

The simulation shows that a sand lens acting as a conduit for saturated flow can have a dominant influence on the entire flow system and can strongly influence the hydrologic character of affected

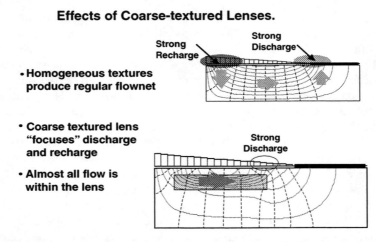

Figure 3.27 A comparison of a landscape with homogeneous flow matrix with a similar landscape containing a sand lens embedded in the homogeneous materials. Under saturated flow the sand lens is far more permeable and conductive than the surrounding materials. Water tends to flow into the sand lens and is transported laterally.

wetlands. Under the same hydraulic gradients, flow occurs primarily within the sand lens, with little flow occurring in the fine-textured matrix within which the sand lens is embedded. Groundwater recharge is associated with the up-gradient portion of the sand lens, and groundwater discharge is associated with the down-gradient portion.

Because of much higher hydraulic conductivity, water can be transported laterally in the sand lens, even under small hydraulic gradients. If the sand lens pinches out and terminates, the hydraulic gradient pushes the water to the surface, resulting in a seep. Such seeps can occur even though the sand lens does not crop out at the surface. The effect is exaggerated if the sand lens terminates at the surface, and high volumes of groundwater discharge can form actual spring-heads at these locations. It is important to realize that under these conditions, the sand lens is the flow system. When modeling groundwater flow in such a system, the flow occurring in the fine-textured matrix can be insignificant. Wetlands are frequently formed above these groundwater discharge areas, and many such wetlands have an artesian source of water (Winter 1989).

Areas associated with the up-gradient portion of the sand lens will be strong recharge sites. Soil in these recharge basins will be leached, and often have strongly developed illuvial horizons such as an argillic horizon. Similarly, wetlands associated with down-gradient portions of the sand lens will be strong groundwater discharge sites. Soils in these discharge basins frequently accumulate salts and nutrients and lack leached illuvial horizons. These soils may be highly organic due to the persistent saturation caused by consistent groundwater discharge.

Saline seeps, which are common in the semiarid west, are excellent examples of wet areas resulting from preferential flow in sand lenses and similar zones of higher conductivity. Saline seeps are typically dry for several years in a row because the conductive coarse-textured zones are above the water table. During a pluvial (wet) cycle, however, the water table rises as the sand lens becomes recharged. Once saturated, groundwater flows to points of discharge where the sand lens outcrops or pinches out near the ground surface. The water carries abundant salts that accumulate on the soil surface as discharging groundwater evaporates. Seeps are often discovered during the pluvial cycle by driving a tractor into the seep area, with uncomfortable consequences. Calcareous fens, an unusual type of wetland dominated by groundwater discharge, represent another type of wetland that is commonly associated with coarse-textured lenses embedded in fine-textured sediments.

The presence of less permeable layers in a more permeable groundwater flow matrix also impacts groundwater flow systems and associated wetlands (Figure 3.28). These restrictive layers may have high clay contents, they may contain a restrictive and impermeable soil structure (e.g., platy type), or high bulk densities may characterize them. Groundwater flow in an idealized landscape with a homogeneous flow matrix is compared in Figure 3.28 to a similar landscape containing a less permeable lens embedded in the homogeneous materials. Hydraulic gradients are the same in both cases. The scenario is applicable to any situation where fine-textured sediments underlie coarser-textured sediments, for example, on outwash plains, where fine-textured lacustrine sediments are overlain by coarser outwash sands. In soils, clay-rich argillic horizons frequently have overlying, coarser-textured, and more permeable E horizons that conduct most of the water in sloping landscapes.

The FLOWNET simulation shows that the layer with the lowest hydraulic conductivity restricts downward groundwater flow and forces water to move around it, directing the flow path through more permeable sediments. The result is slower water removal due to shallow gradients that slope to a depression at the edge of the wetland. Additionally, the direct loss of water by ET from the area, poor internal drainage within the overlying sediments, and the potential development of a groundwater mound above the restrictive lens also occur. If the sediments under the restrictive lens are unsaturated, a perched water table results. If the groundwater mound intersects the soil surface, the resulting wetland is similarly a "perched" wetland with soils that have formed under "epiaquic" conditions, or water that has accumulated above the soil and tends to move down, or recharge, the groundwater. Soils with an epiaquic moisture regime typically have an unsaturated zone underlying a saturated zone.

Effects of Fine-textured Lenses.

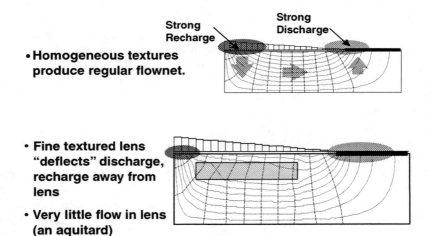

- **Homogeneous textures produce regular flownet.**

- **Fine textured lens "deflects" discharge, recharge away from lens**
- **Very little flow in lens (an aquitard)**

Figure 3.28 The rectangle in the FLOWNET is a fine-textured lens that acts to deflect flow around the lens. Flow in the lens or aquitard is nominal. Recharge occurs before the lens or above the lens and flows laterally. Argillic horizons can act like an aquitard on landscapes.

The effects on groundwater flow of a highly impermeable argillic horizon under a more permeable E horizon in the epiaquic Edina series (Fine, smectitic, mesic Vertic Argialbolls) are discussed in some detail in Chapter 9.

APPLICATIONS: WETLAND HYDROLOGY

Hydrology and Wetland Classifications

Hydrogeomorphic Classification

In order to classify the relationship of landscape and wetlands, we refer to Brinson's (1993) hydrogeomorphic model (HGM). The classes which comprise Brinson's (1993) basic categories in his HGM system separate and group wetlands based on geomorphic setting, dominant source of water, and hydroperiod. These classes reflect wetland processes, such as seasonal depression, because the energy of water is expressed (kinetic energy) or constrained (potential energy) by its soil-geomorphic condition. For example, groundwater in a sloping wetland moves quite differently than groundwater in flats, depressions, fringing, and riverine systems. Depressional wetland systems are the only HGM class covered in the following discussion. The hydrogeomorphic system is discussed in more detail in Chapter 9.

Stewart and Kantrud Depressional Classification

Stewart and Kantrud's (1971) Wetland Classification System defines hydroperiod for the Northern Prairies of the Unites States and Canada. Perhaps these concepts can be extended to nontidal wetlands outside the Northern Prairie region. The Stewart and Kantrud classification divides hydroperiod into three groups based on long-term climatic conditions: (i) normal water levels, (ii) less water than normal, or drought phase, and (iii) more water than normal, or pluvial phase.

Table 3.1   Classes and Zones Related to Ponding Regime and Ponding Duration

Class	Central Vegetation Zone	Ponding Regime	Ponding Duration (Normal Conditions)
I	Low prairie[1]	Ephemeral	Few days in spring
II	Wet meadow[2]	Temporary	Few weeks in spring; few days after heavy rain
III	Shallow marsh	Seasonal	1–3 months; spring early summer
IV	Deep marsh	Semi-permanent	5 months typical
V	Permanent open water	Permanent	Most years except drought
VI	Intermittent alkali	Varies	Varies
VII	Fen	Saturated	Rarely ponded; groundwater saturated

[1] The low-prairie zone is too dry to be considered part of a jurisdictional wetland.
[2] The wet meadow zone is the driest part of a jurisdictional wetland.

From Stewart, R.E. and H.A. Kantrud. 1971. Classification of natural ponds and lakes in glaciated prairie region. U.S. Fish Wildl. Serv., Res. Publ. 92. U.S. Govt. Printing Office. Washington, DC.

Stewart and Kantrud (1971) used their definition of hydroperiod to further classify depressional wetlands based on recognizable vegetation zones that develop in response to normal seasonal variations in hydroperiod. They grouped prairie wetland vegetation into zones characterized (1) by distinctive plant community structure and assemblages of plant species, and (2) ponding regime (Table 3.1).

Wetland classes are based on the type of vegetation zone occupying the pond center; thus the wettest zone defines the class. Class II temporary wetlands, for example, are dominated by a wet meadow plant community but lack vegetation typically found in a shallow marsh community. A Class IV semipermanent wetland characteristically has a central zone dominated by a deep-marsh plant community adapted to semipermanent ponding, and peripheral shallow-marsh, wet meadow, and low-prairie zones, indicating progressively shorter degrees of inundation. Figure 3.29 illustrates a "Class IV semipermanent pond or lake" with the relationship of vegetation zones to each other.

Zonal Classification

The wetland classification system of Cowardin et al. (1979), hereafter referred to as the Cowardin system, is similar in some respects to the Stewart and Kantrud system. The Cowardin system, which is more comprehensive, focuses on vegetation zones rather than on the entire wetland

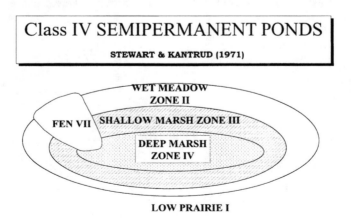

Figure 3.29 Arrangement of vegetation zones in a semipermanent pond or lake with a small fen. The wetland edge is the outer wet-meadow or fen zone. The low-prairie is not part of a jurisdictional wetland. (Adapted from Stewart, R.E. and H.A. Kantrud. 1971. Classification of natural ponds and lakes in glaciated prairie region. U.S. Fish Wildl. Serv., Res. Publ. 92. U.S. Govt. Printing Office. Washington, DC.)

basin. For example, in the Cowardin system, the emergent shallow marsh of Stewart and Kantrud would be separated from the emergent, deep-marsh vegetation zone as a distinct wetland class. Many wetlands characterized under one Stewart and Kantrud class would be characterized under two or more classes in the Cowardin system.

Landscape Hydrology Related to Wetland Morphology and Function

Regional Studies (Macroscale)

Climatology and geomorphology are broad complex disciplines with important applications to understanding hydric soil genesis. Regional wetland characteristics often result from Earth's physical features over broad geographic areas (physiography) interacting with climate differences. For instance, unglaciated areas differ from glaciated areas, and prairie glacial areas differ from forested glaciated areas (Winter and Woo 1990; Winter 1992). Winter and Woo (1990) called divisions at this scale "hydrogeologic physiography" and divided the United States into a few general categories. Climatic criteria, based on gradients between wet–dry and cold–warm extremes, are used by Winter and Woo (1990) to identify a number of varieties of specific regional physiographic types (Figure 3.30). For example, glacial terrains characterized by youthful till landscapes with poorly integrated drainage are further broken down by climate into the eastern glacial terrain, which has high precipitation, and prairie glacial terrain (Prairie Pothole Region or PPR), which is characterized by lower precipitation (Figure 3.31). Both regions are fairly representative of a continental climate with cold winter and warm summers. Snow covers the ground 30 to 50% of the time. The presence of snow cover and frost during a significant portion of the year has a strong impact on wetlands. Even though winter precipitation is usually low, the precipitation that falls is stored in the snow pack, to be released upon spring snowmelt. Because much of the ground is still frozen, runoff is maximized. The period immediately after spring snowmelt is frequently the time of highest water levels for wetlands in these areas, a fact that readily distinguishes cold climate wetlands from those in warmer climates.

It is precipitation, however, that really distinguishes eastern from prairie glacial terrain. The prairie is definitely drier, with average annual precipitation varying from 400 to 600 mm/yr. compared to the eastern region's 600 to 1400 mm/yr.

A more important measure of climate that directly affects wetland hydroperiod, and integrates the effects of temperature and precipitation is the difference between precipitation and pan evapotranspiration. The PPR is characterized by a moisture deficit, whereas the eastern regions have moisture excess (Figure 3.32).

Hydrogeologic Physiography

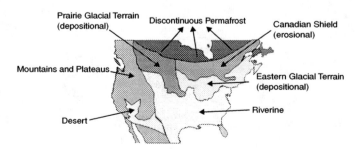

Figure 3.30 Climate discriminates the wetlands in the eastern glacial terrain from wetlands in the prairie glacial terrain. (Adapted from Winter, T.C. and Woo, M-K., 1990. Hydrology of lakes and wetlands, pp. 159–187. *In* Wolman, M.G., and Riggs, H.C. (Eds.) *Surface Water Hydrology. The Geology of North America,* v. 0-1. Geological Society of America, Boulder, CO.)

Precip. and Temperature

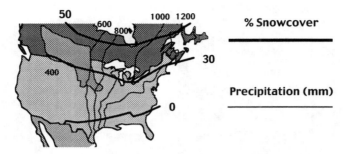

Figure 3.31   Contrasting yearly precipitation values in the prairie and eastern glacial terrains. The prairie glacial terrain is added for perspective in relation to the precipitation. (Adapted from Winter, T.C. and Woo, M-K., 1990. Hydrology of lakes and wetlands, pp. 159–187. *In* Wolman, M.G., and Riggs, H.C. (Eds.) *Surface Water Hydrology. The Geology of North America,* v. 0-1. Geological Society of America, Boulder, CO.)

The existence of a moisture deficit in the PPR and a moisture excess in the eastern glaciated terrains has a great bearing on groundwater recharge and discharge relationships. In the eastern glaciated terrain it spawns the development of an integrated surface drainage system. A precipitation surplus is the driving force that causes wetlands to fill to the point where they spill over the lowest portions of their catchments to form these integrated drainage networks. In the eastern glaciated terrain, characterized by moisture, drainage networks are present but poorly integrated due to the youthful, hummocky nature of the unconsolidated tills draped over the underlying bedrock. The PPR landscape is similar geologically; however, low precipitation coupled with moisture deficits ensures that the wetlands usually will not fill to overflowing. The result is a hummocky landscape that is a mosaic of thousands of undrained catchments placed at varying elevations in thick till. Wetlands, varying in ponding duration from ephemeral to permanent, generally occupy highest to lowest positions, respectively, within the catchment.

Precipitation – Pan Evaporation (cm)

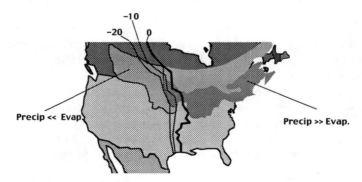

Figure 3.32   The border between the prairie and eastern glacial terrains is characterized by the difference between precipitation and pan evapotranspiration. (Adapted from Winter, T.C. and Woo, M-K., 1990. Hydrology of lakes and wetlands, pp. 159–187. *In* Wolman, M.G., and Riggs, H.C. (Eds.) *Surface Water Hydrology. The Geology of North America,* v. 0-1. Geological Society of America, Boulder, CO.)

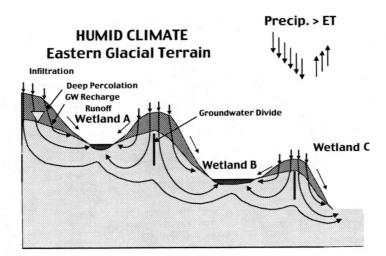

Figure 3.33   Humid glacial terrain with groundwater divides in each minor upland. Recharge occurs in uplands, and their soils are leached. Discharge occurs in adjacent wetlands. Surface drainage is developed, although initially it is deranged.

Groundwater Recharge and Discharge Relationships in Humid, Hummocky Landscapes

Figure 3.33 presents an idealized example of local groundwater relationships in hummocky topography of humid regions characterized by a precipitation surplus. After a precipitation event, a portion of the water falls on the wetland itself (direct interception), a portion is received as runoff from the surrounding catchment, and a portion infiltrates the upland soil and percolates downward or laterally as long as positive hydraulic gradients exist.

Local groundwater flow systems overlay regional systems. Because precipitation events in the humid region are closely spaced in time, a succession of recharge events drives infiltrated water via deep percolation to the water table. Groundwater is thus recharged in the upland (Figure 3.33), resulting in leached soil profiles. If percolating water reaches the water table faster than it can be discharged to low areas, then a groundwater mound develops under topographic highs. Figure 3.33 represents a generally accepted hydrologic model for groundwater recharge for humid regions. The water table is a subdued replica of the surface topography, and wetlands tend to be foci of local discharge. Groundwater divides form at the crests of the groundwater mounds under topographic highs. These divides are "no-flow" boundaries across which streamlines will not flow; hence, they identify the local flow systems that are superimposed on the regional flow systems in hummocky topography.

Over time, runoff, groundwater discharge, and direct interception will flood the pond until the surface water overtops the lowest portions of the catchment. The resulting meandering, relatively disorganized surface flow (deranged drainage) usually connects wetlands to each other in hummocky eastern glaciated terrain.

To summarize groundwater recharge–discharge relationships in humid regions:

1. Groundwater recharge occurs in uplands, and upland soils are typically leached.
2. Wetlands are "usually" loci of groundwater discharge.
3. Surface drainages (initially deranged) develop.
4. Many local flow systems overlay regional flow systems.

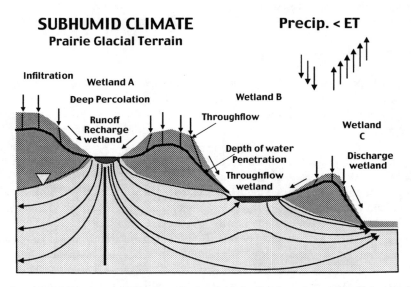

SUBHUMID CLIMATE
Prairie Glacial Terrain

Precip. < ET

Figure 3.34   In subhumid landscapes, the groundwater divide is often in a depression. These landscapes often have flowthrough and discharge wetlands as well as recharge wetlands.

Groundwater Recharge and Discharge Relationships in Subhumid, Hummocky Landscapes

Figure 3.34 is an example of local groundwater relationships in hummocky topography of subhumid regions that are characterized by a moisture deficit. Wetlands are still recharged via direct precipitation and overland flow. The longer intervals between precipitation events and the usually intense nature of the events themselves, however, ensure that deep percolation and groundwater recharge does not regularly occur under topographic highs. The groundwater mound is not present under the high because not enough new water infiltrates or penetrates deep enough to reach the water table. Much of the soil water returns to the atmosphere by evapotranspiration before the next recharge event occurs. The overall lack of precipitation coupled with high evapotranspiration further ensure that wetlands will not fill to overflowing.

Groundwater is recharged frequently at the edges of ponded wetlands and under dry wetlands because groundwater recharge occurs first where the vadose zone is thinnest (Winter 1983). The above factors result is a landscape dominated by closed catchments and nonexistent surface drainage. Because deep percolation is minimized by the lack of frequent precipitation, interdepressional uplands are relatively uninvolved in transfers of water to and from the water table. In the subhumid PPR, therefore, groundwater recharge and discharge are depression focused (Lissey 1971, Sloan 1972). Seasonally ponded wetlands in upland positions (e.g., Wetland A, Figure 3.34) recharge the groundwater with relatively fresh overland flow and snowmelt. A portion of this recharge water moves downward and laterally into and out of intermediate throughflow wetlands (Wetland B, Figure 3.34), and is subsequently discharged into a low-lying discharge-type wetland (Wetland C, Figure 3.34).

To summarize groundwater recharge–discharge relationships in subhumid regions:

1. Groundwater recharge and discharge are depression-focused.
2. Uplands are relatively uninvolved in groundwater recharge and discharge. Upland soils often contain evidence of limited deep percolation (e.g., presence of Ck horizons, Cky horizons).

3. Surface drainages are limited or nonexistent.
4. Wetlands are distinctly recharge, flowthrough, and discharge with respect to groundwater flow.

A Proposed Wetland–Climatic Sequence

A series of hydrology–climatic sequences was constructed based on experiences in studying soils across climatic regions (Richardson et al. 1992, 1994) and on information from *Wetlands of Canada* (National Wetlands Working Group 1988). The hydroclimatic sequences were divided into four zones, moving east to west across the northern region of North America: (1) Zone 1 — perhumid, (2) Zone 2 — humid, (3) Zone 3 — subhumid, and (4) Zone 4 — semiarid. Zones 1 and 2 relate to the humid region eastern and prairie glacial terrains mentioned in the preceding section. Zones 3 and 4 related to drier terrains.

Excess precipitation in perhumid landscapes leaches the soil of easily soluble materials, including nutrients, and tends to favor acid-forming plants that produce tannin. Tannin is an excellent preservative of organic matter, and that is why it is used to "tan" leather. Tannin restricts bacterial decomposition. The slow loss of mor-type humus or organic material from acid bogs may be largely due to the tannin-created preservation. Mor humus does not mix with the mineral soil nor do bacteria consume it. Its slow decomposition is largely from fungi. Large peatlands, extending for several miles, often cover existing landscapes (Moore and Bellamy 1974). In a depression, organic matter or primary peat accumulates in saturated conditions, reducing the size of the water storage. Next to form are secondary peats that fill the depression up to the limit of water retention. Lastly, acid peats usually formed from sphagnum moss by the growth of "tertiary peat" on the existing peat and often on the land surface around the depression covering the landscape out from the depression (Moore and Bellamy 1974). "Tertiary peats are those which develop above the physical limits of groundwater, the peat itself acting as a reservoir holding a volume of water by capillarity above the level of the main groundwater mass draining through the landscape" (Moore and Bellamy 1974). Such a peat blanket is illustrated in Figure 3.35. Blanket peats are more common in areas of low evapotranspiration and a high amount of precipitation, such as eastern Canada and northern Finland. Water flow is restricted primarily to the peat, and stream initiation is prohibited. In peat basins containing only primary peat, water flow occurs into the basin (cf. humid climatic region). Any water that infiltrates the peat mat and reaches the mineral soil will probably flow laterally

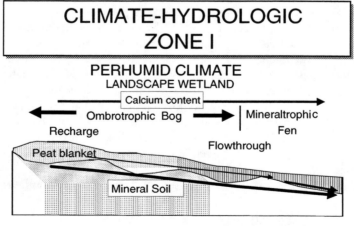

Figure 3.35 Perhumid blanket peatland with tertiary peat covering the landscape. Water flows in the peat or in the mineral soil below the peat. Lower areas are enriched with nutrients. Upper areas are distinctly nutrient deficient.

CLIMATE – HYDROLOGIC ZONE IV

DEPRESSION–FOCUSED RECHARGE WETLANDS

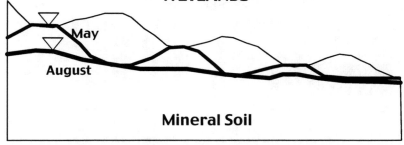

Figure 3.36 In semiarid regions with hummocky topography the depressions are nearly recharge areas. (Adapted from Miller, J.J., D.F. Acton, and R.J. St. Arnaud. 1985. The effect of groundwater on soil formation in a morainal landscape in Saskatchewan. *Can. J. Soil Sci.* 65:293–307.)

below the peat in these landscapes. Secondary peats create a situation that stops or inhibits the growth of stream channels. This lack of channel development results from the fact that water only flows below, on, or in the peat mat. The only water that reaches the peat surface is rainwater and hence is very nutrient poor.

Zone 2 is the same as the humid climate discussed earlier in the section titled *Groundwater Recharge and Discharge Relationships in Humid, Hummocky Landscapes,* and Zone 3 is the same as the subhumid climate discussed in the section dealing with subhumid, hummocky landscapes. Zone 4 (semiarid) contains dominantly recharge wetlands because the lack of precipitation and high ET precludes the integrated groundwater systems of the aforementioned zones. The climate is so dry that only recharge wetlands or low prairies occur, with a few saline ponds (Figure 3.36). Miller et al. (1985) describe this type of landscape in a semiarid climate. Fifteen of sixteen catchments that they studied were characterized by recharge hydrology and corresponding soil morphologies, such as soils with argillic horizons in the wetlands. Wetland soils were leached, and the surrounding wetland edge soils were calcareous and dominated by evaporites. Many of these soils contained natric horizons.

Generalized Landscapes with Soils and Hydrology

Winter (1988) related two generalized landscapes in an effort to unify the hydrodynamics of nontidal wetlands. The following demonstrates that in combination with soil information, his landscapes seem to provide a framework for interpretation. His landscapes consisted of a high landform and a low landform connected by a scarp or steeper slope.

The first of these generalized landscapes consists of a smooth flat upland with a corresponding lowland. This model landscape compares well with the Atlantic Coastal Plain "red-edge" landscapes observed by Daniels and Gamble (1967). These soils in the southeastern states are well drained and hematitic often with a distinct red color. The wetter and more interior soils become progressively yellower first as a function of iron hydration and then gray due to iron losses from the poorly drained soils. We present a modified version here with soil classifications added to demonstrate the landscape–hydrology–soil continuum (Figure 3.37). The actual coastal area used for our model

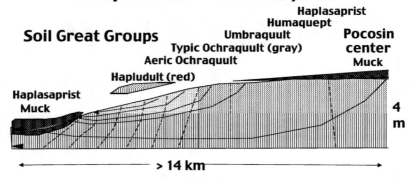

Figure 3.37   Soil distribution and flownet for a high rainfall flat upland typical of the low coastal plains near the Atlantic Ocean.

has a thin aquifer over an aquitard that is several miles wide. The hydraulic gradient is thus very low. The equipotential lines are widely spaced. Most of the recharge actually occurs from the Umbraquults to the Hapludults and not from the pocosin center muck-textured Histosol or organic soils. The pocosin center soils only receive rainwater as a water source (ombrotrophic) but drain the water exceedingly slowly such that the water becomes stagnant (stagno-groundwater recharge). The nutrients and soluble ions are slowly removed over time. The pocosin center soils, therefore, are mostly leached Histosols (organic soils). The Haplosaprist muck in the low landscape position in Figure 3.37 is an example of a mineralotrophic soil (mineral-rich Histosol).

Recharge is highest in the soils on the edge of the upper landform. These soils have argillic horizons and have lost iron due to reduction grading from the Hapludult to the Umbraquult. Colors range from red in the oxidized Hapludults to gray in the more reduced Umbraquults.

Winter's (1988) second generalized landscape, which he called "hummocky topography," is typified by local flow systems centered on depressions and intervening microhighs. We illustrate this type of landscape with a flownet modeled from an area in south central North Dakota (Figure 3.38). The landscape transect that we sampled has seven distinct depressions with many smaller ones that are too small for the scale. The transect distance is about 2 miles (3 km). Equipotential lines occur in 0.5 m (20 inches) head intervals (dashed). There is approximately 6 m (20 feet) of head loss over the entire transect, with head decreasing from the left (south) to the right (north). Bold arrows mark the three largest wetlands. The illustration characterizes a landscape with regional flow being disrupted by complex local flow systems. At a larger scale, with the smaller depressions visible, flow is even more disrupted.

Lissey (1971) described depression-focused recharge and discharge ponds. Water in a ponded condition flows even if the movement is extremely slow. The movement impacts soils by removing or adding dissolved components and translocating clay materials. Discharging groundwater tends to add material to the soils, while recharging groundwater leaches material from the soil. Groundwater flow can reverse or alternate, thereby leading to a reversal in pedogenic processes. Over time, the dominant flow processes will be manifested in a unique pedogenic morphologic signature. An interpretation of the hydrologic regime can, therefore, be made using soil morphology (Richardson 1997).

A major problem with using soil morphology as an indicator of wetland hydrology, however, is that the natural groundwater hydrologic regime has often been altered through anthropogenic disturbance activities. These activities may include ditches and tile lines for removing water from

FLOWNET OF A HUMMOCKY

TOPOGRAPHY IN TILL from Dickey, County,

ND

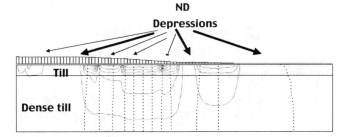

Figure 3.38 A FLOWNET simulation based on a landscape in till topography in south central North Dakota. The equipotential lines are 0.5 m decreasing increments from the high on the left (south) to the low on the right (north).

a wetland, and dams and dikes that prevent water from entering a wetland. (Committee on Characterization of Wetlands 1995). It takes years for soil morphology to equilibrate with the new hydrologic regime. The morphologic indicators may be relict features indicative of the predisturbance hydrologic conditions.

For the examination of the small depressions that were too small to see individually on Figure 3.38, the smooth topography model of Winter (1988) could be utilized on each one because only local flow would be involved. For example in recharge wetlands, water collects in depressions and percolates slowly to the water table (Figure 3.39). Percolating water often forms mounded water tables in topographically low areas (Knuteson et al. 1989). Knuteson et al. (1989) described recharge

WET SEASON EVENTS
DEPRESSION FOCUSED
RECHARGE

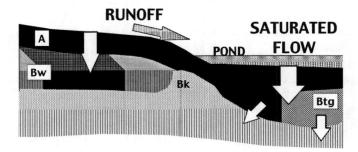

Figure 3.39 Wet season water flow system in depression-focused recharge wetlands. Variations in climate, stratigraphy, and topography alter details of the basic model. (Data from Lissey, A. 1971. Depression-focused transient groundwater flow patterns in Manitoba. *Geol. Assoc. Can. Spec. Paper* 9:333–341; Knuteson, J.A., J.L. Richardson, D.D. Patterson, and L. Prunty. 1989. Pedogenic carbonates in a Calciaquoll associated with a recharge wetland. *Soil Sci. Soc. Am. J.* 53:495–499; Richardson, J.L., J.L. Arndt, and J. Freeland. 1994. Wetland soils of the prairie potholes. *Adv. Agron.* 52:121–171.)

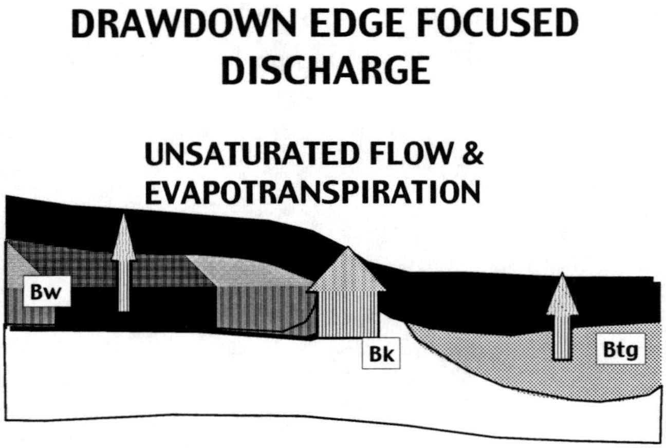

DRAWDOWN EDGE FOCUSED DISCHARGE

UNSATURATED FLOW & EVAPOTRANSPIRATION

Figure 3.40   When the pond dries, upward flow is established by the drying influence at the surface of evapotranspiration and creates an upward wet to dry matric potential that initiates unsaturated upward flow. The edges of the depression have the longest period of time with upward flow and lack much downward flow in the wet periods, hence the thicker Bk horizons.

wetlands formed in a subhumid climate of eastern North Dakota. They observed that the water table mounded under the depression during ponding events. The water table surface also had a steeper relief than existed on the ground surface; the mound disappeared or was lowered during the drying of the wetland. Recharge wetlands are common in subhumid and drier climates, and they usually dry out during the growing season.

During precipitation events, or during spring snow melt, water moves by overland flow or by infiltration and throughflow into the wetland. The soil profiles tend to be leached in the uplands during these events, removing some carbonates and creating a Bw horizon. The Bw horizon is a weakly developed horizon.

The edge of the depression receives water that discharges from throughflow or transient flow during the aforementioned precipitation events (Figure 3.40). In times of low precipitation, these areas dry out and have abundant water moving upward via unsaturated flow through the soil in response to plant uptake and evapotranspiration. Dissolved materials are left as the water evaporates, resulting in the formation of Bk horizons. Carbonate levels in these horizons have been well in excess of 30%. This illustrates the fact that over one quarter of the soil mass of these horizons has formed as an evaporite. Knuteson et al. (1989) examined the rate of formation of these horizons based on unsaturated flow and concluded that a horizon of this type can form in a few thousand years.

The pond area receives much water and temporarily has water above the soil surface nearly every year. The pond centers become inundated earlier and stay wet longer than other portions of the local landscape. Water moves downward through the profile along a hydraulic gradient (Figure 3.39), leaching and translocating material with it. Much of the dissolved material is completely leached from the profile, although some may be returned to the soil as the pond dries. Translocated clays accumulate at depth in the profile forming impermeable Btg horizons. These Btg horizons slow the percolation of water through the wetland bottom and increase the effectiveness of the pond to hold water.

The water flow system illustrated in Figures 3.39 and 3.40 results in soils with Bk horizons (carbonate accumulation) adjacent to soils with Bt horizons (carbonates removed and clays translocated). These soil types are extremely contrasting even though they are separated by only a few centimeters of elevation.

Recharge occurs first where the vadose zone is thinnest.

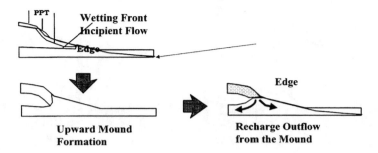

Figure 3.41 The development of a groundwater mound during rain events alters water flow into a wetland. The vadose zone is thinnest here. (Adapted from Winter, T.C. 1989. Hydrologic studies of wetlands in the northern prairies. pp. 16–54. *In* A. Van der Walk (Ed.) *Northern Prairie Wetlands.* Iowa State University Press. Ames, IA.)

Zonation in Wetlands: Edge Effects

The edges of ponds and wetlands often display alternating flow regimes (e.g., saturated–unsaturated) several times per year. Such edge-focused processes were discussed in a preceding section. The wave-action mentioned earlier, for instance, created different landforms and soil types at the wetland edge. We previously mentioned the "red edge" effect and other edge phenomena. We will examine other edge-focused processes further in this section.

Flow reversals are specific hydrologic occurrences that are frequently observed at pond edges (Rosenberry and Winter 1997). Flow reversals occur when recharge flow changes to discharge flow, or vice versa. After rainfall events, infiltration and interflow shunt water to the pond edge and create a mounded water table (Figure 3.41). The water table is already near the soil surface at a pond edge. Groundwater moving as interflow now fills the pores that are not saturated. It is easy to saturate soils when the water table is near the surface both because of the thinness of the unsaturated zone and the large amount of unsaturated pore space present in the unsaturated capillary fringe (Winter 1983). The mounded water table at the wetland edge rises above the pond and acts as somewhat of a miniature drainage divide. The mound is a recharge mound, with groundwater moving both downslope into the pond and into the earth. The mound (Figure 3.41) intercepts interflow and shunts much of it via infiltration into the ground. Some of the interflow also recharges the mound. During these events the soil is leached. This scenario is the opposite of the evaporative discharge often seen during dry periods at the edge, and the usual discharge of groundwater into the pond (Rosenberry and Winter 1997; Figure 3.42).

Plants at the edge of the wetland, such as phreatophytes and hydrophytes, are consumptive water users. Phreatophytes act like large water pumps, and selective plantings of these water users can alter local subsurface hydrology in the same manner as the pumping well in Figure 3.10. They create a depression in the water table, which illustrates that the water table mound is removed by water losses and replaced by a depression in the water table not long after the cessation of rain (Rosenberry and Winter 1997). The flow is reversed, and the water table depression also acts as a barrier to groundwater flowing into the pond. Wetland edges have frequent flow reversals of this type. During mound and depression phases, groundwater is restricted in its movement to the wetland.

Whittig and Janitzky (1963) in their classic paper described a wetland edge effect consisting of the accumulation of sodium carbonate (Figure 3.43). This type of edge effect has been widely known and is used as a model to illustrate salinization and alkalinization in warm climates. Chemical

Evapotranspiration by phreatophytes results in groundwater depressions forming at the pond edge

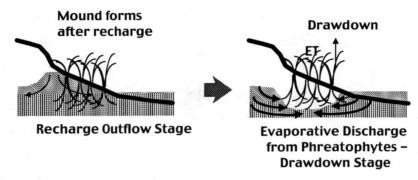

Figure 3.42   The mound dissipates quickly because the vegetation at wetland edges, particularly phreatophytes and hydrophytes, consume large quantities of water. These plants create a drawdown of the water table and disrupt water flow to the pond. Mounds alternating with drawdown depressions at the pond edge represent flow reversals.

reduction via microbial transformations liberates the carbonate anion that then reacts with calcium to form the mineral calcite. Calcite precipitation removes calcium from the system, which increases the relative amounts of carbonate and bicarbonate anions in the soil solution. As the soil dries, matric potentials increase and water moves via capillarity transporting these anions, as well as sodium cations, toward the soil surface. During the evaporation process, the water loses dissolved carbon dioxide, resulting in an increase in pH. When bicarbonate looses carbon dioxide, carbonate forms. Whittig and Janitzky (1963) noted pH values as high as 10 in some of their profiles, with abundant sodium carbonate forming as a surface efflorescence. Inland and at slightly higher elevations, carbon dioxide is not a factor in carbonate formation. The carbon dioxide stays in solution, sulfate is not reduced, and thereby does not precipitate or form either calcium carbonate or sodium carbonate. In these places, the soils become saline with accumulations of sodium and magnesium sulfates.

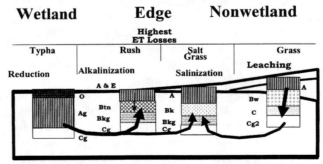

Figure 3.43   Edge-focused evaporative discharge with sodium carbonate development. This edge is more common in mesic and warmer climates. (Adapted from Whittig, L.D. and P. Janitzky. 1963. Mechanisms of formation of sodium carbonate in soils I. Manifestations of biologic conversions. *J. Soil Sci.* 14:322–333.)

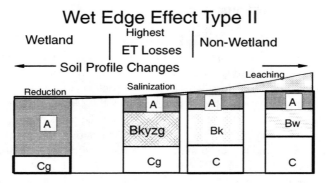

Figure 3.44 Evaporative discharge edge with gypsum and calcite rather than sodium carbonate. This edge is more common in cooler climates. (From Steinwand, A.L. and J.L. Richardson. 1989. Gypsum occurrence in soils on the margin of semipermanent prairie pothole wetlands. *Soil Sci. Soc. Am. J.* 53:836–842. With permission.)

In northern climates, carbon dioxide remains in solution longer because the cool temperature retards sulfate reduction and allows for more dissolved carbon dioxide. In North Dakota and the Prairie Provinces of Canada, abundant sulfate is present and some reduction to sulfide occurs; however, the amount of carbonate in solution is less than the amount of available calcium (Arndt and Richardson 1988, 1989, Steinwand and Richardson 1989). Calcite and gypsum, therefore, are produced in place of sodium carbonate at the edge (Figure 3.44). The pathways of calcite and gypsum production are explained more fully in Chapter 18 and in Arndt and Richardson (1992). The result is that in northern areas, soil salinity is dominated by calcite and has pH levels that seldom exceed 8.3.

WETLAND HYDROLOGY AND JURISDICTIONAL WETLAND DETERMINATIONS

Wetlands are regulated under a variety of federal, state, and local statutes; however, in order to regulate a resource, the resource must be defined. The majority of the regulatory agencies that have jurisdiction over the nation's wetland resource use the 1987 U.S. Army Corps of Engineers (COE) manual to identify wetlands. While this chapter provides the background and context to understand wetland hydrology and assessment, the 1987 manual is the current authority that provides methods to assess the presence/absence of wetland hydrology in jurisdictional wetlands. The following discussion places the concepts of wetland hydrology developed above into a regulatory context. The Corps of Engineers currently maintains an updated version of the 1987 manual on the Internet, complete with user guidance. The reader is directed to the online version of the 1987 manual for more details.

Wetland Hydrology Defined

The U.S. Army COE (1987) *Wetlands Delineation Manual* defines wetland hydrology as follows:

"The term 'wetland hydrology' encompasses all hydrologic characteristics of areas that are periodically inundated or have soils **saturated to the surface** at some time during the growing season. Areas with evident characteristics of wetland hydrology are those where the presence of water has an overriding influence on characteristics of vegetation and soils due to anaerobic and reducing conditions, respectively. Such characteristics are usually present in areas that are inundated or have soils that are **saturated to the surface** for sufficient duration to develop hydric soils and support vegetation typically adapted for life in periodically anaerobic soil conditions" (paragraph 46, emphasis added).

Duration of Inundation or Saturation

Saturation to the surface for some period is an apparent requirement for wetland hydrology to be present. Table 5 of the 1987 manual provides guidance on the duration of inundation or saturation that is required for wetland hydrology to exist. Areas that are intermittently or never inundated or saturated (i.e., less than 5% of the growing season) have such conditions for an insufficient duration to be classified as wetland. For example, in the Minneapolis, Minnesota area, the growing season lasts from about May 1 to October 1, or 153 days based on the soil survey data from the area; 5% of the growing season equates to 7.65 days. Thus, in Minneapolis, inundation or saturation to the surface must be present for an absolute minimum of 8 days during the growing season for wetland hydrology to exist as defined in the 1987 manual.

While the presence or absence of a water table for 8 days during the growing season can be easily determined by monitoring water levels along well transects, the duration criteria are confounded by the requirement that this level of ponding duration and intensity be present "in most years." Recent guidance from the COE has indicated that "in most years" means 51 years out of 100 (March 1992 COE Guidance on the 1987 Manual). Thus, when assessing hydrology using wells, the climatic context is extremely important because the standard could not feasibly be determined experimentally.

Field Methodology for Determining Wetland Hydrology

The U.S. Army COE 1987 manual provides a field methodology for determining if soil saturation is present:

"Examination of this indicator requires digging a soil pit to a depth of 16 inches and observing the level at which water stands in the hole after sufficient length of time has been allowed for water to drain into the hole. The required time will vary depending on soil texture. In some cases, the upper level at which water is flowing into the pit can be observed by examining the wall of the hole. This level represents the depth to the water table. The depth to saturated soils will always be nearer the surface due to the capillary fringe. For soil saturation to impact vegetation, it must occur within a major portion of the root zone (usually within 12 inches of the surface) of the prevalent vegetation" (paragraph 49.b. [2]).

This open borehole methodology indicates that the parameter being measured is whether the water table is within 12 inches of the surface. With a water table at this shallow depth, it is generally assumed that saturation to the surface will periodically occur due to water table fluctuations or capillary action. The reader is directed to caveats discussed in the section titled *Adhesion, Cohesion, and Capillarity* in this chapter on the use of capillary fringe concepts in defining the saturated zone.

Importance of the Wetland Hydrology Parameter to Jurisdictional Wetland Determinations

The U.S. Army COE 1987 manual makes clear that the presence of wetland hydrology *may not* be inferred from the presence of hydric soils and a predominance of hydrophytic plants, particularly when an area has been altered from "normal circumstances." The 1987 manual states that:

"… sole reliance on vegetation or either of the other parameters as the determinant of wetlands can sometimes be misleading. Many plant species can grow successfully in both wetlands and non-wetlands, and hydrophytic vegetation and hydric soils may persist for decades following alteration of hydrology that will render an area a non-wetland" (paragraph 19).

The 1987 Manual also provides further guidance on drained hydric soils:

"A drained hydric soil is one in which sufficient ground or surface water has been removed by artificial means such that the area will no longer support hydrophytic vegetation. Onsite evidence of drained soils includes:

a. Presence of ditches or canals of sufficient depth to lower the water table below the major portion of the root zone of the prevalent vegetation.
b. Presence of dikes, levees, or similar structures that obstruct normal inundation of an area.
c. Presence of a tile system to promote subsurface drainage.
d. Diversion of upland surface runoff from an area.

Although it is important to record such evidence of drainage of an area, a hydric soil that has been drained or partially drained still allows the soil parameter to be met. The area, however, will not qualify as a wetland if the degree of drainage has been sufficient to preclude the presence of either vegetation **or a hydrologic regime** that occurs in wetlands" (paragraph 38, emphasis added).

This analysis in the 1987 manual suggests that the correct assessment of the hydrologic parameter is essential to delineate jurisdictional wetlands, especially in areas where hydrology has been impacted by anthropogenic or natural causes, resulting in possibly relict hydric soils and relict hydrophyte plant communities being present. Such areas would fall into the "Atypical" situation covered by Section F of the 1987 manual. The online version of the 1987 manual provides the following guidance:

"[W]hen such activities occur [reference is to draining, ditching, levees, deposition of fill, irrigation, and impoundments] an area may fail to meet the diagnostic criteria for a wetland. Likewise, hydric soil indicators may be absent in some recently created wetlands. In such cases, an alternative method must be employed in making wetland determinations" (paragraph 12.a).

Application of Basic Hydrologic Concepts to Jurisdictional Wetlands

The U.S. Army COE 1987 manual provides scant guidance regarding what alternative methods are suitable in altered situations, nor does it provide estimates for the extent of anthropogenically altered wetlands that may require alternative methods. The need for alternative methods may be far greater than is generally recognized because few landscapes, especially in agricultural, urban, and suburban landscapes, are in their natural state. Many wetlands have been impacted by agriculture and urbanization, with the result that wetland hydrology, hydrophytic plant communities, and hydric soils are not in equilibrium with each other. Under these conditions a routine delineation may not accurately define the extent of the wetland resource. Many wetland specialists prefer to perform hydrologic studies in these areas because of suspected relict hydric soils and hydrophytic vegetation.

"When hydrologic alteration is suspected, performance of an adequate study should consist of, at a minimum, the following procedures (modified from Section F, the 1987 manual):

1. **Describe the type of alteration.** Anthropogenic impacts to wetland hydrology may be subtle or obvious, and may result in an alteration to wetter or drier conditions. Agricultural drainage ditches, drain tiles, dikes, levees, and filling are obvious attempts to remove water from an area or prevent water from flowing onto an area. Stormwater drains and diversions are obvious indicators that water may be added to an area. The effects of urbanization and agricultural use are more subtle, and may have broad, regional impacts on the groundwater system that are not obvious, yet may result in a continuous, overall decline in the health and magnitude of the wetland resource.

2. **Describe the effects of the alteration.** The effects of several hydrologic alterations can be theoretically addressed by employing many of the concepts examined in this chapter, focusing on an assessment of the effects the alterations have on the water balance of the study area. For example:

- Drainage increases the outputs of the water balance equation at the expense of storage. The result is a decline in water table depth and reduced wetland acreage. Well studies are often the only way to effectively determine if the affected area is partially drained and jurisdictional, or effectively drained and non-jurisdictional.
- Stormwater inputs increase water inputs at the expense of outputs, frequently resulting in an increase in storage and an enlarged wetland. Inverse condemnation (too much land restored to a wetland) lawsuits are frequently lodged by landowners affected by additions of stormwater to existing wetlands in urbanizing areas.
- Stream channelization results in more efficient removal of flood flows, with the result that riparian wetlands at the periphery of channelized streams become drier.
- Urbanization results in an increased area of impervious surfaces that prevent infiltration and reduce groundwater recharge. The management of stormwater off of these surfaces, however, can result in significantly increased runoff to wetlands that are part of the stormwater system.
- Tillage in a wetland's catchment accelerates sedimentation and infilling of the wetland, and has poorly understood effects on groundwater dynamics and the water balance of affected wetlands. Hydrographs along well transects are frequently used to assess the presence of jurisdictional wetland hydrology in hydrologically altered situations. It is difficult and expensive, however, to monitor the wells for a sufficiently long period to interpolate from the data the presence/absence of wetland hydrology for 51 out of 100 years. When hydrographs and well transects are employed, it is particularly important to provide a strong long- and short-term climatic context, to describe the effects of the alteration as well as possible, and to document supporting observations such as the presence of invading upland plant species.

3. **Characterize the preexisting conditions.** This characterization is commonly performed with an interpretation of the existing aerial photo history augmented with map analyses, literature searches, soil survey information, and soils and vegetation documentation. An important change that should be mentioned is the change from phreatophytes, which are heavy water users, to field crops, which use very little water comparatively.

Considerations, Caveats

Jurisdictional wetland delineation has as its focus the dry edge of the wetland. It is an unfortunate reality that wetland delineation does not focus on wetland presence or absence, but instead focuses on the aerial extent of the wetland. The term "unfortunate" is used because wetland delineation takes the most dynamic portion of the wetland that exists as a transition zone and turns it into a two-dimensional line. It is for these reasons that most of the disputes involving jurisdictional wetland boundaries occur at the wetland edge: we take something that exists as a gradient in three dimensions and turn it into two. In many situations this representation of the wetland boundary is unrealistic.

It is also at this dry edge where the soil–landscape–hydrology interactions result in the development of hydric soil morphology that is transitional to upland soil characteristics. In addition to being the location of the jurisdictional boundary, sediment deposition also occurs primarily at the wetland edge. Sediment deposition has significant impacts on wetland longevity, functions, and quality, especially when accelerated by human activities. It is unfortunate that researchers often ignore these transitional areas. Pond interiors are often the only locations that have water level recorders and other instrumentation for measuring hydroperiod. Measuring hydroperiod only in the interiors and not on the wetland edges results in an incomplete picture of hydroperiod. It is only through an understanding of the dynamic hydrology of the transition zone between wetland and upland that we can understand the interactions between hydrology, soils, and vegetation sufficiently to make accurate jurisdictional determinations, and wisely manage the wetland resource.

SUMMARY

A wetland, as suggested by the nature of the name, consists of two natural media interacting: water and soil. Wetland hydrology is dynamic and can change with a single rainstorm event, or a rapid snowmelt, or during a hot windy day. The wetland water balance is the fundamental relationship between inputs, outputs, and storage that dictates the presence or absence of a wetland. The water may come from the landscape where it has been gathered from its catchment basin or fall directly on the wetland via precipitation. Water, once in the wetland, either stays, leaves by evapotranspiration, or it drains away.

To be a hydric soil, the soil must remain saturated for an extended time and be chemically reduced. The chemical and physical processes that occur by water moving into, through, and from the soil alter it in distinct, visible ways. These changes occur slowly over time as a response to the water activity. This visible hydrologic signature is called soil morphology. Recharge dominance, for instance, is the direct movement of water from the wetland to groundwater. The movement of water over time in this manner leaches soluble material and translocates clay in the soil. Discharge dominance, on the other hand, adds materials such as calcium carbonate to hydric soils. Iron is usually chemically reduced in saturated conditions and often alternatively oxidized during drier periods. This creates a distinct morphological pattern that reflects both the soil chemistry and hydrologic conditions. Hydric soil indicators developed from the process.

Landscape, climatological, and biological conditions must exist to get and keep a wetland wet. Hillslope geometry and position, such as the base of long slopes, shed and concentrate water at certain places. Depressions frequently constrain water from flowing freely to a stream. Strata, such as sand lenses, may gather the water from a large catchment and concentrate the water in a wetland. Climatic constraints, such as copious quantities of precipitation or very low evapotranspiration rates, maintain water in the wetland throughout a year or periodically during a wet season. Certain plants may foster the retention of water and aid in wetland creation. All these conditions are reflected in hydric soils. The soils reflect the hydrology of the pedons throughout the wetlands and can be used to determine the hydrology expected over time, the wetland as a whole, or zones within a wetland.

Alteration of the wetland, frequently for an economic purpose, changes wetland hydrology. Sadly, a rather long period of time may occur before the hydric soils equilibrate and reflect the new hydrologic conditions via their soil morphology.

REFERENCES

Arndt, J.L. and J.L. Richardson. 1988. Hydrology, salinity, and hydric soil development in a North Dakota prairie-pothole wetland system. *Wetlands: J. Soc. Wetland Sci.* 8:94–108.

Arndt, J.L. and J.L. Richardson. 1989. Geochemical development of hydric soil salinity in a North Dakota prairie-pothole wetland system. *Soil Sci. Soc. Am. J.* 53:848–855.

Arndt, J.L. and J.L. Richardson. 1992. Carbonate and gypsum chemistry in saturated, neutral pH soil environments. p. 179–187. *In* R.D. Robarts and M.L. Bothwell (Eds.) *Aquatic Ecosystems in Semi-arid Regions: Implications for Resource Management.* N. H. R. I. Symposium Series 7, Environment Canada, Saskatoon, Saskatchewan, Canada.

Arndt, J.L. and J.L. Richardson. 1993. Temporal variation in salinity of shallow groundwater collected from periphery of North Dakota wetlands (U.S.A.). *J. Hydrology.* 141:75–105.

Bailey, R.W. and O.L. Copeland. 1961. Low flow discharges and plant cover relations on two mountain watersheds in Utah. *Intern. Assoc. Sci. Hydrol. Pub.* 51:267–278.

Bouma, J. 1990. Using morphometric expressions for macropores to improve soil physical analyses of field soils. *Geoderma* 46:3–11.

Brady, N.L. and R.R. Weil. 1998. *The Nature and Properties of Soils,* 12th ed. Prentice-Hall, Englewood Cliffs, NJ. 740 p.

Brinson, M.M. 1993. A hydrogeomorphic classification for wetlands. Wetlands Research Program Tech. Rept. WRP-DE-4. U.S. Army Corps of Engineers, Waterways Experiment Station. Vicksburg, MS.

Carroll, D. 1970. *Rock Weathering*. Plenum Press. New York.

Cedargren, H.R. 1967. *Seepage, Drainage, and Flow Nets*. John Wiley & Sons, New York.

Chorley, R.J. 1978. The hillslope hydrological cycle, pp 1–42, *In* M.J. Kirkby (Ed.) *Hillslope Hydrology*. John Wiley & Sons, New York.

Committee on Characterization of Wetlands. 1995. *Wetlands: Characteristics and Boundaries*. W.W. Lewis, (Chair), National Research Council, National Academy of Sciences, Washington, DC.

Cowardin, L.M., V. Carter, F.C. Golet, and E.T. LaRoe. 1979. *Classification of Wetland and Deepwater Habitats of the United States*. FWS/OBS-79/31. U.S. Fish and Wildlife Service, Washington, DC.

Daniels, R.B. and E.E. Gamble. 1967. The edge effect in some Ultisols in the North Carolina coastal plain. *Geoderma* 1:117–124.

Elburg, H. van, G.B. Engelen, and C.J. Hemker. 1990. The Free University, Institute of Earth Sciences, DeBoeleaan 1085, Amsterdam, The Netherlands.

Freeland, J.A. 1996. Soils and sediments as indicators of agricultural impacts on northern prairie wetlands. Ph.D. dissertation, North Dakota State University, Fargo, ND.

Freeland, J.A., J.L. Richardson, and L.A. Foss. 1999. Soil indicators of agricultural impacts on northern prairie wetlands: Cottonwood Lake Research area, North Dakota. *Wetlands* 19:78–89.

Freeze, R.A. and J.A. Cherry. 1979. *Groundwater*. Prentice Hall, Englewood Cliffs, New Jersey.

Freeze, R.A. and P.A. Witherspoon. 1966. Theoretical analysis of regional groundwater flow: 1. Analytical and numerical solutions to the mathematical model. *Water Resource Res.* 2:641–656.

Freeze, R.A. and P.A. Witherspoon. 1967. Theoretical analysis of regional groundwater flow: 2. Effect of water table configuration and subsurface permeability variation. *Water Resource Res.* 3:623–634.

Freeze, R.A. and P.A. Witherspoon. 1968. Theoretical analysis of regional groundwater flow: 3. Quantitative interpretations. *Water Resource Res.* 4:581–590.

Gambrel, R.P. and W.H. Patrick, Jr. 1978. Chemical and microbiological properties of anaerobic soils and sediments. p. 375–423 *In* D.D. Hook and R.M.M. Crawford (Eds.) *Plant Life in Anaerobic Environments*. Ann Arbor Sci. Publ. Inc., Ann Arbor, MI.

Gilley, J.E., B.D. Patton, P.E. Nyren, and J.R. Simanton. 1996. Grazing and haying effects on runoff and erosion from a former conservation reserve program site. *Applied Engineering in Agriculture* 12:681–684.

Greenwood, D.J. 1961. The effect of oxygen concentration on the decomposition of organic materials in soil. *Plant and Soil* 14:360–376.

Hewlett, J.D. and W.L. Nutter. 1970. The varying source area of streamflow from upland basins. pp. 65–83. Paper presented at Symposium on Interdisciplinary Aspects of Watershed Management, Montana State University, American Society of Civil Engineers, New York.

Hubbert, M.K. 1940. The theory of groundwater motion. *J. Geol.* 48:785–944.

Hurt, G.W., P.M. Whited, and R. Pringle (Eds.). 1996. *Field Indicators of Hydric Soils of the United States*. USDA Natural Resources Conservation Service, U.S. Govt. Printing Office, Washington, DC.

Jenny, H. 1941. *Factors of Soil Formation*. McGraw-Hill, New York.

Kirkby, M.J. and R.J. Chorley. 1967. Throughflow, overland flow and erosion. *Bull. Intern. Assoc. Sci. Hydrology* 12:5–21.

Kirkham, D. 1947. Studies of hillslope seepage in the Iowan drift area. *Soil Sci. Soc. Am. Proc.* 12:73–80.

Knuteson, J.A., J.L. Richardson, D.D. Patterson, and L. Prunty. 1989. Pedogenic carbonates in a Calciaquoll associated with a recharge wetland. *Soil Sci. Soc. Am. J.* 53:495–499.

Kramer, J., J. Printz, J. Richardson, and G. Goven. 1992. Managing grass, small grains, and cattle. *Rangelands* 14(4):214–215.

Kutilek, M. and D.R. Nielsen. 1994. *Soil Hydrology*. Catena Verlag, Cremlingen-Destedt, Germany.

Lissey, A. 1971. Depression-focused transient groundwater flow patterns in Manitoba. *Geol. Assoc. Can. Spec. Paper* 9:333–341.

Maasland, M. and H.C. Haskew. 1957. The auger hole method of measuring the hydraulic conductivity of soil and its application to tile drainage problems. Proc. 3rd Cong., International Commission on Irrigation and Drainage (ICID), p. 8:69–8:114.

Mausbach, M.J. 1992. Soil survey interpretations for wet soils. *In* J. M. Kimble (Ed.) Proc. 8th Intern. Soil Correlation Meeting (VIII ISCOM): Characterization, Classification and Utilization of Soils. USDA–SCS. National Soil Survey Center, Lincoln, NE.

Meyboom, P. 1967. Mass-transfer studies to determine the groundwater regime of permanent lakes in hummocky moraine of western Canada. *J. Hydrol.* 5:117–142.

Miller, J.J., D.F. Acton, and R.J. St. Arnaud. 1985. The effect of groundwater on soil formation in a morainal landscape in Saskatchewan. *Can. J. Soil Sci.* 65:293–307.

Mills, J.G. and M. Zwarich. 1986. Transient groundwater flow surrounding a recharge slough in a till plain. *Can. J. Soil Sci.* 66:121–134.

Moore, P.D. and D.J. Bellamy. 1974. *Peatlands.* Springer Verlag, New York.

National Wetlands Working Group. 1988. *Wetlands of Canada.* Ecological Land Classification Series, No. 24 Environment Canada Polyscience Publications. Ottawa, Ontario.

Pauling, L. 1970. *General Chemistry.* Dover Publications. New York. 989 pp.

Pfannkuch, H.O. and T.C. Winter. 1984. Effect of anisotropy and groundwater system geometry on seepage through lakebeds, 1. Analog and dimension analysis. *J. Hydrol.* 75:213–237.

Richardson, J.L. 1997. Soil development and morphology in relation to shallow ground water — an interpretation tool. pp. 229–233 *In* K.W. Watson and A. Zaporozec (Eds.) *Advances in Ground-Water Hydrology: A Decade of Progress.* American Institute of Hydrology, St. Paul, MN.

Richardson, J.L., J.L. Arndt, and J. Freeland. 1994. Wetland soils of the prairie potholes. *Adv. Agron.* 52:121–171.

Richardson, J.L. and R.J. Bigler. 1984. Principle component analysis of prairie pothole soils in North Dakota. *Soil Sci. Soc. Am. J.* 48:1350–1355.

Richardson, J.L. and F.D. Hole. 1978. Influence of vegetation on water repellency in selected western Wisconsin soils. *Soil Sci. Soc. Am. J.* 42:465–467.

Richardson, J.L., L.P. Wilding, and R.B. Daniels. 1992. Recharge and discharge of ground water in aquic conditions illustrated with flownet analysis. *Geoderma* 53:65–78.

Rosenberry, D.O. and T.C. Winter. 1997. Dynamics of water-table fluctuations in a upland between two prairie-pothole wetlands in North Dakota. *J. Hydrology* 191:266–289.

Schoeneberger, P.J., D.A. Wysocki, E.C. Benham, and W.D. Broderson. 1998. *Field Book for Describing and Sampling Soils.* National Soil Survey Center, Natural Resources Conservation Service, USDA, Lincoln, NE.

Schot, Paul. 1991. Solute transport by groundwater flow to wetland ecosystems. Ph.D. thesis, University of Utrecht, Geografisch Instituut Rijksuniversiteirt. 134p.

Schwartzendruber, D., M.J. De Boodt, and D. Kirkham. 1954. Capillary intake rate of water and soil structure. *Soils Sci. Soc. Am. Proc.* 18:1–7.

Seelig, B.D. and J.L. Richardson. 1994. A sodic soil toposequence related to focused water flow. *Soil Sci. Soc. Am. J.* 58:156–163.

Simonson, R.W. 1959. Outline of a generalized theory of soil genesis. *Soil Sci. Soc. Am. Proc.* 23:152–156.

Skaggs, R.L., D. Amatya, R.O. Evans, and J.E. Parsons. 1994. Characterization and evaluation of proposed hydrologic criteria for wetlands. *J. Soil and Water Cons.* 49(5):501–510.

Skaggs, R.W., W.F. Hunt, G.M. Chescheir, and D.M. Amatya. 1995. Reference simulations for evaluating wetland hydrology. *In* K.L. Campbell (Ed.) *Versatility of Wetlands in the Agricultural Landscape.* Hyatt Regency, Tampa, Florida, September 17–20, 1995 Am. Soc. Agric. Engineers, St. Joseph, MI.

Sloan, C.E. 1972, *Ground-water Hydrology of Prairie Potholes in North Dakota.* U.S. Geol. Survey Professional Paper 585-C U. S. Govt. Printing Office, Washington, DC.

Steinwand, A.L. and J.L. Richardson. 1989. Gypsum occurrence in soils on the margin of semipermanent prairie pothole wetlands. *Soil Sci. Soc. Am. J.* 53:836–842.

Stewart, R.E. and H.A. Kantrud. 1971. Classification of natural ponds and lakes in glaciated prairie region. U.S. Fish Wildl. Serv., Res. Publ. 92. U.S. Gvt. Printing Office. Washington, DC.

Tischendorf, W.G. 1968. Tracing stormflow to a varying source area in small forested watershed in the southeastern Piedmont. Ph.D. dissertation, Univ. of Georgia, Athens, GA.

Toth, J. 1963. A theoretical analysis of groundwater flow in small drainage basins. *J. Geophys. Res.* 68:4197–4213.

U.S. Army COE. 1987. *Wetlands Delineation Manual.* Environmental Laboratory, U.S. Army Corps of Engineers Waterways Experiment Station, Vicksburg, MS.

U.S. Army COE St. Paul District. 1996. U.S. Army COE St. Paul District, 17 April 1996, Guidelines for submitting wetland delineations to the St. Paul District Corps of Engineers and Local Units of Government in the State of Minnesota. Public Notice 96-01078-SDE. USA-COE, St. Paul, MN.

Whittig, L.D. and P. Janitzky. 1963. Mechanisms of formation of sodium carbonate in soils I. Manifestations of biologic conversions. *J. Soil Sci.* 14:322–333.

Winter, T.C. 1983. The interaction of lakes with variably saturated porous media. *Water Resour. Res.* 19:1203–1218.

Winter, T.C. 1988. A conceptual framework for assessing cumulative impacts on hydrology of nontidal wetlands. *Environmental Management* 12:605–620.

Winter, T.C. 1989. Hydrologic studies of wetlands in the northern prairies. pp. 16–54. *In* A. Van der Valk (Ed.) *Northern Prairie Wetlands.* Iowa State University Press. Ames, IA.

Winter, T.C. 1992. A physiographic and climatic framework for hydrologic studies of wetlands. pp. 127–148. *In* Roberts, R.D. and M.L. Bothwell (Ed.) *Aquatic Ecosystems in Semi-arid Regions: Implications for Resource Management.* N. H. R. I. Symposium Series 7, Environment Canada Saskatoon.

Winter, T.C. and Woo, M-K. 1990. Hydrology of lakes and wetlands, pp. 159–187. *In* Wolman, M.G. and Riggs, H.C. (Eds.) *Surface Water Hydrology. The Geology of North America,* v. 0-1. Geological Society of America, Boulder, CO.

CHAPTER 4

Redox Chemistry of Hydric Soils

M. J. Vepraskas and S. P. Faulkner

INTRODUCTION

Hydric soils are described in Chapter 2 as soils that formed under anaerobic conditions that develop while the soils are inundated or saturated near their surface. These soils can form under a variety of hydrologic regimes that include nearly continuous saturation (swamps, marshes), short-duration flooding (riparian systems), and periodic saturation by groundwater. The most significant effect of excess water is isolation of the soil from the atmosphere and the prevention of O_2 from entering the soil. The blockage of atmospheric O_2 induces biological and chemical processes that change the soil from an aerobic and oxidized state to an anaerobic and reduced state. This shift in the aeration status of the soil allows chemical reactions to occur that develop the common characteristics of hydric soils, such as the accumulation of organic carbon in A horizons, gray-colored subsoil horizons, and production of gases such as H_2S and CH_4. In addition, the creation of anaerobic conditions requires adaptations in plants if they are to survive in the anaerobic hydric soils.

This chapter discusses the chemistry of hydric soils by focusing on the oxidation–reduction reactions that affect certain properties and functions of hydric soils and form the indicators by which hydric soils are identified (Chapter 7). Both the biological and chemical functions of wetlands are controlled to a large degree by oxidation–reduction chemical reactions (Mitsch and Gosselink 1993). The fundamentals behind these reactions will be reviewed in this chapter along with methods of monitoring these reactions in the field, and the effects of these reactions on major nutrient cycles in wetlands.

In our experience, soil chemistry is probably the subject least understood by students of hydric soils and wetlands in general. Therefore, the following treatment is intended to be simple, and to cover those topics that can be related to the field study of hydric soils. Students wishing more detailed treatments are encouraged to consult the work of Ponnamperuma (1972) in particular, as well as the discussion of redox reactions in McBride (1994) and Sparks (1995).

OXIDATION AND REDUCTION BASICS

Oxidation–reduction (redox) reactions govern many of the chemical processes occurring in saturated soils and sediments (Baas-Becking et al. 1960). Redox reactions transfer electrons among atoms. As a result of the electron transfer, electron donor atoms increase in valence, and the electron acceptor atoms decrease in valence. Such changes in valence usually alter the phase in which the atom occurs in the soils, such as causing solid minerals to dissolve or dissolved ions to turn to gases. The loss of one or more electrons from an atom is known as **oxidation,** because in the early days of chemistry the known oxidation reactions, such as rust formation, always involved oxygen. The gain of one or more electrons by an atom is called **reduction** because the addition of negatively charged electrons reduces the overall valence of the atom. Each complete redox reaction contains an oxidation and a reduction component that are called *half-reactions*. Redox reactions are more easily understood and evaluated when the oxidation and reduction half-reactions are considered separately. This is appropriate because oxidation and reduction processes each produce different effects on the soil.

For example, in aerobic soils organic compounds such as the carbohydrate glucose can be oxidized to CO_2 as shown in the following reaction:

$$C_6H_{12}O_6 + 6O_2 \rightarrow 6CO_2 + 6H_2O \qquad \text{(Equation 1)}$$

This reaction can be broken down into an oxidation half-reaction and a reduction half-reaction:

$$C_6H_{12}O_6 + 6H_2O \rightarrow 6CO_2 + 24e^- + 24H^+ \text{ (Oxidation)} \qquad \text{(Equation 2)}$$

$$6O_2 + 24e^- + 24H^+ \rightarrow 12H_2O \text{ (Reduction)} \qquad \text{(Equation 3)}$$

The basic oxidation half-reactions in soils are catalyzed by microorganisms during their respiration process (Chapter 5). The respiration is responsible for releasing one or more electrons as well as hydrogen ions. Oxidation occurs whenever heterotrophic microorganisms are using organic tissues as their carbon source for respiration, as when organic tissues are being decomposed in soils. For this discussion, bacteria will be considered the major group of organisms initiating the oxidation processes in soil. Organic tissues are the major source of electrons, and when the tissues are oxidized the electrons released are used for reducing reactions. The most important point to remember is that when organic tissues are not present, or when bacteria are not respiring, redox reactions of the type discussed in this chapter will not occur in the soil.

Alternate Electron Acceptors

Electron acceptors are the substances reduced in the redox reactions. Oxygen is the major electron acceptor used in redox reactions in aerobic soils. However, in anaerobic soils, where O_2 is not present, other electron acceptors have to be used by bacteria if they are to continue their respiration by oxidizing organic compounds. The major electron acceptors that are available in anaerobic soils are contained in the following compounds: NO_3^-, MnO_2, $Fe(OH)_3$, SO_4^{2-}, and CO_2 (Ponnamperuma 1972, Turner and Patrick 1968).

Theoretically, the electron acceptors are reduced in anaerobic soils in the order shown above. In an idealized case, when organic compounds are being oxidized, O_2 will be the only electron acceptor used while it is available. When the soil becomes anaerobic upon the complete reduction of most available O_2, then NO_3^- will be the acceptor reduced while it is available. This same sequence is followed by the other compounds shown. Thus, if O_2 is never depleted, the reduction of the other compounds will never occur. While not all bacteria use the same electron acceptors, we will assume that most soils contain all microbial species necessary to reduce each of the electron acceptors noted earlier.

The order of reduction discussed above is idealized and probably does not occur in soil horizons exactly as predicted from theoretical grounds. It has been observed that the reduction of Fe^{3+} and

Table 4.1 Half-Cell Reducing Reactions and the Equations Used to Calculate the Phase Change Lines Shown in Figure 4.1

Half-Cell Reaction	Redox Potential (Eh, mV) =
$1/4O_2 + H^+ + e^- = 1/2H_2O$	$1229 + 59\log(P_{O2})^{1/4} - 59pH$
$1/5NO_3^- + 6/5H^+ + e^- = 1/10N_2 + 3/5H_2O$	$1245 - 59[\log(P_{N2})^{1/10} - \log(NO_3)^{1/5}] - 71pH$
$1/2MnO_2 + 2H^+ + e^- = 1/2Mn^{2+} + H_2O$	$1224 - 59\log(Mn^{2+}) - 118pH$
$Fe(OH)_3 + 3H^+ + e^- = Fe^{2+} + 3H_2O$	$1057 - 59\log(Fe^{2+}) - 177pH$
$FeOOH + 3H^+ + e^- = Fe^{2+} + 2H_2O$	$724 - 59\log(Fe^{2+}) - 177pH$
$1/2Fe_2O_3 + 3H^+ + e^- = Fe^{2+} + 3/2H_2O$	$707 - 59\log(Fe^{2+}) - 177pH$
$1/8SO_4^{2-} + 5/4H^+ + e^- = 1/8H_2S + 1/2H_2O$	$303 - 59[\log(P_{H2S})^{1/8} - \log(SO_4^{2-})^{1/8}] - 74pH$
$1/8CO_2 + H^+ + e^- = 1/8CH_4 + 1/4H_2O$	$169 - 59[\log(P_{CH4}) - \log(P_{CO2})^{1/2}] - 59pH$
$H^+ + e^- = 1/2H_2$	$0.00 - 59[\log(P_{H2})^{1/2} - 59pH$

Mn^{4+} can occur in a soil even though some O_2 is still present (McBride 1994). The theoretical order of reduction requires that the soil's Eh value be an equilibrium value such that all redox half-reactions have adjusted to it. For this to happen, the soil's Eh must remain stable over a certain time period, be the same across the horizon, and all electron acceptors have to be able to react at a similar rate. A soil's Eh is never stable for long if the soil is affected by a fluctuating water table. Furthermore, Eh values will vary across a soil horizon at some periods because organic tissues are not uniformly distributed: roots can be found at cracks or in large channels, but not in some parts of the soil matrix. This means that reducing reactions that are occurring around a dead root will not be the same as those occurring in an air bubble a few centimeters away. In addition, electron acceptors also do not become reduced at similar rates. A discussion of reaction kinetics is beyond the scope of this chapter, but the topic has been reviewed by McBride (1994), who provides a thorough discussion of the order of reduction of the electron acceptors. Despite these inherent problems, the general order of reduction presented above is useful for understanding the general reduction sequence that occurs in hydric soils.

Principal Reducing Reactions in Hydric Soils

Reducing reactions, especially those that use compounds other than O_2, are the ones most responsible for the major chemical processes that occur in hydric soils such as denitrification, production of mottled soil colors, and production of hydrogen sulfide and methane gases. Common reducing reactions found in hydric soils are listed in Table 4.1. Because the electron acceptors most commonly used are compounds that contain oxygen, the basic reducing reactions produce water as a by-product as shown in Equation 3 and Table 4.1. This process removes H^+ ions from solution and causes the pHs of acid soils to rise during the reduction process.

Oxygen reduction occurs when organic tissues are being oxidized in a soil horizon that lies above the water table and in a soil that is not covered by water. Oxygen reduction can also occur in saturated soils where O_2 is dissolved in the soil solution. This frequently occurs when water (rainfall) has recently infiltrated a soil. When oxygen reduction has removed virtually all dissolved O_2, organic tissues decompose more slowly. If anaerobic conditions and slow decomposition are maintained for a long period, then organic C accumulates and organic soils may form (Chapter 6).

Denitrification is the reduction of nitrate to dinitrogen gas by the following reaction:

$$2NO_3^- + 10e^- + 12H^+ \rightarrow N_2 + 6H_2O \qquad \text{(Equation 4)}$$

Other gaseous by-products containing N are also possible. The reaction is similar to oxygen reduction in that both a gas and water are produced. This reaction improves water quality by removing NO_3^-, but it has no direct impact on soil properties such as color or organic C content, which can be used to identify hydric soils in the field.

Manganese reduction occurs after most of the nitrate has been reduced. Manganese exists primarily in valence states of 2+, and 4+. The reducing reaction is:

$$MnO_2 + 2e^- + 4H^+ \rightarrow Mn^{2+} + 2H_2O \qquad \text{(Equation 5)}$$

The MnO_2 is a mineral with a black color. When reduced, the oxide dissolves and Mn^{2+} stays in solution and can move with the soil water.

Iron reduction is the reducing reaction occurring in hydric soils that affects soil color. Iron behaves much like Mn and has two oxidation states — 2+ and 3+. When oxidized, the ferric form of Fe (Fe^{3+}) occurs as an oxide or hydroxide mineral. All of these oxidized forms of Fe impart brown, red, or yellow colors to the soil. The reduced ferrous Fe (Fe^{2+}) is colorless, soluble, and can move through the soil. The reducing reaction that ferric Fe undergoes varies with the type of ferric-Fe mineral present, as shown in Table 4.1. For amorphous Fe minerals the reducing reaction is:

$$Fe(OH)_3 + e^- + 3H^+ \rightarrow Fe^{2+} + 3H_2O \qquad \text{(Equation 6)}$$

Sulfate reduction is performed by obligate anaerobic bacteria (Germida 1998). The basic reaction is similar to that for nitrate reduction, and it too produces a gaseous product:

$$SO_4^{2-} + 8e^- + 10H^+ \rightarrow H_2S + 4H_2O \qquad \text{(Equation 7)}$$

The H_2S gas has a smell like that of rotten eggs. It can be easily detected in the field, but occurs most often near coasts where seawater supplies SO_4^{2-} for reduction.

Carbon dioxide reduction produces methane, or what is commonly called natural gas used in homes. This reaction is also similar to the others that produce a gaseous by-product:

$$CO_2 + 8e^- + 8H^+ \rightarrow CH_4 + 2H_2O \qquad \text{(Equation 8)}$$

Methane is an inflammable gas. It can be identified in the field when it is collected in water-filled plastic bags that are inverted and placed on the surface of a submerged soil for 24 hours. If the bubble of gas trapped in the bag is allowed to escape through a pinhole placed in the bag, and if it ignites in the presence of a flame, it is assumed to be methane (J. M. Kimble, USDA, personal communication). While this technique has been described to the authors, neither of us has actually verified it.

Factors Leading to Reduction in Soils

Four conditions are needed for a soil to become anaerobic and to support the reducing reactions discussed above (Meek et al. 1968, Bouma 1983): (1) the soil must be saturated or inundated to exclude atmospheric O_2; (2) the soil must contain organic tissues that can be oxidized or decomposed; (3) a microbial population must be respiring and oxidizing the organic tissues; and (4) the water should be stagnant or moving very slowly. Saturation or inundation are needed to keep the atmospheric O_2 out of the soil. Exclusion of atmospheric O_2 is probably the major factor that determines when reduction can occur in the soil. Presence of oxidizable organic tissues is probably the most important factor determining whether or not reduction occurs in a saturated soil (Beauchamp et al. 1989). Some soils are known to be saturated yet do not display any signs that reducing reactions such as Fe^{3+} reduction have occurred. In most instances, such soils simply lack the oxidizable organic tissues needed to supply the electrons used in reducing reactions (Couto et al. 1985).

A respiring microbial population is essential to the formation of reduced soils. Bacteria are widespread, abundant, varied, and adapted to function in the climates in which they occur. As reducing chemical reactions are studied more extensively in the field, it is becoming clear that they occur more frequently than originally thought (Megonigal et al. 1996, Clark and Ping 1997). Lastly, stagnant water is needed for reducing reactions to occur (Gilman 1994). Moving water, either in the form of groundwater or flood water, retards the onset of reduction particularly Fe reduction. The moving water apparently carries oxygen through the soil. While the water is in motion, its O_2 is difficult to deplete.

QUANTIFYING REDOX REACTIONS IN SOILS

Thermodynamic Principles

Oxidation–reduction reactions can be expressed thermodynamically using the concept of redox potential (Eh). This discussion begins with a review of thermodynamic principles that can be applied directly in the field to evaluate which redox reactions are occurring in a soil. The theory behind redox potential can be derived by considering the general reducing equation:

$$\text{Oxidized molecule} + m\text{H}^+ + n \text{ electrons} = \text{Reduced molecule} \qquad \text{(Equation 9)}$$

where m is the number of moles of protons, and n is the number of moles of electrons used in the reaction. This reaction can be expressed quantitatively by calculating the Gibbs free energy (ΔG) for the reaction:

$$\Delta G = \Delta G^\circ + RT \ln \frac{(\text{Re}\,d)}{(\text{Ox})(\text{H+})^m} \qquad \text{(Equation 10)}$$

where ΔG° is the standard free energy change, R is the gas constant, T is absolute temperature, and (Red) and (Ox) represent the activities of reduced and oxidized species. This equation can be transformed into one more applicable to us by converting the Gibbs free energy into a unit of voltage using the relationship $\Delta G = -nEF$:

$$Eh = E^\circ - \frac{RT}{nF} \ln \frac{(\text{Re}\,d)}{(\text{Ox})} - \frac{mRT}{nF} \ln(\text{H}^+) \qquad \text{(Equation 11)}$$

where Eh is the electrode potential (redox potential) for the reaction, E° is the potential of the half-reaction under standard conditions (unit activities of reactants under 1 atmosphere of pressure and a temperature of 298°K), and F is the Faraday constant. Equation 11 is called the Nernst equation. Substituting values for R, F, and T of 8.3 J/K mol, 9.65×10^4 coulombs mol^{-1}, and 298°K, respectively, converting the logarithm, and substituting pH for $-\log(\text{H}^+)$ the Nernst equation can be simplified to:

$$Eh(mV) = E^\circ - \frac{59}{n} \log \frac{(\text{Re}\,d)}{(\text{Ox})} + \frac{59\,m}{n} pH \qquad \text{(Equation 12)}$$

The Nernst equation shows that the reduction of an element will create a specific Eh value at equilibrium; however, the exact Eh value will vary with soil pH and the concentration (activity) of oxidized and reduced species in the soil. This equation has practical value for monitoring the development of reducing conditions in hydric soils in the field.

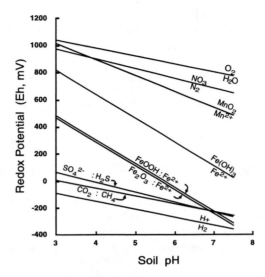

Figure 4.1 An Eh–pH phase diagram for the reducing reactions shown in Table 4.1. The lines were computed for the following conditions: dissolved species were assumed to have an activity of 10^{-5} M, partial pressures for O_2 and CO_2 were 0.2 and 0.8 atmospheres, respectively, and partial pressures of the remaining gases were assumed to be 0.001 atm.

Eh/pH Phase Diagrams

Equation 12 is used in Figure 4.1 to graphically portray the major reducing reactions occurring in hydric soils. The figure was prepared using the equations shown in Table 4.1, which were modified from the half-reactions described earlier. The equations represent the following conditions: dissolved species were assumed to have activities of 10^{-5} M, partial pressures for O_2 and CO_2 were 0.2 and 0.8 atmospheres, respectively, and partial pressures of the remaining gases were assumed to be 0.001 atm, which approximate what might be found in nature (McBride 1994).

The upper and lower lines in Figure 4.1 are the theoretical limits expected for redox potentials in soils because of the buffering effect of water on redox reactions. Eh values above the upper line shown in Figure 4.1 are prevented at equilibrium because water in the soil would oxidize to O_2 and supply electrons which would lower the Eh. Eh values below the lower line are prevented because water (which supplies H^+) would be reduced to H_2, consuming electrons and raising the Eh. The Eh values at which the other reducing reactions occur vary with pH, and also vary with the assumptions regarding the concentrations noted earlier. These theoretical limits vary with pH as described by the Nernst equation.

The order or sequence for which the electron acceptors are reduced is clearly shown in Figure 4.1. The sequence changes somewhat for different pHs. The Fe oxides shown in Table 4.1 each have separate phase lines. The nearly amorphous $Fe(OH)_3$ minerals (ferrihydrite) reduce at a higher Eh value for a given pH than do the crystalline minerals of FeOOH (goethite) or Fe_2O_3 (hematite). Field studies have shown that the $Fe(OH)_3$ minerals occupy 30 to 60% of these Fe minerals in hydric soils (Richardson and Hole 1979).

Reliability of Phase Diagrams for Field Use

Eh/pH phase diagrams are useful for showing how reduction and oxidation of a given species vary with the pH of the solution, and they also show the relationship among the different elements that undergo redox reactions. Once a redox phase diagram is in hand, the next logical step is to measure Eh and pH in the field and use these data to predict the phase a given element is in. It is possible to do this for some redox reactants, but phase diagrams have two potential problems which directly limit field applications. The first deals with mixed redox couples, and the second with the kinetics of redox reactions.

Mixed Redox Couples

The lines on an Eh/pH diagram show the Eh and pH values where a specific redox couple (half-reaction) is expected to undergo a phase change and attain the concentration that was used to develop the diagram. Each line on the phase diagram was computed by assuming that both the Eh and pH values measured in the soil solutions were influenced only by a single redox half-reaction and that equilibrium had been achieved. This will generally not be the case if other substances are present in the soil solution which are also undergoing redox reactions, and if the soil's Eh value is changing over time. In such cases the soil's Eh value would be a *mixed potential*, or an average potential determined by a number of the half-reactions shown in Table 4.1, and not simply the result of a single redox half-reaction. These "average Eh values" complicate the use of phase diagrams for interpretations of redox data because they are not in equilibrium with each other, and therefore the actual Eh at which a phase change will occur cannot be predicted precisely using the equations of Table 4.1.

The presence of mixed redox potentials also creates problems when attempting to adjust Eh values for different pHs. For example, where the ratio of protons to electrons (m/n in Equation 12) is unity in the half cell reaction, the Nernst equation predicts a 59 mV change in Eh per pH unit. This value is sometimes used to adjust measured redox potentials for comparison at a given pH, but as shown in Table 4.1, the ratio of m/n varies for different redox couples and ranges from −59 to −177 mV/pH unit. The Eh/pH slope predicted from the Nernst equation assumes that a specific redox couple controls the pH of the system. While this may be true for controlled laboratory solutions, the pH of natural soils and sediments is buffered by silicates, carbonates, and insoluble oxide and hydroxide minerals which are not always involved in redox reactions (Bohn et al. 1985, Lindsay 1979). Therefore, it is not surprising that measured slopes in natural soils deviate from the predicted values. Applying a theoretical correction factor to adjust Eh values for pH differences among soils may be inappropriate for natural conditions (Bohn 1985, Ponnamperuma 1972). We recommend that Eh values measured in soils not be adjusted to a common pH, but rather that the pH of the soil be measured and reported whenever Eh values are reported.

Mixed redox couples can also alter the apparent slopes of the phase lines shown in Figure 4.1. For instance, a change of +177 mV per pH unit is the predicted slope for the reduction of $Fe(OH)_3$ to Fe^{2+} (Table 4.1) based on the m/n value of 3 (i.e., 24/8). In a series of experiments where he added different kinds of plant organic matter to several different kinds of soils, Zhi-guang (1985) found that this slope varied as a function of the ratio of ferrous iron to organic matter. In sandy soils with almost no Fe^{2+}, the slope matched the theoretical value of 59 mv per pH unit. As the Fe^{2+} concentration increased, the slope also increased but did not reach the theoretical value of −177 mV per pH unit. On the other hand, Collins and Buol (1970) found good agreement between the measured and theoretical Eh/pH relationship for soils containing more Fe minerals. In summary, we feel phase diagrams such as those shown in Figure 4.1 will be most useful for interpreting redox data for elements that are abundant (e.g., Fe) in a soil, and where soil pHs are influenced by the redox reactions and are not buffered by carbonates as would be expected at soil pHs >7.

Reaction Kinetics

Another problem that complicates the use of phase diagrams with natural Eh data is that some redox reactions occur much more slowly than others. This is particularly true for the reduction of O_2, NO_3^-, and MnO_2 (McBride 1994). The effect of this is that the actual Eh at which detectable amounts of reduced species of these compounds occurs tends to be 200 to 300 mV lower than what would be predicted in Figure 4.1. This means that redox potential measurements in soils may not relate well to the chemical composition of soil solutions that are predicted by Figure 4.1. On the other hand, redox reactions related to Fe have been found to begin in soils (using Pt electrodes) near the Eh values specified in Figure 4.1. Reduction of SO_4 and CO_2 also begin at Eh values similar to those predicted in Figure 4.1. In summary, phase diagrams can be useful to interpret data for transformation of specific Fe minerals in soils, but caution is needed for predicting when reduction occurs for O_2, NO_3^-, and MnO_2.

The Concept of pe

Redox reactions written as half-reactions treat electrons (e^-) as a reactive species very similar to H^+. While free electrons do not occur in solution in any appreciable amount, the electrons can be considered as having a specific activity. Electron activity is expressed as *pe*, which has been defined as (Ponnamperuma 1972):

$$pe = -\log(e^-) = \frac{Eh(mV)}{59} \qquad \text{(Equation 13)}$$

Solutions with a high electron activity (low pe) and low Eh value conceptually have an abundance of "free electrons." These solutions will be expected to reduce O_2, NO_3, MnO_2, etc. Solutions that have a low electron activity (high pe) and high Eh value can be thought of as having virtually no "free electrons," and will maintain the elements of O, N, Mn, Fe, etc., in their oxidized forms. The pe can also be used as a substitute for Eh in Equation 12:

$$pe = \frac{E°}{59} - \frac{1}{n}\log\frac{(Re\,d)}{(Ox)} - \frac{mpH}{n} \qquad \text{(Equation 14)}$$

This equation can be used to develop phase diagrams like that shown in Figure 4.1. Although the pe concept is useful for chemical equilibria studies, it is a theoretical concept that cannot be measured directly in nature. We will continue to use redox potential (Eh) as our measure of reducing intensity because this voltage can be measured in the field.

MEASURING REDUCTION IN SOILS

Chemical Analyses

The chemistry of hydric soils can be evaluated in a general sense by measuring the concentrations of reduced species in solution. If for example there is no measurable O_2 in solution, the soil is known to be anaerobic. If Fe^{2+} is detected in solution, we can predict from theoretical grounds that the soil is probably anaerobic, that denitrification has occurred (if NO_3^- was present initially), that manganese reduction has taken place, but that the reduction of SO_4^{2-} and CO_2 may or may not have occurred. Reaction kinetics and microsite reduction can create exceptions to these interpretations. Chemical evaluations of all reduced species in solution is expensive and usually used only for research purposes as described in the "Nutrient Pools, Transformations, and Cycles" section of this chapter.

Dyes

A less expensive alternative to measuring soil solution chemistry is to use a dye that reacts with reduced forms of key elements. The most widely used dyes for field evaluations of reduction react with Fe^{2+}. Childs (1981) discussed the use of α, α'-dipyridyl in the field. Heaney and Davison

(1977) showed that the α, α'-dipyridyl reagent reliably distinguished Fe^{2+} from Fe^{3+}, and that dye results corresponded well with measurements of the concentration of these species. Other dyes such as 1, 10-phenanthroline, are available to detect Fe^{2+} in reduced soils, and all can be used in similar ways (Richardson and Hole 1979). Dyes work quickly in the field and are easy to use. To test for Fe^{2+} in the field, a sample of *saturated* soil is extracted and the dye solution immediately sprayed onto it. If Fe^{2+} is present, it will react with the dye within one minute and change color. Both 1, 10-phenanthroline and α, α'-dipyridyl turn red when they react with Fe^{2+}. It must be remembered that these dyes detect only Fe^{2+}. If a positive reaction occurs after the dye is applied to a soil sample, it can be assumed that the soil is reduced in terms of Fe, and that the soil must also be anaerobic. If no reaction to the dye is found, then all we know is that Fe^{2+} is not present. The soil in this case may be anaerobic, but not Fe-reduced, or it may be aerobic. Either of these two cases will produce a negative reaction to the dye solution.

A 0.2% solution of α, α'-dipyridyl dye is used in the field by soil classifiers of the USDA Natural Resources Conservation Service (Soil Survey Staff 1999). It is prepared by first dissolving 77 g of ammonium acetate in 1 liter of distilled water. Then 2 g of α, α'-dipyridyl dye powder is added and the mixture stirred until the dye dissolves. The dye powder and solution are both sensitive to light and should be kept in brown bottles or in the dark. This solution can be applied with a dropper to freshly broken surfaces of saturated soils. If a pink (low ferrous iron) or red (high ferrous iron) color develops within a minute, ferrous iron is present. This procedure uses a neutral (pH ~ 7.0) solution, which avoids potential errors associated with photochemical reduction of ferric–organic complexes. Avoid spraying onto soils contacted by steel augers or shovels, because these may give false positive tests. For dark-colored soils (Mollisols, Histosols), the use of white filter paper improves the ability to observe color development.

False positive errors from photochemical reduction of ferric–organic compounds can occur when samples to which the dye has been applied are exposed to bright sunlight. In addition, exposure to air can rapidly oxidize Fe^{2+} to Fe^{3+} when pH > 6 (Theis and Singer 1973) and produce a false negative result. Childs (1981) describes the development of the test and the errors associated with the photochemical reduction of ferric–organic complexes.

Redox Potential Measurements

Redox potential (Eh in Equation 12) is a voltage that can be measured in the soil and used to predict the types of reduced species that would be expected in the soil solution. The Eh measurements are evaluated along with soil pH data and an Eh/pH phase diagram such as that shown in Figure 4.1. The redox potential voltage must be measured between a Pt-tipped electrode and a reference electrode that creates a standard set of conditions. Platinum electrodes are sometimes called microelectrodes because they consist of a small piece of Pt wire that is placed in the soil. The Pt wire is assumed to be chemically inert and only conducts electrons. It generally does not react itself with other soil constituents and does not oxidize readily as do Fe, Cu, and Al metals. Reduced soils transfer electrons to the Pt electrode, while oxidized soils tend to take electrons from the electrode. For actual redox potential measurements, the electron flow is prevented. The potential or voltage developed between the soil solution and a reference electrode is measured with a meter that has been designed to detect small voltages. The voltages developed in soil range from approximately +1 to −1 V, and are usually expressed in millivolts (mV).

There are several methods of Pt-electrode construction, but they all follow the same basic design (Faulkner et al. 1989, Patrick et al. 1996). For soil systems, 18-gauge platinum wire (approximately 1 mm in diameter) is preferred because it is more resistant to bending when inserted in the soil. The Pt wire is cut into 1.3-cm segments, with wire-cutting pliers that are used only for cutting platinum, and cleansed in a 1:1 mixture of concentrated nitric and hydrochloric acids for at least 4 hours. This removes any surface contamination that could occur during cutting or handling. The cut wire segments are then soaked overnight in distilled, deionized water.

For field studies of less than 3 years' duration, welding or fusing the platinum directly to a 12- or 14-gauge copper wire or brass rod is the least complicated method to use. All exposed metal except Pt must be insulated with a nonconducting material (e.g., heat-shrink tubing) and a water-proof epoxy. This welded/fused design is most appropriate for studies of less than 3 years, because many epoxy cements are not stable for extended periods under continuous exposure to water. Longer measurement periods are better served by a glass body electrode (see Patrick et al. [1996] for a complete description).

Platinum electrodes can be "permanently" installed in the soil and left in place for up to a year to monitor a complete wetting and drying cycle. After a year, some electrodes should be removed and retested in the laboratory to ensure that problems related to component breakdown are not occurring. The installation process must seal the electrodes from the movement of air or water from the surface to the tip. This can be done by augering a hole, filling it with a slurry made from the extracted soil, and inserting the cleaned Pt electrode to the appropriate depth. The slurry must have the same chemical properties as the soil the Pt tip is placed in.

Redox potential measurements are made in the field using a portable pH/millivolt (mV) meter and a saturated calomel or silver/silver-chloride reference electrode. Commercial voltmeters can be used, but not all of them register millivolts. The reference electrode normally is not permanently installed at the site. To begin readings, the reference electrode is pushed a short distance into wet or moist soil at the surface to ensure a good electrical contact. If the soil is relatively dry, a knife or soil probe is used to excavate a shallow hole to hold the electrode upright. Water should be poured into the hole to provide good electrical contact between the reference electrode and soil solution. If the soil is dry, a dilute salt solution (i.e., 5 g KCl in 100 ml H_2O) can be used to moisten the reference electrode hole and prevent a junction potential from being established between the reference electrode and the soil. The reference electrode is connected to the "common" terminal on the commercial meters. The other terminal (for voltage) is connected to a single Pt electrode that is buried in the soil. To take a measurement after the electrodes are connected to the meter, the meter is turned on and the voltage allowed to stabilize before a single number can be recorded. This stabilization can be immediate, or it may require several minutes until the "drift" in the voltage stops.

Correcting Field Voltages to the Standard Hydrogen Electrode

The voltage measured in the field between the buried Pt wire and a reference electrode is not the redox potential or Eh. True redox potentials are measured against a standard hydrogen electrode which consists of a Pt plate with H_2 gas moving across its surface. Such an electrode is impractical for field use. Correction factors are used to adjust the field voltage measured with one type of reference electrode to the voltage that would have been measured had a standard hydrogen electrode been used. The correction factors for two common reference electrodes are listed in Table 4.2. The correction is simply:

$$\text{Field Voltage} + \text{Correction Factor} = \text{Redox Potential (Eh)} \qquad \text{(Equation 15)}$$

Variability in Redox Potential

Redox potential measurements made at a single point in the soil may change over the course of a year by 1000 mV or more if the soil is periodically saturated or flooded and reducing reactions occur. Less variation is expected in soils that never saturate as well as ones that are permanently inundated. An example of the variation in redox potential for one hydric soil is shown in Figure 4.2, where data for the mean of five redox potential measurements are plotted, along with the minimum and maximum values found for the same depth. Before the soil became saturated in 1998 the redox potential was above 600 mV, and the range in values among the five electrodes was about 100 mV, which is relatively small. Within a few days of the soil saturating due to a rising water

Table 4.2 **Correction Factors Needed to Adjust Voltages Measured in the Field to Redox Potentials (Eh's) for Two Commonly Used Reference Electrodes**

Temperature (°C)	mV	
	Calomel (Hg-containing)	Ag/AgCl
25	244	197
20	248	200
15	251	204
10	254	207
5	257	210
0	260	214

Note: The factors are added to field-measured voltages to correct the values to voltages measured with standard hydrogen electrodes. Correction factors for the Ag/AgCl electrode assume the electrode is filled with a saturated KCl solution.

table, the redox potential fell, but the rate of fall was not the same among all five electrodes. During the period of decrease in redox potential across the horizon the range in values was over 600 mV. By day 60 (in 1998) the range in redox potentials again was approximately 100 mV even though the mean potential was near 0 mV. Later periods of greater redox potential variability were associated with periodic draining and resaturation.

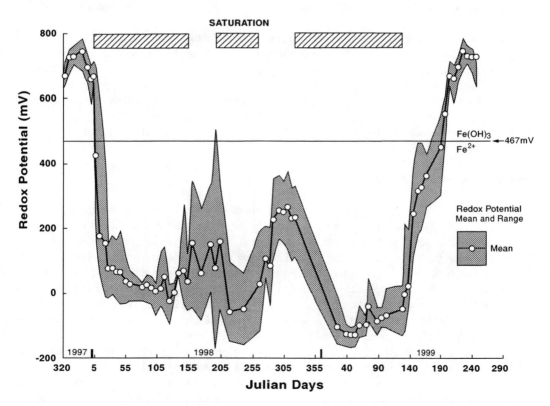

Figure 4.2 Variation in redox potential for a hydric soil at a depth of 30 cm. Data are the mean and range of five Pt electrodes. Variation among electrodes is greatest during periods when soil is either saturating or draining, and less variation occurs when the soil is either saturated or drained for several weeks. Reduction of $Fe(OH)_3$ occurs within weeks of the soil saturating, and reduced Fe can be maintained even during intermittent periods when soil is unsaturated.

The type of variability illustrated in Figure 4.2 is real and must be expected when making redox potential measurements in hydric soils that undergo periodic saturation and drainage. The variability seems to be caused by the oxidation of organic tissues and the corresponding reducing reactions occurring in microsites (Crozier et al. 1995, Parkin 1987). Microsites are simply small volumes of soil on the order of 1 to 5 cm^3 that surround decomposing tissues such as a dead root or leaf. Examples of microsites where reduction occurred are shown in Plates 13 and 14, where the microsites occur within the gray-colored soil. When the redox potentials shown in Figure 4.2 were >600 mV, the soil was unsaturated and O_2 was controlling or poisoning the system. After saturation occurred, the oxidation of organic tissues by bacteria continued. After dissolved O_2 was depleted, alternate electron acceptors were used in the reducing reactions. The Pt electrode that recorded the fastest drop in redox potential following saturation may have been adjacent to the decomposing tissue (near the microsite of reduction), while the electrode that responded most slowly may have been farther away. Although there are broad ranges in Eh following saturation due to the reduction occurring in microsites, over time the range in Eh narrows as the dissolved O_2 in the soil solution is depleted and a greater volume of soil becomes reduced.

To characterize the redox potentials in hydric soils an adequate number of measurements must be made across a horizon to account for the variability expected in the redox potentials. Statistical analyses applied to redox data have usually indicated that 10 or more electrodes per depth are needed for an acceptable level of precision over a complete wetting/drying cycle. This is generally too expensive for routine use. We recommend, however, that at least five Pt electrodes be installed at each depth for which redox potential measurements are desired. Under no circumstances that we can imagine, should a single redox potential measurement be used to assess reducing conditions in the field.

In summary, soil redox potential measurements remain the most versatile tool we currently have for assessing reducing reactions economically for virtually any soil. The method, when properly applied, provides useful data on reducing reactions. The spatial and temporal variability in Eh is magnified during the initial periods of flooding/saturation and draining as the system changes from aerobic to anaerobic and back again. Because of these conditions, it is important to collect data over a period that includes a saturating and draining cycle. The most effective way to partially overcome the problem of spatial heterogeneity of a given soil is through replication of the measurement equipment.

Interpreting Redox Potential Changes in Nature

Redox potential measurements are made to evaluate changes in soil chemistry. Because of the problems created by the mixed potentials and reaction kinetics discussed earlier, it is safest to base the interpretations of redox data on one or two elements that are abundant in soils and react quickly to changes in redox potential. We will use Fe as the element for interpreting changes in redox potential over time, and focus on the reduction of $Fe(OH)_3$. The first step is to identify the redox potential at which $Fe(OH)_3$ reduces to Fe^{2+}. This redox potential is obtained from the Eh/pH diagram shown in Figure 4.1 by using the average pH of the soil measured over time. For the soil shown in Figure 4.2, the average pH was found to be 5.0. From the Eh/pH diagram it can be seen that at this pH $Fe(OH)_3$ reduces to Fe^{2+} when the Eh is below 467 mV.

The phase change for $Fe(OH)_3$ to Fe^{2+} is shown in Figure 4.2 by the horizontal line at an Eh of 467 mV. The data in Figure 4.2 can be interpreted by considering when and for how long Fe^{2+} was in solution. It can be seen that during most of 1998, Fe^{2+} would have been expected to be in solution. We know from our earlier discussion that if Fe^{2+} is present, we can assume that most dissolved O_2 has been reduced to H_2O, that most NO_3^- present has been denitrified, and that most Mn oxides have been reduced to Mn^{2+}. Microsite reduction and reaction kinetics affect the validity of these assumptions as discussed previously. Phase lines for SO_4^{2-} and CO_2 could also be added to interpret whether these materials were reduced as well. Such interpretations are simple and

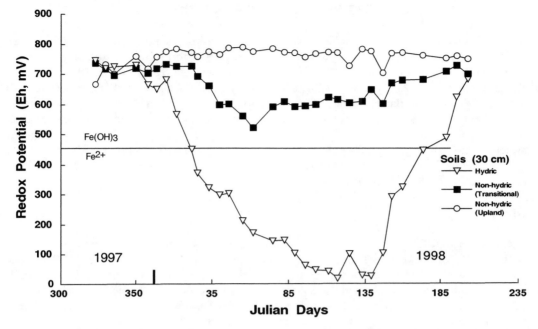

Figure 4.3 Comparison of mean redox potential values among three soils (30 cm depth): a hydric soil, a non-hydric soil in the transition to the upland, and in an upland, non-hydric soil. The hydric soil is the only one where the redox potential fell low enough for Fe reduction to occur. The other two soils either did not saturate or were not saturated for a long enough period for Fe reduction to occur.

straightforward, and can be verified by analyzing soil samples with dyes that react with Fe^{2+} or by analyzing water samples for Fe^{2+}.

Redox potential changes that occurred over time in a landscape consisting of a hydric soil, transition zone, and upland area are shown in Figure 4.3. These redox potential data are the mean of five electrodes at a depth of 30 cm. The soils all had a pH of 5.0, and the $Fe(OH)_3$ phase line has been added to the figure. The occurrence of saturation clearly controls the fluctuation in redox potential among the three landscape positions. The upland soil never became saturated during the study period, and it can be seen that its redox potential remained high and fairly constant. The transitional soil was saturated for short periods (data not shown), but the redox potential never fell to a point where Fe reduction would have been expected. On the other hand, the hydric soil was saturated for an extended period, and Fe reducing conditions occurred for approximately 150 days.

pH Changes in Reduced Soils

Oxidation–reduction reactions in anaerobic soil can cause changes in the soil's pH. As shown in Table 4.1, the reducing reactions consume protons, and a change in pH should be expected as a result. Ponnamperuma (1972) showed that the amount of change varies among soils, but in general, reduction causes the soil pH to shift *toward* 7 but not to necessarily *reach* 7. Reduction in acid soils generally increases the pH, while in alkaline soils it can reduce pH. The amount of pH change can be as high as three pH units following several weeks of submergence, although changes of <2 pH units are probably more typical. The degree of change depends on the amount of reduction taking place and is determined by the amount of oxidizable organic tissue, as well as the amount of reducible electron acceptors. According to Ponnamperuma (1972), pH values remain <6.5 in acid soils containing low amounts of organic matter and reducible Fe oxides or hydroxides. In alkaline soils, pH tends to decrease toward 7, possibly due to the production of CO_2. Most acid organic soils are low in Fe, and submergence would not cause large increases in pH.

PLANT NUTRIENT POOLS, TRANSFORMATIONS, AND CYCLES

Natural wetland systems maintain a wider range of redox reactions than upland ecosystems and transform plant nutrients among solid, solute, and gaseous forms. As a result, they are capable of recycling plant nutrients among the soil, water, and atmosphere. The pools, transformations, and fluxes of the major plant nutrients N, P, and S in freshwater wetlands will be reviewed in this section. Freshwater wetlands will be our focus because they occupy over 90% of the world's wetland area (Giblin and Weider 1992). Representative examples of hydric soils and chemical processes were chosen because it is beyond the scope of this chapter to exhaustively review all wetland systems and possible processing mechanisms for every potential contaminant. More complete reviews can be found in Reddy and D'Angelo (1994), Richardson (1999), and Giblin and Wieder (1992) for the nutrients N, P, and S, respectively.

Nitrogen

Nitrogen transformations in hydric soils are a complex assortment of interrelated processes controlled by microbial activity and the redox status of the soil (Gambrell and Patrick 1978, Reddy and Patrick 1984). These transformations are summarized in the generalized diagram of the nitrogen cycle for oxidized and reduced zones in a flooded soil shown in Figure 4.4. The diagram is simplified and not all intermediary forms of N are shown. We can assume that the nitrogen cycle contains five basic transformations that can occur in hydric soils (Figure 4.4A). The major N pools or forms in which N occurs in natural freshwater wetlands are: (a) total organic N, which consists of N in plants, microbes, and sediment, (b) available inorganic N in water and sediment (primarily NO_3^- and NH_4^+), and (c) N_2 and N_2O gases. The organic N pool is the largest, with 100 to 1000 g N m^{-2} occurring primarily as sediment-N (Howard-Williams and Downes 1993). Sediment-N accounts for 80 to 90% of all the N in wetlands. The total plant N pool is roughly an order of magnitude less than total sediment-N. Inorganic-sediment N is another order of magnitude less than the plant pool. Major N inputs into wetlands come from atmospheric deposition (NO_3^- and NH_4^+), N_2 fixation, and groundwater input of NO_3^-.

As shown in Figure 4.4A, *N-Fixation* is the conversion of N_2 gas into NH_3, which is then incorporated into organic tissues. N-fixation rates in wetlands vary from 0.02 to 90 g N m^{-2} y^{-1} around the world, being lowest in the arctic areas and highest in areas that receive excessive nutrient inputs (Howard-Williams and Downes 1993). N-fixation supplies the majority of N to some wetlands, but adds <5% of the N input to others (DeLaune and Patrick 1990, Howard-Williams and Downes 1993). N-fixation in wetlands is accomplished by both blue-green algae and bacteria. Blue-green algae fix nitrogen through photosynthesis and function primarily in water above the soil surface. Other bacteria can fix nitrogen anaerobically and are most abundant in the upper 5 cm of anaerobic soil. Although diffusion of N_2 through water is slow, hydrophytic plants can transport N_2 to their roots as they transport O_2 (Patrick 1982). For this reason, N-fixation rates in flooded soil are greatest in the zone of highest root growth.

Ammonification or mineralization is the conversion of organic-N into ammonium (NH_4^+) by microorganisms. It occurs when organic tissues are oxidized and the N content of the tissue exceeds the requirements of the microbes. Ammonification can occur aerobically or anaerobically (Figure 4.4B), but aerobic ammonification is much faster. Patrick (1982) reported that corn stalks which are high in N compounds lost 20% of their weight in 17 days when they were decomposed anaerobically, but lost 37% in the same period when decomposed aerobically. Patrick (1982) also reported that rye grass lost 7% of its weight in 84 days under anaerobic decomposition, but lost 17% when decomposed aerobically for 66 days. While anaerobic decomposition is slower, it results in approximately five times the N being maintained as NH_4^+ than occurs under aerobic conditions, in part because nitrification is prevented (Patrick 1982). According to Ponnamperuma (1972) most

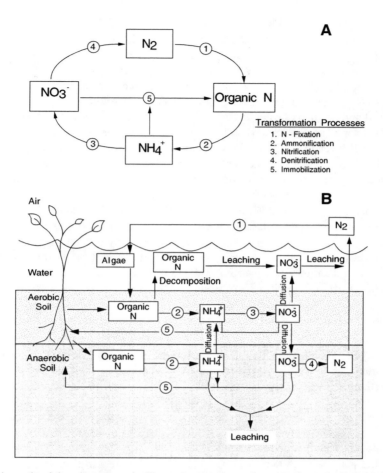

Figure 4.4   Schematic of the nitrogen cycle. The major transformations are shown in (A) while the portions of the soil, as well as water and air in which the transformations occur, are shown in (B). The N cycle contains a gaseous N phase, which allows N to be removed from the wetland.

of the mineralizable N can be converted to ammonium within 2 weeks of submergence in soils with neutral pH, adequate levels of P, and a temperature that is not limiting microbial growth.

Nitrification is the production of NO_3^- from NH_4^+. It occurs in aerobic soils in a two-step process. The ammonium is first converted to nitrite (NO_2^-), and then the nitrite is converted to nitrate (NO_3^-). Both steps in the process are completed by a restricted group of bacteria that include *Nitrosomonas, Nitrosococcus,* and *Nitrobacter* species. Nitrification occurs in the aerobic soil zone of hydric soils and around aerated roots growing in anaerobic soil. Rates for nitrification are variable and depend on the quantities of both NH_4^+ and O_2 in the soil. Rates have been reported to range from 0.01 to 0.16 g N m^{-2} day^{-1} (Reddy and D'Angelo 1994). In hydric soils, nitrification rates are limited primarily by the supply of O_2 (Reddy and D'Angelo 1994).

Denitrification is the reduction of NO_3^- to either N_2 or N_2O gases, which is accomplished by a large number of different bacteria (Ponnamperuma 1972). Denitrification is the major mechanism that returns N, originally fixed from the atmosphere, back to the atmosphere. It is a reducing reaction with a rate dependent on the availability of decomposable organic tissues, as well as a supply of NO_3^-. Because NO_3^- is used as an alternate electron acceptor to O_2, denitrification occurs in anaerobic zones of soils or sediments. The anaerobic zones can be the small microsites described earlier. Denitrification rates range from 0.003 to 1.02 g N m^{-2} day^{-1} (Reddy and D'Angelo 1994).

Immobilization is the conversion of mineral forms of N into plant tissue. Immobilization and ammonification generally occur simultaneously.

Most flooded soils have a thin (<5 cm), oxidized layer at the surface where oxidation reactions occur (Faulkner and Richardson 1989, Moore and Reddy 1994). This layer forms by O_2 diffusing from the atmosphere into the overlying floodwater, and then into the soil (Figure 4.4B) (Howeler and Bouldin 1971). Reduction processes dominate in the anaerobic zone below this oxidized layer. In both reduced and oxidized layers, organic N may be mineralized to NH_4^+. In the reduced layer, NH_4^+ is stable and may be adsorbed to sediment exchange sites or used by both plants and microbes. The thin, oxidized layer in flooded soils is important in N transformations because NH_4^+ may be oxidized to NO_3^- by chemautotrophic bacteria (nitrification) in this layer. Depletion of NH_4^+ in the upper, oxidized layer causes NH_4^+ to diffuse upward in response to the concentration gradient. This diffusion process may be effective from 4 to 12 cm deep.

Nitrate is unstable in reduced zones and is quickly depleted via assimilative reduction (taken up by organisms and used in their tissue), denitrification, or leaching (Figure 4.4B). Redox potential, pH, moisture content, labile (readily oxidizable) C source, and temperature control the rate of NO_3^- reduction. For example, at pHs < 6 the reduction of N_2O to N_2 is strongly inhibited. The sequential processes of mineralization, nitrification, and denitrification dominate wetland N cycling and potentially process 20 to 80 g N m^{-2} y^{-1} (Bowden 1987).

Phosphorus

The soil P cycle is fundamentally different from the N cycle in that it has no substantial gaseous phase to facilitate the removal of P from wetlands. The lack of the gaseous phase means that P entering a wetland is either stored in wetland sediment and plant tissue, or it is in solution and may be carried out of the wetland by flowing water. Phosphorus in soils has a constant valence of +5, and it is not directly affected by redox processes because it does not change valence. The solubility of P in soils and water, however, is affected by redox processes that dissolve or decompose the compounds that have bonded to P.

A simplified P cycle is shown in Figure 4.5A. Soil P exists in three major forms: organic P, fixed mineral P, and orthophosphate (ortho P). Orthophosphate exists as an anion in the forms of $H_2PO_4^{2-}$, HPO_4^{1-}, and PO_4^{3-}, at pHs of 2 to 7, 8 to 12, and >13, respectively. Fixed mineral P consists of ortho-P bound to an oxide or hydroxide containing Al or Fe^{3+}, or bound to cations Ca or Mg, all of which are insoluble at certain pHs. Organic forms of P found in plants occur in compounds such as inositol phosphates, phospholipids, and nucleic acids. Insoluble organic P compounds can also be found in sediment in the form of partially decomposed plant tissue. Organic forms of P also include orthophosphate anions that are electrostatically bound to an organic compound such as humic or fulvic acid. These acids are negatively charged, as is the ortho-P, and a bonding of the two is accomplished by a cation bridge which can be Ca, Mg, Al, or Fe. Soluble P may consist of ortho-P anions or as certain organic P forms.

Organic-P and fixed mineral-P comprise approximately 80 to 90% of the P in a wetland (Figure 4.5B). Living plants store most of the remaining P, leaving very little in the water column (Richardson 1999). Whether P in the soil is in the organic form or fixed mineral form depends largely on the type of soil present. Most soil P (>95%) in organic soils is in the organic form with cycling among P-forms controlled by biological forces (i.e., microbes and plants). In mineral soils, most of the soil P may be bound to minerals containing Al, Fe, Ca, or Mg.

The transformation of fixed mineral P into soluble orthophosphate is controlled by the interaction of redox potential and pH (Holford and Patrick 1979, Sah and Mikkelsen 1986). In acid soils (pHs 4 to 7) ortho-P is preferentially adsorbed onto Fe and Al oxides and hydroxides. At pHs <4, the Al and Fe oxide and hydroxides dissolve, and any ortho-P bonded to them can be released to the soil solution (Lindsay 1979). Likewise, Fe^{3+} phosphates such as strengite ($FePO_4 \cdot 2H_2O$)

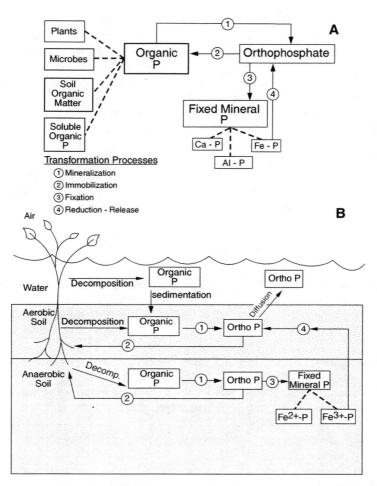

Figure 4.5 Schematic of the phosphorus cycle. The major transformations are shown in (A) while the portions of the soil, as well as water in which the transformations occur, are shown in (B). The P cycle does not have a gaseous phase for P, and this causes P to either build up in the wetland soil, or to be carried out of the wetland with moving water.

can also dissolve if the redox potential falls low enough to reduce the ferric Fe to Fe^{2+} (Moore and Reddy 1994). Aluminum does not change valence in soils and is unaffected by redox potential. Precipitation as insoluble Ca or Mg-phosphates or adsorption to carbonates are the dominant transformations at pHs > 7. These are the predominant forms of mineral P forms in arid soils. Just as Al, Ca and Mg do not participate in redox reactions, so the solubility of Ca or Mg-P complexes is determined by soil or water pH rather than Eh (Moore and Reddy 1994).

The amount of P that a soil can adsorb is directly related to the amount of oxalate-extractable (amorphous) Al and Fe^{3+} oxides and hydroxides (Richardson 1985, Reddy and D'Angelo 1994). The P sorption capacity of an oxidized soil may increase during flooding and reduction due to formation of amorphous ferrous hydroxides, which have a greater surface area and more sorption sites than the more crystalline, oxidized, ferric forms (Holford and Patrick 1979). While organic materials can also absorb P, the amount of organic material in the soil is generally not as good a predictor of P adsorbing capacity as the amounts of amorphous Fe and Al oxides and hydroxides. Because soils have a finite amount of Fe and Al oxides and hydroxides, their P sorbing capacity is limited. Once it is exceeded, no more P can be retained in the wetland, and it is then kept in solution and carried out of the wetland with water.

In addition to the amount of P a wetland can hold, the rate at which P is added to a wetland also determines whether P is retained within or exported from a wetland. Richardson (1999) presented an analysis of 125 wetland sites and showed that the approximate maximum rate at which a wetland can absorb P is <1 g P m^{-2} y^{-1}. When this rate of P input is exceeded, wetlands do not absorb P fast enough to remove it from solution, and the excess P is exported from the wetland with flowing water. Sediment accretion processes control the long-term P removal capability of wetland ecosystems. Sediment accretion rates for peats have been estimated to store <1 g P m^{-2} y^{-1} (Richardson 1999).

There is little direct uptake of phosphate from the water column by emergent wetland vegetation because the soil is the major source of nutrients (Richardson 1999). Growing vegetation is a temporary nutrient-storage compartment resulting in seasonal exports following plant death. The long-term role of emergent vegetation is to transform inorganic P to organic forms. Microorganisms play a definite role in P cycling in wetlands, but the microbial pool is small in terms of P storage (10 to 20% of total P) (Richardson and Marshall 1986).

Sulfur

The S cycle has been studied less than those for N and P in freshwater wetlands. The major forms of S found in these wetlands are shown in Figure 4.6A. Most ($>70\%$) soil S occurs in an organic form (Wieder and Lang 1986, 1988). The remaining inorganic S is distributed as *reduced inorganic sulfides* (RIS) such as pyrite ($<10\%$), *dissolved SO_4^{2-}* ($<20\%$), and *gases* consisting of H_2S and dimethyl sulfide ($<5\%$) (Giblin and Weider, 1992).

Sulfur transformations are biologically mediated and, like P and N transformations, are affected by the interaction between redox potential and pH. In aerated portions of the hydric soil (Figure 4.6B) immobilization, oxidation of inorganic sulfide and elemental sulfur, and mineralization of organic S to inorganic SO_4^{2-} are the dominant processes (Giblin and Weider 1992). Under reducing conditions, sulfate reduction transforms SO_4^{2-} to H_2S during respiration by obligate anaerobic bacteria. The H_2S formed by sulfate reduction can be released to the atmosphere or can react with organic matter, providing another pathway for converting inorganic-S to organic-S. With SO_4^{2-} reduction and sufficient Fe^{2+}, iron sulfides (FeS, FeS_2) can form; pyrite (FeS_2) formation requires alternating (either temporally or spatially) anaerobiosis with limited aeration.

Despite the small size of the inorganic pool, this fraction is the most important for S cycling, retention, and mobility. Fluxes through the inorganic pool dominate S cycling in wetlands that have high SO_4^{2-}, inputs such as those usually associated with wastewater additions. Wieder and Lang (1988) calculated that 3.5 to 4 times as much inorganic S was processed compared to the organic S pool through alternating SO_4^{2-} reduction and sulfide/sulfur oxidation. This has important implications for wetland S cycles because S inputs are primarily SO_4^{2-} from atmospheric deposition and either natural or amended hydrologic sources. Sulfate retention by aerobic, mineral soils is dominated by the same adsorption mechanisms involved in PO_4^{3-} retention. However, adsorbed SO_4^{2-} is displaced by PO_4^{3-} on the exchange sites, but PO_4^{3-} is not displaced by SO_4^{2-}.

Significant fluxes of S to the atmosphere cause the wetland to function as a transformer as opposed to a true sink. Studies reviewed by Giblin and Weider (1992) show that H_2S emission from freshwater wetlands are generally <200 mg S m^{-2} y^{-1}, and losses from dimethyl sulfide emission are approximately the same. Sulfate reduction rates in saltwater marshes can be 10 times higher than those in freshwater wetlands due to a greater amount of SO_4^{-2} input (Giblin and Weider 1992).

SUMMARY

Hydric soils differ from upland soils in that they are anaerobic in their upper 30 cm for some period during most years. The anaerobic conditions develop when oxidation–reduction reactions

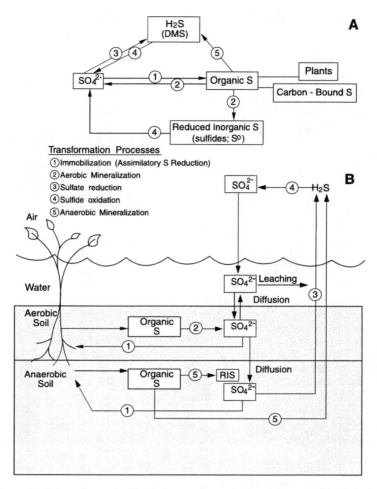

Figure 4.6 Schematic of the sulfur cycle. The major transformations are shown in (A), while the portions of the soil, as well as water and air in which the transformations occur, are shown in (B). The sulfur cycle includes a gaseous phase as does the nitrogen cycle, allowing S to be removed from the wetland.

occur in the soil that transfer electrons from donor to acceptor atoms. These reactions require that a soil: (1) be saturated with slowly moving or stagnant water, (2) have oxidizable organic C, and (3) have an active microbial population. The soils must be saturated to exclude atmospheric oxygen gas, which is a strong electron acceptor. Organic C is needed to supply the electrons used in the reduction process. An active microbial population transfers the electrons from donors to acceptors as they respire and oxidize organic tissues.

The major electron acceptors include the elements O, N, Mn, Fe, S, and C. The elements are reduced in this order. The presence of oxygen in the form of O_2 gas can keep all other elements from being reduced. The reduction of these elements can be monitored in soils by measuring the oxidation–reduction (redox) potential or by using dye solutions. Redox potential is an electrical measurement, where the voltage developed between two electrodes can be related to the chemistry of the soil solution. Dyes such as α, α'-dipyridyl react with reduced forms of Fe and allow its detection in the field.

Oxidation–reduction reactions affect soil color by causing organic C to accumulate and Fe or Mn oxides to become concentrated or depleted in portions of the soil. The hydric soil field indicators that are used to identify hydric soils are all formed by oxidation–reduction reactions. These reactions are also responsible for the cycling of N, P, and S in soils by transforming these elements into organic and inorganic forms. This cycling has important implications for maintaining the quality of fresh waters.

REFERENCES

Baas-Becking, L.G.M., I.R. Kaplan, and D. Moore. 1960. Limits of the natural environment in terms of pH and oxidation–reduction potentials. *J. Geology* 68:243–284.

Beauchamp, E., J.T. Trevors, and J.W. Paul. 1989. Carbon sources for bacterial denitrification. *Adv. Soil Sci.* 10:113–142.

Bohn, H.L., B.L. McNeal, and G.A. O'Connor. 1985. *Soil Chemistry.* John Wiley & Sons, New York.

Bouma, J. 1983. Hydrology and soil genesis of soils with aquic moisture regimes. pp. 253-281. *In* L. P. Wilding, N.E. Smeck, and G.F. Hall (Eds.). *Pedogenesis and Soil Taxonomy. I. Concepts and Interactions.* Elsevier Science Publishers B. V., Amsterdam, The Netherlands.

Bowden, W.B. 1987. The biogeochemistry of nitrogen in freshwater wetlands. *Biogeochemistry* 4:313–348.

Childs, C.W. 1981. Field test for ferrous iron and ferric-organic complexes (on exchange sites or in water-soluble forms) in soils. *Aust. J. Soil Res.* 19:175-180.

Clark, M.H. and C.L. Ping. 1997. Hydrology, morphology, and redox potentials in four soils of South Central Alaska. pp. 113–131. *In* M.J. Vepraskas and S.W. Sprecher (Eds.) *Aquic Conditions and Hydric Soils: The Problem Soils.* SSSA Spec. Publ. No. 50, Soil Sci. Soc. Am., Madison, WI.

Collins, J.F. and S.W. Buol. 1970. Effects of fluctuations in the Eh–pH environment of iron and/or manganese equilibria. *Soil Sci.* 110:111-118.

Couto, W., C. Sanzonowicz, and A. De O. Bacellos. 1985. Factors affecting oxidation–reduction processes in an Oxisol with a seasonal water table. *Soil Sci. Soc. Am. J.* 49:1245–1248.

Crozier, C.R., I. Devai, and R.D. DeLaune. 1995. Methane and reduced sulfur gas production by fresh and dried wetland soils. *Soil Sci. Soc. Am. J.* 59:277–284.

DeLaune, R.D. and W.H. Patrick, Jr. 1990. Nitrogen cycling in Louisiana Gulf Coast brackish marshes. *Hydrobiologia* 199:73–79.

Faulkner, S.P. and C.J. Richardson. 1989. Physical and chemical characteristics of freshwater wetland soils. p. 41–72. *In* D.A. Hammer (Ed.) *Constructed Wetlands for Wastewater Treatment.* Lewis Publishers, Chelsea, MI.

Faulkner, S.P., W.H. Patrick, Jr., and R.P. Gambrell. 1989. Field techniques for measuring wetland soil parameters. *Soil Sci. Soc. Am. J.* 53:883-890.

Gambrell, R.P. and W.H. Patrick, Jr. 1978. Chemical and microbiological properties of anaerobic soils and sediments. pp. 375-423. *In Plant Life in Anaerobic Environments.* Ann Arbor Science Publishers, Inc. Ann Arbor, Michigan.

Germida, J.J. 1998. Transformations of sulfur. pp. 346–368. *In* D.M. Sylvia, J.J. Furhmann, P.G. Hartel, and D.A. Zuberer (Eds.) *Principles and Applications of Soil Microbiology.* Prentice Hall, Upper Saddle River, NJ.

Giblin, A.E. and R.K. Wieder. 1992. Sulphur cycling in marine and freshwater wetlands. pp. 85–117. *In* R.W. Howarth, J.W.B. Stewart, and M.V. Ivanov (Eds.) *Sulphur Cycling on the Continents. Wetlands, Terrestrial Ecosystems, and Associated Ecosystems.* SCOPE 48. John Wiley & Sons, New York.

Gilman, K. 1994. *Hydrology and Wetland Conservation.* John Wiley & Sons, New York.

Heaney, S.I. and W. Davison. 1977. The determination of ferrous iron in natural waters with 2,2′-bipyridyl. *Limnol. Oceanogr.* 22:753–60.

Holford, I.C.R. and W.H. Patrick, Jr. 1979. Effects of reduction and pH changes on phosphate sorption and mobility in an acid soil. *Soil Sci. Soc. Am. J.* 43:292–297.

Howard-Williams, C. and M.T. Downes. 1993. Nitrogen cycling in wetlands. pp. 141–167. *In* T.P. Burt, A.L. Heathwaite, and S.T. Trudgill (Eds.) *Nitrate: Processes, Patterns, and Management.* John Wiley & Sons, New York.

Howeler, R.H. and D.R. Bouldin. 1971. The diffusion and consumption of oxygen in submerged soils. *Soil Sci. Soc. Amer. Proc.* 35:202–208.

Lindsay, A.L. 1979. *Chemical Equilibria in Soils.* John Wiley & Sons, New York.

McBride, M.B. 1994. *Environmental Chemistry of Soils.* Oxford Univ. Press, New York.

Meek, B.D., A.J. McKenzie, and L.B. Gross. 1968. Effects of organic matter, flooding time, and temperature on the dissolution of iron and manganese from soil *in situ. Soil Sci. Am. Proc.* 32:634–638.

Megonigal, J.P., S.P. Faulkner, and W.H. Patrick, Jr. 1996. The microbial activity season in southeastern hydric soils. *Soil Sci. Soc. Am. J.* 60:1263–1266.

Mitsch, W.J. and J.G. Gosselink. 1993. *Wetlands.* 2nd ed. Van Nostrand Reinhold, New York.

Moore, P.A. and K.R. Reddy. 1994. Role of Eh and pH on phosphorus geochemistry in sediments of Lake Okeechobee, Florida. *J. Envir. Qual.* 23:955–964.

Parkin, T.B. 1987. Soil microsites as a source of denitrification variability. *Soil Sci. Soc. Am. J.* 51:1194–1199.

Patrick, W.H. 1982. Nitrogen transformations in submerged soils. *In* F.J. Stevenson (Ed.) *Nitrogen in Agricultural Soils.* Agron. Monogr. 22. ASA, CSSA, and SSSA, Madison, WI.

Patrick, W.H., Jr., R.P. Gambrell, and S.P. Faulkner. 1996. Redox measurements of soils. pp. 1255-1276. *In* D.L. Sparks (Ed.) *Methods of Soil Analysis: Part 3 — Chemical Methods.* 3rd ed. Soil Sci. Soc. Am., Madison, WI.

Ponnamperuma, F. N. 1972. The chemistry of submerged soils. *Adv. Agron.* 24:29–96.

Reddy, K.R. and E.M. D'Angelo. 1994. Soil processes regulating water quality in wetlands. pp. 309–324. *In* W.J. Mitsch (Ed.) *Global Wetlands. Old World and New.* Elsevier Sci. B.V., Amsterdam, The Netherlands.

Reddy, K.R. and W.H. Patrick, Jr. 1984. Nitrogen transformations and loss in flooded soils and sediments. *CRC Critical Reviews in Environmental Control* 13:273–309.

Richardson, C.J. 1985. Mechanisms controlling phosphorus retention capacity in freshwater wetland. *Science* 228:1424–1427.

Richardson, C.J. 1989 Wetlands as transformers, filters, and sinks for nutrients. *In Freshwater Wetlands: Perspectives on Natural, Managed and Degraded Ecosystems.* Univ. of Georgia. Savannah River Ecology Lab. Ninth Symposium. Charleston, SC.

Richardson, C.J. 1999. The role of wetlands in storage, release, and cycling of phosphorus on the landscape: a 25-year retrospective. pp. 47–68. *In* K.R. Reddy, G.A. O'Connor, and C.L. Schelske (Eds.) *Phosphorus Biogeochemistry in Subtropical Ecosystems.* Lewis Publ., Boca Raton, FL.

Richardson, C.J. and P.E. Marshall. 1986. Processes controlling movement, storage, and export of phosphorus in a fen peatland. *Ecol. Monogr.* 56:279–302.

Richardson, J.L. and F.D. Hole. 1979. Mottling and iron distribution in a Glossoboralf Haplaquoll hydrosequence on a glacial moraine in northwestern Wisconsin. *Soil Sci. Soc. Am. J.* 43:552–558.

Sah, R.N. and D.S. Mikkelsen. 1986. Transformations of inorganic phosphorus during the flooding and draining cycles of soil. *Soil Sci. Soc. Am. J.* 50:62–67.

Soil Survey Staff. 1999. *Soil Taxonomy: A Basic System of Soil Classification for Making and Interpreting Soil Surveys.* 2nd ed. USDA, Natural Resources Conservation Service, Agr. Handbook No. 436. U.S. Govt. Printing Office, Washington, DC.

Sparks, D.L. 1995. *Environmental Soil Chemistry.* Academic Press, San Diego, CA.

Theis, T.L. and P.C. Singer. 1973. The stabilization of ferrous iron by organic compounds in natural waters. p. 178–192. *In* P.C. Singer (Ed.) *Trace Metals and Metal–Organic Interactions in Natural Waters.* Ann Arbor Science Publ., Ann Arbor, MI.

Turner, F.T. and W.H. Patrick, Jr. 1968. Chemical changes in waterlogged soils as a result of oxygen depletion. *Trans. 9th Intern. Cong. of Soil Sci.* 4:53–65.

Wieder, R.K. and G.E. Lang. 1986. Fe, Al, Mn, and S chemistry of sphagnum peat in four peatlands with different metal and sulfur input. *Water, Air, and Soil Pollut.* 29:309–320.

Wieder, R.K. and G.E. Lang. 1988. Cycling of inorganic and organic sulfur in peat from big run bog, West Virginia. *Biogeochemistry* 5:221–242.

Zhi-guang, L. 1985. Oxidation-reduction potential. pp. 1–26. *In* Y. Tian-red (Ed.) *Physical Chemistry of Paddy Soils.* Springer-Verlag, Berlin.

CHAPTER 5

Biology of Wetland Soils

Christopher B. Craft

INTRODUCTION

Hydric soil characteristics are the product of microorganisms, vegetation, and animal activity in soils saturated with water. Microorganisms are key agents regulating organic matter accumulation and other hydric soil characteristics, such as the formation of redoximorphic features and H_2S evolution (Chapter 7). Microbial decomposition of organic matter in combination with plant productivity determines whether and to what extent organic matter accumulates in the soil. The combination of biomass production by plants and reduced microbial decomposition in waterlogged environments leads to accumulation of soil organic matter, producing the surficial organic-rich horizons common to many hydric soils. Both plants and animals contribute to redoximorphic features, such as Fe pore linings (oxidized rhizospheres) as well as reduced soil matrix. This chapter discusses the environmental factors that affect distribution and abundance of the wetland soil organisms that produce physical and chemical characteristics unique to hydric soils. The role of microorganisms, vegetation, and animals in regulating the nature, extent, and intensity of hydric soil characteristics is also discussed.

MICROBIAL PROCESSES

Organic Matter Decomposition

Microorganisms such as bacteria and fungi are responsible for the decomposition of organic matter. Many of these organisms are heterotrophic in that they obtain energy from organic matter to support growth, metabolism, and reproduction. Bacteria numerically are the most abundant decomposers with more than a billion per gram of soil, while fungi are most abundant in terms of biomass (Atlas and Bartha 1987, Brady 1990, Paul and Clark 1996). Bacteria are important agents of both aerobic and anaerobic decomposition and are primarily responsible for organic matter decomposition in anaerobic environments (Benner et al. 1984, 1986b). Fungi are predominantly aerobes or facultative anaerobes and are the dominant agents of decomposition in aerobic environ-

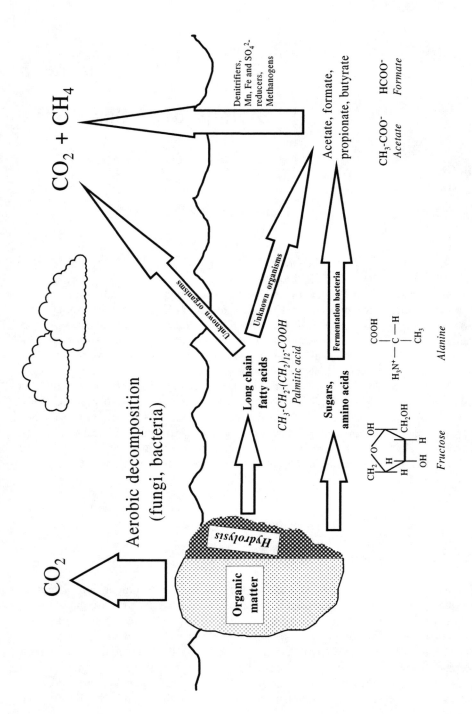

Figure 5.1   Aerobic and anaerobic pathways of organic matter decomposition in soils. Interterrestrial soils, aerobic decomposition predominates as organic matter is oxidized to CO_2 during respiration. In wetland soils, decomposition occurs via fermentation bacteria and bacteria that use terminal electron acceptors (Fe, Mn, NO_3^-, SO_4^{2-}) other than oxygen.

ments (Pritchett 1979, Paul and Clark 1996), including standing dead stems of wetland vegetation (Newell 1989, Newell et al. 1993).

In terrestrial soils, decomposition is an aerobic process because oxygen is used as a terminal electron acceptor to convert organic carbon (CH_2O) to carbon dioxide (CO_2) as shown below (see also Figure 5.1):

$$[CH_2O]n + nH_2O \rightarrow nCO_2 + 4n\ e^- + 4n\ H^+ \qquad \text{(Equation 1)}$$

$$O_2 + 4\ e^- + 4\ H^+ \rightarrow 2\ H_2O \qquad \text{(Equation 2)}$$

When wetland soils are inundated or saturated, respiring microorganisms quickly consume O_2 that is dissolved in the water or trapped in the soil. This causes decomposition to occur via anaerobic pathways using other terminal electron acceptors, such as oxidized iron (Fe^{3+}) and manganese (Mn^{4+}), nitrate (NO_3^-) and sulfate (SO_4^{2-}) (Figure 5.2). Aerobic decomposition is more efficient than anaerobic decomposition, yielding more energy (ATP) and more rapid decomposition of organic matter than anaerobic decomposition (Atlas and Bartha 1981, Mitsch and Gosselink 1993). Whereas aerobic decomposition produces 38 moles of ATP for every 1 mole of glucose metabolized, fermentation bacteria, which are important agents of anaerobic decomposition, produce only 2 moles of ATP for every mole of glucose metabolized (Atlas and Bartha 1981). Because anaerobic bacteria operate at a much lower energy level than aerobic bacteria, decomposition proceeds much more slowly in anaerobic and oxygen-limited environments such as wetlands (Ponnamperuma 1972). An alternative explanation for slow decomposition proposed by Lee (1992) is based on the idea that anaerobic soils and sediments contain a paucity of soil fauna including organisms that fragment organic matter and bacterial grazers. The absence of soil fauna, especially bacterial grazers, leads to the accumulation of bacterially derived organic compounds that are preserved in anaerobic soils and sediments. In either case, the accumulation of organic matter in wetland soils is the end result of reduced fragmentation and decomposition of organic matter.

Many types of fermentation bacteria exist, including those of the genus *Bacillus*, *Clostridium*, and *Lactobacillus* (Atlas and Bartha 1981, Lovley 1991). During fermentation, complex organic molecules are broken down into simple compounds, such as formate, acetate, and ethanol (Figure 5.1; Ponnamperuma 1972). There are numerous fermentation products, including simple alcohols, low-molecular-weight organic acids, fatty acids, and CO_2 (Ponnamperuma 1972, Atlas and Bartha 1981, Nedwell 1984, Capone and Kiene 1988). These compounds serve as substrates for other facultative and obligate anaerobic bacteria, including denitrifiers, Fe, Mn and SO_4^{2-} reducers, and methanogenic bacteria (Ponnamperuma 1972, Svorensen and Jorgensen 1987, Capone and Kiene 1988, Lovley 1991). Thus, the reduction of Fe^{3+}, Mn^{4+}, and SO_4^{2-} depends on the production of low-molecular-weight organic compounds during fermentation, as shown in Figure 5.1.

In addition to oxygen, organic matter decomposition in wetlands is regulated by other factors, including temperature, pH, organic matter quality, nutrients, and the availability of other terminal electron acceptors. Different microorganisms have different tolerances to temperature, but microbial activity and decomposition generally increase with increasing temperature up to approximately 35 to 40°C (Atlas and Bartha 1987, Benner et al. 1986a, Paul and Clark 1996). Most bacteria cannot survive extreme acidity or alkalinity, and optimum pH and decomposition occur in the range of pH 6 to 8 (Atlas and Bartha 1987, Benner et al. 1985, Brady 1990). Fungi are moderately acidophilic, and optimal growth usually is in the range of pH 4 to 6 (Brady 1990, Paul and Clark 1996). Organic matter quality also affects the rate of decomposition by bacteria and fungi. Organic compounds, such as sugars and amino acids, decompose more rapidly than other constituents of organic matter, such as cellulose and hemicellulose, while the lignin component is the slowest to decompose (Benner et al. 1984, Nedwell 1984, Moran et al. 1989, Paul and Clark 1996). Van Veen et al. (1984) developed a model that divides "fresh" organic matter into three components: (1) easily decom-

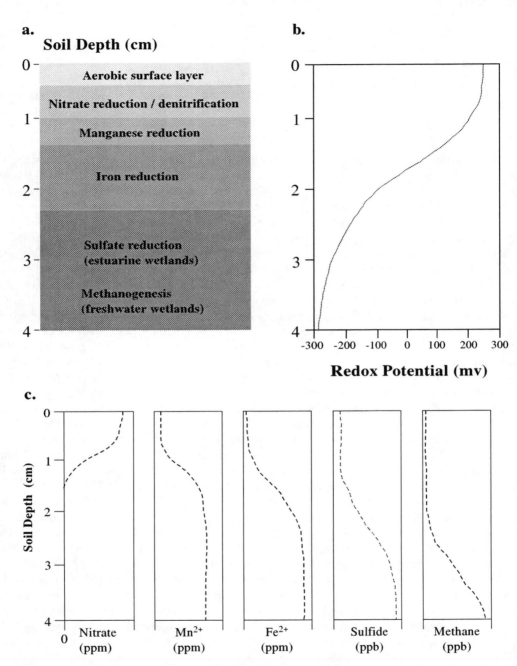

Figure 5.2 (a) Generalized diagram of the relative importance of different microbial reducing reactions, (b) soil redox potential and (c) concentrations of reducing reactants (NO_3^-) and products (Mn^{2+}, Fe^{2+}, SO_4^{2-}, CH_4) with depth in a wetland soil profile. (Part "c" is adapted from Patrick, W.H., Jr. and R.D. DeLaune. 1972. Characterization of the oxidized and reduced zones in flooded soil. *Soil Sci. Soc. Am. Proc.* 36:573–576.)

posable sugars and amino acids, (2) slowly decomposable cellulose, and (3) hemicellulose and resistant lignin. Estimated decomposition rates (quantity per day) were 0.2, 0.08, and 0.01 for sugars–amino acids, cellulose and hemicellulose, and lignin, respectively (Voroney et al. 1981).

Microbially Mediated Reducing Reactions: Denitrification, Mn, Fe and SO_4^{2-} Reduction, and Methanogenesis

When a soil is flooded, inundated, or saturated with water, aerobic decomposing microorganisms quickly consume the available oxygen, and facultative anaerobic microorganisms begin to use Fe^{3+}, Mn^{4+}, and SO_4^{2-} and other compounds as terminal electron acceptors to oxidize organic compounds to obtain energy for growth. These biochemical reactions occur in a predictable sequence corresponding to the availability of electron donors and acceptors in the soil, as discussed in Chapter 3. Upon flooding or inundation, aerobic decomposers quickly utilize the available O_2. Once most of the O_2 is consumed, denitrifying bacteria use nitrate as an energy source, followed by Mn and Fe reducers, SO_4^{2-} reducers, and finally methanogens. The same sequence of reducing reactions can be observed within an inundated wetland soil profile (Figure 5.2). Typically, a surface oxidized surface layer of varying thickness (mm to cm) exists at the soil surface. Oxygen penetration deeper into the soil is facilitated by roots and burrowing by animals, so the spatial and temporal variation in this zone can be quite dynamic. Below this oxidized layer is a gradient of decreasing redox potential and increasingly reduced subsurface soil horizons. The oxidized surface zone is maintained by diffusion of oxygen from the overlying water column. But, because oxygen diffuses 10,000 times more slowly through water than through air, it is consumed in this oxidized zone and does not penetrate into the subsoil. Below the oxidized zone, the redox potential decreases, and progressively reduced zones dominated by denitrification, Fe/Mn reduction, and SO_4^{2-} reduction/methanogenesis occur that correspond to increasing depth in the soil (Figure 5.2). These gradients of reducing reactions depend on the availability of reducible species (e.g., NO_3^-, oxidized Fe and Mn, SO_4^{2-}) and can be found in both freshwater and estuarine wetlands. A major distinction between freshwater and estuarine wetlands, however, is the predominance of sulfate reduction over methanogenesis in estuarine wetlands (Figure 5.2). The abundant SO_4^{2-} in seawater provides a continuous source of sulfate that supports high rates of sulfate reduction in salt marshes and other saline estuarine wetlands (Capone and Kiene 1988). In contrast, freshwater wetlands receive SO_4^{2-} primarily from atmospheric sources. In these wetlands, sulfate reduction is only a minor reducing reaction because methanogenesis predominates at low redox potentials (Capone and Kiene 1988).

Denitrification

Denitrification, the reduction of NO_3^- to gaseous N (NO, N_2O, N_2), is the first microbial reducing reaction to occur when O_2 is depleted. Denitrification is dependent on the redox potential, NO_3^- supply, and available C as well as other factors, such as temperature and pH, that affect all microbial processes (Seitzinger 1988, Groffman 1994). Wetland soils typically are chemically reduced and often contain ample organic C, especially compared to terrestrial ecosystems (Craft 1997). As such, denitrification in wetlands generally is limited by the availability of NO_3^- (Ambus and Lowrance 1991, Johnston 1991). In wetlands receiving loadings of NO_3^- from nonpoint runoff or other sources, denitrification can remove significant amounts of NO_3^-, up to 60 g N/m²/yr (Table 5.1). In most wetlands, however, denitrification is much lower (<4 g N/m²/yr) and is limited by the amount of NO_3^- produced by nitrifying bacteria (Table 5.1). Nitrifiers, which convert NH_4^+ to NO_3^-, are aerobic organisms and are confined to the narrow aerobic surface zone of the soil. Thus, denitrification in most wetland soils is closely coupled to nitrification (Capone and Kiene 1988, Groffman 1994).

Table 5.1   Rates of Denitrification, Mn, Fe, and SO_4^{2-} Reduction and Methanogenesis in Freshwater and Estuarine Wetlands

	Freshwater	Estuarine
Denitrification[1] (unenriched)	0–4	0–3
(g N/m²/yr) (nutrient enriched)	30–60	
SO_4^{2-} reduction[2] (g S/m²/yr)	0–540	800–2400
Methanogenesis[3] (g C/m²/yr)	0.5–74	0.5–10

[1] From Johnston 1991, Lindau and DeLaune 1991, Groffman 1994, DeLaune et al. 1996, and Boustany et al. 1997.
[2] From Howarth and Teal 1979, Howarth and Giblin 1983, Howarth 1984, Howes et al. 1984, Castro and Dierberg 1987, Wieder and Lang 1988, and Wieder et al. 1990.
[3] From Bartlett et al. 1987, Wieder et al. 1990, Moore 1994.

Mn and Fe Reduction

The hydric soil characteristics associated with redoximorphic features are the products of dissimilatory reduction of Fe. Microorganisms use Fe^{3+} as an external electron acceptor to oxidize organic C to provide energy and substrate for cell metabolism. The process reduces Fe^{3+} minerals to Fe^{2+}. It is the formation and leaching of reduced iron (Fe^{2+}) in the soil that results in the characteristic gray color found in many hydric soils (Bloomfield 1950, 1951, Ottow 1971, Van Breeman 1988). Numerous studies indicate that Fe and Mn reduction in soils and sediments is microbially mediated, as evidenced by inhibition of Fe^{2+} and Mn^{2+} production after additions of toxins such as cyanide (Kamura et al. 1963), mercury (DeCastro and Ehlich 1970, Jones et al. 1983), and chloroform (Bromfield 1954). Laboratory studies indicate that optimal microbial Fe reduction occurs at temperatures near 30 to 35°C and pH of 7 (Jones et al. 1983, Lovely and Phillips 1988). Excellent reviews of microbial Fe and Mn reduction are presented by Lovley (1987, 1991).

The availability of other terminal electron acceptors, particularly NO_3^-, can inhibit Fe reduction and the development of redoximorphic features. Several studies have demonstrated that Fe reduction is inhibited by NO_3^- (Yuan and Ponnamperuma 1966, Svorensen 1982, Jones et al. 1983), although not all Fe reducers are inhibited (Svorensen 1982, Jones et al. 1983). It is believed that NO_3^- inhibits Fe reduction because NO_3^- reducers are able to out-compete Fe reducers for electron donors (Lovley 1991). In contrast, Mn reduction can occur in the presence of NO_3^- (Billen 1982), as evidenced by the simultaneous reduction of NO_3^- and Mn^{4+} (Burdige and Nealson 1985).

The forms of oxidized Fe and Mn in the soil also dictate the rate and extent of Fe and Mn reduction (Lovley and Phillips 1986a, Lovley 1991). Most oxidized Fe and Mn in soils exist as insoluble oxides that vary in physical (particle size, surface area) and chemical composition (crystalline structure, reactivity, and oxidation state) (Bromfield and David 1978, Murray 1979, Schwertmann 1988). Highly crystalline forms of oxidized Fe are not susceptible to Fe reduction, while soluble chelated Fe oxides are readily reduced (Munch and Ottow 1980, Jones 1983, Jones et al. 1983, Lovley and Phillips 1986b, Lovley 1991). Most of the reducible Fe in soils consists of poorly crystalline Fe^{3+} oxides (Lovley 1991), with amorphous $Fe(OH)_3$ and ferrihydrite (a poorly crystalline Fe^{3+} oxide) likely being the predominant forms of Fe reduced (Ponnamperuma et al. 1967, Brannon et al. 1984, Lovley 1987). Lovley and Phillips (1987) found a strong relationship between hydroxylamine extractable Fe^{3+} (which extracts poorly crystalline Fe) and Fe reduction in freshwater and estuarine sediments.

Sulfate Reduction

In estuarine wetland soils and in freshwater environments with appreciable SO_4^{2-}, microbial reduction of sulfate to S^- or H_2S occurs. The characteristic "rotten egg" odor of tidal and estuarine wetland soils is the result of H_2S production by sulfate reducers. Numerous species of bacteria, including *Desulfovibrio, Desulfobacter, Desulfococcus,* and others, can use SO_4^{2-} as a terminal electron acceptor (Ponnamperuma 1972, Widdel and Hansen 1992). In wetland soils and sediments, most SO_4^{2-} reducing bacteria are mesophiles, existing at optimum temperatures of 20 to 40°C (Wieder and Lang 1988, Widdel and Hansen 1992). Like Fe and Mn reduction, SO_4^{2-} reduction may be inhibited by the presence of other terminal electron acceptors. Appreciable SO_4^{2-} reduction does not occur in the presence of oxidized Fe and Mn or nitrate (Ponnamperuma 1972, Yoshida 1975) because Fe/Mn reducers and denitrifiers maintain the concentration of electron donors at concentrations that are too low to support populations of SO_4^{2-} reducers (Lovley 1991).

A major difference between freshwater and estuarine wetlands is the predominance of sulfate reduction over methanogenesis in estuarine wetlands. The concentration of SO_4^{2-} in most freshwater ecosystems is only 100 to 200 μM as compared to 20 to 30 mM in seawater (Capone and Kiene 1988). As such, increased SO_4^{2-} availability in estuarine wetlands leads to higher rates of sulfate reduction than in freshwater wetlands (Castro and Dierberg 1987, Capone and Kiene 1988). Estimates of sulfate reduction in salt marsh sediments are on the order of 1000 to 2000 g $S/m^2/yr$, with approximately 50 to 90% of the total organic matter decomposition occurring via this pathway (Table 5.1; Howarth 1984, Howes et al. 1984, Howarth and Giblin 1983). In freshwater wetlands, sulfate reduction is generally limited by the availability of SO_4^{2-} (Nedwell 1984). Although sulfate reduction generally is lower in freshwater than in estuarine wetlands, significant sulfur inputs from acid deposition may lead to rates of sulfate reduction that are comparable to those measured in estuarine soils and sediments (Table 5.1).

Methanogenesis

Methane-producing bacteria occur under the most anaerobic and chemically reduced conditions. Methanogenesis yields less energy than other reducing reactions and, thus, does not predominate until other reducible substrates in the soil (NO_3^-, oxidized Fe and Mn, SO_4^{2-}) have been consumed. Like other reducing reactions, methanogenesis is regulated by temperature (Bridgham and Richardson 1992, Moore 1994, Updegraff et al. 1995), redox potential (Kludze and DeLaune 1994), availability of other reducible substrates (Bartlett et al. 1987, Capone and Kiene 1988), and labile organic carbon sources (Yavitt et al. 1987, Capone and Kiene 1988, Bridgham and Richardson 1992). The presence of other reducible substrates tends to suppress methanogenesis because of (1) the reduced energy yield associated with methanogenesis and (2) the inability to compete with denitrifiers and Fe, Mn, and SO_4^{2-} reducers for organic substrates (Svorensen et al. 1981, Bartlett et al. 1987, Capone and Kiene 1988, Oremland 1988, Yavitt and Lang 1990, Lovley 1991). Other factors, including water table depth, net primary production, and organic matter quality, also affect methanogenesis. Moore (1994) observed a positive relationship between seasonal water table depth and methane flux in Canadian peatlands. Likewise, Whiting and Chanton (1993) reported that methane flux from various North American freshwater wetlands increased with wetland net ecosystem productivity. Many freshwater peatlands exhibit low rates of methanogenesis because of the absence of low-molecular-weight carbon sources such as acetate, formate, and methanol (Yavitt et al. 1987, Capone and Kiene 1988, Bridgham and Richardson 1992). Methane fluxes from freshwater wetlands range from 0 to 74 g $C/m^2/yr$, with much lower amounts reported for estuarine wetlands, 0 to 10 C $g/m^2/yr$ (Table 5.1).

WETLAND VEGETATION

Like microorganisms, wetland vegetation is critical to the development of hydric soil characteristics. Accumulation of net primary production (NPP) (leaves, twigs, stems, wood, roots) from emergent and woody vegetation leads to the formation of surficial layers high in organic matter and, sometimes, to the development of histic epipedons and Histosols (Chapter 6). Translocation of dissolved organic matter from surface to subsurface soil horizons produces organic rich subsurface horizons characteristic of wet Spodosols (Chapter 15).

In this section, wetland ecosystems are grouped into general classes based on water source (freshwater, estuarine), dominant vegetation (herbaceous, woody), and soil type (mineral, organic) (Figures 5.3 and 5.4). While this type of classification omits much information on hydrology, water chemistry, and other attributes, it provides a framework for understanding the relationship between the different classes of wetlands and hydric soil characteristics that depend on accumulation of organic matter. Freshwater wetlands are dominated by either herbaceous emergent vegetation (marshes), woody vegetation (swamp forests), or peatmosses (*Sphagnum* spp.) and other acid-tolerant plants (bogs) (Figure 5.3). The estuarine wetlands, except for mangroves, are dominated by herbaceous vegetation (Figure 5.4). Mangrove forests are found along tropical and subtropical shorelines. These estuarine wetlands occupy areas of the landscape that are dominated by salt, brackish, and freshwater tidal marshes in temperate regions.

Net Primary Production, Decomposition and Organic Matter Accumulation

Organic matter accumulation in wetland soils represents the balance between NPP and decomposition (Schlesinger 1991, Mitsch and Gosselink 1993). Both processes are regulated by hydroperiod (duration of inundation or saturation), nutrient availability, and abiotic stressors (salinity, H_2S, acidity) as well as other factors (solar radiation, air/soil temperature). Hydroperiod regulates primary producers and decomposers by limiting the availability of O_2 for aerobic respiration. The availability of nutrients (nitrogen [N], phosphorus [P], calcium [Ca]) can enhance or limit productivity and decomposition, while stressors such as salinity, H_2S, and acidity can inhibit or halt both processes.

Hydroperiod

Flooding, inundation, or saturation of wetland soils initiates the development of anaerobic soil conditions that may be periodic or continuous, depending on the wetland. Many plant species, especially those adapted to terrestrial environments, become stressed when O_2 becomes limiting. When the O_2 supply to the roots is restricted, aerobic metabolism shuts down, and processes critical to growth (cell extension and division) and maintenance (nutrient absorption) slow dramatically, or even stop. With prolonged inundation, toxic fermentation products (acetaldehyde, ethanol) accumulate, causing cytoplasmic acidosis, destruction of cell organelles (mitochondria), and plant death (Roberts 1988, Vartapetian 1988). Many wetland species, however, possess morphological and metabolic adaptations that enable them to survive and even thrive in anaerobic soils. Such adaptations include aerenchyma (air-filled spaces in shoot tissue that transport O_2 to the roots where respiration occurs), pneumatophores (aerial roots in cypress [knees] and mangroves), swollen or buttressed trunks (which increase surface area for O_2 uptake), lenticles (pores in woody stems and trunks for increased O_2 diffusion), and anaerobic metabolic pathways that produce nontoxic end products (e.g., malate) (Mendelssohn et al. 1981, Mitsch and Gosselink 1993).

In wetlands, NPP is affected by both hydroperiod and hydrodynamics (Figures 5.5a and 5.5b). Hydrodynamics, the direction and velocity of water flow, regulates NPP by controlling oxygen and nutrient inputs from floodwaters or tides (Moore and Bellamy 1974, Brinson et al. 1981a, b, Mitsch and Gosselink 1993). Generally, the more "open" a wetland is, the greater the NPP (Lugo and Snedaker 1974, Gosselink and Turner 1978, Brinson et al. 1981a, b, Hopkinson 1992). "Open" wetlands, such as tidal marshes and alluvial forests, are among the most productive wetlands (Figure 5.5b; Brinson et al. 1981a, Good et al. 1982) and, in fact, are among the most productive ecosystems in the world (Mitsch and Gosselink 1993). In these wetlands, predictable periodic inundation brings in O_2 and nutrient-rich sediment (Odum 1980, Brinson et al. 1981b) and also flushes toxins from

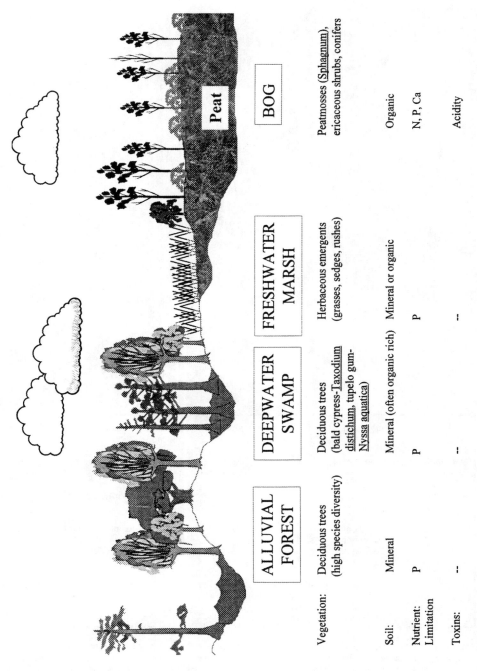

ALLUVIAL FOREST

DEEPWATER SWAMP

FRESHWATER MARSH

BOG

Peat

	ALLUVIAL FOREST	DEEPWATER SWAMP	FRESHWATER MARSH	BOG
Vegetation:	Deciduous trees (high species diversity)	Deciduous trees (bald cypress-_Taxodium distichum_, tupelo gum-_Nyssa aquatica_)	Herbaceous emergents (grasses, sedges, rushes)	Peatmosses (_Sphagnum_), ericaceous shrubs, conifers
Soil:	Mineral	Mineral (often organic rich)	Mineral or organic	Organic
Nutrient: Limitation	P	P	P	N, P, Ca
Toxins:	--	--	--	Acidity

Figure 5.3 Four classes of freshwater wetlands (bog, marsh/fen, alluvial swamp forest, and deepwater swamp forest). The dominant vegetation and soil type along with potential limiting nutrients and toxins are presented in the table at the bottom of the figure. (Nutrient limitation is from Lugo and Snedaker 1974, Valiela and Teal 1974, Sullivan and Daiber 1974, Broome et al. 1975, Brown 1981, Broome et al. 1988, Odum and McIvor 1990, Mitsch and Gosselink 1993, Craft et al. 1995, Bridgham et al. 1996. Toxins are from Lugo and Snedaker 1974, Mendelssohn et al. 1982, Odum and McIvor 1990, Mitsch and Gosselink 1993, Bridgham et al. 1996.)

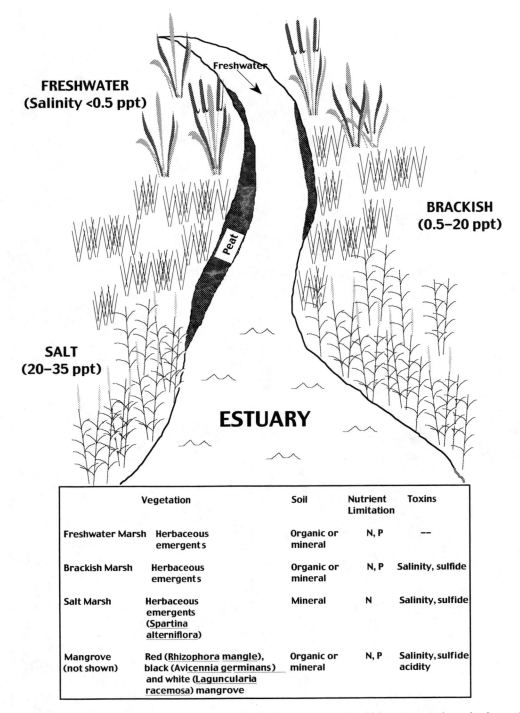

	Vegetation	Soil	Nutrient Limitation	Toxins
Freshwater Marsh	Herbaceous emergents	Organic or mineral	N, P	--
Brackish Marsh	Herbaceous emergents	Organic or mineral	N, P	Salinity, sulfide
Salt Marsh	Herbaceous emergents (Spartina alterniflora)	Mineral	N	Salinity, sulfide
Mangrove (not shown)	Red (Rhizophora mangle), black (Avicennia germinans) and white (Laguncularia racemosa) mangrove	Organic or mineral	N, P	Salinity, sulfide acidity

Figure 5.4   Four classes of estuarine wetlands (tidal freshwater marsh, brackish water marsh, and salt marsh. In tropical and subtropical regions, mangrove forests (not shown) replace the salt- and brackish-water marshes that dominate in temperate regions. The dominant vegetation and soil type along with potential limiting nutrients and toxins are presented in the table at the bottom of the figure. See Figure 5.3 for the literature citations of nutrient limitation and toxins.

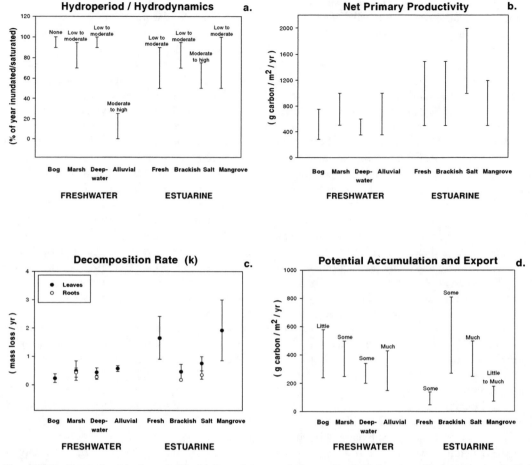

Figure 5.5 Generalized hydroperiod/hydrodynamics, net primary production, decomposition rate and potential soil organic matter accumulation and export in various freshwater and estuarine wetlands. (NPP is from Bernard and Gorham 1978, Reader 1978, Richardson et al. 1978, Van der Valk and Davis 1978, Whigham et al. 1978, Brinson et al. 1981a, Brown and Lugo 1982, Conner and Day 1982, Mann 1983, Odum et al. 1984, Stout 1988, Mitsch and Gosselink 1993, Szumigalski and Bayley 1996. NPP of alluvial wetlands is aboveground productivity only. Decomposition rate (k) is from Reader and Stewart 1972, Reimold et al. 1975, Brinson 1977, Cruz 1978, Chamie and Richardson 1978, Odum and Heywood 1978, Van der Valk and Davis 1978, Puriveth 1980, Hackney and Cruz 1980, Brinson et al. 1981a, Day 1982, Valiela et al. 1982, Yates and Day 1983, Odum et al. 1984, Benner et al. 1991, Mitsch and Gosselink 1993.) Potential accumulation and export are calculated as the difference between NPP and decomposition using the median or lower range (freshwater tidal marsh, 0.91; mangrove forest, 0.85) of the decomposition rate.

the system (Steever et al. 1976, Gosselink and Turner 1978). In contrast, "closed" wetlands such as bogs and pocosins, receive most of their water and nutrient inputs from precipitation. These "closed" wetlands have low NPP as compared to other wetlands (Figure 5.5b; Richardson et al. 1978, Mitsch and Gosselink 1993).

The duration of inundation or saturation also affects NPP of wetland vegetation, although probably to a lesser extent than hydrodynamics. Maximum productivity generally occurs under intermediate rather than either short or continuous hydroperiod (Figures 5.5a and 5.5b; Mitsch and Gosselink 1993). In southern deepwater swamps (cypress–gum forests), NPP is greatest in seasonally inundated forests as compared to drained or permanently flooded forests (Mitsch and Ewel 1979, Conner and Day 1982, Wharton et al. 1982). The relationship between hydrology and NPP of emergent wetland vegetation is similar although less clear. For example, *S. alterniflora* growing adjacent to tidal creeks typically has much higher NPP than *S. alterniflora* in back marsh areas, leading to the distinction between "tall" and "short" forms of *Spartina* (Wiegert and Freeman 1990). However, part of the growth difference in these forms is likely due to other factors, such as increased nutrient inputs and tidal flushing of toxins (salinity, sulfides). Likewise, in freshwater wetlands, NPP of Everglades sawgrass (*Cladium*) marshes is greatest under intermediate (inundated 50% of the time) rather than short or continuous inundation (Urban et al. 1993, David 1996).

Accumulation of soil organic matter depends not only on plant productivity but also on the rate of decomposition of dead and senescing plant material. Like productivity, decomposition tends to be lowest in wetlands with extended hydroperiod and low hydrodynamics (Figure 5.5c; Bridgham et al. 1991). Bogs have among the lowest rate of decomposition, which is due, in part, to lack of flushing and biotic production of hydrogen (H^+) ions and soil acidity (Moore and Bellamy 1974). Decomposition is also slower in other low-energy wetlands, such as deepwater swamps and brackish marshes, whereas wetlands characterized by high hydrodynamics and/or short hydroperiod (tidal marshes, mangroves, alluvial forests) typically have high rates of decomposition (Figure 5.5c).

It is believed that in most wetlands, soil organic matter accumulation is the result of root production and decomposition. In many wetlands, below-ground production is comparable to or greater than above-ground production (Valiela et al. 1982, Broome et al. 1986, Megonigal and Day 1992, Nyman et al. 1995). Like above-ground production, root production is greater under periodic rather than continuous inundation (Megonigal and Day 1992). Decomposition of roots, however, is generally slower than above-ground material (Figure 5.5c). In a mesocosm study, Day et al. (1989) found that only 20 to 30% of cypress roots decomposed over a 12-month period as compared to 30 to 60% for above-ground material. Day et al. (1989) also observed that decomposition also proceeded more quickly under periodic rather than continuous inundation. In a brackish-water marsh, only 16 to 20% of the roots of *Juncus* and *Spartina* decomposed in a year, with essentially no decomposition occurring below 10 cm (Hackney and Cruz 1980). As with woody vegetation, salt and freshwater marsh roots and rhizomes decompose more slowly than above-ground material such as leaves (Figure 5.5c).

Hydroperiod and hydrodynamics interact to regulate soil organic matter accumulation and lead to the development of hydric soil characteristics related to organic C (Chapters 7 and 8). Organic matter accumulation tends to be greatest under extended hydroperiod and low to moderate hydro-dynamics (Figure 5.5d; Gosselink and Turner 1978, Craft et al. 1993). Wetlands with low hydro-dynamics (bogs, fens, irregularly inundated tidal marshes, deepwater swamps) accumulate organic matter more readily than "open" wetlands (alluvial swamps, salt marshes) (Figure 5.5d). In "open" wetlands, much of the organic matter is exported by flooding and tides, and does not accumulate to the extent found in more "closed" low-energy wetlands (Gosselink and Turner 1978, Hopkinson 1992, Craft et al. 1993).

Soil organic matter accumulation generally is highest in wetlands (marshes, fens, bogs) dominated by herbaceous vegetation (marshes, fens) and *Sphagnum* (bogs) (Figure 5.5d). Many of these wetlands are characterized by relatively long hydroperiod and low hydrodynamics. Wetland herbaceous vegetation (grasses, rushes, sedges) typically has greater NPP than woody vegetation (Brinson et al. 1981a). The abundant NPP and often stable hydroperiod results in substantial accumulation of organic matter over time in these wetlands. Subtropical wetlands, including the Everglades in Florida and the Okefenokee in southern Georgia, represent some of the largest expanses of organic soils in the United States (Mitsch and Gosselink 1993). These organic soils formed primarily from deposition and accumulation of emergent plant detritus (Schlesinger 1978, Gleason and Stone 1994). The vast areas of organic soils of the northern United States and Canada are also the product of organic matter accumulation from nonwoody vegetation, in this case, peatmosses (*Sphagnum* spp.) (Moore and Bellamy 1974, Clymo 1983). In contrast to southern freshwater marshes, organic matter accumulation in northern bogs is due primarily to the low rate of decomposition (Figure 5.5c) caused by low pH, cold temperatures, and low nutrient availability (Moore and Bellamy 1974).

Nutrients

In many wetlands, NPP and decomposition are limited by the availability of nutrients. Nitrogen (N) and phosphorus (P) usually are the primary limiting nutrients (Valiela and Teal 1974, Gambrell and Patrick 1978, Klopatek 1978, Mitsch et al. 1979, Brown 1981, Aerts et al. 1992, Bridgham et al. 1996), although calcium (Ca) and potassium (K) may sometimes limit plant productivity (Steward and Ornes 1975, Bridgham et al. 1996). Many freshwater wetlands are P limited (Figure 5.3; Brown 1981, Vitousek and Howarth 1991, Craft et al. 1995, Bridgham et al. 1996). This is because P is cycled primarily through sedimentary compartments with no significant biological (e.g., N fixation) or atmospheric inputs (Schlesinger 1991). The largest reservoir of P is the soil, with most sequestered as recalcitrant organic compounds or bound with aluminum (Al), iron (Fe), or Ca (Richardson 1985, Richardson and Marshall 1986, Qualls and Richardson 1995), which are not readily available to plants and microorganisms (Brady 1990, Mitsch and Gosselink 1993). Estuarine wetlands, in contrast, tend to be N limited (Figure 5.4; Valiela and Teal 1974, Sullivan and Daiber 1974, Broome et al. 1975). Salt marshes and mangroves are periodically inundated by a mixture of seawater (which is relatively high in P, 0.01 to 0.1 mg/L P) diluted with freshwater (Long and Mason 1983), which apparently provides sufficient P to support biologic activity. Bogs and pocosins are usually acidic (pH < 4), and additions of lime, which increase soil pH and Ca, may increase NPP (Bridgham et al. 1996).

When nutrients are limiting, additions of these elements stimulate plant growth and decomposition. Many studies have documented the increase in NPP of wetland plant communities in response to nutrient additions (Sullivan and Daiber 1974, Valiela and Teal 1974, Broome et al. 1975, Brown 1981, Richardson and Marshall 1986, Craft et al. 1995, Bridgham et al. 1996). The effect of nutrients in accelerating decomposition has also been demonstrated (Valiela et al. 1982, Maltby 1985, Davis 1991, Amador and Jones 1995, Bridgham et al. 1996). It appears, however, that nutrient enrichment stimulates NPP to a greater extent than decomposition, as evidenced by enhanced organic matter accumulation in eutrophic environments. In the Everglades, peatlands receiving nutrient-enriched agricultural drainage respond with increased NPP (Davis 1989), decomposition (Davis 1991), organic matter accumulation, and accretion of organic material (Craft and Richardson 1993). At eutrophic sites, 24% of the NPP was buried as peat (210 to 220 g C/m²/yr) as compared to only 16% at unenriched sites (80 to 85 g C/m²/yr) (Craft and Richardson 1993).

Toxins

Periodic to continuous flooding, inundation, or saturation creates an anaerobic soil environment that is unfavorable for the growth and survival of many plant species. In addition to an O_2 limiting environment, other stressors (soil acidity, metal toxicity, salinity, and sulfides) affect the ability of wetland vegetation to survive and thrive (Figures 5.3 and 5.4). Soil acidity affects plant growth by saturating cation exchange sites with hydrogen (H^+) and aluminum (Al^{3+}), thereby displacing cations (Ca^{2+}, Mg^{2+}) essential for plant growth (Brady 1990). In strongly acidic soils (pH <4), H^+ and Al saturation of soil cation exchange sites leads to reduced Ca uptake, inhibited root elongation, and reduced growth and mortality (Brady 1990). However, soil acidity is not usually a problem for most wetland soils because prolonged inundation and saturation tend to drive the pH toward neutrality (pH = 7) (Ponnamperuma 1972). In acid soils, microbial reduction of Fe^{3+} to Fe^{2+} consumes H^+ and raises the pH, while in high pH (>7) soils, the production of organic acids during anaerobic decomposition lowers the pH (Ponnamperuma 1972, Faulkner and Richardson 1989). Some peatlands (bogs, pocosins), however, are acidic because they are "closed" or precipitation-driven wetlands with no water or nutrient inputs from surface or groundwaters. The soil pH of these peatlands often approaches 4 or less (Bridgham and Richardson 1993, Bridgham et al. 1996), and the plant communities are dominated by acidophilic (acid-loving) vegetation, such as conifers (pines, black spruce), ericaceous shrubs (heath), and peatmosses (*Sphagnum*) (Heinselman 1970, Moore and Bellamy 1974, Sharitz and Gibbons 1982). Soil acidity indirectly inhibits plant growth by increasing the concentration, sometimes to toxic levels, of reduced Fe and Mn. In acid soils, Fe and Mn toxicity have been reported for a variety of flood-intolerant terrestrial species (Jones and Ethington 1970, Wu 1981, Hodson et al. 1981, Rozema et al. 1985). Microbial reduction in wetland soils can lead to higher levels of Fe and Mn as compared to terrestrial soils. However, many wetland plant species have a higher tolerance for Fe and Mn than terrestrial species (Rozema et al. 1985, Ernst 1990).

In estuarine wetlands, salinity and sulfides are two additional stressors that affect the growth and distribution of vegetation (Figure 5.4). Salinity increases the osmotic potential of the soil water, adversely affecting vegetation in three ways, osmotically, nutritionally, and by a direct toxic effect (Long and Mason 1983). The effects of increased soil salinity include reduced water uptake and transpiration, increased Na uptake, decreased ammonium (NH_4^+) and K^+ uptake (Haines and Dunn 1976, Long and Mason 1983, Mendelssohn and Burdick 1988), and, eventually, reduced growth and survival of salt-intolerant plant species. Estuarine wetland vegetation such as cordgrass (*Spartina* spp.) and mangroves are adapted to saline conditions and, as a result, thrive and outcompete salt-intolerant freshwater wetland vegetation. Some of these adaptations include salt exclusion at the root surface, specialized salt glands on leaves, compartmentalization in cell vacuoles and reduced transpiration (by plants using the C_4 photosynthesis pathway) (Long and Mason 1983, Mitsch and Gosselink 1993). Different species, however, are adapted to different salinity levels (Broome et al. 1995, Flynn et al. 1995, Hackney et al. 1996) so that estuarine marsh and mangrove communities often exhibit distinct spatial patterns of zonation that are due to differences in salinity and other factors (tidal regime, soil redox, nutrient availability) (Ball 1980, Wiegert and Freeman 1990, Hackney et al. 1996).

Like salinity, sulfide toxicity is a problem for estuarine wetland vegetation (Figure 5.4; Koch and Mendelssohn 1989, Mitsch and Gosselink 1993). As described earlier, sulfide or H_2S is a product of microbial SO_4^{2-} reduction. In coastal wetlands, the high SO_4^{2-} concentration in seawater and soil leads to increased SO_4^{2-} reduction and H_2S production as compared to freshwater wetlands (Castro and Dierberg 1987). Hydrogen sulfide inhibits both aerobic and anaerobic (fermentative) metabolic pathways (Pearson and Havill 1988, Koch et al. 1990), carbon assimilation (Pezeshki et al. 1991), and nitrogen (NH_4^+) (Bradley and Morris 1990) and trace metal (Zn, Cu) uptake (Ponnamperuma 1972). Like salinity, sulfide tolerance probably varies from one species to another. Adaptations to excess sulfide include storage of SO_4^{2-} in cell vacuoles, conversion to gaseous compounds (H_2S, dimethylsulfide, carbon disulfide) that can diffuse out, and different metabolic tolerances to sulfide concentrations (Ernst 1990).

SOIL FAUNA

Wetlands are an important habitat for a variety of animals, including detritus-feeding invertebrates, larval and adult insects, amphibians and reptiles, waterfowl and wading birds, furbearers (beaver, muskrat, *Nutria*), and large mammals (bear). Many animals, especially insects and amphibians, are dependent on wetlands and aquatic habitats for reproduction (egg laying) to complete their life cycles. The abundant primary production of wetlands supports a large and diverse assemblage of consumers, especially invertebrates and birds. Wetlands are detritus-based ecosystems, producing and accumulating large amounts of organic matter. This organic matter serves as an energy and nutrient source for detritus-feeding organisms, especially invertebrates. Many animals, such as annelid and oligochaete worms, crayfish, and crabs, create burrows that serve as conduits

Table 5.2   Common Soil and Surface Dwelling Invertebrates Found in Different Wetland Classes. (Macrofauna include those organisms greater than 500 μm in diameter. Meiofauna are organisms 63 to 500 μm in diameter.)

	Meiofauna	Macrofauna
Freshwater		
Marsh/Fen[1]	Amphipods, copepods	Insect larvae (chironomids, a.k.a. midges), gastropods (snails), isopods
Swamp forest		
(Deepwater)	Nematodes, amphipods, copepods, oligochaete worms	Crayfish, clams, oligochaete worms, snails, isopods, midge larvae
(Alluvial)[2]	Nematodes, oligochaete worms, "terrestrial" invertebrates (mites–acari, springtails–collembola)	Oligochaete worms including earthworms — *Lumbricus* spp., crayfish
Estuarine		
Tidal marsh[3]		
(Freshwater)	Nematodes, amphipods, oligochaete worms	Oligochaete and polychaete worms, midge larvae
(Brackish)[4]	Nematodes, copepods, oligochaete worms	Oligochaete worms, fiddler crabs — *Uca* spp., snails, ribbed mussel — *Geukensia demissa*
(Salt)[5]	Nematodes, copepods, oligochaete worms	Oligochaete worms, fiddler crabs, mud crabs — *Sesarma* spp., periwinkle (gastropod), snails, ribbed mussel, oyster (*Crassostrea virginica*), hard clam (*Mercenaria mercenaria*)
Mangrove forest[6]	Nematodes, amphipods, copepods, polychaete worms	Fiddler crabs, oysters, barnacles, a.k.a. *Balanus* spp.

[1] Voigts 1976, Krieger 1992, Mitsch and Gosselink 1993, Rader and Richardson 1994.
[2] Zizer 1978, Wharton et al. 1981, 1982, Sklar 1985, Duffy and LaBar 1994.
[3] Diaz et al. 1978, Odum et al. 1984, 1988, Findley et al. 1989, Mitsch and Gosselink 1993.
[4] Stout 1984.
[5] Stout 1984, Montague et al. 1981, Wiegert and Freeman 1990.
[6] Odum and Heald 1972, Kuenzler 1974, Odum et al. 1982, Mitsch and Gosselink 1993.

for oxygen and water movement, oxygenating the soil and contributing to the development of redoximorphic features. Burrow walls are sites of reprecipitation of Fe and Mn, and oxidation of other reduced compounds (HS^-, NH_4^+) (Aller 1988). Table 5.2 describes some of the common invertebrates found in wetland soils (see also Figures 5.6–5.8). The same factors (oxygen limitation/hydroperiod, nutrients, toxins) that regulate wetland plant communities also affect the distribution and abundance of soil fauna. However, unlike plants, many animals are motile and can avoid adverse environmental conditions by migrating elsewhere.

Studies of freshwater wetland invertebrate communities reveal that hydrology is an important determinant of the abundance and distribution of soil fauna. In bottomland swamps, invertebrate assemblages vary with the seasonal inundation pattern (spring flooding vs. summer dry down). Wharton et al. (1982) describe three assemblages of invertebrates in these systems. The first includes inundation fauna that live in the soil and litter during periods of flooding. Oligochaete worms, copepods, isopods, nematodes, and gastropods (snails) are the dominant organisms, although amphipods, tubellarians, ostracods, lumbricid worms, and midge (Chironomid) larvae may be present (Figure 5.6; Zizer 1978, Sklar and Conner 1979, Wharton et al. 1982, Sklar 1985, Duffy and LaBar 1994, Golladay et al. 1997). Oligochaetes and midge larvae tolerate low levels of dissolved oxygen and, thus, are common in swamps as well as other freshwater wetlands (Mitsch and Gosselink 1993).

The second assemblage consists of litter invertebrates (mites [*Acari*], springtails [*Collembola*], and earthworms) that are found in floodplain leaf litter during dry periods (Figure 5.5). These

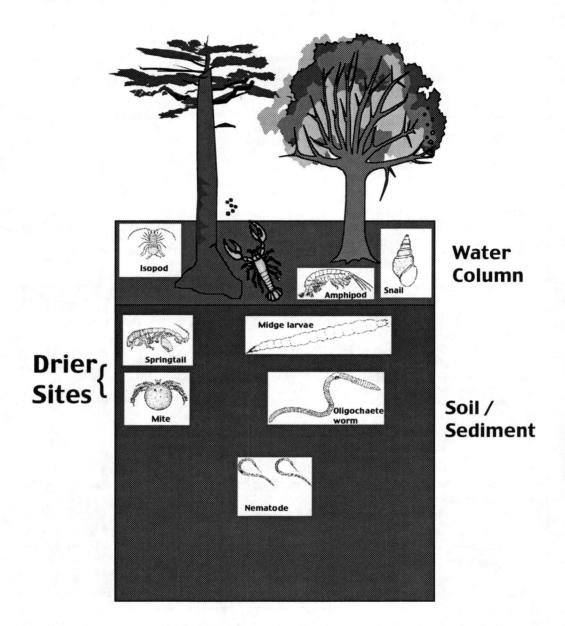

Figure 5.6   Common soil and aquatic invertebrates of a deepwater/alluvial swamp.

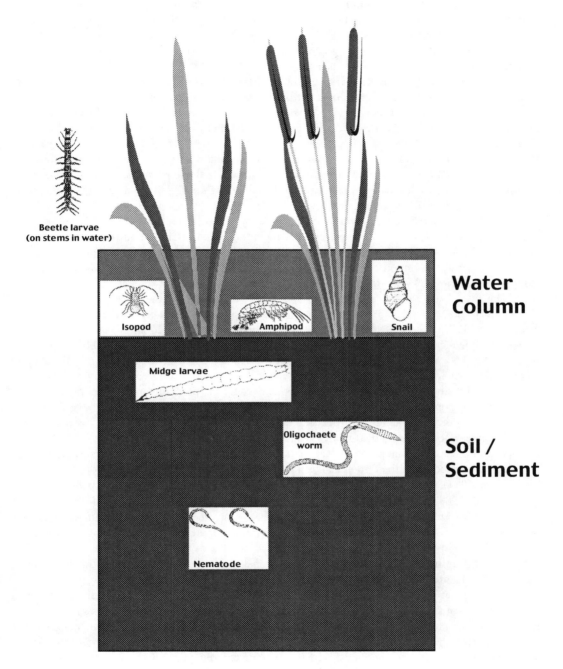

Figure 5.7   Common soil and aquatic invertebrates of a freshwater marsh.

organisms are more representative of terrestrial soils, existing on drier sites or utilizing the floodplain during the seasonal summer–autumn dry period. Mites and springtails account for >90% of the invertebrates in swamp forest litter during dry periods (Wharton et al. 1982). On even drier sites, the invertebrate assemblage becomes predominantly terrestrial, with numerous millipedes, crickets, and beetles in addition to mites and springtails (Zizer 1978, Wharton et al. 1982).

The third assemblage includes persistent fauna that exist throughout the year in the floodplain, in flooded or saturated soils. These organisms include crayfish and insect larvae. More than 20 species of chimney-building crayfish (*Procambarus*, *Cambarus*) are found east of the Mississippi River (Wharton et al. 1981). In floodplain soils, crayfish burrows may be abundant and elevated as much as 12 inches above the soil surface, giving the appearance of chimneys. Crayfish densities in floodplain soils can range from 21 to 46/m² and contribute up to 33% of the faunal biomass (Konikoff 1977, Wharton et al. 1982). Studies of dwarf crayfish (*Cambarellus* sp.) along the Mississippi gulf coast indicate a preference for wetland habitats containing abundant litter and submerged and emergent vegetation (Peterson et al. 1996).

In freshwater marshes, the invertebrate community is dominated by insect larvae (Table 5.2, Figure 5.7). Many of these immature insects, including midge (Chironomid) larvae or "blood-worms," are found in submerged soils and detritus (Krieger 1992, Mitsch and Gosselink 1993). Other invertebrates common to freshwater marshes include amphipods, copepods, and isopods (Voigts 1976, Krieger 1992, Rader and Richardson 1994). Many freshwater marshes occasionally dry down as a result of periodic droughts. As a result, patterns of the invertebrate community change over time, corresponding to changes in marsh water levels and plant community composition. Voigts (1976) recorded changes in prairie pothole invertebrate communities during a 5- to 20-year drydown cycle. During the initial drydown, the community was dominated by isopods. As the water level began to rise, emergent vegetation colonized the marsh, and amphipods, chironomid larvae, and other insect larvae dominated. As water levels continued to rise, emergent vegetation gave way to floating aquatic plants, and copepods became the predominant invertebrates.

Invertebrate communities of tidal wetlands have been extensively studied because of the importance of salt marshes and mangroves for sustaining the abundant secondary productivity of estuarine and nearshore waters. Studies of tidal freshwater marshes along the east coast (New York, Virginia, South Carolina) indicate that amphipods, copepods, oligochaete and polychaete worms, chironomid larvae, and nematodes are the dominant soil fauna (Table 5.2; Diaz et al. 1978, Findley et al. 1989, Odum et al. 1984). Nearly all of the organisms in these wetlands are freshwater-dependent species (Odum et al. 1988), although salinity-tolerant estuarine invertebrates, such as fiddler (*Uca*) and blue crabs (*Callinectes sapidus*), occasionally migrate into these wetlands (Odum et al. 1984). Species diversity is lower in tidal freshwater marshes compared to nontidal wetlands upstream (Odum et al. 1984) or higher-salinity salt marshes downstream (Lopez 1988, Odum et al. 1988). Freshwater macroinvertebrates, such as crayfish, probably are common in these wetlands, although there is little data to support or refute this (Odum et al. 1984).

Brackish-water and salt marsh wetlands typically have higher densities and diversity of soil invertebrates than freshwater wetlands. In fact, secondary production in salt marshes is among the highest (per unit area) of any ecosystem on earth (Daiber 1982, Wiegert and Freeman 1990). The meiofauna (63 to 500 μm) of brackish water and salt marshes are dominated by organisms similar to those found in freshwater wetlands. Nematodes, harpacticoid copepods, and oligochaetes are the dominant meiofauna in *Juncus*-dominated brackish-water marshes and salt marshes dominated by *Spartina alterniflora* (Figure 5.8; Daiber 1982, Stout 1984, 1988). The macrofauna (>500 μm) community, however, is much larger and more diverse in estuarine as compared to freshwater wetlands. Salt and brackish marshes contain many more gastropods (snails) and bivalves than freshwater tidal and nontidal wetlands. Commercially important bivalves, such as the ribbed mussel (*Geukensia demissa*), eastern oyster (*Crassostrea virginica*), and hard clam (*Mercenaria mercenaria*), as well as numerous noncommercial species, are common residents of coastal wetlands, especially in salt marshes (Figure 5.8; Daiber 1982, Wiegert and Freeman 1990). Likewise, peri-

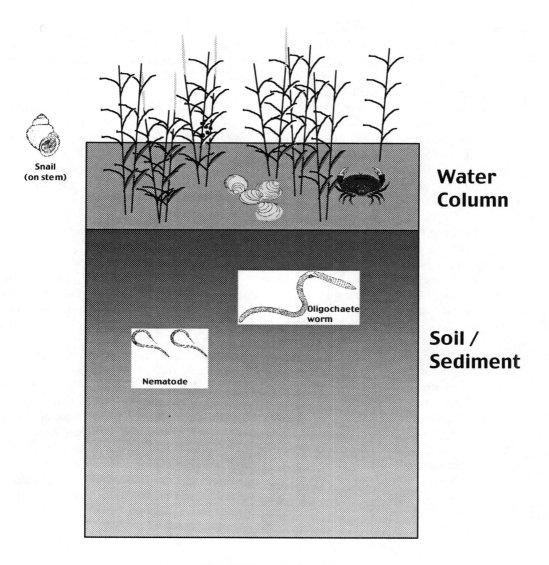

Figure 5.8   Common soil and benthic invertebrates of a salt marsh.

winkle (*Littorina irrorata*) and mud (*Ilyanassa obsoleta, Melampus bidentatus*) snails are abundant in salt marshes (Figure 5.8). In a Georgia salt marsh there were more than 700 periwinkle and mud snails per m^2 of marsh surface (Montague et al. 1981) with 50 to >400 periwinkle snails per m^2 (Newell 1993). Another unique feature of coastal wetlands is the preponderance of burrowing crabs. In these wetlands, fiddler (*Uca* spp.) and marsh (*Sesarma* spp.) crabs are the primary agents of soil aeration and water movement. *Uca* is important in creating burrows in brackish, salt, and mangrove wetlands (Teal 1958, Montague 1980). In addition to oxygenation of the soil, *Uca* burrows enhance *S. alterniflora* productivity, microbial decomposers, and benthic algae by reducing salinity and sulfide and enhancing nutrient availability (Montague 1980, 1982, Bertness 1985). Montague et al. (1981) reported that *Uca* densities in a Georgia salt marsh exceeded 80 to 200/m^2, with total community biomass of 15 g C/m^2. Deposit feeding estuarine invertebrates, such as snails and annelid worms, process large amounts of mineral matter during feeding and, in doing so, also contribute to bioturbation (Lopez and Levinton 1987). A comparison of macroinvertebrates in regularly inundated "low" and irregularly inundated "high" marsh zones of a *Juncus* marsh revealed higher invertebrate density and diversity in the low marsh zone (Subrahmanyan et al. 1976). These differences were probably due to a combination of factors, including greater accessibility for estuarine organisms, more detritus in the low marsh zone, and differences in soil physicochemical characteristics (salinity, pH, drainage, nutrients [N, P]) (Coultas 1969, Long and Mason 1983, Stout 1984).

Mangrove forests contain many of the same soil invertebrates found in other estuarine wetlands, including nematodes, polychaete worms, and fiddler crabs (Odum and Heald 1972, Kuenzler 1974). Surface-dwelling invertebrates include amphipods, harpacticoid copepods, and bivalves (mussels, oysters) (Odum and Heald 1972). Additionally, the prop roots of mangroves (aerial roots that provide support) are important faunal habitat in these systems (Odum et al. 1982, Mitsch and Gosselink 1993). Large numbers of barnacles and other filter feeders attach themselves to the prop roots (Mitsch and Gosselink 1993). Fiddler crabs utilize prop roots during high tide and burrow into the sediments during low tide (Kuenzler 1974).

Oxygen limitation, nutrients, and toxins affect soil fauna in the same manner as plants. However, many animals have at least some ability to migrate to more favorable environments. One difference between freshwater and estuarine soil fauna is that freshwater animals frequently have developed the physiological adaptations needed to withstand anaerobic conditions for longer periods (days to months) than is the case with estuarine animals (Lopez 1988). In contrast, estuarine benthic fauna typically are adapted to relatively short (hours to days) periods of anoxia, which may be an evolutionary adaptation to tidal mixing (Lopez 1988). Nutrient enrichment appears to increase the density and diversity of wetland invertebrates. Rader and Richardson (1994) reported higher densities and diversity of macroinvertebrates in nutrient-enriched areas of the Everglades that receive agricultural drainage, with 5 times as many invertebrates at enriched sites ($53,000/m^2$) than at unenriched locations ($10,000/m^2$). At both sites, the dominant invertebrates were dipteran larvae, gastropods, and amphipods (Rader and Richardson 1994). As with wetland vegetation, salinity and sulfide can be toxic to soil fauna. Invertebrates, especially those that live in estuarine environments, possess adaptations to deal with these stressors. Many estuarine wetland invertebrates are able to regulate the osmotic potential of their body through various mechanisms, including salt secreting glands, reduced exoskeleton permeability, and other specialized organs (Long and Mason 1983, Mitsch and Gosselink 1993). Other invertebrates become "inactive" during periods of stress caused by low oxygen or excess salinity or sulfide.

CONCLUSIONS

Soil organisms are critical to the development and evolution of wetland soil characteristics. Microorganisms and plant roots consume the limited O_2 in saturated and waterlogged soils and sediments. The depletion of O_2 leads to a shift from aerobic microbial metabolism to less efficient anaerobic metabolism fueled by terminal electron acceptors other than O_2. In the absence of O_2, microorganisms use NO_3^-, Mn^{4+} and Fe^{3+}, SO_4^{2-} and CO_2, producing end products such as N_2O, NO, reduced forms of Mn and Fe, H_2S, and CH_4. The formation of redoximorphic features and H_2S are metabolic products of microbial reducing reactions. Accumulation of organic matter is another important characteristic of wetland soils regulated by microorganisms as well as vegetation. The combination of high plant productivity and reduced decomposition of organic matter results in the development of organic-rich surface soil horizons and sometimes the formation of peat and muck soils. Organic matter accumulation represents the balance between productivity and decomposition. As such, the rate and extent of organic matter accumulation depends on factors such as hydrology, nutrients (N, P), and toxins (acidity, salinity, H_2S) that affect the ability of plants and microorganisms to survive and grow. Plant roots and soil fauna are key agents of bioturbation, oxygenation, and water movement in wetland soils. Root channels and burrow structure serve as conduits for water and air movement, producing redoximorphic features (Chapter 7) and enhancing the productivity of plant and microbial communities. Clearly, the unique characteristics of wetland soils and sediments are a reflection of the microorganisms, plant communities, and soil fauna that inhabit these periodically to permanently wet environments.

ACKNOWLEDGMENTS

I am grateful for the insightful reviews of an earlier draft of the manuscript by Carolyn Currin and Julie Fontaine. Steve Golladay provided thoughtful comments that greatly improved the Soil Fauna section. Many thanks to Bill Casey and Connie Chiang for reviewing and proofreading a later draft of the manuscript.

REFERENCES

Aerts, R., B. Wallen, and N. Malmer. 1992. Growth limiting nutrients in *Sphagnum*-dominated bogs subject to low and high atmospheric nitrogen supply. *J. Ecol.* 80:131–140.

Aller, R.C. 1988. Benthic fauna and biogeochemical processes in marine sediments: the role of burrow structures. pp. 301–338. *In* T.H. Blackburn and J. Sorenson (Eds.) *Nitrogen Cycling in Coastal Marine Environments.* SCOPE (Scientific Advisory Committee on Problems of the Environment), John Wiley & Sons, New York.

Ambus, P. and R. Lowrance. 1991. Comparison of denitrification in two riparian soils. *Soil Sci. Soc. Am. J.* 55:994–997.

Amador, J.A. and R.D. Jones. 1995. Carbon mineralization in pristine and phosphorus-enriched peat soils of the Everglades. *Soil Sci.* 159:129–141.

Atlas, R.M. and R. Bartha. 1981. *Microbial Ecology: Fundamentals and Applications.* Addison-Wesley Publishing Co., Reading, MA.

Atlas, R.M. and R. Bartha. 1987. *Microbial Ecology: Fundamentals and Applications.* 2nd edition. The Benjamin/Cummings Publishing Co., Reading, MA.

Ball, M.C. 1980. Patterns of secondary succession in a mangrove forest in south Florida. *Oecologia* 44:226–235.

Bartlett, K.B., D.S. Bartlett, R.C. Harriss, and D.I. Sebacher. 1987. Methane emissions along a salt marsh salinity gradient. *Biogeochem.* 4:183–202.

Benner, R., M.L. Fogel, and E.K. Sprague. 1991. Diagenesis of belowground biomass of *Spartina alterniflora* in salt-marsh sediments. *Limnol. Oceanogr.* 36:1358–1374.

Benner, R., A.E. Maccubbin, and R.E. Hodson. 1986a. Temporal relationships between the deposition and microbial degradation of lignocellulosic detritus in a Georgia salt marsh and the Okefenokee Swamp. *Microb. Ecol.* 12:291–298.

Benner, R., M.A. Moran, and R.E. Hodson. 1985. Effects of pH and plant source on lignocellulose biodegradation rates in two wetland ecosystems, the Okefenokee swamp and a Georgia salt marsh. *Limnol. Oceanogr.* 30:489–499.

Benner, R., M.A. Moran, and R.E. Hodson. 1986b. Biogeochemical cycling of lignocellulosic carbon in marine and freshwater ecosystems: relative contributions of procaryotes and eucaryotes. *Limnol. Oceanogr.* 31:89–100.

Benner, R., S.Y. Newell, A.E. Maccubbin, and R.E. Hodson. 1984. Relative contributions of bacteria and fungi to rates of degradation of lignocellulosic detritus in salt-marsh sediments. *Appl. Environ. Microbiol.* 48:36–40.

Bernard, J.M. and E. Gorham. 1978. Life history aspects of primary production in sedge wetlands. pp. 39–52. *In* R.E. Good, D.F. Whigham, and R.L. Simpson (Eds.) *Freshwater Wetlands: Ecological Processes and Management Potential.* Academic Press, New York.

Bertness, M.D. 1985. Fiddler crab regulation of *Spartina alterniflora* production on a New England salt marsh. *Ecol.* 66:1042–1055.

Billen, G. 1982. Modelling the processes of organic matter degradation and nutrient recycling in sedimentary environments. pp. 15–52. *In* D.B. Nedwell and C.M. Brown (Eds.) *Sedimentary Microbiology*. Academic Press, New York.

Bloomfield, C. 1950. Some observations of gleying. *J. Soil Sci.* 1:205–211.

Bloomfield, C. 1951. Experiments on the mechanism of gley formation. *J. Soil Sci.* 2:196–211.

Bradley, P.M. and J.T. Morris. 1990. Influence of oxygen and sulfide concentration on nitrogen uptake kinetics in *Spartina alterniflora*. *Ecol.* 71:282–287.

Brady, N.C., 1990. *The Nature and Properties of Soils*. Macmillan Publishing, New York.

Brannon, J.M., D. Gunnison, R.M. Smart, and R.L. Chen. 1984. Effects of added organic matter on iron and manganese redox systems in sediment. *Geomicrobiol. J.* 3:319–341.

Bridgham, S.D. and C.J. Richardson. 1992. Mechanisms controlling soil respiration (CO_2 and CH_4) in southern peatlands. *Soil Biol. Biochem.* 24:1089–1099.

Bridgham, S.D. and C.J. Richardson. 1993. The biogeochemistry of North Carolina freshwater peatlands: hydrology and nutrient gradients. *Wetlands* 13:207–218.

Bridgham, S.D., J. Pastor, J.A. Jannsens, C. Chapin, and T.J. Malterer. 1996. Multiple limiting gradients in peatlands: a call for a new paradigm. *Wetlands* 16:45–65.

Bridgham, S.D., C.J. Richardson, E. Maltby, and S.P. Faulkner. 1991. Cellulose decay in natural and disturbed peatlands in North Carolina. *J. Environ. Qual.* 20:695–701.

Brinson, M.M. 1977. Decomposition and nutrient exchange of litter in an alluvial swamp forest. *Ecol.* 58:601–609.

Brinson, M.M., A.E. Lugo, and S. Brown. 1981a. Primary productivity, decomposition and consumer activity in freshwater wetlands. *Ann. Rev. Ecol. Syst.* 12:123–161.

Brinson, M.M., B.L. Swift, R.C. Plantico, and J.S. Barclay. 1981b. Riparian ecosystems: their ecology and status. FWS/OBS-81/17. USFWS, Biological Services Program, Washington, DC.

Bromfield, S.M. 1954. Reduction of ferric compounds by soil bacteria. *J. Gen. Microbiol.* 11:1–6.

Bromfield, S.M. and D.J. David. 1978. Properties of biologically formed manganese in relation to soil manganese. *Aust. J. Soil Res.* 16:79–89.

Broome, S.W., I.A. Mendelssohn, and K.L. McKee. 1995. Relative growth of *Spartina patens* (Ait.) Muhl. and *Scirpus olneyi* Gray occurring in a mixed stand as affected by salinity and flooding depth. *Wetlands* 15:20–30.

Broome, S.W., E.D. Seneca, and W.W. Woodhouse, Jr. 1986. Long-term growth and development of transplants of the salt marsh grass *Spartina alterniflora*. *Estuaries* 9:63–74.

Broome, S.W., E.D. Seneca, and W.W. Woodhouse, Jr. 1988. Tidal salt marsh restoration. *Aquat. Bot.* 32:1–22.

Broome, S.W., W.W. Woodhouse, Jr., and E.D. Seneca. 1975. The relationship of mineral nutrients to the growth of *Spartina alterniflora* in North Carolina. II. The effects of N, P and Fe fertilizers. *Soil Sci. Soc. Am. Proc.* 39:301–307.

Brown, S.L. 1981. A comparison of the structure, primary productivity and transpiration of cypress ecosystems in Florida. *Ecol. Monogr.* 51:403–427.

Brown, S.L. and A.E. Lugo. 1982. A comparison of structural and functional characteristics of saltwater and freshwater forested wetlands, p. 109–130. *In* B. Gopal, R.E. Turner, R.G. Wetzel, and D.F. Whigham (Eds.) *Wetlands: Ecology and Management*. National Institute of Ecology and International Scientific Publications, Jaipur, India.

Boustany, R.G., C.R. Crozier, J.M. Rybczyk, and R.R. Twilley. 1997. Denitrification in a south Louisiana wetland forest receiving treated sewage effluent. *Wetlands Ecology and Management* 4:273–283.

Burdige, D.J. and K.H. Nealson. 1985. Microbial manganese reduction by enrichment cultures from coastal marine sediments. *Appl. Environ. Microbiol.* 50:491–497.

Capone, D.G. and R.P. Kiene. 1988. Comparison of microbial dynamics in marine and freshwater sediments: contrasts in anaerobic carbon metabolism. *Limnol. Oceanogr.* 33:725–749.

Castro, M.S. and F.E. Dierberg. 1987. Biogenic hydrogen sulfide emissions from selected Florida wetlands. *Water Air Soil Poll.* 33:1–13.

Chamie, J.P. and C.J. Richardson. 1978. Decomposition in northern wetlands. pp. 115–130. *In* R.E. Good, D.F. Whigham, and R.L. Simpson (Eds.) *Freshwater Wetlands: Ecological Processes and Management Potential*. Academic Press, New York.

Clymo, R.S. 1983. Peat. pp. 159–224. *In* A.J.P. Gore (Ed.) *Mires, Swamp, Bog, Fen and Moor, Ecosystems of the World 4A*. Elsevier Scientific Publishing, New York.

Conner, W.H. and J.W. Day, Jr., 1982. The ecology of forested wetlands in the southeastern United States. pp. 69–87. *In* B. Gopal, R.E. Turner, R.G. Wetzel, and D.F. Whigham (Eds.) *Wetlands: Ecology and Management.* National Institute of Ecology and International Scientific Publications, Jaipur, India.

Coultas, C.L. 1969. Some saline marsh soils in north Florida, part 1. *Soil Crop Sci. Soc. Florida Proc.* 29:111–123.

Craft, C.B. and C.J. Richardson. 1993. Peat accretion and phosphorus accumulation along a eutrophication gradient in the Northern Everglades. *Biogeochem.* 22:133–156.

Craft, C.B., E.D. Seneca, and S.W. Broome. 1993. Vertical accretion in microtidal regularly and irregularly flooded estuarine marshes. *Estuar. Coastal Shelf Sci.* 37:371–386.

Craft, C.B., J. Vymazal, and C.J. Richardson. 1995. Response of Everglades plant communities to nitrogen and phosphorus additions. *Wetlands* 15:258–271.

Craft, C.B. 1997. Dynamics of nitrogen and phosphorus retention during wetland ecosystem succession. *Wetlands Ecology and Management* 4:177–187.

Cruz, A.A. de la. Production and transport of detritus in wetlands. 1978. pp. 162–174. *In* P.E. Greeson, P.E., J.R. Clark, and J.E. Clark (Eds.) *Wetland Functions and Values: The State of Our Understanding.* American Water Resources Association, Minneapolis, MN.

Daiber, F.C. 1982. *Animals of the Tidal Marsh.* Van Nostrand Reinhold, New York.

David, P.G. 1996. Changes in plant communities relative to hydrologic conditions in the Florida Everglades. *Wetlands* 16:15–23.

Davis, S.M. 1989. Sawgrass and cattail production in relation to nutrient supply in the Everglades. pp. 325–341. *In* R.R. Sharitz and J.W. Gibbons (Eds.) *Freshwater Wetlands and Wildlife.* CONF-8603101, DOE symposium series no. 61, USDOE Office of Scientific and Technical Information, Oak Ridge, TN.

Davis, S.M. 1991. Growth, decomposition and nutrient retention of *Cladium jamaicense* Crantz and *Typha domingensis* Pers. in the Florida Everglades. *Aquat. Bot.* 40:203–224.

Day, F.P., Jr. 1982. Litter decomposition rates in the seasonally flooded Great Dismal swamp. *Ecol.* 63:670–678.

Day, F.P., Jr., J.P. Megonigal, and L.C. Lee. 1989. Cypress root decomposition in experimental wetland mesocosms. *Wetlands* 9:263–282.

DeCastro, A.F. and H.L. Ehlich. 1970. Reduction of iron oxide minerals by a marine *Bacillus*. *Antonie van Leeuwenhoek* 36:317–327.

DeLaune, R.D., R.R. Boar, C.W. Lindau, and B.A. Kleiss. 1996. Denitrification in bottomland hardwood wetland soils of the Cache River. *Wetlands* 16:309–320.

Diaz, R.J., D.F. Boesch, J.L. Haver, C.A. Stone, and K. Munson. 1978. *Habitat Development, Field Investigations,* Windmill Point *Marsh Development Site,* James River, Virginia. part II: *Aquatic Biology-Benthos.* U.S. Army Waterways Experiment Station, Technical report D-77-23. pp. 18–54.

Duffy, W.G. and D.J. LaBar. 1994. Aquatic invertebrate production in southeastern USA wetlands during winter and spring. *Wetlands* 14:88–97.

Ernst, W.H.O. 1990. Ecophysiology of plants in waterlogged and flooded environments. *Aquat. Bot.* 38:73–90.

Faulkner, S.P. and C.J. Richardson. 1989. Physical and chemical characteristics of freshwater wetland soils. p. 41–72. *In* D.A. Hammer (Ed.) *Constructed Wetlands for Wastewater Treatment.* Lewis Publishers, Chelsea, MI.

Findley, S., K. Schoeberl, and B. Wagner. 1989. Abundance, composition and dynamics of the invertebrate fauna of a tidal freshwater wetland. *J. N. Am. Benthol. Soc.* 8:140–148.

Flynn, K.M., K.L. McKee, and I.A. Mendelssohn. 1995. Recovery of freshwater marsh vegetation after a saltwater intrusion event. *Oecologia* 103:63–72.

Gambrell, R.P. and W.H. Patrick, Jr. 1978. Chemical and microbiological properties of anaerobic soils and sediments. pp. 375–423. *In* D.D. Hook and R.M.M. Crawford (Eds.) *Plant Life in Anaerobic Environments.* Ann Arbor Science Publishers, Ann Arbor, MI.

Gleason, P.J. and P. Stone. 1994. Age, origin and landscape evolution of the Everglades peatland. pp. 149–198. *In* S.M. Davis and J.C. Ogden (Eds.) *Everglades: The Ecosystem and Its Restoration.* St. Lucie Press, Delray Beach, FL.

Golladay, S.W., B.W. Taylor, and B.J. Palik. 1997. Invertebrate communities of forested limesink wetlands in southwest Georgia, USA: habitat use and influence of extended inundation. *Wetlands* 17:383–393.

Good, R.E., N.F. Good, and B.R. Frasco. 1982. A review of primary production and decomposition dynamics of the belowground marsh component. pp. 139–157. *In* V. Kennedy (Ed.) *Estuarine Comparisons.* Academic Press, New York.

Gosselink, J.G. and R.E. Turner. 1978. The role of hydrology in freshwater wetland ecosystems. pp. 63–78. *In* R.E. Good, D.F. Whigham, and R.L. Simpson (Eds.) *Freshwater Wetlands: Ecological Processes and Management Potential*. Academic Press, New York.

Groffman, P.M. 1994. Denitrification in freshwater wetlands. *Curr. Top. Wetland Biogeochem.* 1:15–35.

Hackney, C.T. and A.A. de la Cruz. 1980. *In situ* decomposition of roots and rhizomes of two tidal marsh plants. *Ecol.* 61:226–231.

Hackney, C.T., S. Brady, L. Stemmy, M. Boris, C. Dennis, T. Hancock, M. O'Bryon, C. Tilton, and E. Barbee. 1996. Does intertidal vegetation indicate specific soil and hydrologic conditions? *Wetlands* 16:89–94.

Haines, B.L. and E.L. Dunn. 1976. Growth and resource allocation responses of *Spartina alterniflora* Loisel to three levels of NH_4-N, Fe and NaCl in solution culture. *Bot. Gazette* 137:224–230.

Heinselman, M.L. 1970. Landscape evolution and peatland types, and the Lake Agassiz Peatlands Natural Area. 1970. *Ecol. Monogr.* 40:235–261.

Hodson, M.J., M.M. Smith, S.J. Wainwright, and H. Opik. 1981. Cation cotolerance in a salt-tolerant clone of *Agrostis stolonifera* L. *New Phytol.* 90:253–261.

Hopkinson, C.S., Jr. 1992. A comparison of ecosystem dynamics in freshwater wetlands. *Estuaries* 15:549–562.

Howarth, R.W. 1984. The ecological significance of sulfur in the energy dynamics of salt marsh and coastal marine sediments. *Biogeochem.* 1:5–27.

Howarth, R.W. and A. Giblin. 1983. Sulfate reduction in the salt marshes at Sapelo Island, Georgia. *Limnol. Oceanogr.* 28:70–82.

Howarth, R.W. and J.M. Teal. 1979. Sulfate reduction in a New England salt marsh. *Limnol. Oceanogr.* 24:999–1013.

Howes, B.L., J.W.H. Dacey, and G.M. King. 1984. Carbon flow through oxygen and sulfate reduction pathways in salt marsh sediments. *Limnol. Oceanogr.* 29:1037–1051.

Johnston, C.A. 1991. Sediment and nutrient retention by freshwater wetlands: effects on surface water quality. *Crit. Rev. Environ. Control* 21:491–565.

Jones, H.E. and J.R. Ethington. 1970. Comparative studies of plant growth and distribution in relation to waterlogging. I. The survival of *Erica cinerea* L. and *E. tetralix* L. and its apparent relationship to iron and manganese uptake in waterlogged soil. *J. Ecol.* 58:487–496.

Jones, J.G. 1983. A note on the isolation and enumeration of bacteria which deposit and reduce ferric iron. *J. Appl. Bacteriol.* 54:305–310.

Jones, J.G., S. Gardiner, and B.M. Simon. 1983. Bacterial reduction of ferric iron in a stratified eutrophic lake. *J. Gen. Microbiol.* 129:131–139.

Kamura, T., Y. Takai, and K. Ishikawa. 1963. Microbial reduction mechanism of ferric iron in paddy soils. Part 1. *Soil Sci. Plant Nutr.* 9:171–175.

Klopatek, J.M. 1978. Nutrient dynamics of freshwater riverine marshes and the role of emergent macrophytes. p. 195–216. *In* R.E. Good, D.F. Whigham, and R.L. Simpson (Eds.) *Freshwater Wetlands: Ecological Processes and Management Potential*. Academic Press, New York.

Kludze, H.K. and R.D. DeLaune. 1994. Methane emissions and growth of *Spartina patens* in response to soil redox intensity. *Soil Sci. Soc. Am. J.* 58:1838–1845.

Konikoff, M. 1977. Studies of the life history and ecology of the red swamp crawfish, *Procambarus clarkii*, in the lower Atchafalaya Basin floodway. Final report to the U.S. Fish and Wildlife Service, Dept. of Biology, Univ. Southwestern Louisiana, Lafayette, LA.

Koch, M.S. and I.A. Mendelssohn. 1989. Sulphide as a soil phytotoxin: differential responses in two marsh species. *J. Ecol.* 77:565–578.

Koch, M.S., I.A. Mendelssohn, and K.L. McKee. 1990. Mechanism for the hydrogen sulfide-induced growth limitation in wetland macrophytes. *Limnol. Oceanogr.* 35:399–408.

Krieger, K.A. 1992. The ecology of invertebrates in Great Lakes coastal wetlands: current knowledge and research needs. *J. Great Lakes Res.* 18:634–650.

Kuenzler, E.J. 1974. Mangrove swamp systems. pp. 346–371. *In* H.T. Odum, B.J. Copeland, and E.A. McMahan (Eds.) *Coastal Ecological Systems of the United States, Volume 1*. The Conservation Foundation, Washington, DC.

Lee, C. 1992. Controls on organic carbon preservation: the use of stratified water bodies to compare intrinsic rates of decomposition in oxic and anoxic systems. *Geochim. Cosmochim. Acta* 56:3323–3335.

Lindau, C.W. and R.D. DeLaune. 1991. Dinitrogen and nitrous oxide emission and entrapment in *Spartina alterniflora* salt marsh soils following addition of N-15 labeled ammonium and nitrate. *Estuar. Coastal Shelf Sci.* 32:161–172.

Long, S.P. and C.F. Mason. 1983. *Salt Marsh Ecology.* Chapman and Hall, New York.

Lopez, G.R. 1988. Comparative ecology of the macrofauna of freshwater and marine muds. *Limnol. Oceanogr.* 33:946–962.

Lopez, G.R. and J.S. Levinton. 1987. Ecology of deposit feeding animals in marine sediments. *Q. Rev. Biol.* 62:235–260.

Lovley, D.R. 1987. Organic matter mineralization with the reduction of ferric iron: a review. *Geomicrobiol. J.* 5:375–399.

Lovley, D.R. 1991. Dissimilatory iron (III) and manganese (IV) reduction. *Microbiol. Rev.* 55:259–287.

Lovley, D.R. and E.J.P. Phillips. 1986a. Organic matter mineralization with reduction of ferric iron in anaerobic sediments. *Appl. Environ. Microbiol.* 51:683–689.

Lovley, D.R. and E.J.P. Phillips. 1986b. Availability of ferric iron for microbial reduction in bottom sediments of the freshwater tidal Potomac River. *Appl. Environ. Microbiol.* 52:751–757.

Lovley, D.R. and E.J.P. Phillips. 1987. Competitive mechanisms for inhibition of sulfate reduction and methane production in the zone of ferric iron reduction in sediments. *Appl. Environ. Microbiol.* 53:2636–2641.

Lovley, D.R. and E.J.P. Phillips. 1988. Novel mode of microbial energy metabolism: organic carbon oxidation coupled to dissimilatory reduction of iron or manganese. *Appl. Environ. Microbiol.* 54:1472–1480.

Lugo A.E. and S.C. Snedaker. 1974. The ecology of mangroves. *Ann. Rev. Ecol. Syst.* 5: 39–64.

Maltby, E. 1985. Effects of nutrient loadings on decomposition profiles in the water column and submerged peat in the Everglades. pp. 450–464. *In Tropical Peat Resources — Prospects and Potential.* Proceedings of the IPS symposium. Kingston, Jamaica. February 25 — March 1, 1985.

Mann, K.H. 1983. *Ecology of Coastal Waters: A Systems Approach.* University of California Press, Los Angeles.

Megonigal, J.P. and F.P. Day. 1992. Effects of flooding on root and shoot production of bald cypress in large experimental enclosures. *Ecol.* 73:1182–1193.

Mendelssohn, I.A. and D.M. Burdick. 1988. The relationship of soil parameters and root metabolism to primary production in periodically inundated soils. pp. 398–428. *In* D.D. Hook and others (Eds.) *The Ecology and Management of Wetlands. Volume 1: Ecology of Wetlands.* Timber Press, Portland OR.

Mendelssohn, I.A, K.L. McKee, and W.H. Patrick, Jr. 1981. Oxygen deficiency in *Spartina alterniflora* roots: metabolic adaptation to anoxia. *Science* 214:439–441.

Mendelssohn, I.A., K.L. McKee, and T.R. Postek. 1982. Sublethal stresses controlling *Spartina alterniflora* productivity. pp. 223–242. *In* B. Gopal, R.E. Turner, R.G. Wetzel, and D.F. Whigham (Eds.) *Wetlands: Ecology and Management.* National Institute of Ecology and International Scientific Publications, Jaipur, India.

Mitsch, W.J. and K.C. Ewel. 1979. Comparative biomass and growth of cypress in Florida wetlands. *Am. Midl. Nat.* 101:417–426.

Mitsch, W.J. and J.G. Gosselink. 1993. *Wetlands.* Van Nostrand Reinhold, New York.

Mitsch, W.J., C.L. Dorge, and J.R. Wiemhoff. 1979. Ecosystem dynamics and a phosphorus budget of an alluvial cypress swamp in southern Illinois. *Ecol.* 60:1116–1124.

Montague, C.L. 1980. A natural history of temperate western atlantic fiddler crab (Genus *Uca*) with reference to their impact on the salt marsh. *Contrib. Mar. Sci.* 23:25–54.

Montague, C.L. 1982. The influence of fiddler crab burrows and burrowing on metabolic processes in salt marsh sediments. pp. 283–301. *In* V.S. Kennedy (Ed.) *Estuarine Comparisons.* Academic Press, New York.

Montague, C.L., S.M. Bunker, E.B. Haines, M.L. Pace, and R.L. Wetzel. 1981. Aquatic macroconsumers. Chapter 4. *In* L.R. Pomeroy and R.G. Wiegert (Eds.) *The Ecology of a Salt Marsh.* Springer-Verlag, New York.

Moore, P.D. and D.J. Bellamy. 1974. *Peatlands.* Springer-Verlag, New York.

Moore, T.R. 1994. Trace gas emissions from Canadian peatlands and the effect of climate change. *Wetlands* 14:223–228.

Moran, M.A., R. Benner, and R.E. Hodson. 1989. Kinetics of microbial degradation of vascular plant material in two wetland ecosystems. *Oecologia* 79:158–167.

Munch, J.C. and J.C.G. Ottow. 1980. Preferential reduction of amorphous to crystalline iron oxides by bacterial activity. *Soil Sci.* 129:15–21.

Murray, J.W. 1979. Iron oxides. pp. 47–98. *In* R.G. Burns (Ed.) *Marine Minerals.* Mineralogical Society of America, Washington, DC.

Nedwell, D.B. 1984. The input and mineralization of organic carbon in anaerobic aquatic sediments. *Adv. Microb. Ecol.* 7:93–131.

Newell, S.Y. 1993. Decomposition of shoots of a salt marsh grass: methodology and dynamics of microbial assemblages. *Adv. Microb. Ecol.* 13:301–326.

Newell, S.Y., R.D. Fallon, and J.D. Miller. 1989. Decomposition and microbial dynamics for standing naturally positioned leaves of the salt-marsh grass, *Spartina alterniflora. Mar. Biol.* 101:471–481.

Nyman, J.A., R.D. DeLaune, S.R. Pezeshki, and W.H. Patrick, Jr. 1995. Organic matter fluxes and marsh stability in a rapidly submerging estuarine marsh. *Estuaries* 18:207–218.

Odum, E.P. 1980. The status of three ecosystem-level processes regarding salt marsh estuaries: tidal subsidy, outwelling and detritus-based food chains. pp. 485–495. *In* V.S. Kennedy (Ed.) *Estuarine Perspectives.* Academic Press, New York.

Odum, W.E. and E.J. Heald. 1972. Trophic analyses of an estuarine mangrove community. *Bull. Mar. Sci.* 22:671–738.

Odum, W.E. and M.A. Heywood. 1978. Decomposition of intertidal freshwater marsh plants. pp. 89–97. *In* R.E. Good, D.F. Whigham, and R.L. Simpson (Eds.) *Freshwater Wetlands: Ecological Processes and Management Potential.* Academic Press, New York.

Odum, W.E. and C.C. McIvor. 1990. Mangroves. pp. 517–548. *In* R.L. Meyers and J.J. Ewel (Eds.) *Ecosystems of Florida.* University of Central Florida Press, Orlando, FL.

Odum, W.E., C.C. McIvor, and T.J. Smith, III. 1982. *The Ecology of the Mangroves of South Florida: A Community Profile.* Tech. Report FWS/OBS/81-24. U.S. Fish and Wildlife Services, Division of Biological Services, Washington, DC.

Odum, W.E., L.P. Rozas, and C.C. McIvor. 1988. A comparison of fish and invertebrate community composition in tidal freshwater and oligohaline marsh systems. pp. 561–569. *In* D.D. Hook and others (Eds.) *The Ecology and Management of Wetlands. Volume 1: Ecology of Wetlands.* Timber Press, Portland OR.

Odum, W.E., T.J. Smith, III, J.K. Hoover, and C.C. McIvor. 1984. *The Ecology of Tidal Freshwater Marshes of the United States East Coast: A Community Profile.* FWS/OBS-84/17. U.S. Fish and Wildlife Service, Washington, DC.

Oremland, R.S. 1988. Biogeochemistry of methanogenic bacteria. pp. 641–705. *In* A.J.B. Zehnder (Ed.) *Biology of Anaerobic Microorganisms.* John Wiley & Sons, New York.

Ottow, J.C.G. 1971. Iron reduction and gley formation by nitrogen fixing *Clostridia* bacteria. *Oecologia* 6:164–175.

Patrick, W.H., Jr. and R.D. DeLaune. 1972. Characterization of the oxidized and reduced zones in flooded soil. *Soil Sci. Soc. Am. Proc.* 36:573–576.

Paul, E.A. and F.C. Clark. 1996. *Soil Biology and Biochemistry.* Academic Press, New York.

Pearson, J. and D.C. Havill. 1988. The effect of hypoxia and sulfide on culture grown wetland and nonwetland plants. 2. metabolic and physiological changes. *J. Exp. Bot.* 39:431–439.

Peterson, M.S., J.F. Fitzpatrick, Jr., and S.J. Vanderkooy. 1996. Distribution and habitat use by dwarf crayfishes (Decapoda: Cambaridae: *Cambarellus). Wetlands* 16:594–598.

Pezeshki, S.R., R.D. DeLaune, and S.Z. Pan. 1991. Relationship of soil hydrogen sulfide level to net carbon assimilation of *Panicum hemitomon* and *Spartina patens. Vegetatio* 95:159–166.

Ponnamperuma, F.N. 1972. The chemistry of submerged soils. *Adv. Agron.* 24:29–96.

Ponnamperuma, F.N., E.M. Tianco, and T. Loy. 1967. Redox equilibria in flooded soils: I. the iron hydroxide system. *Soil Sci.* 103:374–382.

Pritchett, W.L. 1979. *Properties and Management of Forest Soils.* John Wiley & Sons, New York.

Puriveth, P. 1980. Decomposition of emergent macrophytes in a Wisconsin marsh. *Hydrobiologia* 72:231–242.

Qualls, R.G. and C.J. Richardson. 1995. Forms of soil phosphorus along a nutrient enrichment gradient in the northern Everglades. *Soil Sci.* 160:183–198.

Rader, R.B. and C.J. Richardson. 1994. Response of macroinvertebrates and small fish to nutrient enrichment in the northern Everglades. *Wetlands* 14:134–146.

Reader, R.J. 1978. Primary production in northern bog marshes. pp. 53–62. *In* R.E. Good, D.F. Whigham, and R.L. Simpson (Eds.) *Freshwater Wetlands: Ecological Processes and Management Potential.* Academic Press, New York.

Reader, R.J. and J.M. Stewart. 1972. The relationship between net primary production and accumulation for a peatland in southeastern Manitoba. *Ecol.* 53:1024–1037.

Reimold, R.J., J.L. Gallagher, R.A. Linthurst, and W.J. Pfeiffer. 1975. Detritus production in coastal salt marshes. pp. 217–227. In L.E. Cronin (Ed.) *Estuarine Research I.* Academic Press, New York.

Richardson, C.J. 1985. Mechanisms controlling phosphorus retention capacity in freshwater wetlands. *Science* 228:1424–1427.

Richardson, C.J. and P.E. Marshall. 1986. Processes controlling movement, storage and export of phosphorus from a fen peatland. *Ecol. Monogr.* 56:279–302.

Richardson, C.J., D.A. Tilton, J.A. Kadlec, P.M. Chamie, and W.A. Wentz. 1978. Nutrient Dynamics of Northern Wetland Ecosystems. pp. 217–241. *In* R.E. Good, D.F. Whigham, and R.L. Simpson (Eds.) *Freshwater Wetlands: Ecological Processes and Management Potential.* Academic Press, New York.

Roberts, J.K.M. 1988. Cytoplasmic acidosis and flooding in crop plants. pp. 392–397. *In* D.D. Hook and others (Eds.) *The Ecology and Management of Wetlands. Volume 1: Ecology of Wetlands.* Timber Press, Portland OR.

Rozema, J., P. Bijwaard, G. Prast, and R. Broekman. 1985. Ecophysiological adaptations of coastal halophytes from foredunes and salt marshes. *Vegetatio* 62:499–521.

Schlesinger, W.H. 1978. Community structure, dynamics and nutrient cycling in the Okefenokee cypress swamp-forest. *Ecol. Monogr.* 48:43–65.

Schlesinger, W.H. 1991. *Biogeochemistry: An Analysis of Global Change.* Academic Press, New York.

Schwertmann, U. 1988. Occurrence and formation of iron oxides in various pedoenvironments. pp. 267–308. *In* J.W. Stucki, B.A. Goodman, and U. Schwertmann (Eds.) *Iron in Soil and Clay Minerals.* D. Reidel Publishing Co., Boston.

Seitzinger, S.P. 1988. Denitrification in freshwater and coastal marine ecosystems: ecological and geochemical significance. *Limnol. Oceanogr.* 33:702–724.

Sharitz, R.R. and J.W. Gibbons. 1982. *The Ecology of Southeastern Shrub Bogs (Pocosins) and Carolina Bays: A Community Profile.* FWS/OBS-82/04. U.S. Fish and Wildlife Service, Division of Biological Services, Washington, DC.

Sklar, F.H. 1985. Seasonality and community structure of the backswamp invertebrates in a Louisiana cypress–tupelo wetland. *Wetlands* 5:69–86.

Sklar, F.H. and W.H. Conner. 1979. Effects of altered hydrology on primary production and aquatic animal populations in a Louisiana swamp forest. pp. 191–208. *In* J.W. Day, Jr., D.D. Culley, Jr., R.E. Turner, and A.T. Humphrey (Eds.) *Proceedings of the Third Coastal Marsh and Estuary Management Symposium.* Louisiana State University, Division of Continuing Education, Baton Rouge, LA.

Steever, E.Z., R.S. Warren, and W.A. Niering. 1976. Tidal energy subsidy and standing crop production of *Spartina alterniflora. Estuarine Coastal Mar. Sci.* 4:473–478.

Steward, K.K. and W.H. Ornes. 1975. The autecology of sawgrass in the Florida Everglades. *Ecol.* 56:162–171.

Stout, J.P. 1984. *The Ecology of Irregularly Flooded Salt Marshes of the Northeastern Gulf of Mexico: A Community Profile.* Biological Report 85(7.1). U.S. Fish and Wildlife Service, Washington, DC.

Stout, J.P. 1988. Irregularly flooded salt marshes of the Gulf and Atlantic Coasts of the United States. pp. 511–525. *In* D.D. Hook and others (Eds.) *The Ecology and Management of Wetlands. Volume 1: Ecology of Wetlands.* Timber Press, Portland, OR.

Subrahmanyan, C.B., W.L. Kruczynski, and S.H. Drake. 1976. Studies of animal communities in two north Florida salt marshes. Part II. Macroinvertebrate communities. *Bull. Mar. Sci.* 25:445–465.

Sullivan, M.J. and F.C. Daiber. 1974. Response in production of cordgrass, *Spartina alterniflora,* to inorganic nitrogen and phosphorus fertilizer. *Chesapeake Sci.* 15:121–123.

Svorensen, J. 1982. Reduction of ferric iron in anaerobic marine sediment and interaction with reduction of nitrate and sulfate. *Appl. Environ. Microbiol.* 43:319–324.

Svorensen, J. and B.B. Jorgensen. 1987. Early diagenesis in sediments from Danish coastal waters: microbial activity and Mn–Fe–S geochemistry. *Geochim. Cosmochim. Acta* 51:1583–1590.

Svorensen, J., D. Christensen, and B.B. Jorgensen. 1981. Volatile fatty acids and hydrogen as substrates for sulfate reducing bacteria in anaerobic marine sediment. *Appl. Environ. Microbiol.* 42:5–11.

Szumigalski, A.R. and S.E. Bayley. 1996. Net aboveground primary production along a bog-rich fen gradient in central Alberta, Canada. *Wetlands* 16:467–476.

Teal, J.M. 1958. Distribution of fiddler crabs in Georgia salt marshes. *Ecol.* 39:18–19.

Updegraff, K., J. Pastor, S.D. Bridgham, and C.A. Johnston. 1995. Environmental and substrate controls over carbon and nitrogen mineralization in northern wetlands. *Ecol. Appl.* 5:151–163.

Urban, N.H., S.M. Davis, and N.G. Aumen. 1993. Fluctuations in sawgrass and cattail density in Everglades Water Conservation Area 2A under varying nutrient, hydrologic and fire regimes. *Aquat. Bot.* 46:203–223.

Valiela, I. and J.M. Teal. 1974. Nutrient limitation in salt marsh vegetation. pp. 547–563. *In* R.J. Reimold, R.J. and W.H. Queen (Eds.) *Ecology of Halophytes.* Academic Press, New York.

Valiela, I., B. Howes, R. Howarth, A. Giblin, K. Foreman, J.M. Teal, and J.E. Hobbie. 1982. Regulation of primary production and decomposition in a salt marsh ecosystem. pp. 151–168. *In* B. Gopal, R.E. Turner, R.G. Wetzel, and D.F. Whigham (Eds.) *Wetlands: Ecology and Management.* National Institute of Ecology and International Scientific Publications, Jaipur, India.

Van Breemen, N. 1988. Long-term chemical, mineralogical and morphological effects of iron redox processes in periodically flooded soils. pp. 811–823. *In* J.W. Stucki, B.A. Goodman, and U. Schwertmann (Eds.) *Iron in Soil and Clay Minerals.* D. Reidel Publishing Co., Boston.

Van der Valk, A.G. and C.B. Davis. 1978. Primary production of prairie glacial marshes. pp. 21–37. *In* R.E. Good, D.F. Whigham, and R.L. Simpson (Eds.) *Freshwater Wetlands: Ecological Processes and Management Potential.* Academic Press, New York.

Van Veen, J.A., J.N. Ladd, and M.J. Frissel. 1984. Modeling C and N turnover through the microbial biomass in soil. *Plant Soil* 76:257–274.

Vartapetian, B.B. 1988. Ultrastructure studies as a means of evaluating plant tolerance to flooding. pp. 452–456. *In* D.D. Hook and others (Eds.) *The Ecology and Management of Wetlands. Volume 1: Ecology of Wetlands.* Timber Press, Portland OR.

Vitousek, P.M. and R.W. Howarth. 1991. Nitrogen limitation on land and in sea: how can it occur? *Biogeochem.* 13:87–115.

Voigts, D.K. 1976. Aquatic invertebrate abundance in relation to changing marsh vegetation. *Am. Midl. Nat.* 95:312–322.

Voroney, R.P., J.A. Van Veen, and E.A. Paul. 1981. Organic C dynamics in grassland soils. 2. Model validation and simulation of long term effects of cultivation and rainfall erosion. *Can. J. Soil Sci.* 61:211–224.

Wharton, C.H., V.W. Lambou, J. Newson, P.V. Winger, L.L. Gaddy, and R. Mancke. 1981. The fauna of bottomland hardwoods in southeastern United States. pp. 87–100. *In* J.R. Clark and J. Benforado (Eds.) *Wetlands of Bottomland Hardwood Forests.* Elsevier, Amsterdam.

Wharton, C.H., W.M. Kitchens, E.C. Pendleton, and T.W. Sipe. 1982. *The Ecology of Bottomland Hardwood Swamps of the Southeast: A Community Profile.* FWS/OBS-81/37. U.S. Fish and Wildlife Service, Biological Services Program, Washington, DC.

Whigham, D.F., J. McCormick, R.E. Good, and R.L. Simpson. 1978. Biomass and primary production in freshwater tidal wetlands of the middle Atlantic coast. pp. 3–20. *In* R.E. Good, D.F. Whigham, and R.L. Simpson (Eds.) *Freshwater Wetlands: Ecological Processes and Management Potential.* Academic Press, New York.

Whiting, G.J. and J.P. Chanton. 1993. Primary production control of methane emission from wetlands. *Nature* 364:794–795.

Widdel, F. and T.A. Hansen. 1992. The dissimilatory sulfate and sulfur reducing bacteria. pp. 583–624. *In* A. Balows, H.G. Truper, M. Dworkin, W. Harder, and K.H. Schleifer (Eds.) *The Prokaryotes: A Handbook on the Biology of Bacteria: Ecophysiology, Isolation, Identification, Applications.* Springer-Verlag, New York.

Wieder, R.K. and G.E. Lang. 1988. Cycling of inorganic and organic sulfur in peat from Big Run Bog, West Virginia. *Biogeochem.* 5:221–242.

Wieder, R.K., J.B. Yavitt, and G.E. Lang. 1990. Methane production and sulfate reduction in two Appalachian peatlands. *Biogeochem.* 10:81–104.

Wiegert, R.G. and B.J. Freeman. 1990. *Tidal Salt Marshes of the Southeast Atlantic Coast: A Community Profile.* Biological report 85(7.29). U.S. Fish and Wildlife Service, Washington, DC.

Wu, L. 1981. The potential for evolution of salinity tolerance in *Agrostis stolonifera* L. and *Agrostis tenuis* Sibth. *New Phytol.* 89:471–486.

Yates, R.F.K. and F.P. Day, Jr. 1983. Decay rates and nutrient dynamics in confined and unconfined leaf litter in the Great Dismal swamp. *Am. Midl. Nat.* 110:37–45.

Yavitt, J.B. and G.E. Lang. 1990. Methane production in contrasting wetland sites: response to organic-chemical components of peat and to sulfate reduction. *Geomicrobiol. J.* 8:27–46.

Yavitt, J.B., G.E. Lang, and R.K. Wieder. 1987. Control of carbon mineralization to CH_4 and CO_2 in anaerobic, *Sphagnum*-derived peat from Big Run Bog, West Virginia. *Biogeochem.* 4:141–157.

Yoshida, T. 1975. Microbial metabolism of flooded soils. pp. 83–123. *In* E.A. Paul. and A.D. MacLaren (Eds.) *Soil Biochemistry.* Marcel Dekker Inc., New York.

Yuan, W.L. and F.N. Ponnamperuma. 1966. Chemical retardation of the reduction of flooded soils and the growth of rice. *Plant Soil* 25:347–360.

Zizer, S.W. 1978. Seasonal variations in water chemistry and diversity of the phytophilic macroinvertebrates of three swamp communities in southeastern Louisiana. *The Southwestern Naturalist* 23:545–562.

CHAPTER 6

Organic Matter Accumulation and Organic Soils

Mary E. Collins and R.J. Kuehl

INTRODUCTION

Almost all life in the soil depends on organic matter for energy and nutrients. Organic matter contributes to plant growth through modification of the physical, chemical, and biological properties of the soil. The formation of soil organic matter (SOM) is extremely complex, and because of the importance of SOM in soil-forming processes and soil fertility, as well as in wetland delineation, ecology, and management, the study of SOM formation and accumulation has spawned worldwide comprehensive research. The dynamics of SOM play a major role in ecosystems, and SOM and organic soils have a significant effect on the management of wetlands and soil and water quality.

De Saussure is usually credited with introducing the term *humus* (Latin equivalent of soil) in the early 1800s to describe the dark-colored organic material in soil (Stevenson 1994). He was the first to show that humus is richer in C and poorer in H and O than the plant material from which it is derived. Investigators in the early years developed many controversial and confusing ideas on the nature and properties of humus and its role in soil formation. The resulting large volume of literature presents a complicated history of the studies on humus. Because of a lack of knowledge about soil microorganisms and their role in the transformation of organic substances, the formation of humus was considered by early workers to be purely a chemical process (Kononova 1961). Some authors try to make a distinction between SOM and the term "humus." The Soil Science Society of America (1997) indicates that the terms are most often used synonymously.

In this chapter, the basic processes and dynamics involved in the accumulation and decomposition of SOM will be discussed, with emphasis on soils developed under anaerobic conditions. The classification of wetland soils with the highest content of organic matter (organic soils, also known as Histosols) will be presented, and organic soils will be defined in terms of their degree of decomposition. The range shows the complexity of the processes in geographic and climatic distribution of organic soils. The majority of organic soils occur in the frigid temperature regime in Alaska, but the largest contiguous area of Histosols is in the hyperthermic region of Florida. The physical and chemical properties of SOM, the importance of nutrient cycling, the methods of analysis for organic soils, and the management of them are also discussed.

SOIL ORGANIC MATTER

Soil organic matter (SOM) refers to the sum total of all substances in soils containing elements in organic form (e.g., organic C, organic P, organic N) and may be defined simply as plant and animal residues in various stages of decomposition. The Soil Science Society of America (1997) defines SOM as the organic fraction of the soil exclusive of undecayed plant and animal residues. Soils inherit organic matter from living organisms, and thus, SOM is partly alive and partly dead. Living plant roots are usually excluded from SOM, but both fresh and partly decomposed residues of plants and animals (macroorganisms) are included, along with the tissue of living and dead microorganisms (microbes).

To simplify the terminology, SOM is usually subdivided into (a) humic and (b) nonhumic substances. Humic substances, a complex mixture containing thousands of organic compounds, are the relatively stable portions of SOM. The bulk of SOM consists of humic substances (Schnitzer 1978). These are amorphous, dark-colored, partly aromatic, chemically complex substances. Humic substances no longer exhibit specific chemical and physical characteristics normally associated with well-defined organic compounds and are resistant to chemical and biological degradation.

Nonhumic substances include those with still-recognizable chemical characteristics, such as amino acids, carbohydrates, fatty acids, pigments, proteins, resins, and waxes. These compounds are generally easily degraded in soils and have short life spans.

Organic matter accumulation occurs in soils when the production of organic materials is high, or when conditions are not optimal for the decomposition of organic materials (Chapter 5). When conditions are optimal for microbial activity, organic matter decomposes so rapidly that accumulation does not occur to a significant extent. Cycles of soil saturation followed by drying appear to stimulate the decomposition of SOM. However, in wetland soils, decomposition is generally limited and organic matter accumulates. Decay processes in a wetland ecosystem can be thought of as a continuum, which begins with the input of plant litter that eventually leads to the formation of SOM. The general patterns of early decay appear to be related to the initial litter quality as well as environmental factors such as moisture and temperature. In the boreal wetlands, organic matter decomposition is controlled by moisture, fertility, and organic matter quality, in addition to soil temperature (Trettin et al. 1995).

Moran et al. (1989) showed that the rate of decay of each SOM fraction decreased as decomposition proceeded. A model developed by Moran et al. (1989) describes the decreasing decomposition rate as material ages, and the preferential decomposition of the more labile components and the accumulation of the more refractory components (e.g., lignin) over time. The model also accommodates the transformation of the more refractory components (humification) during the decomposition process.

Organic materials accumulate in wetland environments that are saturated with water at the surface or above for most or nearly all of the year. The organic materials were derived mainly from the remains of vascular plants such as the sedges, rushes, and grasses, but the organic materials also contain woody materials from trees in some areas.

Carbohydrates, composed of carbon, oxygen, and hydrogen, are the primary components of plant materials that are added to the soils as organic materials. Carbohydrates include simple sugars, simple polysaccharides such as starch, and more complex polysaccharides such as cellulose. Other polysaccharide components of plant materials are hemicellulose, pectins, and gums. The polysaccharide group is important because it forms as much as 70% of the dry matter of plant cells and provides a significant source of energy for microorganisms (Everett 1983).

Proteins are another important component of living plant materials, which are important nitrogen sources in the organic soil materials. Other components of organic soil materials are lipids, tannins, and lignins. Lipids include fats, waxes, and plant pigments. Tannins impart the brown, yellowish-brown, and black colors to the plant materials and the water surrounding them (Everett, 1983).

Lignin is a component of the cell walls of plants. It is one of the most common constituents of organic soil materials because it is one of the most chemically resistant plant compounds.

During the development of soils, SOM is accumulated through the formation of biomass and organic detritus. Sources of organic soil material include relatively coarse plant debris that is deposited within or upon the soil as litter, and colloidal and soluble organic compounds from root exudates, microbial products, or leachate from the litter. The colloidal fraction contains the microbial population.

Plant material enters the soil as particulates that can undergo biological/biochemical transformations. Soluble litter components can be distributed through the soil matrix where they can be adsorbed to mineral surfaces, precipitated with other organic compounds, or immobilized in the biomass.

ORGANIC SOILS AND ORGANIC SOIL MATERIAL

Many terms have been used to describe organic soils. The term "peat" has been used for many years as a general term to describe the soils with various amounts of undecomposed plant remains. Early emphasis was placed on the botanical source of the plant remains, and several types of peats (limnic peats, telmatic peats, terrestic peats, etc.) were recognized.

Specific terms have been defined for organic soils and organic material. The majority of the terms depend on the degree of decomposition of the organic matter. These terms are used throughout the world (Table 6.1). In Alaska and other northern areas of North America, as well as areas in Europe with large expanses of organic materials, peat is still used today as a general textural term indicating a high fiber content. The term "peatland" is used as a general landscape term to describe the expansive area of their occurrence. Peatlands form in depressions, slopes, and raised bogs and occur over a range of climatic conditions from boreal to arctic. Peatlands are often divided into bogs, fens, mires, and swamps. These are particular landscapes or ecosystems and are often given the general description of peat-accumulating wetlands (Mitsch and Gosselink 1993). Bogs are areas of peat that are acid. *Bogs* formed primarily from shrub or moss vegetation. The tannin content of bogs is high. *Fens* are sedge and grass-like plant dominated areas of organic materials that contain considerable bases, such as calcium. *Swamps* are formed under woody vegetation with variable amounts of tannin. *Mires* is a common but confusing term used in Europe to indicate the ecosystems in which waterlogged peat has accumulated, usually in raised areas. Mitsch and Gosselink (1993) suggest that *mire,* as used in Europe, is synonymous with any peat-accumulating wetland. Moore and Bellamy (1974) also used *mires* to cover all wetland ecosystems in which peat accumulates.

Table 6.1 Terms and Definitions Used to Describe Organic Soils and Organic Matter

Term	Definition
Peat	Organic soil material in which the original plant remains are recognizable (fibric material).
Muck	Organic soil material in which the original plant remains are not recognizable. Contains more mineral matter and is usually darker in color than peat.
Fen	A peat-accumulating wetland that receives some drainage from surrounding mineral soils and usually supports marsh-like vegetation. These areas are richer in nutrients and less acidic than bogs. The soils under fens are peat (Histosol) if the fen has been present for a while.
Mull	A forest humus type characterized by intimate incorporation of organic matter into the upper mineral soil (i.e., a well-developed A horizon) in contrast to accumulation on the surface.
Mor	A type of forest humus characterized by an accumulation of organic matter on the soil surface in matted Oe (F) horizons, reflecting the dominant mycogenous decomposers.
Pocosin	Temperate zone evergreen shrub bogs dominated by pond pine, ericaceous shrubs and *Sphagnum*.
Bog	A peat-accumulating wetland that has no significant inflows or outflows and supports acidophilic mosses, particularly *Sphagnum*.
Moder	A type of forest humus transitional between mull and mor.

From Soil Science Society of America. 1997. *Glossary of Soil Science Terms 1996.* Soil Science Society of America, Inc., Madison, WI.

By definition, organic soil material is saturated with water for long periods (or artificially drained). Excluding live roots, organic soil material has an organic C content (by weight) that ranges from 12% or more if the mineral part of the soil contains no clay to 18% or more organic carbon if the mineral part of the soil contains 60% or more clay (Chapter 1; Soil Survey Staff 1998). This definition includes organic soil materials that are commonly called *peat* or *muck*. By definition, organic soil materials that are never saturated with water for more than a few days must contain 20% or more (by weight) organic carbon (Soil Survey Staff 1998). This definition is intended to include leaf litter or a root mat, which accumulates on the surfaces of many undisturbed soils under varying moisture conditions. It can be seen by this latter definition that not all organic soil materials accumulate under water or in anaerobic conditions.

Morphology of organic soils is distinct because the parent material is biological in origin (i.e., organic rather than mineral). The O master horizon is designated for layers dominated by organic material. O horizons have three subordinate horizon designations, a (sapric), e (hemic), and i (fibric), that indicate the degree of organic matter decomposition (Soil Survey Staff 1996b). In the field, the degree of decomposition of the organic matter is determined by estimating the rubbed and unrubbed fiber content of the soil materials (Table 1.1 of Chapter 1). Fibers are defined as pieces of plant tissue (excluding live roots) large enough to be retained on a 100-mesh sieve (openings 0.15 mm in diameter) when the materials are screened after dispersion in sodium hexametophosphate (Soil Survey Staff 1998). Fibers show evidence of cellular structure of the plants from which they were derived and should be decomposed enough so that they can be crushed or shredded with the fingers. Pieces of wood that are larger than 2 cm in cross section and are so undecomposed that the material cannot be crushed with the fingers are not considered fibers (Soil Survey Staff 1998). It is recommended that the soil materials be rubbed with firm pressure between the thumb and fingers 10 times to determine the rubbed fiber content (Soil Survey Staff 1998). The rubbed material may be examined under a hand lens with 10 power or higher to more accurately estimate the fiber content. The rubbed soil materials can also be examined in the laboratory by dispersing the materials with sodium pyrophosphate and washing on a screen.

Sapric material (Gr. *sapros*, rotten) is the most highly decomposed organic material and has a rubbed fiber content of < one sixth of the soil volume. The fiber content is generally < one third of the volume before rubbing. The individual organic fibers cannot be identified. Because sapric soil materials have the lowest content of plant fibers, the bulk density can be expected to be higher and the water content lower (on a dry weight basis at saturation) than hemic or fibric soil materials. Sapric soil materials are common in many Histosols that have been drained and cultivated. These materials are usually black or very dark gray in color. The bulk density of sapric soil materials is usually > 0.2 g cm^{-3}, and the maximum water content when saturated is normally < 450% on an ovendry basis (Soil Survey Staff 1998). A general horizon sequence for a soil formed in organic soil materials made up of highly decomposed plant materials, such as a Typic Haplosaprist, is: Oa1, Oa2, etc. The Oa designations indicate that the organic materials are highly decomposed (sapric). A photograph of a Histosol (Limnic Haplosaprists) in Minnesota is shown in Plate 3. The thickness of the organic materials (sapric) in this soil is 100 cm. The underlying materials consist of both organic and inorganic materials that likely were deposited in water by the action of aquatic animals, or were derived from underwater and floating plants and subsequently modified by aquatic animals (Soil Survey Staff 1996b). Snail shells can be seen in the photograph. These underlying materials are called limnic (Gr. *limne*, lake). One of the materials that commonly occur in limnic materials is coprogenous earth (sedimentary peat). A complete description of the soil shown in Plate 3, including the underlying organic and mineral soil horizons, is presented in Table 6.2.

Hemic material (Gr. *hemi*, half) is intermediate in degree of organic matter decomposition and has a rubbed fiber content of one sixth to two fifths by volume. In an unrubbed sample of hemic material the fibers can be seen and range from one third to two thirds of the volume. Colors of hemic soil materials commonly range from dark grayish-brown to dark reddish-brown. Bulk density of hemic soil materials is commonly between 0.07 and 0.18 g cm^{-3}. Their maximum water content

Table 6.2   An Example of the Morphology for a Histosol in Minnesota

Series:	Muskego
Location:	Sibley County, Minnesota
Classification:	Coprogenous, euic, mesic Limnic Haplosaprists (formerly Medisaprists)
Landscape position:	Depression on till plains
Drainage class:	Very poorly drained
Parent material:	Highly decomposed organic material overlying silty coprogenous earth.

Horizon	Depth (cm)	Description
Oap	0–22	Black (10YR 2/1) muck; very dark gray (10YR 3/1) dry; weak fine granular structure; friable; moderately acid; abrupt smooth boundary.
Oa1	22–60	Black (10YR2/1) muck, very dark gray (10YR 3/1) dry; weak thin platy structure parting to weak fine granular; friable; moderately acid; gradual smooth boundary.
Oa2	60–100	Black (10YR 2/1) muck; very dark grayish brown (10YR 3/2) dry; weak fine granular structure; friable; moderately acid; clear smooth boundary.
C1	100–112	Black (5Y2.5Y/1) mucky silt loam (coprogenous earth); common fine prominent dark yellowish brown (10YR 4/6) and common olive gray (5Y 5/2) mottles; massive; friable; slight effervescence; slightly alkaline; clear smooth boundary.
C2	112–180	Olive gray (5Y 5/2) silt loam (coprogenous earth); common fine prominent dark yellowish brown (10YR 4/6) mottles; massive; friable; strong effervescence; moderately alkaline.

From Domeier, M.J. 1997. *Soil Survey of Sibley County, Minnesota.* USDA–NRCS, in cooperation with the Minn. Ag. Exp. Sta., U.S. Govt. Printing Office, Washington, DC.

when saturated commonly ranges from 450 to 850% or more. In a soil description, the layers of hemic soil materials are designated Oe.

Fibric soil material (L. *fibra*, fiber) is slightly decomposed organic soil material which contains two fifths to three fourths or more fibers by volume after rubbing. Even after being rubbed 10 times or more, individual fibers should still be seen, and it should be possible to determine the botanic origin of the plant remains that make up the soil. The Oi horizon designations are used for fibric materials. The relatively undecomposed fibric materials are most common in the colder climates and occur in all parts of Alaska (Riger et al. 1979). Fibric soil materials are widespread in the raised bogs of the boreal forest zones. Fibric soil materials have very low bulk densities (<0.1 g cm^{-3}) and the water content, when the soil is saturated, ranges from about 850 to $> 3000\%$ of the weight of the ovendry material (Soil Survey Staff 1998). The colors of fibric soil materials are commonly brown, light yellowish-brown, dark brown, or reddish-brown. A root mat or leaf litter on the soil surface is considered fibric materials, and the horizon or layer would be designated as Oi. A complete description of a soil formed in fibric materials in Alaska is presented in Table 6.3. This soil is classified as Sphagnic Fibristels and by definition, three fourths of more of the fibric material in the upper 50 cm, or to a depth of a lithic content, whichever is shallower, is derived from *Sphagnum* (Soil Survey Staff 1998).

DISTRIBUTION OF ORGANIC SOIL MATERIALS

Histosols cover approximately 2.9% of the U.S. (Natural Resources Conservation Service 1997). Clearly the largest acreage of these organic soils is in the colder climates (Alaska, Minnesota, and Michigan), but areas of the southeastern coastal plain (Florida, Louisiana, and North Carolina) contain large areas of Histosols (Chapter 16). The broad range of temperatures in which Histosols exist makes it difficult to quantify the processes involved in production and accumulation of SOM. Histosols extend from cryic to isohyperthermic soil temperature regimes. All Histosols except for Folists are considered hydric soils in the United States (Natural Resources Conservation Service 1996).

Table 6.3 An Example of the Morphology of the Salamatof Soil Series, Formed in Fibric Materials in Alaska

Classification:	Dysic, frigid Sphagnic Fibristels (formerly Sphagnic Borofibrists)
Landscape position:	Muskegs
Drainage class:	Very poorly drained
Parent material:	Thick beds of sphagnum moss and thin layers of sedges.

Horizon	Depth (cm)	Description
Oi1	0–22	Brown (10YR 4/3) when wet; pale brown (10YR 6/3) when squeezed dry; raw, undecomposed sphagnum moss peat; many roots; many dark-colored, coarse, woody particles; extremely acid; gradual boundary.
Oi2	22–150	Dark brown (7.5YR 3/2) when wet; brown (7.5YR 4/2) when squeezed dry; coarse moss peat; a few thin strata of coarse sedge peat; a few woody particles; a few live roots to a depth of 46 cm; peat material is slightly finer below a depth of 61 cm; extremely acid.

From Riger, S., D.B. Schoephorster, and C.E. Furbush. 1979. *Exploratory Soil Survey of Alaska.* USDA–SCS, Washington, DC.

CLASSIFICATION OF ORGANIC SOILS

Early classification systems, beginning with the Russian system in the late 1890s and continuing to the *Keys to Soil Taxonomy* (Soil Survey Staff 1998), recognized the importance of organic soils by establishing categories and classes for them (Table 6.4). Sibirtsev's 1895 soil classification system considered organic soils (e.g., Moor and Bog soils) as having more or less well-developed soil characteristics that reflected the result of some local factor of soil development (Baldwin et al. 1938). They were classified as Intrazonal. Marbut's system of soil classification in 1935 (Baldwin et al. 1938), recognized organic soils by creating the classes "Peat and Swamp soils," both under Pedalfers (acidic, iron- and aluminum-accumulating soils in the eastern U.S.) and Pedocals (high base soils in the western U.S.). The 1938 system used Sibirtsev's terminology and classified organic soils as Intrazonal soils at the Order level and as Hydromorphic soils at the suborder level (Baldwin et al. 1938). The "1st Approximation" to *Soil Taxonomy* proposed in 1951 greatly changed both the terminology and importance of organic soils. Fronobods (Order level) were soils that had greater than "30% organic matter in peat or G (gleyed) horizon" (Cline 1979). In this system, organic soils had classes down to the Great Soil Group level. The uniqueness of organic soils continued throughout the seven approximations of the new soil classification system (*Soil Taxonomy*) in the United States.

U.S. *Soil Taxonomy*

In the U.S. *Soil Taxonomy* system there are currently 12 soil orders, and these were described in Chapter 1. Orders are the broadest grouping of the soils (highest categorical level in *Soil Taxonomy*), based on the presence of selected soil properties. Histosols is the Order used to classify organic soils (Soil Survey Staff 1975). Histosols extend from Alaska to Florida. Orders are subdivided into numerous Suborders. The Suborders for Histosols are Folists, Fibrists, Hemists, and Saprists. Folists are organic soils that are not permanently wet. These are never saturated with water except for a few days following heavy rains. The other Suborders represent organic soils that are permanently wet under natural conditions and are defined on the degree of decomposition. Fibrists are Histosols that have the least decomposed organic matter (fibric material). Organic matter in Hemists is intermediately decomposed (hemic material). Saprists have organic matter that is the most decomposed (sapric material). The Great Groups for Histosols are defined on the presence of sphagnum moss or humilluvic materials, soil temperature regime, or a sulfuric horizon. Each Great Group has a number of Subgroups which are differentiated by the thickness of the organic layers, fiber content of the organic materials, depth to mineral soil or bedrock, etc. A diagram showing the subdivision of Histosols into Suborders, Great Groups, and Subgroups is presented in

Table 6.4 Classification of Organic Soils from the 1890s to 1996

Soil Classification System	Sibirtsev' 1895 System	Marbut's 1935 System	1938 System	1st Approximation (1951)	Soil Taxonomy (1975)	Keys to Soil Taxonomy (1998)
Highest level	Intrazonal Soils (Division B)	Pedalfers (Category VI)	Pedocals (Category VI)	Intrazonal Soils (Order)	Fronobods (Order)	Histosols (Order)
Lower level	Moor and bog soils (Class 8)	Peat soils (Category III)	Peat soils (Category III)	Hydromorphic soils (Suborder)	Fronosols (Unnamed)	Histosols Folists (Suborder)
		Swamp soils (Category III)	Swamp soils (Category III)			Hemists (Suborder)
						Fibrists (Suborder)
						Saprists (Suborder)

From Cline, M.G. 1979. Soil Classification in the United States. *Agronomy Mimeo* No. 79-12. 2nd print. Agron. Dept., Cornell Univ., Ithaca, NY; Soil Survey Staff. 1998. *Keys to Soil Taxonomy.* 8th ed. USDA–NRCS. U.S. Govt. Printing Office, Washington, DC.

Example of the Classification of the Muskego Soil Series – Limnic Haplosaprists

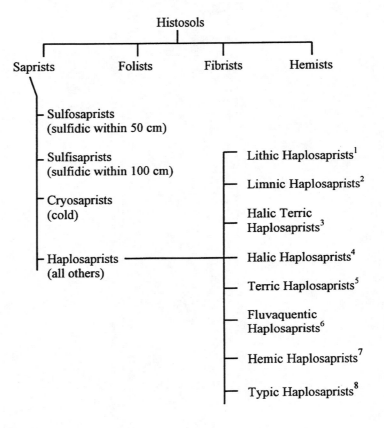

[1] Bedrock within a depth of 130 cm
[2] Contain limnic layers within a depth of 130 cm
[3] Have a layer 30 cm or thicker with an electrical conductivity of 30 dS/m or more for 6 months or more and have a mineral layer 30 cm or thicker above a depth of 130 cm.
[4] Have a layer 30 cm or thicker with an electrical conductivity of 30 dS/m or more for 6 months or more
[5] Have a mineral layer 30 cm or thicker above a depth of 130 cm.
[6] Contain thin mineral layers within the organic materials
[7] Have hemic or fibric layers below the surface layer, 25 cm or more in total thickness
[8] Other Haplosaprists

Figure 6.1 Diagram showing the Histosol suborders and the great groups and subgroups within the Saprists. The pedon of the Muskego soil series, shown in Plate 3, classifies as Limnic Haplosaprists.

Figure 6.1. The Histosol Order is subdivided into four Suborders; the Saprists Suborder is subdivided into four Great Groups; and as shown in Figure 6.1, the Haplosaprists Great Group is subdivided into eight Subgroups.

The classes of Histosols have decreased because in the revised *Keys to Soil Taxonomy* (Soil Survey Staff 1998), organic soils in the very cold climates with permafrost are now classified in a new Suborder, Histels. At the Great Group level, there were formerly 123 classes, but some, such as Luvifibrists, were "tentatively established for use in other countries if needed" (Soil Survey Staff 1975). *Keys to Soil Taxonomy* (Soil Survey Staff 1998) has the same Suborders as in 1975, but the number of Great Groups has decreased from 20 to 16, and the number of Subgroups from 123 to

60. The decrease is primarily due to placement of the Histosols in the very cold climates into the Gelisols Order. The Suborder Histels is within the Gelisols Order (cold climate) and includes soils that contain large amounts of organic carbon that accumulates under anaerobic conditions and cold climates (see Chapter 16). Histels have permafrost within 100 cm of the soil surface, or gelic materials (materials that show evidence of frost churning or ice segregation) within 100 cm of the soil surface and permafrost within 200 cm of the soil surface (Soil Survey Staff 1998). Histels also contain 80% or more by volume organic materials from the soil surface to a depth of 50 cm, or to a dense or bedrock layer, if less than 50 cm.

Soils that lack the thick organic surface layer of Histosols or Histels are considered mineral soils even though soils may have a surface layer with a very high organic matter content. For the mineral soils that contain a thin organic surface layer, a histic (Gr. *histos*, tissue) epipedon was defined. The term "epipedon" is used to designate a surface layer. The main difference between a Histosol (an organic soil) and a mineral soil that has a histic epipedon is the thickness of the organic materials. Histosols have a minimum thickness; the organic layers must be at least 40 cm thick within the upper 80 cm of soil. The soils that are shallow (< 50 cm) over limestone or gravel must have organic layers that constitute at least two thirds of the depth of the soil (Soil Survey Staff 1996b). Therefore, if a soil has organic material that does not meet the required thickness for a Histosol, the soil can have a histic epipedon. The histic epipedon is defined as a layer that has saturation and reduction for some time during normal years, or is artificially drained. The presence of the histic epipedon is noted in the classification of the soil. For example, a Histic Alaquod is a mineral soil in the Spodosol Order, with a histic epipedon. A histic epipedon is 20 to 40 cm thick, and requires 12 to 18% organic C, depending upon the clay content of the soil. This is the same amount of organic C as required for Histosols. A histic epipedon can be 20 to 60 cm thick in soils containing 75% or more sphagnum fibers, or having a bulk density < 0.1 g/cm^3 (Soil Survey Staff 1996b). The histic epipedon, by definition and current interpretation, cannot occur in Histosols because the organic soil materials can be no thicker than 40 cm and still meet the requirements of a histic epipedon. However, the histic epipedon can occur in many of the other soil Orders in the soil classification system in the U.S.

Some soil materials have a distinctly higher SOM content, but not enough to classify as a histic epipedon or a Histosol. A modifier may be added to the mineral soil texture (i.e., mucky sand) to describe the increased SOM. When rubbing a mucky sand between the fingers, the soil should feel greasy, but upon the third rub, the gritty texture of the sand should be evident (Watts and Sprecher 1995). The amounts of clay and organic C needed to qualify for "mucky-modified mineral textures" are given in Table 1.2 of Chapter 1.

Other Soil Classification Systems

Soil classification systems in other countries also separate soils into organic and mineral. A comparison of selected classification systems is presented in Table 6.5.

The Food and Agriculture Organization of the United Nations/Educational, Scientific, and Cultural Organization (FAO/UNESCO) legend of the Soil Map of the World (FAO 1974) also has classified organic soils as Histosols. These Histosols are defined similar to U.S. Histosols. At a lower level, the pH of the Histosol is used to classify the soil, as well as the soil temperature (e.g., gelic, very cold).

The French system classifies organic soils at the highest level as Hydromorphic soils, with the lower levels defined based on the amount of organic matter. The Canadian system is similar to the U.S. system in that the lower level is defined by the relative decomposition of the organic matter (Bentley 1978). Fibrisols correspond to Fibrists; Mesisols to Hemists; Humisols to Saprists; and Folisols to Folists.

In the Kubiena (1953) system, which is used mostly in Europe, organic soils can be separated as "Subaqueous or Underwater Soils" and "Division of Semiterrestrial or Flooding and Groundwater

Soils" at the highest level. At lower levels, under each division there are specific classes ("Peat-forming subaqueous soils") and types ("Fen") of organic soils. The English and Wales system of classification places organic soils under "Peat (Organic) soils" at its highest level, with a subdivision loosely describing the degree of organic matter decomposition.

USE OF *SOIL TAXONOMY* CLASSIFICATIONS TO ESTIMATE SOM LEVELS OF SOILS

Soil organic matter is about 56% soil organic C (SOC) (Nelson and Sommers 1982). Kern (1994) used SOC data on soils throughout the contiguous U.S. and categorized the soils into 10 majors group (soil Orders) according to the classification in *Soil Taxonomy* (Soil Survey Staff 1975). As expected, the results of his grouping showed that to a depth of 1 m, Histosols contained the highest levels of SOC (84.3 kg m^{-2}) and contained nearly six times as much SOC as the soils with the next highest amount (Spodosols).

Kern (1994) concluded that there was too much variation within the groupings at the Order level. Additional smaller groupings were made at the Suborder (e.g., Aquents) and Great Group levels (e.g., Hydraquents) of *Soil Taxonomy*. These groupings allowed the soils to be differentiated based on characteristics such as wetness and presence of layers with a higher clay content or higher salt content. In nearly all the groupings at the Great Group level, the wet soils had the highest level of SOC. For example, the Hydraquents contained a very high amount of SOC (28.8 kg m^{-2}). These are wet soils of tidal areas and show the accumulation of organic matter as affected by the saturated conditions as well as the higher clay contents. Other soils with a very high SOC content (34.4 kg m^{-2}) were the Umbraquults, which by definition have a higher organic matter content in the surface layer. These examples show the usefulness of the *Soil Taxonomy* classifications in the prediction of soils in which the higher organic matter levels can be expected to occur in the wet soil groupings. One exception was in the Spodosol order. The wet Spodosols (Aquods) did not have a high SOC content. This lower organic matter accumulation may substantiate the current guidelines that indicate many of the Spodosols, although poorly drained, are not saturated near the surface for long periods of time and do not have a significant organic matter accumulation in the surface layer. These soils are not hydric soils (Hurt and Puckett 1992).

ORGANIC SOIL FORMATION

Organic soils in North America that developed under anaerobic conditions occur in areas where precipitation exceeds evaporation, where soils receive runoff water such as in depressional areas and low-lying areas along streams, and in low-lying coastal areas where the groundwater table is high. Organic soils that formed along a tidal creek in New Hampshire are shown in Plate 4 (Kelsea and Gove 1994). The soils in this area are Sulfihemists. These are organic soils in an intermediate level of decomposition that have sulfidic materials within a depth of 100 cm (Soil Survey Staff 1996b).

In many freshwater ecosystems, slope gradients are low and drainage inflows are diffuse and meandering, which permits sedimentation and allows for the accumulation of SOM under anaerobic conditions. During organic soil development, SOM accumulates through the formation of biomass and organic detritus.

Basic Processes and Rates

The basic processes involved in the production and accumulation of organic matter were discussed in Chapter 5. In summary, organic matter accumulates in saturated soils due primarily to a lack of O$_2$ during waterlogging, and the inhibitory effect of protonated organic acids (Pon-

Table 6.5 Classification of Organic Soils in Different Soil Classification Systems

Classification System — Relative Level	U.S. Soil Taxonomy	FAO/UNESCO	French System	Canadian System	Kubiena System	England and Wales
Highest level	Histosols	Histosols	Hydromorphic soils	Organic	Subaqueous or underwater soils	Peat (organic) soils
					Division of semiterrestrial or flooding and groundwater soils	
Lower level	Folist Fibrist Hemist Saprist	Eutric Histosol Dystric Histosol Gelic Histosol	Organic hydromorphic soils Medium organic hydromorphic soils Low organic matter hydromorphic	Fibrisol Mesisol Humisol Folisol	Peat-forming subaqueous soils Semiterrestrial peat soils Gley soils with humus formation	Raw peat soils Earthy peat soils

Adapted from Buol, S.W., F.D. Hole, and R.J. McCracken. 1989. *Soil Genesis and Classification*, 3rd ed. Iowa State Univ. Press, Ames, IA.

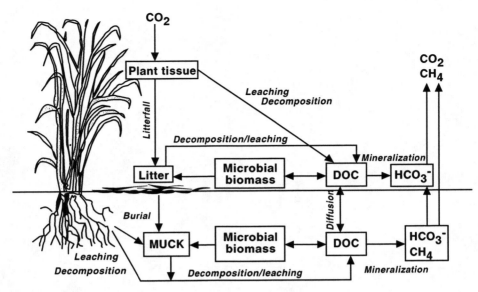

Figure 6.2 Wetland carbon cycle showing the transformations that take place in the production and accumulation of SOM. (From DeBusk, W.F. 1996. Organic matter turnover along a nutrient gradient in the Everglades. Ph.D. dissertation. Univ. of Florida, Gainesville.)

namperuma 1972; Gambrell and Patrick 1978; Bohn et al. 1985; Tate 1987) and organic matter grows (Ponnamperuma 1972).

DeBusk (1996), in a study of organic matter turnover in the Florida Everglades, diagrammed a wetland carbon cycle (Figure 6.2) to describe the transformations that occur in the production and accumulation of SOM. In the wetland carbon cycle, the transformations and flows that occur are represented as a collection of compartments. Simultaneous transfer of mass occurs among the compartments. In the vegetation compartment, macrophytic and algal species transform inorganic carbon to organic carbon through photosynthesis. Organic carbon is stored in the system in living (vegetation and microbial biomass) and nonliving (dead plant tissue, litter, SOM) forms. Nonliving storage of organic carbon is very high in wetlands in relation to other ecosystems; this storage is an energy reserve for the ecosystem (Wetzel 1992).

The microbial biomass is very important in the wetland carbon cycle because most of the organic carbon fixed in wetland systems by both phytoplankton and macrophytes is processed and recycled entirely by bacteria without involvement of the food web of higher animals (Wetzel 1984). Microbial biomass comprises only a small fraction of the nonliving organic matter, and yet most of the ecosystem production passes through the microbial component several times (Elliott et al. 1984). The microbial decomposers in the biomass derive their energy and carbon for growth from organic carbon, and thus facilitate the recycling of energy and carbon within and outside the wetland ecosystem (Wetzel 1992). The significance of dissolved organic carbon (DOC) in the wetland carbon cycle has not been clearly defined (DeBusk 1996). The role of DOC has not been well established because DOC represents such a broad spectrum of organic compounds (Wetzel 1984).

McDowell et al. (1969), using radiocarbon measurements on organic soils in the Upper Everglades of Florida, showed that formation of the organic materials began about 4400 years B.P. (before present). Development of organic materials was initially very slow but increased greatly about 3500 to 4000 years ago as a result of a rapid rise in sea level. McDowell et al. (1969) showed that organic soil development proceeded rapidly from 3500 to about 1200 years B.P. During this time, plant growth increased rapidly, and the organic materials were preserved as fibrous peats with very little mineral soil mixing.

Schlesinger (1977) measured the organic matter accumulation in ecosystems throughout the world by estimating the total detritus (kg cm^{-2}) in soils. The highest accumulations by his estimates were in swamps and marshes (mean of 68.6 kg cm^{-2}). Tundra had the second highest amount (mean of 21.6 kg cm^{-2}). The amount of detritus in tropical forests was considerably lower, with an estimated mean of 10.4 kg cm^{-2}. Schlesinger (1977) pointed out that decomposition in tropical forests is rapid, and even though the production of litter is high in these areas, accumulation of organic matter is low.

The highest amounts of organic matter generally accumulate in the surface layers of wetland soils (see Chapter 15 for a discussion of the organic matter accumulation in the subsoils of Spodosols). Richardson and Bigler (1984), in a study of wetland soils of the Prairie Pothole region of North Dakota, observed a steady decrease in SOM levels with depth. They concluded that the incorporation of plant materials to the soil surface continually adds fresh materials to the surface of mineral soils, and organic matter is lost in waterlogged soils at depth by slow microbial decomposition.

A few processes affecting the decomposition and rate of accumulation of SOM will be illustrated using as examples the cypress swamps in the southeastern U.S., a small lake in northwest Florida, and the northern wetlands in Canada.

Cypress Swamps of the Southeastern U.S. Coastal Plain

The cypress swamps in the cypress/pine flatwoods ecosystem are a unique type of forested wetland that covers much of the Atlantic and Gulf coastal plain of Alabama, Florida, Georgia, Louisiana, South Carolina, and Mississippi. The typical landscape is relatively flat with scattered small-to-large, closed depressions. Many are only one to two hectares or less in size, but some are considerably larger. The depressions generally are karstic features and often support a bald-cypress (*Taxodium distichum* L.) vegetation. Slash pine (*Pinus elliottii* L.) and loblolly pine (*Pinus taeda* L.) commonly dominate the pine flatwoods. The depressions on the landscape, to and through which the surface water and groundwater flow, contain water on the surface for much of the year, and thus are commonly referred to as ponds. Water depth in the ponds varies throughout the year depending on the balance between rainfall and evapotranspiration (Plate 5).

The ponds, in their initial development, receive surface runoff and subsurface flow of nutrient-rich water. This water provides an excellent medium for the growth of microorganisms such as algae, bacteria, and diatoms, which reproduce, die, and settle to the bottom of the ponds. The deposits of these microorganisms as well as other sediments slowly reduce the depth of water in the ponds. As the ponds become shallower, plants begin to grow. Submergent plants grow initially, followed by floating plants. These plants contribute organic matter to the bottom of the ponds, and microbial respiration reduces the oxygen in the bottom of the ponds. Water-loving plants become established, and more organic matter is added to the bottom of the ponds. Eventually, the O_2 is depleted in the bottom of the ponds. Floating plants continue to flourish on the surface of the ponds and begin to support mosses, such as the *Sphagnum* sp. As the organic materials continue to be added, the ponds continue filling with organic matter. When the pond basin is filled, increasing amounts of *Sphagnum* sp. grow on the surface.

The rate of accumulation of organic materials is highly variable and depends on several factors, including the rate of biomass production, the amount of decomposition that takes place, and the number of fires that occur in the forested areas. Even though the ponds may contain water for much of the year, during dry periods it is common for forest fires to occur. As a result of the forest fires, trees die and are removed. When trees are removed, the evapotranspiration rates decrease and the water table level in the pond rises. With the higher water table, the rate of decomposition of the organic materials decreases and thus the accumulation of organic materials increases.

The history of several ponds and lakes has been studied in Florida to try to determine the rate of accumulation of organic materials. One small pond (40 m long and 15 m wide) in north-central Florida was investigated by Collins and Hanninen (personal communication, 1993) with ground-penetrating radar (GPR) to determine the thickness of the organic materials and the depth to the underlying limestone. Pollen samples were also collected from the organic materials to learn more about the vegetative history of the pond and the rate of accumulation of the organic materials. Results of this study showed that the rate of organic matter accumulation differed during the period 0 to 4250 yr B.P., ranging from 0.07 to 0.03 cm yr^{-1}. The organic sediments in the pond accumulated at a rate of about 0.043 cm yr^{-1} during the period 2500 to 3000 yr B.P. But, during the period 0 to 2500 yr B.P., the organic matter in the pond accumulated at a rate of about 0.03 cm yr^{-1}. The increased rate of organic matter accumulation reflects a change in the vegetation and climate during that period. Pollen analyses showed that in the past 2500 years, the vegetation in the area of the pond changed from a drier environment dominated by *Pinus* to a wetter vegetation accommodating *Taxodium*.

Small Lake in Northwest Florida

A small lake (about 500 m in diameter) in northwest Florida was studied by Watts et al. (1992), and the vegetation history of the past 40,000 years was recorded. This study showed substantial climatic changes during the period as reflected in the vegetative changes. The fluctuations in vegetative cover indicate substantial differences in the rate of organic matter accumulations over the years. The vegetative history indicated that from 40,000 to 29,000 yr B.P. was a time for forests with abundant pines and oaks and diverse mesic tree species. From 29,000 to 14,000 yr B.P. was a period of species-poor pine forests because of climatic variations. Spruce trees were present from 14,000 to 12,000 yr B.P. during a cold climatic phase (similar to southern Quebec today). At about 12,000 yr B.P., the forests were oak dominated and contained many dried-out lake basins. Sometime after 7760 yr B.P., there was a rise of about 4 m in the water table, and the cypress swamps developed.

Northern Wetlands of Canada

Ovenden (1990) observed that peatlands began to develop from 5000 to 3000 yr B.P. on the southern part of Glacial Lake Agassiz. During that time period, pollen records indicate increasing precipitation and declining summer temperatures. Ovenden (1990) showed rates of peat accumulation in different types of peat layers in the northern wetlands of Canada that ranged from about 0.01 to 0.06 cm yr^{-1}. Accumulation rates as high as 0.2 cm yr^{-1} were noted in organic materials described as woody sedge peat.

EFFECT OF LANDSCAPE POSITION AND HYDROLOGY ON SOM ACCUMULATION

The effect of the landscape position and hydrology on the development of organic materials is apparent when considering the extensive areas of organic soils in areas such as Alaska and the Everglades in Florida. The effect of plant types on SOM decomposition is also significant.

Landscape Position

Landscape position affects the formation and accumulation of SOM through its influence on rainfall, runoff, and soil moisture retention. Soils in depressions usually have high SOM because retention of water and the resulting anaerobic conditions help preserve the organic materials. Organic soils form on landscapes that accumulate water faster than it can drain away. Flat topography, such as is common in the coastal plains, lowers flow gradients. In depressional basins, water retention can actually alter the landscape geomorphology. As water retention increases, the water table can rise above the levels of the original landform. The higher water table allows for organic matter to form above the surface of the original landform. So, over time, surface- or groundwater-flow wetlands become primarily precipitation driven.

Hydrology

The relationship of hydrology and the development of organic soils largely results from the interaction of landforms and groundwater levels, with dependence on precipitation and temperature effects (Moore and Bellamy 1974). Mausbach and Richardson (1994) stated that the hydrology of wetlands is generally governed by the amount of precipitation.

Two major types of landscapes are discussed with respect to hydrology and organic soils: depressional wetlands and sloping wetlands. Depressional wetlands occur in a confined landscape, and rainfall cannot leave because of the landscape constraints. Sloping wetlands receive surface or groundwater flow.

In sloping wetlands, water flow usually occurs below the surface because of the permeable underlying mineral materials. Sloping wetlands that have water discharge areas may develop fens. Holte (1966) described elevated mounds of organic materials as fens in Iowa, and termed the fens "discharge cones" to note the strong groundwater discharge from these areas. Mausbach and Richardson (1994) noted that one feature common to all fen wetlands is the predominance of groundwater discharge from areas higher on the landscape.

Siegel (1983) investigated the hydrogeological setting for the organic soils in Glacial Lake Agassiz. He found that groundwater circulated in flow paths several km long in the organic materials and underlying mineral soils, and hypothesized that groundwater mounds create the raised bogs or mires that are common in the area. The groundwater discharge contributes the needed wetness for the organic materials to form.

Organic materials form in regions with an excess of precipitation over evaporation, and thus, Histosols are extensive in the cooler climates, such as Alaska. Histosols also occur in areas with high precipitation and low outflow, such as the Everglades of Florida. Richardson et al. (1994) noted that Histosols are not common from North Dakota westward except in fens. They attribute the lack of Histosols in this area to the frequent water drawdown and drying which does not allow organic sediments to accumulate. This is the result of the exposure of sediments to air during drawdown and carbon mineralization during microbially mediated SO_4^{2-} reduction. Komor (1992) noted that in the prairie potholes of North Dakota, oxidation and consumption of organic matter occur as a result of aeration during frequent drawdown and sulfate reduction. Malterer and Farnham (1985) and Richardson et al. (1987) pointed out that fens seldom have a water drawdown and are low in SO_4^{2-}. Fens contain Histosols and occur in North Dakota and throughout the pothole region of the glaciated areas in central North America (Holte 1996, van der Valk 1975, Vitt et al. 1975, Malterer and Farnham, 1985).

Plant Types

Plant-associated properties that contribute to decomposition rates include the carbon:nitrogen ratio, lignin content, materials rich in tannins, and humified materials (Neue 1985, Tate 1987, Oades 1988). These researchers have shown that plant species are more resistant to decomposition if they are higher in lignins or decompose to materials that are rich in tannins. Many studies indicate that fresh organic matter in the form of leaves and other plant debris is readily decomposable and provides a quick source of energy for microorganisms.

The decomposition of plant fibers is less of a function of *in situ* decomposition than the plant type (Moore and Bellamy, 1974). Some plants do not preserve many fibers. Sphagnum moss produces fibrous, difficult to decompose organic materials.

The contrast in forest vs. prairie vegetation is commonly used to illustrate the effect of vegetation on the accumulation of organic matter. In most grassland soils, the organic matter content is highest in the surface and declines gradually deeper into the soil. The organic matter content closely follows the distribution of grass roots because much of the organic matter is the result of the annual death and regrowth of the grass roots. A forested soil may contain as much organic matter in the surface layer as a grassland soil, but the surface layer (A horizon) of a forested soil is usually much thinner. Tree roots live longer than grass and add little organic matter to the soil each year. Much of the organic matter in the surface layer in forested soils is the result of deposits of leaves and twigs, which are mixed into the upper layer of the soil.

ENVIRONMENTAL FACTORS

The environmental factors of temperature and precipitation play major roles in the accumulation and decomposition of SOM. The role of soil pH appears to be less well defined.

Temperature and Moisture

Jenny (1950) studied the effect of climate on soil nitrogen and SOM levels in soils in the midwestern U.S. His study showed that for each 10°C rise in mean annual temperature, the soil nitrogen and SOM content of the soil decreased 2 to 3 times. Therefore, it would be expected that warmer climatic zones would have the lowest levels of SOM. However, increasing rainfall in an area can have a tremendous effect on SOM production because the moisture promotes greater plant growth and the production of large quantities of plant materials for organic matter synthesis. Harsh climatic conditions can lead to the preservation of SOM. Thus, even with warm temperatures, under anaerobic conditions in soils, such as in the Florida Everglades, the decomposition of organic matter by microorganisms decreases, resulting in significant increases in SOM accumulation. Similarly, in the saturated soils of the Great Lakes region, the cold winters result in a decrease in the activity of microorganisms and the preservation of SOM. Some researchers have suggested that at 4°C or less, microorganisms cease functioning (Tate 1987). It is apparent that, under saturated soil conditions, O_2 deficiencies for prolonged periods prevent the complete microbial decay of plant materials over a wide temperature range.

pH

Some researchers have shown that organic matter decomposition is slower in acid soils (Jenkinson 1971, Neue 1985), while others suggest that decomposition is increased in acid soils (Oades 1988). This apparent conflict shows the complex nature and multiple interactions of the processes at work in the decomposition of SOM.

The results of a study by Moran et al. (1989) showed that the decomposition of *Spartina alterniflora* L. in a salt marsh was quite rapid and only 15% of the original material remained after 2 months. In contrast, in a freshwater swamp under acid conditions, 65% of the *Carex walteriana* L. material remained after a year of decomposition. This difference was assumed to be related to environmental differences in the pH of the soils at the two sites, rather than differences in the composition of the plants.

Many of the organic soils that have been analyzed in the U.S., particularly in the southeastern U.S., have low pH values. Bridgham and Richardson (1993), in a study of Medisaprists (now Haplosaprists) in the pocosins and gum swamps of North Carolina, observed that the organic soils had pH values < 4.0. Collins et al. (1997) reported that the mean pH value of 27 Histosol pedons (Medisaprists) in Florida was 4.2. On the other hand, organic soils in fens may be calcareous and have pH values of 7.4 or higher. Histosols in Alaska range from alkaline (pH >7.4) on the arctic coastal plain to slightly acid (pH < 6.6) on the arctic foothills, and range from slightly to strongly acid (pH 6.5 to 5.1) in the boreal zone (Riger et al. 1979). Richardson et al. (1987) noted that calcareous fens with histic epipedons have been reported in Iowa, Minnesota, North Dakota, and Wisconsin. The more acid organic soils occur in areas with higher rainfall and leaching.

PHYSICAL PROPERTIES OF ORGANIC MATERIALS

Bulk density, mineral content, hydraulic conductivity, and water-holding capacity or water retention are important physical properties that can be routinely analyzed in organic soils. These properties are related to the porosity and pore-size distribution of the organic materials, which are related to the degree of decomposition of those materials. The least decomposed organic materials have many large pores which permit rapid water movement, while the more decomposed organic materials have finer pores which may permit only slow water movement.

In general, the following relationships can be expected in organic soils: increasing unrubbed fiber content results in decreasing bulk density values, increasing hydraulic conductivity values, and an increase in the saturated water content (% by volume) of the soils. Boelter (1969) showed a curvilinear relationship of water content at saturation to unrubbed fiber content and bulk density. Bishel-Machung et al. (1996), in a study of wetlands in Pennsylvania, showed a negative relationship between organic matter and bulk density (Pearson correlation coefficient = –0.804) as well as organic matter and pH (r = –0.534). Collins et al. (1997), in a statistical analysis of 27 Saprists in Florida, showed a similar negative correlation coefficient (r = –0.697) between organic carbon and bulk density, as well as organic carbon and pH (r = –0.505).

Bulk Density

Bulk density generally is very low in organic soils and has been used to quantify the state of decomposition of the organic materials. Farnham and Finney (1965) reported values as low as 0.06 g cm^{-3}, on a dry weight basis. Bulk density tends to increase with increasing decomposition. Hurt et al. (1995) stated that the bulk density of peat (fibric materials), mucky peat (hemic materials), and muck (sapric materials) is <0.1 g cm^{-3}, 0.1 to 0.2 g cm^{-3}, and >0.2 g cm^{-3}, respectively. The upper limit of the bulk density of sapric materials has not been defined. Collins et al. (1997) reported that the mean bulk density of 27 Medisaprists (now Haplosaprists) in Florida was 0.65 g cm^{-3}.

A simple test to measure bulk density, mineral content, and water content on a single core sample was described by Lynn et al. (1974). Their procedure is to trim a core of undisturbed organic soil into a tared (T), 130-ml aluminum moisture can. The can volume (V) is used for bulk density. Cover the sample and weigh (A). Place the sample in an oven at 110°C overnight. Cool the sample and weigh (B). Again, heat the sample at 400°C overnight, cool, and weigh (C). Compute the water content, mineral content, and bulk density with the following equations:

$$\text{Water Content (\%)} = (A - B)/(B - T) \times 100 \qquad \text{(Equation 1)}$$

$$\text{Mineral Content (\%)} = (C - T)/(B - T) \times 100 \qquad \text{(Equation 2)}$$

$$\text{Bulk Density (g/cc)} = (B - T)/V \qquad \text{(Equation 3)}$$

Hydraulic Conductivity

Hydraulic conductivity is difficult to measure in organic soils. Differences in pore-size distribution of the organic materials, which are related to the stage of decomposition of the materials, can result in major differences in the rate of movement of saturated water. Boelter (1965) measured hydraulic conductivity of organic soils in northern Minnesota. The study showed that the hydraulic conductivity of layers of an organic soil described as sphagnum moss peat ranged from very rapid water movement in a surface layer (described as undecomposed mosses) to much slower movement in a subsurface layer (described as moderately decomposed moss peat). Some of the rates of movement were too rapid to measure with the techniques used in the Boelter (1965) study, while some of the more decomposed organic materials (well-decomposed peat) had hydraulic conductivity values slower than often expected in soils with a high clay content. Collins et al. (1997) reported high hydraulic conductivity values (68 cm hr^{-1}) for 27 Histosols (Medisaprists, now Haplosaprists) analyzed in Florida.

Water-Holding Capacity

The water-holding capacity of organic soils is generally estimated from the % organic C, bulk density, and water-retention data. Organic soils are commonly saturated and have very high water-holding capacities, both on a weight and volume basis. Boelter (1969) reported that the water content of saturated peat in northern Minnesota ranged from nearly 100% by volume in the undecomposed sphagnum moss peat in the surface layer to about 80% by volume in the more decomposed subsurface layers. Water-retention data from Boelter (1969) showed, however, that the undecomposed sphagnum moss peat loses a large portion of its saturated water content (about 25%) at suctions of only 5 cm of water. Water-retention determinations are made by placing undisturbed soil cores in Tempe pressure cells. The cells are saturated and then sequentially extracted under a number of different suctions of water (Soil Survey Staff 1996a).

CHEMICAL PROPERTIES OF SOIL ORGANIC MATTER

Carbon is the dominant component of SOM, and the determination of the organic C content of soils is probably the most important and widely used analysis of soils with a high content of SOM. Organic C is commonly measured in soils by the Walkley–Black method (wet combustion) (Walkley 1934), or by ignition (dry combustion). The determination of organic C by loss on ignition is a taxonomic criterion for organic soil materials (Soil Survey Staff 1996a). The percentage of organic matter lost on ignition is used to determine the organic C content of soils with a high content of SOM.

In the ignition method, a dry sample is placed in a cold muffle furnace and raised to a temperature of 400°C for a period of 16 hours. The difference in mass before and after ignition is the organic matter content of the sample because organic matter is oxidized and CO_2 is lost. The measurement of organic carbon can be used as an indirect determination of SOM through the use of a conversion factor in which

$$\% \text{ Organic C} \times 1.724 = \% \text{ SOM} \qquad \text{(Equation 4)}$$

Although the literature indicates that the proportion of organic carbon in SOM for a range of soils is highly variable, the "Van Bemmelen factor" of 1.724 has been used for many years (Soil Survey Staff 1996a). The assumption made in using this factor is that SOM contains about 58% organic C.

The Walkley–Black wet combustion method for the determination of organic carbon is generally considered invalid if the organic C content is > 8% (Soil Survey Staff 1996a). Although the Walkley–Black method converts the most active forms of organic C in soils, it does not yield complete oxidation of these compounds and is not an appropriate procedure for soils with a high content of SOM.

Table 6.6 Chemical and Physical Properties of Organic and Mineral Soils

Property	Organic	Mineral
Organic carbon content (%)	>12–18	<12–18
Bulk density (g/cm^3)	<0.6	1.0–2.0
Porosity (%)	>80	45–55
pH	<4.5*	3.5–8.5
Base saturation	Low	Low to high
Cation exchange capacity	High	Low to high
Plant available water	Often low**	Low to high
Hydraulic conductivity	Moderate to rapid	Very low to very rapid

 * Although organic soils are commonly quite acid, pH values of organic soils in fens may be > 7.0.
**Organic soils have a very high water holding capacity. However, much of the water is held either in the larger pores (gravitational water) or in very small pores and is unavailable for plant growth (Boelter and Blake, 1964).

Comparison of Organic and Mineral Soil Chemical and Physical Properties

It is difficult to compare the properties of organic and mineral soil materials because most soils contain at least traces of both materials in each horizon, and the soil properties are highly dependent upon the sand, silt, and clay content of the soils. Mineral soils as well as organic soil materials can have a very wide range of soil properties. In general though, it is known that organic soils have lower bulk densities and higher water-holding capacities, higher porosity, and higher cation exchange capacities than mineral soils. A general comparison of organic and mineral soils is shown in Table 6.6.

TYPES OF ORGANIC SOIL COMPONENTS

SOM is a complex mixture of substances that are the result of the addition of organic residues of plant and animal origin into the soil and their continual transformation through biological, chemical, and physical factors. The organic fraction in soils contains various substances that represent either organic residues undergoing decomposition, metabolic products of microorganisms using organic residues as a source of energy, or products of resynthesis in the form of bacterial cells.

The major organic components of SOM are the humic substances. Humic substances are formed from the chemical and biological degradation of plant and animal residues and the synthetic activities of microorganisms. The components of plant and animal residues that serve as a source of humic substances is not completely clear. Several pathways have been proposed for the formation of humic substances during the decay of plant and animal residues. Investigators have confirmed the transformation of the nitrogen of plant residues into bacterial plasma, the nonlignin origin of humic substances in the early stages of humification of plant residues, etc. It has also been shown that water-soluble organic compounds of plant tissues participate in humus formation, and that tannin-like substances are sources of humic substances during early stages of humification.

The basic cell wall construction of vascular plants includes a framework component of cellulose, a matric component of linear polysaccharides (hemicellulose), and an encrusting component composed of lignin (Zeikus 1981). Lignin occurs in cells of conductive and supportive tissue, and thus is not found in algae and mosses. The presence of lignin is the ultimate limiting factor in decomposition of vascular plant tissues (Zeikus 1981).

Organic residues undergoing decomposition include compounds such as fats, carbohydrates, proteins and their decomposition products such as amino acids, lignins, tannic substances, resins and terpenes, and various organic acids, aromatic materials, alcohols, and hydrocarbons. Collectively, these compounds form about 10 to 15% of the total amount of SOM. A separate group of substances, the origin and nature of which are not fully understood, are termed *humic substances*. This group comprises about 85 to 90% of soil humus. "Humic substances" is a term used to describe the colored substances formed by secondary synthesis reactions or its fractions obtained on the basis of solubility characteristics. The main groups of humic substances are humic acids, fulvic acids, and humin.

Humic Substances: Humic Acids, Fulvic Acids, and Humin

The classical method of fractionating SOM is based on the different solubilities of the components in alkali and acidic solutions. Humic acid is the dark-colored organic material that can be extracted from soil by dilute bases and other reagents, and it is insoluble (precipitated) in dilute acid. Humin is the alkali insoluble fraction of SOM or humus. The fulvic acids are the fraction of SOM that are soluble in both base and acid.

Humic acids have a complex structure: At least two main components participate in the formation of their molecule: (i) compounds of phenolic or quinoid nature, and (ii) nitrogen-containing compounds (amino acids and peptides). Humic acid molecules may also contain a third component, which are substances of a carbohydrate nature. Humic acids have been divided into two groups: the brown humic acids, which are not coagulated from the basic solution in the presence of electrolyte, and the gray humic acids, which are coagulated in the presence of electrolyte.

Humic substances not extracted from soils during treatment with basic solutions are placed in the humin group. In soils, the humin group is partly represented by carbonized plant residues, which may be present in soils with restricted aeration.

The term *fulvic acids* was first introduced for a group of acids occurring in peat waters. At very low pH values (<3.0), these acids are "straw yellow" in color. This was the basis for the introduction of the term *fulvic acids* because fulvus = yellow. With increasing pH values, the fulvic acids change from orange to a "wine red" color. Fulvic acids, like humic acids, have structural elements of an aromatic nature, nitrogen-containing substances, and reducing substances.

Soil Carbohydrates

Carbohydrates are significant in the formation of SOM because they serve as an energy source for microorganisms, and also as building blocks for humus formation. Carbohydrates have been estimated to constitute 5 to 25% of SOM, and thereby are the second most abundant component of humus (Stevenson 1994). Plant remains contain carbohydrates in the form of simple sugars, hemicellulose, and cellulose, but these are generally decomposed rapidly by soil microorganisms. The microbial products make up the major part of the carbohydrates in SOM, with the exception of the carbohydrates in the soil leaf litter. In soils containing undecayed or partially decayed plant remains, a portion of the carbohydrates exists as cellulose.

The major elements in humic and fulvic acids are carbon and oxygen. The carbon content of humic acids generally ranges from about 54 to 59% (Steelik 1985), while the oxygen content ranges from 33 to 39%. Fulvic acids have lower carbon (40 to 50%), but higher oxygen (40 to 50%) contents. Humic and fulvic acids contain smaller percentages of hydrogen, nitrogen, and sulfur.

NUTRIENT CYCLES INVOLVING SOM

The role of SOM in nutrient cycling involves the conservation and recycling of nutrients between the microbial–plant–soil communities. Nutrient cycling involves two major processes: immobilization and mineralization. Immobilization is the uptake of inorganic nutrient ions by microorganisms. The immobilization of nutrients refers to the use and incorporation of nutrients into living matter by microorganisms and living plants. Mineralization is the conversion of nutrients in organic matter into inorganic ions by microbial decomposers.

The organic matter that is added to the soil consists of a variety of compounds. As microorganisms decompose the organic matter, the most easily digested and readily available materials, such as the carbohydrates and proteins, are utilized. While digesting the plant residues, microorganisms use the carbon and other nutrients for their own growth, and thus are temporarily unavailable or immobilized in microbial biomass. The synthesized tissues of the microorganisms eventually are mineralized and become available for further decomposition after the microorganisms die. The net effect is the release of energy as heat, the formation of carbon dioxide and water, and the conversion of organically bound nutrients (N, P, S) to ammonium, phosphate, sulfate, and many other nutrients as simple metallic ions, which are made available to other organisms for growth.

Under anaerobic soil conditions, different microorganisms become active. The oxidized forms of plant nutrients are reduced. Sulfur is reduced to hydrogen sulfide, and the characteristic "rotten egg" odor is produced. One of the more important reactions in soils under anaerobic conditions is denitrification, in which NO_3^- is reduced to nitrogen gas.

As discussed by Bray and Andrews (1924), the transformation of lignin during the humification process is quite complex. During humification, plant residues become brown in color, and with sufficient moisture, a brown liquid, an aqueous solution of humic substances, is formed. Humification occurs most rapidly in decomposing leaves of grasses, and much more slowly in roots and needles of pine trees. Woody residues generally have the slowest decomposition rate, while residues of legume–grass mixtures, which are rich in nitrogen and low in lignin, are the most readily decomposed. Leaves and roots of different grasses as well as different trees humify at different rates. The higher the content of easily mobile substances and the lower the content of lignin, the more rapidly the plant residues are humified. Tree leaves are generally a fairly good source of humus. On the other hand, conifer needles are decomposed slowly, apparently because their humification is retarded by the high content of resins in the needles. Observations have shown that the formation of humic substances is possible at the early stages of humification and prior to decomposition of lignified tissues.

It has been shown that microorganisms can wholly decompose any plant substance. Thus, it is evident that the formation of humic substances takes place under intense biological activity. In the first stage of the process, microorganisms decompose the original plant or animal residues to simpler compounds. These compounds later serve as components for the formation of molecules of humic substances (structural units of lignin, tannic substances, etc.). In the second stage, products of the resynthesis and metabolism of microorganisms (amino acids, uronic acids, various substances of aromatic structure) participate and serve as biocatalysts in the formation of humic substances, the end products of their activity, as components of their molecules.

Special conditions may exist in poorly and very poorly drained soils as the wet conditions alter the activities of macro- and microorganisms. In addition to the organic matter decomposing at a reduced rate, the end products of metabolism are different (Stevenson 1994). Fermentation products from incomplete oxidation can lead to the production of CH_4, organic acids, amines, mercaptans, aldehydes, and ketones.

In summary, many different types of reactions can lead to the production of the dark-colored pigments of the humic and fulvic acids in SOM. A multiple origin is likely, but the major process appears to involve polyphenols and quinones in condensation reactions. Polyphenols, derived from lignin, are enzymatically converted to quinones that combine with amino compounds to form N-containing polymers.

MANAGEMENT OF ORGANIC SOILS

Conventional cultivation practices as well as management practices in forested wetlands, such as drainage and harvesting of forested wetlands, affect the organic soils in the wetlands. Disturbance of wetland vegetation and organic soils in the wetlands increases the rate of organic matter decomposition. Trettin et al. (1995) noted that increased organic matter decomposition following soil drainage is a consequence of that silvaculture practice in northern-forested wetlands. Trettin et al. (1996) also found that tree harvesting and site preparation increased soil temperature, which increased the rate of organic matter decomposition.

Silvacultural practices (site preparation for timber harvesting and regeneration) in wetlands can affect the organic matter decomposition rate by altering soil temperature or moisture, or by increasing substrate availability to soil microorganisms (Armentano and Menges 1986, Trettin et al. 1995). These practices may have long-term negative effects on organic soils. Removal of the litter layer, the rhizosphere, and the decay of logs reduces the elevation of the soil surface. Bedding, a site preparation practice that is used to create an elevated planting bed, has been shown to increase organic matter decomposition as a result of drier site conditions and increased aerobic decomposition. In the southeastern U.S., Mader (1990) measured increased organic matter decomposition following clear-cutting in bottomland hardwoods.

One of the major problems related to the use of organic soils is subsidence. There are several factors that play important roles in subsidence of organic soils: biological oxidation, depth to the water table, leaching of soluble organic materials, characteristics of the organic material, compaction, burning, wind erosion, and water erosion. The depth to the water table is certainly one of the major factors involved in subsidence.

Subsidence of organic soils occurs in two stages: primary and secondary. Primary subsidence occurs relatively fast (4 to 10 years) and involves the loss of buoyant force and compaction of the organic soils due to drainage. Secondary subsidence, which is much slower, includes biological oxidation, wind and water erosion, and leaching of organic materials (Everett 1983).

Biological oxidation occurs when there is an increase in the depth to the water table. Subsidence of the organic soils in the Everglades Agricultural Area (EAA) of South Florida has been observed and recorded since 1914. Water table test plots have shown that water table depth and subsidence are directly related. The relationship was expressed by the equation $x = (y - 2.45)/14.77$, in which x equals the subsidence rate in inches/year, and y equals the average depth of the water table in inches (Stephens 1956). In the EAA, lowering the water table by 30, 60, and 90 cm resulted in subsidence rates of 1.6, 3.6, and 5.7 cm yr^{-1}, respectively, of these organic soils (Lucas 1982). This lowering of the water table provided ideal conditions for aerobic microbial decomposition. Microbial activity and decomposition rates are greatest in the upper 10 cm. It has been estimated that by the year 2000, only about 13% of the EAA will have organic materials >1m in thickness due to biological oxidation (Lucas 1982). Continued subsidence over a long period of time has changed the classification of some of the Florida Everglades soils that were originally considered organic. Some soils may no longer have a thick enough organic layer to classify as a Histosol (Collins et al. 1986). An example of subsidence at a concrete monument in the EAA is shown in Plate 6. The top of the monument was set flush with the existing ground level in 1924. Stephens (1956) reported that the concrete monument showed a subsidence of nearly 1 m between 1924 and 1954. The photograph shown in Plate 6 was taken in 1996 and indicates that the soil had subsided a total of about 1.65 m at that time.

A study in Indiana (Jongedyk et al. 1950) showed annual subsidence rates of 1.1, 1.8, and 3.0 cm where water tables were maintained at depths of 42, 68, and 98 cm, respectively. This rate of subsidence is considerably less than that reported by Stephens (1956), and is likely related to the colder and freezing winter temperatures that occur in Indiana. In a Minnesota study over a 5-year period (Roe 1963), subsidence of 15 cm and 60 cm was noted with water table levels of 30 and 135 cm, respectively.

Burning is another factor that can increase or decrease subsidence of organic soils. Before 1950, fires accounted for the majority of the loss of organic materials in the U.S. Some fires were the result of deliberate burning to increase soil pH and soil nutrient levels; others were natural fires caused by lightning (Lucas 1982). In pioneer days, spring burning was a common practice. Railroad locomotives and early farm tractors also contributed to fires. Burning in this manner increased the subsidence of the organic soils. On the other hand, burning has also been used as a management technique to reduce subsidence. Wetlands are often invaded by surrounding trees. As the trees grow, the accumulation of organic matter decreases because trees add little biomass compared to grasses and other organic matter sources. In addition, trees use more water during their growth, which results in increased depths to the water table. Thus, burning the trees in a wetland may actually prolong the life of the organic soils.

Extremely hot fires, which often occur after periods of severe drought, can burn deeply into the organic layers and destroy the forest stand. Severe fires were likely the cause of the formation of the many lakes in the Okefenokee Swamp (Cypert 1961), and the deep, treeless pools in some of the cypress swamps (Ewel and Mitsch 1978).

Wind erosion occurs as the surface of the organic soil dries and decomposes. Even under the most careful management, the resulting loose, powdery surface is easily removed by wind. This type of erosion can be a serious problem during severe windstorms. Constructing and maintaining shrub windbreaks can reduce wind erosion, as can maintaining a cover on the soil surface at all times.

Water erosion generally is not a major cause of loss of organic materials because very minimal runoff occurs in the low-lying, relatively flat areas in which organic soils are typically located.

Proper management techniques can reduce, although not totally stop, the subsidence of Histosols. Management of organic soils involves protection from development followed by an active management program to counteract the effects of the lowering of the water tables. Maintaining a high water table is one of the primary means of reducing biological oxidation and subsidence.

REFERENCES

Armentano, T.V. and E.S. Menges. 1986. Patterns of change in the carbon balance of organic soil wetlands of the temperate zone. *J. Ecol.* 74:755–774.

Baldwin, M., C.E. Kellogg, and J. Thorp. 1938. Soil classification. pp. 979–1001. *In* G. Hambridge (Ed.) *Soils and Men.* U.S. Dept. of Agriculture Yearbook. U.S. Govt. Printing Office, Washington, DC.

Bentley, C.F. (Ed.) 1978. *Photographs and Descriptions of Some Canadian Soils.* Publ. No. B 79-1. The Univ. of Alberta, Edmonton, Alberta.

Bishel-Machung, L., R.P. Brooks, S.S. Yates, and K.L. Hoover. 1996. Soil properties of reference wetlands and wetland creation projects in Pennsylvania. *Wetlands* 16 (4):532–541.

Boelter, D.H. 1965. Hydraulic conductivities of peat. *Soil Sci.* 100:227–231.

Boelter, D.H. 1969. Physical properties of peats as related to degree of decomposition. *Soil Sci. Soc. Amer. Proc.* 33:606–609.

Boelter, D.H. and G.R. Blake. 1964. Importance of volumetric expression of water contents in organic soils. *Soil Sci. Soc. Am. Proc.* 28:176–178.

Bohn, H., B.L. McNeal, and G.A. O'Connor. 1985. *Soil Chemistry,* 2nd ed., John Wiley & Sons, New York.

Bouma, J. 1983. Hydrology and genesis of soils with aquic moisture regimes. pp. 235–281. *In* L.P. Wilding and N.E. Smeck (Eds.) *Pedogenesis and Soil Taxonomy. I. Concepts and Interactions.* Elsevier, Amsterdam, The Netherlands.

Bray, M. and T. Andrews. 1924. Chemical changes of ground wood during decay. *J. Industr. Chem.* 16:137.

Bridgham, S.D. and C.J. Richardson. 1993. Hydrology and nutrient gradients in North Carolina peatlands. *Wetlands* 13:207–218.

Buol, S.W., F.D. Hole, and R.J. McCracken. 1989. *Soil Genesis and Classification.* 3rd ed. Iowa State Univ. Press, Ames, IA.

Cline, M.G. 1979. Soil Classification in the United States. *Agronomy Mimeo* No. 79-12. 2nd print. Agron. Dept., Cornell Univ., Ithaca, NY.

Collins, M.E., G.W. Schellentrager, J.A. Doolittle, and S.F. Shih. 1986. Using ground-penetrating radar to study changes in soil map unit composition in selected Histosols. *Soil Sci. Soc. Am. J.* 50:408–412.

Collins, M.E., F.A. Ovalles, and R.J. Kuehl. 1997. A quantitative description of Florida soils. pp. 248. *In Agronomy Abstracts,* ASA-CSSA-SSSA, Anaheim, CA.

Cypert, E. 1961. The effects of fires in the Okefenokee Swamp in 1954 and 1955. *Am. Midl. Nat.* 66:485–503.

DeBusk, W.F. 1996. Organic matter turnover along a nutrient gradient in the Everglades. Ph.D. dissertation. Univ. of Florida, Gainesville.

Domeier, M.J. 1997. *Soil Survey of Sibley County, Minnesota.* USDA–NRCS, in cooperation with the Minn. Ag. Exp. Sta., U.S. Govt. Printing Office, Washington, DC.

Elliott, E.T., D.C. Coleman, R.E. Ingham, and J.A.Trofymow. 1984. Carbon and energy flow through microflora and microfauna in the soil system of terrestrial ecosystems. pp. 424–433. *In* M.J. Klug, and C.A. Reddy (Eds.) *Current Perspectives in Microbial Ecology.* American Society of Microbiology, Washington, DC.

Everett, K.R. 1983. Histosols. pp. 1–53. *In* L.P. Wilding, N.E. Smeck, and G. F. Hall (Eds.) *Pedogenesis and Soil Taxonomy, II. The Soil Orders.* Elsevier, Amsterdam.

Ewell, K.C. and W.J. Mitsch. 1978. The effects of fire on species composition in cypress dome ecosystems. *Florida Scientist* 41:25–32.

Farnham, R.S. and H.R. Finney. 1965. Classification and properties of organic soils. *Adv. Agron.* 17:115–162.

FAO. 1974. FAO/UNESCO Soil Map of the World, 1:5,000,000. Paris, France.

Gambrell, R.P. and W.H. Patrick, Jr. 1978. Chemical and microbiological properties of anaerobic soils and sediments. P. 375–423. *In* Hook, Donald D. and R.M.M. Crawford (Eds.) *Plant life in Anaerobic Environments.* Ann Arbor Science Publ. Inc., Ann Arbor, MI.

Holte, K.E. 1966. A floristic and ecological analysis of the Excelsior fen complex in northwest Iowa. *In* Ph.D. dissertation. Univ. of Iowa, Iowa City.

Hurt, G.W. and W.E. Puckett. 1992. Proposed hydric soil criteria and their identification. pp. 148–151. *In* J.M. Kimble (ed.) *Proc. of the Eighth International Soil Correlation Meeting (VIII ISCOM): Characterization, Classification, and Utilization of Wet Soils. Louisiana and Texas.* Oct. 6–21, 1990. USDA–SCS Nat. Soil Survey Center, Lincoln, NE.

Hurt, G.W., R.F. Pringle, H.C. Smith, A.J. Tugel, M.P. Whited, and D.Williams. 1995. NRCS recommended field indicators of hydric soils in north and south Florida. pp. 99–114. *In* V.W. Carlisle (Ed.) *Hydric Soils of Florida Handbook,* 2nd ed. Florida Assoc. of Environmental Soil Scientists, Gainesville, FL.

Jenkinson, D.S. 1971. Studies on the decomposition of C^{14} labeled organic matter in soil. *Soil Sci.* 111:64–70.

Jenny, H. 1950. Causes of the high nitrogen and organic matter content of certain tropical forest soils. *Soil Sci.* 69:63–69.

Jongedyk, H.A., R.B. Hickok, I.D. Mayer, and N.K. Ellis. 1950. Subsidence of muck soil in northern Indiana. Sta. Circ. 366, Purdue Univ., Ag. Exp. Sta., West Lafayette, IN.

Kelsea, R.J. and J.P. Gove. 1994. *Soil Survey of Rockingham County, New Hampshire.* USDA–SCS in cooperation with the New Hampshire Ag. Exp. Sta. U.S. Govt. Printing Office, Washington, DC.

Kern, J.S. 1994. Spatial patterns of soil organic carbon in the contiguous United States. *Soil Sci. Soc. Am. J.* 58:439–455.

Komor, S.C. 1992. Bidirectional sulfate diffusion in saline-lake sediments: evidence from Devils Lake, northeast North Dakota. *Geology* 20:314–322.

Kononova, M.M. 1961. *Soil Organic Matter. Its Nature, Its Role in Soil Formation and in Soil Fertility.* The Academy of Sciences of the USSR. The V.V. Dokuchaev Institute. Pergamon Press, New York.

Kubiena, W.L. 1953. *The Soils of Europe.* Thomas Murby & Co., London.

Lucas, R.E. 1982. Organic soils (Histosols) formation, distribution, physical and chemical properties and management for crop production. Mich. Ag. Exp. Sta. Res. Rep. 435.

Lynn, W.C., W.E. McKinzie, and R.B. Grossman. 1974. Field laboratory tests for characterization of Histosols. p. 11–20. *In* M. Stelly (Ed.) *Histosols: Their Characteristics, Classification, and Use.* SSSA Spec. Pub. 6. Madison, WI.

Mader, S.F. 1990. Recovery of ecosystem functions and plant community structure by a tupelo–cypress wetland following timber harvesting. Ph.D. dissertation. North Carolina State Univ., Raleigh (Diss. Abstr. 90-25641).

Malterer, T.J. and R.S. Farnham. 1985. The genesis and characterization of some calcareous peat lands in west-central Minnesota. pp. 195. *In Agronomy Abstracts,* ASA-CSSA-SSSA, Chicago, IL.

Mausbach, M.J. and J.L. Richardson, 1994. Biogeochemical processes in hydric soils. *In* p. 68–127. *Current topics in Wetland Biogeochemistry, Volume 1.* Wetland Biogeochemistry Inst. Louisiana State Univ., Baton Rouge.

McDowell, L.L., J.C. Stephens, and E.H. Stewart. 1969. Radiocarbon chronology of the Florida Everglades peat. *Soil Sci. Soc. Am. Proc.* 33:743–745.

Mitsch, W.J. and J.G. Gosselink. 1993. *Wetlands.* 2nd ed. Van Nostrand Reinhold, New York.

Moran, M.A., R. Benner, and R.E. Hodson. 1989. Kinetics of microbial degradation of vascular plant material in two wetland ecosystems. *Oecologia* 79:158–167.

Moore, P.D. and D.J. Bellamy. 1974. *Peatlands.* Springer-Verlag, New York.

Natural Resources Conservation Service. National Soil Survey Center. 1997. Acreage of each soil order in each state in the U.S., USDA–NRCS, Lincoln, NE., unpublished.

Natural Resources Conservation Service. 1996. Field indicators of hydric soils in the United States. G.W. Hurt, P.M. Whited, and R.F. Pringle (Eds.) USDA, NRCS, Fort Worth, TX.

Nelson, D.W. and L.E. Sommers. 1982. Total carbon, organic carbon, and organic matter. Methods of soil analysis. Part 2. 2nd ed. *In* A.L. Page et al. (Eds.), Agron. Monogr. 9. ASA and SSSA, Madison, WI.

Neue, H.U. 1985. Organic matter dynamics in wetland soils. pp. 109–122. *In Wetland Soils: Characterization, Classification, and Utilization.* Proc. of workshop, March 26–April 5, 1984, Manila, Philippines, International Rice Institute, Los Banos, Philippines.

Oades, J.M. 1988. The retention of organic matter in soils. *Biogeochemistry* 5:35–70.

Ovenden, L. 1990. Peat accumulation in northern wetlands. *Quatr. Res.* 33:377–386.

Ponnamperuma, F.N. 1972. The chemistry of submerged soils. *Adv. Agron.* 24:29–96.

Richardson, J.L. and R.J. Bigler. 1984. Principal component analysis of prairie pothole soils in North Dakota. *Soil Sci. Soc. Am. J.* 48: 1350–1355.

Richardson, J.L., T.J. Malterer, A. Gienke, J.L. Arndt, M.J. Rosek, and A.J. Duxbury. 1987. Classification problems associated with histic soils of calcareous fens. *Soil Survey Horizons* 28:53–55.

Richardson, J.L., J.L. Arndt, and J. Freeland. 1994. Wetland soils of the prairie potholes. *Adv. Agron.* 52:121–171.

Riger, S., D.B. Schoephorster, and C.E. Furbush. 1979. *Exploratory Soil Survey of Alaska.* USDA–SCS, Washington, DC.

Roe, H. B. 1963. A study of influence of depth of groundwater level on yields of crops grown on peat lands. Minn. Ag. Exp. Sta. Bull. 330, St. Paul, MN.

Schlesinger, W.H. 1977. Carbon balance in terrestrial detritus. *Annual Rev. Ecol. Syst.* 8:51–81.

Schnitzer, M. 1978. Humic substances: chemistry and reactions. *Dev. Soil Sci.* 8:1–64.

Siegel, D.I. 1983. Groundwater and the evolution of patterned mires, Glacial Lake Agassiz peatlands, northern Minnesota. *J. Ecol.* 71:913–921.

Soil Science Society of America. 1997. *Glossary of Soil Science Terms 1996.* Soil Science Society of America, Inc., Madison, WI.

Soil Survey Staff. 1975. *Soil Taxonomy: A Basic System of Soil Classification for Making and Interpreting Soil Surveys.* USDA–SCS Agr. Handbook 18. U.S. Govt. Printing Office, Washington, DC.

Soil Survey Staff. 1996a. *Soil Survey Laboratory Methods.* USDA–NRCS Soil Surv. Invest. Rep. 42, 3rd ed., U.S. Govt. Printing Office, Washington, DC.

Soil Survey Staff. 1996b. *Keys to Soil Taxonomy.* 7th ed. USDA–NRCS. U.S. Govt. Printing Office, Washington, DC.

Soil Survey Staff. 1998. *Keys to Soil Taxonomy.* 8th ed. USDA–NRCS. U.S. Govt. Printing Office, Washington, DC.

Steelik, C. 1985. Elemental characteristics of humic substances. pp. 457–476. *In* G.R. Aiken, D.M. McKnight, R.L. Wershaw, and P. MacCarthy (Eds.) *Humic Substances in Soil, Sediment, and Water.* John Wiley & Sons, New York.

Stephens, J.C. 1956. Subsidence of organic soils in the Florida Everglades. *Soil Sci. Soc. Am. Proc.* 20:77–80.

Stevenson, F.F. 1994. *Humus Chemistry: Genesis, Composition, Reactions.* 2nd ed. John Wiley & Sons, New York.

Tate, R.L. III. 1987. *Soil Organic Matter — Biological and Ecological Effects.* John Wiley & Sons, New York.

Trettin, C.C., M.F. Jurgensen, M.R. Gale, and J.W. McLaughlin. 1995. Soil carbon in northern forested wetlands: impacts of silvercultural practices. pp. 437–461. *In* W.W. McFee and J.M. Kelley (Eds.) *Carbon Forms and Functions in Forest Soils.* SSSA, Madison, WI.

Trettin, C.C., M. Davidian, M.F. Jurgensen, and R. Lea. 1996. Organic matter decomposition following harvesting and site preparation of a forested wetland. *Soil Sci. Soc. Am. J.* 60:1994–2003.

van der Valk, A.G. 1975. Floristic composition and structure of fen communities in northwest Iowa. *Proc. Iowa Acad. Sci.* 82 (2):113–118.

Vitt, D.H., P. Achuff, and R.E. Andrus. 1975. The vegetation and chemical properties of patterned fens in the Swan Hills, north central Alberta. *Can. J. Bot.* 53:2776–2795.

Walkley, A. 1934. A critical examination of a rapid method for determining organic carbon in soils. *Soil Sci.* 63:251–263.

Watts, W.A., B.C.S. Hansen, and E.C. Grimm. 1992. Camel Lake: A 40,000-Yr record of vegetational and forest history from northwest Florida. *Ecology* 73:1056–1066.

Watts, F.C. and S.W. Sprecher. 1995. Concepts and formation of hydric soils. pp. 33–41. *In* V.W. Carlisle (Ed.) *Hydric Soils of Florida Handbook,* 2nd ed., Florida Assoc. of Environmental Soil Scientists, Gainesville, FL.

Wetzel, R.G. 1984. Detrital dissolved and particulate organic carbon functions in aquatic systems. *Bull. of Marine Sci.* 35:503–509.

Wetzel, R.G. 1992. Gradient-dominated ecosystems: sources and regulatory functions of dissolved organic matter in freshwater ecosystems. *Hydrobiol.* 229:181–198.

Zeikus, J.G. 1981. Lignin metabolism and the carbon cycle. pp. 211–243. *In* M. Alexander (Ed.) *Advances in Microbial Ecology, Volume 5.* Plenum Press, New York.

CHAPTER 7

Morphological Features of Seasonally Reduced Soils

M.J. Vepraskas

INTRODUCTION

Hydric soils are created by oxidation–reduction (redox) chemical reactions that occur when a soil is anaerobic and chemically reduced. The redox reactions produce signs in the soil that they have occurred, and these signs are described in this chapter as *morphological features of reduction* and *hydric soil field indicators* (Hurt et al. 1998). A "reduced" soil is one in which redox reactions have caused reduced forms of O, N, Mn, Fe, or S to be present in the soil solution. "Reduced" is a general term that implies that some elements in addition to O_2 are present in their reduced form. Common reduced forms of elements or compounds that are found in hydric soils include: H_2O, N_2, Mn^{2+}, Fe^{2+}, and H_2S, while their oxidized counterparts are O_2, NO_3^{2-}, MnO_4, $FeOOH$, and SO_4^{2-}, respectively. This chapter will focus on morphological features that form in soils that have been reduced periodically or seasonally. Hydric soil field indicators are the subject of Chapter 8.

Morphological features of seasonally reduced soils include specific color patterns, odors, color changes that occur on exposure to air, or a specific kind of organic material. These features can occur at any depth in the soil, and the abundance of a given feature is variable. The features are direct indicators that the soil was reduced at some point in its history, and therefore, they will be referred to as morphological features of reduction in this chapter. Morphological features of reduction have also been used to estimate which part of the soil is seasonally saturated with free water (Cogger and Kennedy 1992, Franzmeier et al. 1983).

On the other hand, most hydric soil field indicators are soil layers with precisely defined colors, thicknesses, and depths that contain morphological features of reduction in specific amounts. As their name suggests, the field indicators were developed solely to identify hydric soils on-site. Morphological features of reduction are components of hydric soil field indicators, but the two terms are not interchangeable. For example, when Fe hydroxides accumulate around root channels in sufficient quantities to be visible, they form the morphological feature of reduction called an Fe *pore lining*. If these pore linings occupy 3% of a sandy loam soil layer whose matrix has a Munsell color of 5YR 3/1, is 10 cm or more thick and lies entirely within the upper 30 cm of the soil, then the layer qualifies as a Hydric Soil Field Indicator termed a *Redox Dark Surface* (Hurt et al. 1998).

A soil that contains morphological indicators of reduction may not be a hydric soil if the features occur too deeply, yet they still indicate that the soil has experienced chemical reduction at some point in its history.

The purpose of this chapter is threefold: (i) to discuss the principal redox reactions and soil conditions needed to form morphological features of reduced soils and field indicators of hydric soils; (ii) to identify the principal types of morphological features of reduction and review their formation; and (iii) to discuss the ways these features can be interpreted.

IMPORTANT CHEMICAL REACTIONS

Principle Elements Involved

In order to understand how morphological features of reduction form, it is useful to simplify oxidation and reduction processes and consider them to be separate reactions even though the two types of reactions occur simultaneously in complex biochemical processes. As explained in Chapters 4 and 5, oxidation–reduction reactions in most soils begin when bacteria oxidize organic compounds to release electrons and protons in the form of H^+ cations. The basic oxidizing reaction that breaks down organic compounds was presented in Chapter 4. The electrons and protons released by the oxidation of organic compounds react with electron acceptors to complete the microbial respiration process.

The principal reducing reactions that form the morphological features of reduction involve four elements (O, Mn, Fe, and S), which are used as electron acceptors in bacterial respiration. As shown in Table 7.1, there are also four basic groups of features that are associated with the principal reducing chemical reactions. These feature groups will be identified by the major element related to their formation: (a) organic-C based features, (b) Mn-based features, (c) Fe-based features, and (d) S-based features. Selected examples of morphological features of reduction are given for illustration in Table 7.1. The reactions for Mn and Fe are reversible and produce different features as the reactions proceed in either direction.

There is a tendency among some wetland scientists to consider one of the four groups shown in Table 7.1 as inherently better or more reliable in identifying soils that have been reduced. This tendency must be avoided, because the preferred group is usually only the one with which they are most familiar. Each group of features is equivalent in showing that reducing reactions have occurred in the soil.

Table 7.1 Three of the Major Reducing Reactions Related to the Development of Hydric Soils

Reducing Reaction	Approximate Eh (pH 7)* mV	Morphological Feature Formed	
		Group Name	Examples
$O_2 + 4e^- + 4H^+ \rightarrow 2H_2O^{**}$	600	Organic C-based features	Oe, Oa, and some black A horizons
$MnO_2 + 2e^- + 4H^+ \leftrightarrow Mn^{2+} + 2H_2O$	300	Mn-based features	Mn masses and some depletions (black and gray mottles)
$2FeOOH + 4e^- + 6H^+ \leftrightarrow 2Fe^{2+} + 4H_2O$	100	Fe-based features	Fe masses and Fe depletions (red, yellow, and gray mottles)
$SO_4^{2-} + 8e^- + 10H^+ \rightarrow H_2S + 4H_2O$	−200	S-based features	Odor of rotten eggs

* Data from McBride, M.B. 1994. *Environmental Chemistry of Soils*. Oxford Univ. Press, New York.
**This reaction occurs under aerobic conditions, and it is not until the O_2 is depleted that organic C accumulates.

It is useful to place features in these groups because some soils are more likely to have one group of features than others. For example, some sands have virtually no Fe in them because the minerals found in the parent material simply did not contain Fe. The morphological features of reduction that will be found in such soils will consist of organic-C based features, Mn-based features, and occasionally the S-based feature. It makes no sense to search these Fe-poor soils for Fe-based signs of redox reactions. Remember that the features in the four groups are equivalent in showing that reduction has occurred in the soil.

Relation of Features to Eh

The formation of one of the groups of features shown in Table 7.1 requires that the redox potential fall to a certain Eh value. When an aerated soil becomes saturated, the reducing reactions proceed in the order shown and progress from higher Eh to lower Eh. An example of Eh fluctuation over a portion of the year is shown in Figure 7.1 to illustrate the relationship between seasonal Eh fluctuations and feature formation. When the soil is unsaturated and molecular O_2 is present in soil pores, the Eh is relatively high (>500 mv at soil pH 7), and none of the morphological features related to reduction shown in Table 7.1 are forming. Soils in this condition are described as oxidized or aerated. When the soils saturate to the surface, the movement of molecular O_2 from the atmosphere into the soil stops. Bacteria that are still respiring by oxidizing organic compounds can reduce dissolved O_2 in the water. As shown in Figure 7.1, once all the dissolved O_2 has been depleted, the redox potential falls below 500 mv, and bacteria must use other electron acceptors to survive. Morphological features formed by the buildup of organic material can begin to develop at this point because decomposition of organic tissues slows down under anaerobic conditions (Chapters 5 and 6).

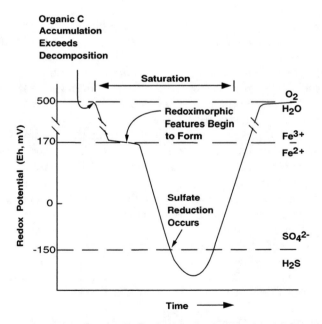

Figure 7.1 Hypothetical changes in redox potential (Eh) over the course of a single "wet" season where the water table rises, inducing reduction in an aerated soil, and then falls, causing the soil to reoxidize. The morphological features of reduction that would be expected to form at each change of Eh are also shown. For this scenario to occur, the soil must contain respiring bacteria that are oxidizing organic C materials.

The Eh continues to fall as long as saturation is maintained and bacteria continue to respire. When the Eh reaches approximately 170 mv (pH 7), the Fe^{3+} ions in some minerals will reduce to Fe^{2+} and dissolve into the soil solution. The soluble Fe^{2+} may diffuse through the soil and concentrate or it may move with the water and be taken out of the soil horizon. As long as the Eh stays below 170 mv, the Fe^{2+} will remain reduced in most cases. The immediate change in the soil that occurs following Fe reduction is that the portion of the soil where reduction occurred will become grayer in color. The gray color occurs because Fe^{2+} is colorless, and the actual color of the soil is determined by the color of the sand, silt, and clay particles in it.

Once all the Fe has been reduced, the Eh will continue to fall, and when it reaches −150 mv, SO_4^{2-} anions may be reduced to H_2S gas. This usually requires a relatively long period of saturation and anaerobic respiration. The gas is produced only while the Eh is below −150 mv.

When the soil drains, O_2 enters the soil and the Eh increases. The production of H_2S ceases, and reduced Fe is oxidized to Fe oxide or hydroxide minerals, which produce the red, yellow, or brown colors seen in many subsoil horizons. Above an Eh of 500 mv, aerobic organisms respire and oxidize undecomposed tissues to CO_2 and water.

BASIC KINDS OF FEATURES

The widespread features that have been found in reduced soils will be described for each of the three principal groups of features shown in Table 7.1. Additional information on features related to organic matter are included in Chapters 6 and 8, while those related to H_2S gas formation were discussed in Chapter 5.

Organic C-Based Features

All organic C-based features consist of one of three kinds of materials that were defined in Chapter 1: organic soil material, mucky mineral soil material, or mineral soil material with a black color. These materials form either distinct horizons (O or A horizons) or occur as aggregates of organic-rich material. O horizon thickness and state of decomposition must be considered when identifying hydric soils. O horizons having a thickness >20 cm and a black or very dark gray color, regardless of the state of decomposition, are in most cases found only in soils that were periodically reduced (Hurt et al. 1998). Thinner layers can also be found in reduced soils, but their thickness and state of decomposition requirements vary for different textural groups and Land Resource Regions, as described in Chapter 8. A horizons consist of mineral (occasionally mucky mineral) soil material, and those that formed in reduced soils have moist Munsell colors with a value of 3 or less and a chroma of 3 or less. The dark color is a direct result of relatively high amounts of organic matter that accumulated under reduced conditions. However, not all soils having dark colors are necessarily saturated and reduced. This is particularly true of the Mollisols found in the midwestern U.S. (Bell and Richardson 1997).

Another group of organic C-based morphological features related to reduction are aggregates of organic materials called *organic bodies* that form around roots. They may be found in mineral horizons or within or just below O horizons lying on the surface. These features consist of organic material or mucky mineral materials as described in Chapter 8.

Iron-Based Morphological Features Related to Reduction

Redoximorphic features are formed by the reduction, movement, and oxidation of Fe and Mn compounds. These features form the gray, red, yellow, brown, or black mottled color patterns that are normally associated with saturated and reduced soils. Redoximorphic features are the most widespread morphological features formed by reduction. There are three basic kinds of redoximorphic features: *redox concentrations*, *redox depletions*, and the *reduced matrix*.

Redox reactions affect Fe and Mn similarly, and the two elements frequently occur together, as noted in Chapter 4. Iron is usually in greater abundance than Mn, but small quantities of Mn can cause some redox concentrations to appear black (Gallaher et al. 1974, Rhoton et al. 1993). Black-colored redox concentrations may be confused with decomposed organic tissue. However, Mn can be detected by spraying the redox concentration with a solution of H_2O_2 (3% concentration). A rapid bubbling of the H_2O_2 solution confirms that Mn is present. The reaction is:

$$MnO_4 + H_2O_2 \rightarrow Mn(OH)_2 + 2O_2$$

where Mn is reduced by the peroxide (Jackson 1965). While a 3% solution of H_2O_2 will react with soil organic matter, the reaction is slower than for Mn, although it can be sped up by heating. Manganese can be abundant in certain soils, such as those having pHs > 7, or in some clays having Munsell hues of 5YR or redder (e.g., Moreland series reported in Hudnall et al. 1990). When Mn is abundant, it can prevent the reduction of Fe and formation of gray soil colors because it is reduced before Fe (McBride 1994). Such Mn-rich soils are probably of small extent, but can be important in certain regions. The remainder of this discussion will focus on Fe, but Mn should be assumed to be included as well.

Redox Concentrations

Redox concentrations are features formed when Fe oxides or hydroxides have accumulated at a point or around a large pore such as a root channel. They have been defined as "bodies of apparent accumulation of Fe–Mn oxides and hydroxides" (Vepraskas 1996). This means that they appear to have formed by Fe or Mn moving into an area, oxidizing, and precipitating. Redox concentrations contain more Fe^{3+} oxides and hydroxides than were found in the soil matrix originally. Three kinds of redox concentrations have been defined: Fe masses, Fe pore linings, and Fe nodules and concretions. These differ in their hardness and also in where they occur in the soil.

Iron masses (Plate 7) are simply soft accumulations of Fe^{3+} oxides and hydroxides that occur in the soil matrix, away from cracks or root channels. They can be of any shape. The masses are soft and easily crushed with the fingers because the concentration of Fe is not great enough to cement the soil particles into a solid mass. Sizes of Fe masses range from 1 mm to over 15 cm in diameter. Because they are found in the matrix, the size of the Fe masses is usually determined by the size of the peds or structural aggregates in the soil which fix the maximize size for the features.

The color of the Fe masses is variable and can be any shade of red, orange, yellow, or brown. The color varies with the type of Fe mineral present. The most common Fe minerals found in Fe masses are goethite, ferrihydrite, and lepidocrocite (Schwertmann and Taylor 1989). These minerals impart hues of 10YR, 7.5YR, and 5YR, respectively. Common value/chroma combinations include 5/6 and 5/8, but other combinations can be found.

Pore linings (Plates 8 and 9) are accumulations of Fe oxides and hydroxides that lie along ped surfaces or root channels. These features are in the soil and not directly on the root. They are similar to oxidized rhizospheres, but whereas oxidized rhizospheres are thought to form on root tissue while the root is alive (Mendelssohn et al. 1995), pore linings do not need a live root in order to form. The distinction between pore linings and oxidized rhizospheres is not important for identifying hydric soils. However, if one needs to identify wetland hydrology, which currently requires the soil to be saturated during the growing season when plants are growing (Environmental Laboratory 1987), then only oxidized rhizospheres can be used because pore linings could develop outside the growing season when soils are reduced and become oxidized as the water table falls (Megonigal et al. 1996).

Pore linings differ from iron masses only in where they occur in the soil: masses occur in the matrix, while the pore linings must be along root channels or cracks. The colors of the two features are similar. Pore linings are generally soft, but in extreme cases the Fe content has reached a level that cements the soil particles together around a root channel. The cemented feature has been called a *pipestem* because it is usually cylindrical and has a small channel running down its axis resembling the shaft of a smoker's pipe (Bidwell et al. 1968).

Nodules and concretions (Plate 10) are hard, generally spherical-shaped bodies made of soil particles cemented by Fe oxides or hydroxides. They range in size from less than 1 mm to over 15 cm in diameter. When broken in half and examined, the concretions are seen to consist of concentric layers like an onion, while no layers are seen in nodules. Most people seem to use the two terms interchangeably, and there is no special significance attached to the layered structure other than it shows that the concretion formed in episodes over time.

The nodules and concretions are difficult to destroy because of their hardness. When they are found in soils, it is never clear whether these features formed in place or were brought into the soil by flooding or by deposition of material eroded from upslope. For this reason, nodules and concretions cannot be considered as reliable indicators of the processes that still occur seasonally in the soil.

Redox Depletions

Redox depletions are zones formed by loss of Fe and other components. They have been defined as "bodies of low chroma (<2), having values of 4 or more where Fe–Mn oxides alone have been stripped out or where both Fe–Mn oxides and clay have been stripped out" (Vepraskas 1996). This definition is used to meet the soil classification requirements for aquic conditions as set forth in *Soil Taxonomy* (Soil Survey Staff 1998). Redox depletions in principle could form with chromas >2 as long as they developed in a soil horizon whose matrix lost Fe by reduction processes.

Two different kinds of redox concentrations have been defined, *Fe depletions* (Plates 11 and 12) and *clay depletions*, and these differ only in whether their texture is similar to that of the matrix or not. Iron depletions form simply by a loss of Fe (and Mn) from a portion of the soil. They have been defined as "low chroma bodies (chromas <2) with clay content similar to that of the adjacent matrix" (Vepraskas 1996). Similar features have also been called gray mottles, gley mottles, albans, and neoalbans (Veneman et al. 1976). Iron depletions frequently occur along root channels and ped surfaces in B horizons. They can occur in the matrix, particularly in A horizons as shown in Plate 13. In some cases the entire matrix is an Fe depletion, such as in E and B horizons of soils that are reduced for long periods.

Clay depletions form by a loss of both Fe and clay. These features have both a low chroma and a coarser texture than found in the adjacent soil matrix. Clay depletions have also been described as silt coatings, skeletans, and neoskeletans (Brewer 1964, Vepraskas and Wilding 1983). They almost always occur along ped surfaces or large root channels. Similar features that occur in the soil matrix were probably formed by silt or sand falling down into and filling a channel. These are technically not clay depletions because their formation is not related to oxidation–reduction chemical reactions. Clay depletions have not been reported within the upper 30 cm of hydric soils and are less important than Fe depletions for hydric soil identification.

Reduced Matrix

The reduced matrix has been defined as a soil matrix that has a "low chroma color *in situ* because of the presence of Fe^{2+}, but whose color changes in hue or chroma when exposed to air as the Fe^{2+} is oxidized to Fe^{3+}" (Vepraskas 1996). The time needed for the color change to occur was set for practical reasons at 30 min, because waiting longer would interfere with the completion of field work. However, there are little data on how long such color changes require.

In principle, the reduced matrix could also be detected by spraying a field moist sample of soil with a dye such as α, α'-dipyridyl that reacts with Fe^{2+}. If a positive reaction with the dye occurs, then it might be assumed that a reduced matrix is also present. The developers of *Soil Taxonomy* (Soil Survey Staff 1998) incorporated this technique as a substitute for redox concentrations if no such features are present. It is also a test to show that the soil is reduced.

Soils that have a reduced matrix during the wet season may develop other morphological features during other periods of the year when the soils drain and oxygen enters some pores. At the time that the reduced matrix occurs in a soil, it is likely that some or all of the redox concentrations (especially pore linings and Fe masses) that were present in the soil prior to the time it became reduced will have dissolved. The dissolved redox concentrations may reform after the water table falls and the Fe^{2+} cations are oxidized. Some hydric soil field indicators, such as the Redox Dark Surface (Hurt et al. 1998), require, in addition to a black matrix color, that a certain percentage of redox concentrations be present before the indicator is met. If the soil is examined at a time when only the reduced matrix is present, the soil may not meet a hydric soil field indicator even though it is actually saturated and reduced at the time of observation. To overcome this problem, it is recommended that a positive test for Fe^{2+} be used as a substitute for redox concentrations in soil descriptions, particularly when hydric soils are being examined.

Gley Soil Colors

Gley colors as defined in Chapter 1 can be one of two types of redoximorphic features, either a reduced matrix or depleted matrix, depending on whether the color changes when exposed to air. Gley colors consist of Munsell hues found on the "Gley Pages" of a Munsell Color Chart.

The gley hues can, in some cases, be unique minerals that contain a reduced form of Fe that is combined with an anion of phosphate, sulfate, carbonate, or other compound (Schwertmann and Taylor 1989). These hues are a morphological indicator of reduction when their values are 4 or more. When values are less than 4, the colors are probably too dark to accurately separate gley hues from other kinds. If the color of the soil does not change upon air drying, the material with the gley hue is probably a redox depletion (Fe depletion). If a color change (e.g., reddening of the matrix) occurs upon oven drying, the material is probably a reduced matrix even though the color change did not occur within 30 minutes of the sample being removed from the soil.

FORMATION OF REDOXIMORPHIC FEATURES

Formation of the organic features related to reduction were described in detail in Chapter 6 and the production of H_2S was described in Chapters 4 and 5, and these processes will not be repeated here. This section will focus only on the formation of redoximorphic features. Redoximorphic features form after one or more of the following three processes have occurred: (1) Fe^{3+} cations in oxides or hydroxides have been reduced, (2) the solubilized Fe^{2+} has moved to another portion of the soil, and (3) the Fe^{2+} has been oxidized to form an Fe mass, pore lining, or nodule. A reduced matrix requires that only the first step occur. Iron depletions require that the first two steps occur, while formation of redox concentrations require that all three steps occur.

Redox Depletions

The process of redox depletion formation is shown in Figure 7.2 for an Fe depletion. To begin the process, assume that the soil matrix had a uniform brown color throughout before an Fe depletion formed. The color came from Fe^{3+} oxide or hydroxide minerals that coated soil particles, such as sand, silt, and clay grains. Each of these grains had a coating of an Fe^{3+} compound that effectively painted the particle surface brown.

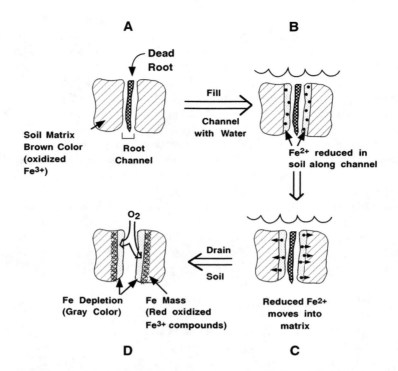

Figure 7.2 Formation of redox depletions (Fe depletions) around a root channel. Initially (A) the soil is uniformly red throughout its matrix. The channel contains a dead root, which is being decomposed by bacteria. Upon flooding (B), the channel is filled with water, and all O_2 dissolved in the water is reduced. When the water is anaerobic, the bacteria reduce Fe oxides and hydroxides in the soil surrounding the channel. The Fe^{2+} dissolves and moves away from the channel (C) leaving the soil particles around the channel stripped of the Fe coatings. If the stripped grains are largely composed of quartz and/or uncoated clay minerals, they will be gray in color (D). The Fe^{2+} may subsequently oxidize to form an Fe mass. (Adapted from Vepraskas, M.J. 1996. *Redoximorphic Features for Identifying Aquic Conditions.* Tech. Bull. 301. NC Agric. Exp. Sta., Raleigh, NC.)

The soil shown in Figure 7.2A has a channel containing a dead root that is being decomposed by bacteria. If the channel is filled with air, the oxidation of the organic matter releases electrons which are used to reduce O_2 to water. As long as the channel contains air, the only reduction that takes place is that of O_2 to water, and there is no change in the color of the soil around the root.

If the channel fills with water, however, the supply of O_2 from the atmosphere is cut off (Figure 7.2B). Bacterial respiration still occurs, and the organic tissue continues to be oxidized. The O_2 dissolved in the soil water is quickly depleted because it is the primary electron acceptor used in the respiration process. After the water becomes anaerobic, the bacteria must use another element to accept electrons to sustain their respiration process. As noted in Chapter 4, the elements used for reduction occur in a sequence, but for this discussion we will focus on Fe.

When the electrons are transferred to Fe^{3+} atoms in minerals, the reduction of Fe causes two changes in the minerals: they lose their color and they dissolve. The dissolved Fe^{2+} moves off the particle surfaces and may diffuse through the soil matrix or may be carried to other parts of the soil in moving water (Figure 7.2C). As Fe^{2+} leaves the particle surfaces, the color of the soil around the channel changes and the soil gradually becomes gray in color. In Munsell terminology the soil's chroma decreases and its value increases until all Fe has been removed from the surfaces of particles lying near the channel. When the soil drains and O_2 enters the soil, the newly formed Fe depletion will retain its gray color because it is the color of the uncoated minerals. Oxygen penetrating into the matrix may oxidize the Fe^{2+} and cause a pore lining or Fe mass to form (Figure 7.2D).

Stripping the soil particles of Fe^{3+} mineral coatings changes the soil's color to the natural color of the soil minerals. Normally this is a gray color when the particles consist of quartz or a clay mineral such as kaolinite. This gray color is relatively permanent and is not affected appreciably by additional periods of saturation and reduction. The only way the color of the gray soil particles could become brown again is if Fe^{2+} were moved onto the particle surfaces and reoxidized to recoat the particle surfaces with more of the "paint" composed of Fe^{3+} minerals.

This is the basic process that forms redox depletions. It is easiest to see in the field when the organic matter occurs as a root that is not near other roots as in Plate 12. In such cases the gray depletions form cylindrical features around root channels. However, if roots are closely spaced along a crack or ped surface, then the depletions will have a planar shape as they coat the surface of the crack, or they may occupy entire layers.

In A horizons (Plate 13) the organic matter that starts depletion formation can be a piece of leaf tissue or a fragment of some other part of the plant. This is an isolated source of organic tissue. The depletions that form around this tissue tend to be spherical. The process that forms the depletion is the same as that shown in Figure 7.2.

Redox Concentrations

Redox concentrations can occur both in the matrix as Fe masses or Fe nodules, and around macropores such as root channels in the case of Fe pore linings. Redox concentrations form when Fe^{2+} in solution moves through the soil toward points of oxidation and precipitates. Points of oxidation can occur in reduced soil where: (1) O_2 enters the soil after a soil drains, (2) when entrapped air is present, or (3) when roots release O_2 to the soil matrix when Fe^{2+} is present. Figure 7.2D illustrates a case where Fe^{2+} diffused into a soil matrix and oxidized where the Fe^{2+} encountered entrapped oxygen, which may occur as an air bubble in the saturated soil. Figure 7.3 illustrates two cases, the first (Figure 7.3A) being where points of oxidation occur around roots that have O_2 transported to them. This process is one way in which pore linings form. The second example (Figure 7.3B) illustrates a case in which O_2 penetrates along macropores such as cracks or root channels and forms both Fe pore linings and Fe masses as Fe^{2+} diffuses to points where O_2 occurs. The formation of Fe pore linings and masses illustrated above has been modeled in laboratory experiments that simulated field conditions (Vepraskas and Bouma 1976). To this author's knowledge no experiments have simulated the formation of Fe nodules and concretions. Apparently these features form slowly over time by repeated episodes of Fe oxidation at the same points in the soil, such as at the interiors of peds where oxygen is entrapped when the soil saturates.

Reduced Matrix

The reduced matrix occurs in soils by a process similar to that shown in Figure 7.2. Once reduced, the Fe^{2+} may remain in place or it may move to portions of the soil and concentrate. This author has seen reduced matrices along cracks in clay soils (Vertisols) where Fe^{2+} produced along a crack was not able to diffuse away from the crack. The time required to form a reduced matrix has not been determined.

Effects of Texture on Redoximorphic Feature Appearance

Despite forming by similar processes, Fe depletions in sands usually appear different from similar features formed in a loam or clay. This can be seen by comparing the soils in Plates 12 and 14. Plate 12 shows a redox depletion around a root channel that formed by the process described in Figure 7.2. Plate 14 shows a redox depletion in a sand that formed by the same process but appears as a roughly circular gray area that is sometimes described as a "splotchy pattern." The Fe

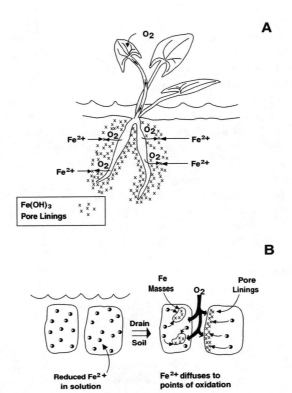

Figure 7.3   Formation of redox concentrations by two different processes. In A, the Fe^{2+} is oxidized around roots which are bringing O_2 into the flooded and reduced soil. Pore linings are the type of redox concentration formed. In B, the Fe^{2+} is oxidized in the matrix after the soil has drained, and O_2 has been able to penetrate into reduced portions of the matrix. Iron masses are formed in this case. It is also possible for Fe masses to form where air has been trapped inside peds. (Adapted from Vepraskas, M.J. 1996. *Redoximorphic Features for Identifying Aquic Conditions.* Tech. Bull. 301. NC Agric. Exp. Sta., Raleigh, NC.)

depletions that appear in Plates 12 and 14 both formed by the process shown in Figure 7.2. They look different because sands do not have large, stable root channels or cracks that remain open for long periods to allow large features to form around the same points in the soil. Root channels in sand remain open while they contain a root, but collapse shortly after the root dies and decomposes. As a result, redox depletions develop around a single root, and their shape is determined by the arrangements of stripped sand grains (i.e., those free of Fe coatings) that fell into the collapsing channel after the root decomposed. Furthermore, the low amounts of Fe oxides in sands causes the contrast between the matrix and the depletions to be less than in more Fe-rich loams and clays.

Time Needed to Form Redoximorphic Features

In order to become reduced, a soil must: (1) be saturated with water to exclude O_2 from the atmosphere, (2) contain actively respiring bacteria, and (3) be depleted of dissolved O_2. If any of these conditions are not met, the reduction of Fe will not occur. To achieve these conditions, an adequate supply of decomposable organic C must be available, the soil water should be stagnant, and the soil temperatures must be above approximately 5°C. If organic C levels are too low, there may not be sufficient microbial respiration to deplete the soil water of oxygen even when the soil is saturated. Moving water tends to carry oxygen into the soil and retards the onset of Fe reduction (Cogger and Kennedy 1992, Gilman 1994). Furthermore, it is generally believed that at temperatures

Table 7.2   Period Required for Saturated Soil Cores of Different Organic C Percentages to Develop Fe-Reducing Conditions Under Three Different Soil Temperatures

Soil Core	Organic C (%)	Soil Temperature		
		23°C	9°C (days)	4°C
X	7.5	6(1–20)*	37(22–95)	74(43–120)
Y	2.5	30(10–43)	97(40–140)	160(151–164)
Z	0.8	53(37–72)	97(80–147)	160(47–>180)

* Mean (range).

Adapted from Cogger, C.G. and P.E. Kennedy. 1992. Seasonally saturated soils in the Puget Lowland. I. Saturation, reduction, and color patterns. *Soil Sci.* 153(6):421–433.

<5°C microbial respiration will be too slow to deplete the soil water of oxygen (Megonigal et al. 1996). This 5°C threshold is a general one that works best for plants whose roots have their maximum elongation rate at a soil temperature of 20 to 30°C (Russell 1977). The 5°C threshold is less applicable to organisms adapted to life in colder soils.

The time required for Fe reduction to occur after initiation of saturation or inundation depends on soil conditions. Meek et al. (1968) detected Fe^{2+} in solution after one day of ponding in field plots (1.2 by 1.2 m) to which chopped alfalfa had been added. The amount of Fe^{2+} in solution reached its peak of 5 to 30 mg/L at approximately 4 to 5 days following the initial ponding. Ponnamperuma (1972) reported that in acid soils "high in organic matter" the peak in Fe^{2+} occurs with 1 to 3 weeks of ponding. Other field studies have shown that Fe reduction may be delayed by up to 4 weeks following saturation, and may not occur at all, depending upon soil conditions (Hayes 1998).

The effect that organic matter content and temperature have on the time it takes for Fe reduction to occur is shown in Table 7.2 (Cogger and Kennedy 1992). These data show that there is a lag between the onset of saturation and the onset of Fe reduction, and that the length of the lag period depends on both soil temperature and organic matter percentage. These two factors directly influence the rate of microbial activity. The data in Table 7.2 illustrate why two soils that are saturated for the same length of time could develop widely different amounts of low chroma or gray color as a result of Fe reduction. For example, assume that two soils (represented by cores X and Y in Table 7.2) were saturated for 100 days each year. Further assume that soil horizon X became saturated when its soil temperature was 23°C, while soil horizon Y became saturated when its soil temperature was 9°C. Iron reduction would be expected to last for 94 days in soil X, but for only 3 days in soil Y before the water tables fell in each soil. If the amount of gray color produced in each soil is directly proportional to the length of time they are reduced, then we would expect the amount of gray color seen in horizon X to be over 30 times that seen in horizon Y, despite both horizons being saturated for identical lengths of time.

The soluble Fe^{2+} can move through the soil with moving water or by diffusion. Vepraskas and Guertal (1992) modeled the formation of Fe depletions and found that diffusion is responsible for most of the Fe loss. This is because in many cases water in wetland soils tends to be stagnant.

The oxidation of Fe^{2+} can occur quickly. Ahmad and Nye (1990) showed in laboratory experiments that after 8 h approximately 78% of the Fe^{2+} in both solution and suspension had oxidized at 20°C. Additional experiments with suspensions kept at a pH of 5.75 showed that approximately 60% of the Fe^{2+} had oxidized within 3 h. These results agree with field observations. For example, the reduced matrix is detected by a visible color change that is expected to occur within 30 minutes of exposure to air (Soil Survey Staff 1998). Movement of sufficient Fe^{2+} through the soil and its oxidation to form visible features has been found to occur around the roots of rice seedlings growing in a flooded field in 7 d (Chen et al. 1980).

Table 7.3 Changes in the Quantity of Redox Depletions (features having Munsell values of 4 or more and chromas of 2 or less) in Soils Along and in a Created Deep Marsh (The hydrology was controlled by pumping)

Soil depth (cm)	Saturation (% of year)	Organic C (% by weight)	Redox Depletions		
			Original Soil	After 3 yrs.	After 5 yrs.
			(% by volume)		
Marsh					
18 to 25	100	1.1	50	90	88
25 to 58	100	0.5	50	80	89
Edge of Marsh					
10 to 30	100	1.3	0	40	75
30 to 53	100	0.8	50	50	70
Transition to Upland					
13 to 23	30*	1.5	0	85	70
23 to 41	30	1.0	0	75	60

* Estimated from bimonthly water table data.

Constructed Wetlands

Rates of redoximorphic feature formation were studied under field conditions across the edge of a created deep marsh near Chicago, Illinois, by Vepraskas et al. (1999), and results are shown in Table 7.3. The hydrology of the site was controlled by pumping which brought water to the marsh. Soil horizons were described at three locations: in the marsh, at the edge of the marsh, and in a transition zone bordering the upland. The amount of redox depletions increased over the 5-year period (Table 7.3), but rapid changes occurred within the first 3 years. The transition zone developed over 70% redox depletions within 3 years because the original soil matrix had a chroma of 3, and developing the redox depletions required losing enough Fe to produce a chroma of 2. Redox depletions decreased slightly by the 5th year because the water table had dropped in this transition zone after two relatively dry years. The data in Table 7.3 show that detectable changes in redox depletions occur quickly following changes in the soil hydrology.

A companion study to that of the deep marsh was conducted along a created floodplain adjacent to a created channel that had dams to control water entry and exit. This allowed the number and duration of floods to be controlled. The topsoil applied to plots on the created floodplain was mixed and applied after the floodplain contours had been graded. The purpose of the study was to determine whether redox depletions could form in A horizons that were inundated by short-term floods. The soils were Mollic Endoaquents containing 2% organic C in the A horizon.

Results of the study are shown in Table 7.4. The first induced flood lasted 8 days and produced redox depletions like that shown in Plate 13 that occupied approximately 2% of the horizon's volume. Over the next 3 years, more depletions were formed as the number of floods increased. This study showed that such depletions can form after even a single inundation. On the other hand, after 3 years the flooding was stopped and the depletions began to disappear.

Ditching Effects

Hayes (1998) evaluated changes in soil morphology in a Coastal Plain landscape (interstream divide) at four different distances from a ditch. He found that after 30 years the Bt horizons of soils within 30 m of the ditch had significantly (0.10 level) greater amounts of redox concentrations than soils farther from the ditch (Table 7.5). The near doubling of redox concentrations was related

Table 7.4   Formation of Redox Depletions in the A Horizon of Soils Along a Created Floodplain as a Function of Flood Frequency and Duration (The soil was classified as a Typic Endoaquoll and A horizon had a pH of 7 with 2.3% organic C)

	Year of Study		
	1992	**1993**	**1994**
No. of floods	2	5	2
Flood durations (days)	7 to 11	4 to 44	13 to 14
Redox depletion characteristics			
Abundance (%)	2	7	27
Color (moist)	2.5Y 4/1	5Y 4/1	2.5Y 4/1
Size (mm)	2 to 10	2 to 35	2 to 20

to the soils nearest the ditch being reduced for significantly shorter periods of time. Reduced Fe in groundwater flowing toward the ditch precipitated in the soils near the ditch. The duration of saturation was not affected by the ditch, because while the ditch removed groundwater from the soils within 30 m, the drained pore space was apparently filled by aerated surface water flowing laterally through the A and O horizons toward the ditch. These results show that relatively large changes in soil morphology can occur following even small changes in hydrology.

INTERPRETING MORPHOLOGICAL FEATURES OF REDUCTION

Morphological features of reduction simply show that the soil has been reduced at some point in its past. For example, many organic C-based indicators show where reduction has occurred, as do redox depletions. The reduced matrix and an odor of H_2S indicate that the soil is currently reduced at the place these features are detected. On the other hand, redox concentrations indicate where oxidation has occurred in the past. By themselves, these features give no indication of how long the soils were saturated and reduced.

Occasionally more information is desired, particularly an estimate as to whether and for how long the soils become saturated in a year of normal rainfall. Assessment of the duration of saturation is necessary for some uses, such as onsite waste disposal using septic systems. This information can be inferred by using morphological features of reduction that have been correlated to measurements of saturation.

Table 7.5   Effect of a Drainage Ditch on Durations of Saturation, Fe Reduction, and Redox Concentrations for a Typic Paleaquult in the Coastal Plain Region of North Carolina

Distance from ditch (m)	Duration of:		Redox Concentrations (% by volume)
	Saturation	**Fe Reduction**	
	(% of year)		
7	41a*	13a	39a
30	44a	22a	38a
60	44a	39b	16b
80	45a	34b	20b

* Numbers within the same column that are followed by the same letter were not significantly different at the 0.10 level, as determined by Tukey's *w* procedure.

Note: Data for saturation and reduction were determined at 60 cm for one year, while the abundance of redox concentrations was determined for the depths between 40 and 60 cm by Hayes (1998).

Table 7.6   Relationship of Durations of Saturation and Fe Reduction to the Percentage of Redox Depletions in Two Soils Along a Hillslope

| Landscape Position* | Depth (cm) | Durations of: | | Redox Depletions (% by weight) |
| | | Saturation | Fe Reduction | |
		(% of year)		
Backslope	143–170	38	5	18
Toeslope	118–143	46	28	79

* Soil on the backslope is classified as a Plinthic Paleudalf, and the soil in the toeslope position is classified as a Fragic Glossudalf.

Note: The data show that the percentage of redox depletions is more directly related to the duration of Fe reduction than to the duration of saturation.

Data from Vepraskas, M.J. and L.P. Wilding. 1983. Albic neoskeletans in argillic horizons as indices of seasonal saturation and iron reduction. *Soil Sci. Soc. Am. J.* 47:1202–1208.

Relating Feature Abundance to Duration of Saturation and Reduction

Morphological features that form in reduced soils range widely in their abundance. Abundance is directly related to how long the soils have been reduced, but indirectly related to how long soils have been saturated. This is because soils do not become reduced as soon as saturation begins (Table 7.2). A comparison of the abundance of redox depletions to periods of saturation and reduction is shown in Table 7.6. The data were obtained for two soils, one of which was on the backslope position and another in the toeslope position. The amount of redox depletions varied fourfold between the two soils. The soils were saturated for similar lengths of time, but reduced for longer periods in the toeslope position, which had the most redox depletions. Similar results were reported by Evans and Franzmeier (1986), Cogger and Kennedy (1992), and Couto et al. (1985), who showed that saturation by itself did not produce redox depletions.

Seasonal High Water Table Determinations

The preceding discussion suggests that it is impossible to develop a single relationship for all soils using the amount of gray color (redox depletions) in a soil to predict the specific length of time the soil is saturated at a given depth. An alternative approach has been to simply estimate the approximate height of the "seasonal high water table" from the presence of any redox depletions. This is the simplest way to relate redoximorphic features to saturation. Normally, it is assumed that a water table rises to the level at which redox depletions occur that have chromas of 2 or less and values of 4 or more. Note that, if the value is less than 4 and the chroma is 2 or less, then the color is black or dark gray and not necessarily related to saturation or reduction. The depth at which the redox depletions begin marks the level the seasonal high water table reaches in the soil. It is assumed that the water table will rise to this level in most years of "normal rainfall." It stays at the level of the redox depletions long enough for reduction to occur. This interpretation implies that the water table rises no farther, but this is not known unless detailed records of water table fluctuation are available. All that can be said is that the water table does not stay above the level of the low-chroma colors long enough to cause the reduction of Fe.

The advantage of using redox depletions to determine seasonal high water table is that the determination can be made in virtually any soil, at low cost, and without any additional information on hydrology or rainfall. The disadvantage is that there is no information on saturation frequency or duration. Nevertheless, determining the depth to the seasonal high water table in this way has proven useful and in general reliable for making onsite assessments as to whether a soil was suitable for septic systems. Case studies using this approach have been reported by Cogger and Kennedy (1992), Franzmeier et al. (1983), and Zobeck and Ritchie (1984), among others.

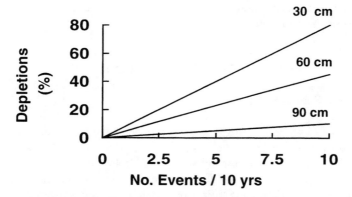

Figure 7.4 Relation of the frequency of saturation lasting 3 weeks or more to the percentage of redox depletions at various depths in a catena of three Ultisols in North Carolina. The saturation frequency was determined by simulating water table fluctuations over a 32-year period using the hydrologic model DRAINMOD.

New Approaches

Hydrologic models that simulate water table levels have provided another tool for developing specific relationships between soil saturation, water table fluctuation, and abundance of morphological features. Simonson and Boersma (1972) may have been the first to develop such relationships. Relationships between saturation frequency and soil color have not been developed widely, but in the few cases where they have it is clear that simulation modeling provides a powerful tool when making interpretations from soil morphology.

The results from one study conducted by the author are shown in Figure 7.4. The hydrologic model used was DRAINMOD (Skaggs 1978), and daily water table levels were computed for a 32-year period using historic rainfall data and onsite calibration of the model. Figure 7.4 shows that abundance of redox depletions was related to periods of saturation lasting 3 weeks or longer. Relationships between abundance and saturation duration changed with depth in this example. The reason for this is related to the decrease in decomposable organic materials with depth. At 90 cm, a given saturation frequency produced fewer depletions than at a depth of 30 cm. At 90 cm, roots are the major source of organic C, and they are oriented vertically and spaced 25 to 50 mm apart on the outside of soil peds. Redox depletions form primarily around these widely spaced roots and occupy less volume than at 30 cm where roots are more abundant and are closely spaced.

The data in Figure 7.4 also show that the abundance of depletions at a given depth can be related to events that do not occur every year. The depletions probably occur during the wetter years and are preserved. When abundance of redox depletions falls below 2%, the events they are related to occur rarely or about once in 20 years.

The advantage of using hydrologic models to predict historic water table levels is that the data produced are very specific and allow prediction of both saturation frequency and duration. Even the occurrence of rare events can be detected. The disadvantage is that the models can be expensive to use due to their need for a variety of soil measurements. In addition, hydrologic models are generally developed for specific kinds of landscapes. For example, DRAINMOD was developed to predict how deep and far apart ditches or tile drains need to be placed in fields to lower the water table a specific amount. It works best in level, coastal plain-type landscapes where groundwater moves laterally to streams or ditches.

The results shown in Figure 7.4 are for illustration only because they are site specific. Different relationships will have to be developed for most other soils because the same duration of saturation will not necessarily produce the same amount of redox depletions as those shown. These differing amounts of redox depletions are caused by differences in organic matter levels, pH, temperature, and so on, which cause a given amount of saturation to produce different durations and levels of reduction.

PROBLEM SITUATIONS

Identifying Relict Features of Reduction

A "relict feature" of reduction is one that has formed in the past and persists in the soil where it can no longer form today. Relict features of reduction make the soil appear to be wetter than it really is. They are useful in identifying soils whose hydrology has changed. A relict feature may persist in soils that were formerly saturated and reduced, but have since been drained by natural or artificial means such that reducing conditions no longer occur. Using relict features to detect altered hydrology can be faster than monitoring water table levels. Soils in which hydrology has changed may mean an area no longer qualifies as a jurisdictional wetland (National Research Council 1995) and is suitable for onsite waste disposal.

Redoximorphic features that are either redox depletions or redox concentrations are the most likely morphological features to be relict features. The reduced matrix must be kept reduced and can never be relict. Carbon-based organic features probably decompose too quickly for them to be preserved for more than 30 years. The single sulphur-based feature known (i.e., H_2S gas) is only found in reduced soils.

Identification of redoximorphic features that might be relict cannot be done with certainty using morphology alone, because hydrologic data are necessary to confirm that the hydrology is different than the features suggest. However, some guidelines can be given for when relict features should be suspected.

Relation to Root Channels and Cracks

In loams and clays, redoximorphic features sometimes form around root channels or cracks, as shown in Figure 7.2 and Plate 12. Redox depletions tend to form along root channels where the organic C occurred and fueled the reduction process. These are frequently the first pores to fill with water following a heavy rain. Redox concentrations that are Fe pore linings must also occur along root channels or ped surfaces. Any time a morphological feature, which had to form along a macropore, is found in the matrix or away from a pore, it can be assumed that it did not form recently and should be considered relict. Examples of this concept are shown schematically in Figure 7.5. Even features that occur in the matrix need to have a consistent relationship to the soil structure and large pores that is consistent with how they form. For example, Fe nodules normally form in the soil matrix. If these are found in Fe depletions on ped surfaces, then it is likely that these nodules are relict features.

Diffuse vs. Sharp Boundaries

Redox concentrations form by accumulation of Fe at certain points in the soil. The amount of Fe in these features is not expected to be the same throughout the feature. Normally Fe quantities decrease from the center of the Fe concentration toward the soil matrix. The zone of decreasing Fe concentration is frequently described as a *diffuse boundary* (see Plate 7). It is sometimes seen as a ring or halo around the Fe concentration that has a slightly different color than the main part of the Fe concentration itself. Diffuse boundaries are assumed to indicate that the feature is forming or has formed in the recent past. In other words, it is reflecting current hydrologic conditions.

When redox concentrations begin to dissolve, or are mixed into the matrix, they acquire *sharp boundaries* with the matrix. In this case the features are no longer forming and are relict. If virtually

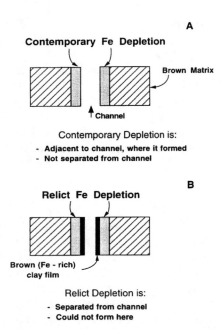

Figure 7.5   Relationship between the location of a redox depletion to a root channel in a soil where the depletion has formed recently and one where it is considered relict. Where the features are thought to be contemporary (A) the depletion abuts the channel, which is the position required for it to form by the process shown in Figure 7.2. In the second case (B), the depletion is separated from the channel by an Fe-rich clay coating. This coating suggests the depletion had to form before the clay was deposited, otherwise the Fe in the clay coating would have been reduced and removed from the coating.

all the redox concentrations in a horizon have such sharp boundaries, then it is likely that the hydrology has changed to make the soils oxidized year round. However, the underlying horizons should also be examined to find features that may be forming and to determine the exact appearance of Fe concentrations with diffuse boundaries in that soil.

When No Indicators Are Present

Occasionally soils that are suspected of being seasonally saturated and reduced do not show any of the common morphological features of reduced soils. This can happen with soils on floodplains in particular (Lindbo 1997). The reasons morphological features of reduction do not form are not completely understood but probably relate to the fact that little Fe reduction occurs. Iron reduction is limited by the soil having low amounts of organic-C at the time of saturation, a high pH, which makes Fe reduction occur only at very low Eh values as discussed in Chapter 4, high levels of Mn oxides in the soil, or large amounts of dissolved O_2 in the water. Identifying hydric soils in areas where no morphological indicators of reduction can be found requires direct measurements of saturation and reduction. Reducing conditions will have to be documented using dyes that react with Fe^{2+} or redox electrodes (Childs 1981; Chapter 4).

FALSE REDOXIMORPHIC FEATURES

Gray Parent Materials

Some soil parent materials have virtually no Fe minerals coating the particle surfaces and contain no Fe-bearing minerals. These materials have a gray color and will remain gray regardless of whether the soils that develop in them become reduced or not. Soils that develop in these deposits will have an A horizon that formed by accumulation of organic debris and a C horizon. Such soils can be well drained, but because of their gray color, which resembles an Fe depletion, they will have the appearance of being seasonally saturated and reduced. This condition should be suspected whenever the parent material (C horizon) is gray in color due to a naturally low amount of Fe. Soils whose parent materials consisted of gray sands will remain gray, and no Fe-based morphological features of reduction will develop. The only morphological indicators of reduction that will develop are organic-C based features or S-based features. Landscape position should also be examined for signs of it being where seasonally saturated soils would be expected. Such positions include the base of steep slopes and concave positions on flat or gently sloping surfaces.

E Horizons

E horizons are layers in the subsoil that developed a gray color through soil-forming processes that may or may not include Fe reduction. E horizons form by eluviation or movement of Fe, clay, and organic matter out of the soil layer that becomes the E. They usually occur below A horizons and must overlie zones of accumulation, such as Bt horizons. E horizons frequently have a chroma of 2 or less and value of 4 or more when the sand and silt grains have been stripped of Fe oxide coatings. The loss of Fe can occur by reduction processes (as in hydric soils), or it can occur because organic acids produced in A or O horizons leach into E horizons and dissolve Fe^{3+} minerals off particle surfaces. In the latter case, the soils are not considered to be hydric. The E is similar to an A horizon in texture and chemical composition except that it does not contain organic matter in the amounts found in the A.

Grayish-colored E horizons that formed in reduced soils are identified by redox concentrations in an abundance that usually exceeds 2%. The concentrations show that Fe has been reduced, moved, and reoxidized into Fe masses or pore linings. In addition, the horizons below the E should also be examined for morphological features of reduction. When redoximorphic features occur both within and below the E horizon, it is likely that the E was formed under reducing conditions.

Geological Materials Below the Rooting Zone

Morphological features of reduction need a source of organic carbon in order to form. The carbon is in greatest concentration near the surface and decreases in concentration with depth. Below a depth of approximately 1 m, the organic carbon is usually found around roots, and it becomes more scarce with increasing depth. C horizons below 1 m can contain features that are, or appear to be, redoximorphic features. While the features may have formed by oxidation–reduction processes, they must be interpreted cautiously because it is not always clear when the features formed. For example, Schoeneberger et al. (1992) described redoximorphic features along fractures in saprolite. Similar features could also be formed by hydrothermal fluids moving upward to the C horizon (see Figures 11–33 in Guilbert and Park 1986). Heated hydrothermal fluids can become strongly reducing when they pass through layers of graphite (Guilbert and Park 1986). Passage through the carbon-rich graphite produces solutions capable of reducing Fe. While such formation may be rare on a global scale, it does happen. When such features are formed deeply in the soil, they can probably be preserved for thousands of years.

SUMMARY

Morphological features of reduced soils form by oxidation–reduction reactions that occur when the soils are anaerobic. The reduction of O_2, Mn^{4+} and Fe^{3+} minerals, and SO_4^{2-} are responsible for the formation of most of the morphological features of reduction. When O_2 is reduced, organic matter accumulation exceeds decomposition, and this leads to the development of features rich in organic C, such as layers of peat or muck. When Fe or Mn minerals are reduced, redoximorphic features develop. The reduction of SO_4^{2-} leads to the production of H_2S gas. Morphological features of reduction show only that a soil was reduced at some point in its history. Their abundance can be related to the frequency of saturation, but such relationships are expected to be site specific and are not widely understood at this time. Hydrologic alterations caused by soil drainage or wetland construction produce changes in morphological features of reduction which can be detected in soils within 3 yrs or less of wetland construction and in less than 30 yrs following ditching. Such rates of formation will vary by region due to differences in temperature, organic carbon levels, as well as other factors. Relict features are those that occur in soils that have been drained, but which formed prior to drainage. They may be identified by their relationship to root channels, cracks, and lack of diffuse boundaries. False redoximorphic features also occur in soils with gray E horizons or gray parent materials. These do not develop under reducing conditions and cannot be used to identify hydric soils.

REFERENCES

Ahmad, N.R. and P.H. Nye. 1990. Coupled diffusion and oxidation of ferrous iron in soils. I. Kinetics of oxygenation of ferrous iron in soils. *J. Soil Sci.* 41:395–409.

Bell, J.C. and J.L. Richardson. 1997. Aquic conditions and hydric soil indicators for Aquolls and Albolls. pp. 23–40. *In* M.J. Vepraskas and S.W. Sprecher (Eds.) *Aquic Conditions and Hydric Soils: The Problem Soils.* SSSA Spec. Pub. No. 50. Soil Sci. Soc. Am., Inc., Madison, WI.

Bidwell, O.W., D.A. Gier, and J.C. Cipra. 1968. Ferromagnesian pedotubules on roots of *Bromus inermis* and *Andropogon gerardii*. *Intern. Congress Soil Sci. Trans.*, Adelaide, Australia, IV:683–692.

Brewer, R. 1964. *Fabric and Mineral Analysis of Soils.* John Wiley & Sons, New York.

Chen, C.C., J.B. Dixon, and F.T. Turner. 1980. Iron coatings on rice roots: morphology and models of development. *Soil Sci. Soc. Am. J.* 44:1113–1119.

Childs, C.W. 1981. Field test for ferrous iron and ferric-organic complexes on exchange sites (in water soluble forms) in soils. *Austral. J. Soil Res.* 19:175–180.

Cogger, C.G. and P.E. Kennedy. 1992. Seasonally saturated soils in the Puget Lowland. I. Saturation, reduction, and color patterns. *Soil Sci.* 153(6):421–433.

Couto, W., C. Sanzonowicz, and A. de O. Barcellos. 1985. Factors affecting oxidation–reduction processes in an Oxisol with a seasonal water table. *Soil Sci. Soc. Am. J.* 49:1245–1248.

Environmental Laboratory. 1987. *Corps of Engineers Wetlands Delineation Manual.* Technical Report Y-87-1. U.S. Army Engineers Waterways Experiment Station, Vicksburg, MS.

Evans, C.V. and D.P. Franzmeier. 1986. Saturation, aeration, and color patterns in a toposequence of soils in north-central Indiana. *Soil Sci. Soc. Am. J.* 50:975–980.

Franzmeier, D.P., J.E. Yahner, G.C. Steinhardt, and H.R. Sinclair. 1983. Color patterns and water table levels in some Indiana soils. *Soil Sci. Soc. Am. J.* 47:1196–1202.

Gallaher, R.N., H.F. Perkins, and K.H. Tan. 1974. Classification, composition, and mineralogy of iron glaebules in a Southern Coastal Plain soil. *Soil Sci.* 117:155–164.

Gilman, K. 1994. *Hydrology and Wetland Conservation.* John Wiley & Sons, New York.

Guilbert, J.M. and C.F. Park, Jr. 1986. *The Geology of Ore Deposits.* W.H. Freeman and Co., New York.

Hayes, W.A., Jr. 1998. *Effect of Ditching on Soil Morphology, Saturation, and Reduction in a Catena of Coastal Plain Soils.* M.S. thesis, North Carolina State University, Raleigh.

Hudnall, W.H., A. Szogi, B.A. Touchet, J. Diagle, J.P. Edwards, and W.C. Lynn. 1990. *Guidebook for Louisiana.* VIII International Soil Correlation Meeting: Classification and Management of Wet Soils. Agronomy Department, Louisiana State University, Baton Rouge.

Hurt, G.W., P.M. Whited, and R.F. Pringle (Eds.) 1998. Field indicators of hydric soils in the United States. Version 4.0. U.S.D.A. Nat. Res. Cons. Serv., Fort Worth, TX.

Jackson, M.L. 1969. *Soil Chemical Analysis–Advanced Course.* 2nd ed, 10th printing. Published by the author, Madison, WI.

Lindbo, D.L. 1997. Entisols: Fluvents and Fluvaquents: problems recognizing aquic and hydric conditions in young, flood plain soils. pp. 133–151. *In* M. J. Vepraskas and S. W. Sprecher (Eds.) *Aquic Conditions and Hydric Soils: The Problem Soils.* SSSA Special Publication No. 50, Soil Sci. Soc. Am., Madison, WI.

McBride, M.B. 1994. *Environmental Chemistry of Soils.* Oxford Univ. Press, New York.

Meek, B.D., A.J. McKenzie, and L.B. Gross. 1968. Effects of organic matter, flooding time and temperature on the dissolution of iron and manganese from soil *in situ. Soil Sci. Soc. Am. Proc.* 32:634–638.

Megonigal, J.P., S.P. Faulkner, and W.H. Patrick, Jr. 1996. The microbial activity season in southeastern hydric soils. *Soil Sci. Soc. of Am. J.* 60:1263–1266.

Mendelssohn, I.A., B.A. Kleiss, and J.A. Wakeley. 1995. Factors controlling the formation of oxidized root channels: a review. *Wetlands* 15(1):37–46.

National Research Council. 1995. *Wetlands: Characteristics and Boundaries.* National Academy Press, Washington, DC.

Ponnamperuma, F.N. 1972. The chemistry of submerged soils. *Adv. Agron.* 24:29–96.

Rhoton, F.E., J.M. Bigham, and D.G. Schultze. 1993. Properties of iron–manganese nodules from a sequence of eroded fragipan soils. *Soil Sci. Am. J.* 57:1386–1392.

Russell, R.S. 1977. *Plant Root Systems.* McGraw-Hill Book Co., London.

Schoeneberger, P.J., S.B. Weed, A. Amoozegar, and S.W. Buol. 1992. Color zonation associated with fractures in a felsic gneiss saprolite. *Soil Sci. Soc. Am. J.* 56:1855–1859.

Schwertmann, U. and R.M. Taylor. 1989. Iron oxides. pp. 379–438. *In* J.B. Dixon and S.W. Weed (Eds.) *Minerals in Soil Environments,* 2nd ed., Soil Sci. Soc. Am., Madison, WI.

Simonson, G.H. and L. Boersma. 1972. Soil morphology and water table relations: II. Correlation between annual water table fluctuations and profile features. *Soil Sci. Soc. Am. J.* 36:649–653.

Skaggs, R.W. 1978. *A Water Management Model for Shallow Water Table Soils.* Technical Report 134. Water Res. Inst., Raleigh, NC.

Soil Survey Staff. 1998. *Keys to Soil Taxonomy.* 8th ed. U.S.D.A., Nat. Res. Cons. Serv. Washington, DC.

Veneman, P.L.M., M.J. Vepraskas, and J. Bouma. 1976. The physical significance of soil mottling in a Wisconsin toposequence. *Geoderma* 15:103–118.

Vepraskas, M.J. 1996. *Redoximorphic Features for Identifying Aquic Conditions.* Tech. Bull. 301. NC Agric. Exp. Sta., Raleigh, NC.

Vepraskas, M.J. and J. Bouma. 1976. Model experiments on mottle formation simulating field conditions. *Geoderma* 15:217–230.

Vepraskas, M.J. and W.R. Guertal. 1992. Morphological indicators of soil wetness. pp. 307–312. *In* J. M. Kimble (Ed.) *Proc. Eighth International Soil Correlation Meeting* (VIII ISCOM): *Characterization, Classification, and Utilization of Wet Soils.* U.S.D.A. Soil Cons. Serv., Nat. Soil Surv. Center, Lincoln, NE.

Vepraskas, M.J., J.L. Richardson, J.P. Tandarich, and S.J. Teets. 1999. Dynamics of hydric soil formation across the edge of a created deep marsh. *Wetlands.* 19(1):78–89.

Vepraskas, M.J. and L.P. Wilding. 1983. Albic neoskeletans in argillic horizons as indices of seasonal saturation and iron reduction. *Soil Sci. Soc. Am. J.* 47:1202–1208.

Zobeck, T.M. and A. Ritchie. 1984. Analysis of long-term water table depth records from a hydrosequence of soils in central Ohio. *Soil Sci. Soc. Am. J.* 48:119–125.

CHAPTER 8

Delineating Hydric Soils

G.W. Hurt and V.W. Carlisle

INTRODUCTION

For centuries wetlands were regarded as little more than habitat for mosquitoes, snakes, and other pests. Today, in addition to recognizing wetlands as habitats for a variety of wildlife species (including mosquitoes and snakes), we are aware that wetlands are the nursery grounds for our fisheries, and that they filter pollutants, reduce flooding, protect against erosion, provide timber products, recharge groundwater reserves, and furnish society with educational, scientific, recreational, and aesthetic benefits. Local, state, and federal governments have enacted laws that regulate the use of wetlands to preserve these public benefits.

To be regulated, wetlands must first be identified and delineated. Most regulated wetlands must have three essential components: (1) hydrophytic vegetation, (2) hydric soils, and (3) wetland hydrology (Cowardin et al. 1979, Tiner and Burke 1995, Environmental Laboratory 1987, Hammer 1992). Technical criteria for each of these characteristics must be met before an area can be identified as a wetland (Environmental Laboratory 1987, U.S. Department of Agriculture 1994). When anaerobic conditions prevail in wetland soils for long enough periods during the growing season a predominance of hydrophytic vegetation is favored. Undrained hydric soils with natural vegetation should support a dominant population of ecologically facultative wetland and obligate wetland plant species; conversely, drained hydric soils without natural vegetation have the ability to support a dominant population of ecologically facultative wetland and obligate wetland plant species once hydrologic modifications are removed or are not maintained.

This chapter presents approaches and methods for identifying and delineating hydric soils for purposes of implementing Section 404 of the Clean Water Act (CWA) and Food Security Act of 1985 as amended by the Food, Agriculture, Conservation, and Trade Act of 1990 and the Federal Agriculture Improvement and Reform Act of 1996 (FSA). It is designed to assist readers in making wetland determinations and delineations using hydric soils as the primary parameter of choice. Separate sections are devoted to preliminary offsite investigations and detailed examination and delineation procedures, with a special section on problem hydric soil delineations. This chapter also includes our observations and recommendations on delineating hydric soils that have been developed over 30 years of studying wetlands.

WETLAND COMPONENTS

Hydrology

It is recognized that the influence of water is the key parameter in the presence or absence of wetlands. Unfortunately, annual and seasonal variations in hydrology make the direct use of this parameter for delineating wetlands in the field very difficult, time consuming, and costly. In addition, requirements for recognition of wetland hydrology vary greatly among regulating agencies. Inundation and/or saturation of the soil surface for 5 to 12.5% of the growing season is a requirement by one agency (Environmental Laboratory 1987). Requirements by another agency are inundation for 7 days, or saturation for 14 days, or inundation for 15 days, or inundation for 10% of the growing season (U.S. Department of Agriculture 1994). Faulkner et al. (1991) found that more than 14 days to as many as 28 days of surface saturation annually may be required to induce sufficient anaerobic conditions for developing hydric soil morphology.

Hydrological records of several years are required to accurately assess the hydrology of a site. Skaggs et al. (1994) documented a site that required 48 years of data to determine that wetland hydrology was present in 24 of those years and therefore met wetland requirements. However, wetland hydrology was not present for the other 24 years, and several consecutive years lacked wetland hydrology. These types of data (years to decades) are desirable but rarely available for borderline sites and most delineation edges. Determination of the hydrologic status of a site must be made based on secondary indicators and, sometimes, short-term saturation records. The reliability of short-term saturation monitoring to determine whether wetland hydrology exists for a given site is suspect (Skaggs et al. 1991). Most often the assumption is made that wetland hydrology exists if hydric soils and hydrophytic vegetation are present.

Vegetation

Presence or absence of wetland vegetation is based on a list of plants (Reed 1988) that grow in areas with insufficient concentration of oxygen for root respiration. Reed (1988) recognized four types of indicator plants that occur in wetlands: (1) obligate wetland plants that occur in wetlands 99% of the time, (2) facultative wetland plants that occur in wetlands 67 to 99% of the time, (3) facultative plants that occur in wetlands 34 to 66% of the time, and (4) facultative upland plants that occur in wetlands 1 to 33% of the time. Failure of this plant list to accurately correlate with wetlands has been well documented. In north-central Florida six sites were investigated (Best et al. 1990); four were wetlands (based on the three parameter approach) and two were uplands. Based on agency methodology for implementation of CWA and FSA, vegetation at all six sites was hydrophytic. In coastal Mississippi six sites were investigated (Erickson and Leslie 1989); five were wetlands and one upland. Vegetation at all six sites was hydrophytic. Similar results have been reported in California (Eicher 1988), North Carolina (Christensen et al. 1988), and Massachusetts (Veneman and Tiner 1990). In addition, vegetation is difficult to assess in dormant seasons, and seasonal and annual variation exists. Situations exist where vegetation cannot be used as a means of identifying a wetland. For example: (1) areas with vegetative cover that no longer consists of natural plant species, (2) site-prepared areas denuded of vegetation, (3) ecotones, and (4) dredge and/or fill areas.

Hydric Soils

Soils provide a reliable method of delineating wetlands, especially in areas with unreliable or unavailable hydrology and in areas of transitional vegetation, or in areas where use of the plant list does not provide delineation assistance (Florida Soil Survey Staff 1992, Hurt and Brown 1995, Segal et al. 1995). According to *Field Indicators of Hydric Soils in the United States* (Hurt et al.

Table 8.1   Criteria for Hydric Soils

1. All Histosols except Folists, or
2. Soils in Aquic Suborders, Great Groups, or Subgroups, Albolls Suborder, Aquisalids, Pachic Subgroups, or Cumulic Subgroups that are:
   a. somewhat poorly drained with a water table equal to 0.0 foot (ft.) from the surface during the growing season, or
   b. poorly drained or very poorly drained and have either:
      (1) water table equal to 0.0 ft. during the growing season if textures are coarse sand, sand, or fine sand in all layers within 20 inches (in.), or for other soils
      (2) water table at less than or equal to 0.5 ft. from the surface during the growing season if permeability is equal to or greater than 6.0 in./hour (h.) in all layers within 20 in., or
      (3) water table at less than or equal to 1.0 ft. from the surface during the growing season, if permeability is less than 6 in./h. in any layer within 20 in., or
3. Soils that are frequently ponded for long or very long duration during the growing season, or
4. Soils that are frequently flooded for long or very long duration during the growing season.

From *Federal Register*, February 24, 1995.

1998): "Nearly all hydric soils exhibit characteristic morphologies that result from repeated periods of saturation and/or inundation for more than a few days. Soil saturation or inundation activates microbiological activity that results in a depletion of oxygen. The resulting anaerobiosis promotes biogeochemical processes such as the accumulation of organic matter and the reduction, translocation, and/or accumulation of iron and other reducible elements." These processes are responsible for the formation of characteristic soil morphologies that persist during both wet and dry periods, making them particularly useful for identifying hydric and other wet soils (Vepraskas 1994, Mausbach and Richardson 1994).

The Hydric Soil Definition (*Federal Register,* July 13, 1994) is: "A hydric soil is a soil that formed under conditions of saturation, flooding, or ponding long enough during the growing season to develop anaerobic conditions in the upper part." Criteria for hydric soils (*Federal Register,* February 24, 1995) have been established (Table 8.1). Relationships and limitations of the hydric soil definition, criteria, and hydric soil indicators must be thoroughly understood to facilitate accurate identification and delineation of hydric soils in the field. All hydric soils must satisfy the requirements of the hydric soil definition. This means the soils must be saturated or inundated during the growing season, and the soil must be anaerobic. Hydric soil criteria were established to generate hydric soil lists, as was discussed in Chapter 2. The criteria by themselves are not intended for onsite application, although the National Technical Committee for Hydric Soils has approved that data proving criteria 1, 3, or 4 can be used to document the presence of a hydric soil. Presence of one (or more) hydric soil indicators (Tables 8.2 and 8.3) is evidence that the definition has been met because indicators form in soils that are saturated or inundated and become anaerobic within 30 cm of the surface. The "growing season" is considered to be that part of the year during which the soil temperature and moisture conditions permit microbial activities (Chapter 5).

PRELIMINARY OFF-SITE INVESTIGATIONS

Prior to any onsite identification or delineation of hydric soils, all available offsite information should be evaluated. Offsite information available for most non-federal lands in the United States and Puerto Rico includes the published soil surveys of the National Cooperative Soil Survey, National Wetlands Inventory Maps produced by the U.S. Fish and Wildlife Service, the topographic quadrangle series of maps produced by the U.S. Geological Survey (USGS), and maps of areas subject to flooding produced by the Federal Emergency Management Agency (FEMA). Reviewing these sources before attempting to identify or delineate hydric soils can significantly reduce time spent in the field. It will also facilitate most onsite identification and delineation procedures.

Table 8.2 Hydric Soil Indicators of the United States*

A. INDICATORS FOR ALL SOILS REGARDLESS OF TEXTURE:

A1 — Histosol. Soil that classifies as a Histosol, except Folists.

A2 — Histic Epipedon. Soil that has a histic epipedon.

A3 — Black Histic. Soils with a layer of peat, mucky peat, or muck 20 cm (8 in.) or more thick starting within the upper 15 cm of the soil surface having hue 10YR or yellower, value 3 or less, and chroma 1 or less.

A4 — Hydrogen Sulfide. Soils with a hydrogen sulfide odor within 30 cm of the soil surface.

A5 — Stratified Layers. Soils with several stratified layers starting within the upper 15 cm of the soil surface. One or more of the layers has value 3 or less with chroma 1 or less and/or it is muck, mucky peat, peat, or mucky modified mineral texture. The remaining layers have value 4 or more and chroma 2 or less.

A6 — Organic Bodies. Soils with a layer that has 2% or more organic bodies of muck or a mucky modified mineral texture, approximately 1 to 3 cm in diameter, starting within 15 cm of the soil surface.

A7 — 5 cm Mucky Mineral. Soils with a mucky modified mineral layer 5 cm or more thick starting within 15 cm of the soil surface.

A8 — Muck Presence. Soils with a layer of muck with value 3 or less and chroma 1 or less within 15 cm of the soil surface.

A9 — 1 cm Muck. Soils with a layer of muck 1 cm or more thick with value 3 or less and chroma 1 or less starting within 15 cm of the soil surface.

A10 — 2 cm Muck. Soils with a layer of muck 2 cm or more thick with value 3 or less and chroma 1 or less starting within 15 cm of the soil surface.

S. INDICATORS FOR SOILS WITH SANDY SOIL MATERIALS:

S1 — Sandy Mucky Mineral. Soils with a mucky modified mineral layer 5 cm or more thick starting within 15 cm of the soil surface.

S2 — 3 cm Mucky Peat or Peat. Soils with a layer of mucky peat or peat 2.5 cm or more thick with value 4 or less and chroma 3 or less starting within 15 cm of the soil surface.

S3 — 5 cm Mucky Peat or Peat. Soils with a layer of mucky peat or peat 5 cm or more thick with value 3 or less and chroma 2 or less starting within 15 cm of the soil surface.

S4 — Sandy Gleyed Matrix. Soils with a gleyed matrix which occupies 60% or more of a layer starting within 15 cm of the soil surface.

S5 — Sandy Redox. Soils with a layer starting within 15 cm of the soil surface that is at least 10 cm thick that has a matrix with 60% or more of its volume chroma 2 or less and 2% or more distinct or prominent redox concentrations as soft masses and/or pore linings.

S6 — Stripped Matrix. Soils with a layer starting within 15 cm of the soil surface in which iron/manganese oxides and/or organic matter have been stripped from the matrix exposing the primary base color of soil materials. The stripped areas and translocated oxides and/or organic matter form a diffuse splotchy pattern of two or more colors. The stripped zones are 10% or more of the volume; they are rounded and approximately 1 to 3 cm in diameter.

S7 — Dark Surface. Soils with a layer 10 cm or more thick starting within the upper 15 cm of the soil surface with a matrix value 3 or less and chroma 1 or less. At least 70% of the visible soil particles must be covered, coated, or similarly masked with organic material. The matrix color of the layer immediately below the dark layer must have chroma 2 or less.

S8 — Polyvalue Below Surface. Soils with a layer that has value 3 or less and chroma 1 or less starting within 15 cm of the soil surface underlain by a layer(s) where translocated organic matter unevenly covers the soil material forming a diffuse splotchy pattern. At least 70% of the visible soil particles in the upper layer must be covered, coated, or masked with organic material. Immediately below this layer, the organic coating occupies 5% or more of the soil volume and has value 3 or less and chroma 1 or less. The remainder of the soil volume has value 4 or more and chroma 1 or less.

S9 — Thin Dark Surface. Soils with a layer 5 cm or more thick within the upper 15 cm of the surface, with value 3 or less and chroma 1 or less. At least 70% of the visible soil particles in this layer must be covered, coated, or masked with organic mer(s) with value 4 or less and chroma 1 or less to a depth of 30 cm or to the spodic horizon, whichever is less.

S10 — Alaska Gleyed. Soils with a layer that has a dominant hue N, 10Y, 5GY, 10GY, 5G, 10G, 5BG, 10BG, 5B, 10B, or 5PB, with value 4 or more in the matrix, within 30 cm of the mineral surface, and underlain by hue 5Y or redder in the same type of parent material.

F. INDICATORS FOR SOILS WITH LOAMY AND CLAYEY SOIL MATERIAL

F1 — Loamy Mucky Mineral. Soils with a mucky modified mineral layer 10 cm or more thick starting within 15 cm of the soil surface.

F2 — Loamy Gleyed Matrix. Soils with a gleyed matrix that occupies 60% or more of a layer starting within 30 cm of the soil surface.

F3 — Depleted Matrix. Soils with a layer at least 15 cm thick with a depleted matrix that has 60% or more chroma 2 or less starting within 25 cm of the surface. The minimum thickness requirement is 5 cm (2 in.) if the depleted matrix is entirely within the upper 15 cm of the mineral soil.

Table 8.2   Hydric Soil Indicators of the United States* *(continued)*

F4 — Depleted Below Dark Surface. Soils with a layer at least 15 cm thick with a depleted matrix that has 60% or more chroma 2 or less starting within 30 cm of the surface. The layer(s) above the depleted matrix have value 3 or less and chroma 2 or less.

F5 — Thick Dark Surface. Soils with a layer at least 15 cm thick with a depleted matrix that has 60% or more chroma 2 or less (or a gleyed matrix) starting below 30 cm of the surface. The layer(s) above the depleted or gleyed matrix have hue N and value 3 or less to a depth of 30 cm and value 3 or less and chroma 1 or less in the remainder of the epipedon.

F6 — Redox Dark Surface. Soils with a layer at least 10 cm thick entirely within the upper 30 cm (12 in.) of the mineral soil that has:

a. matrix value 3 or less and chroma 1 or less and 2% or more distinct or prominent redox concentrations as soft masses or pore linings, or

b. matrix value 3 or less and chroma 2 or less and 5% or more distinct or prominent redox concentrations as soft masses or pore linings.

F7 — Depleted Dark Surface. Soils with redox depletions, with value 5 or more and chroma 2 or less, in a layer at least 10 cm thick entirely within the upper 30 cm of the mineral soil that has:

a. matrix value 3 or less and chroma 1 or less and 10% or more redox depletions, or

b. matrix value 3 or less and chroma 2 or less and 20% or more redox depletions.

F8 — Redox Depressions. Soils in closed depressions subject to ponding with 5% or more distinct or pore linings in a layer 5 cm or more thick entirely within the upper 15 cm of the soil surface.

F9 — Vernal Pools. Soils in closed depressions subject to ponding with a depleted matrix in a layer 5 cm thick entirely within the upper 15 cm of the soil surface.

F10 — Marl. Soils with a layer of marl with a value 5 or more starting within 10 cm of the soil surface.

F11 — Depleted Ochric. Soils with a layer(s) 10 cm or more thick that has 60% or more of the matrix with value 4 or more and chroma 1 or less. The layer is entirely within the upper 25 cm of the soil surface.

F12 — Iron/Manganese Masses. Soils on floodplains with a layer 10 cm or more thick with 40% or more chroma 2 or less, and 2% or more distinct or prominent redox concentrations as soft iron/manganese masses with diffuse boundaries. The layer occurs entirely within 30 cm of the soil surface. Iron/manganese masses have value 3 or less and chroma 3 or less; most commonly they are black. The thickness requirement is waived if the layer is the mineral surface layer.

F13 — Umbric Surface. Soils on concave positions of interstream divides and in depressions, a layer 15 cm or more thick starting within the upper 15 cm of the soil surface with value 3 or less and chroma 1 or less immediately underlain by a layer 10 cm or more thick with chroma 2 or less.

F14 — Alaska Redox Gleyed. Soils with a layer that has dominant matrix hue 5Y with chroma 3 or less, or hue N, 10Y, 5GY, 10GY, 5G, 10G, 5BG, 10BG, 5B, 10B, or 5PB, with 10% or more redox concentrations as pore linings with value and chroma 4 or more. The layer occurs within 30 cm of the soil surface.

F15 — Alaska Gleyed Pores. Soils with a layer that has 10% hue N, 10Y, 5GY, 10GY, 5G, 10G, 5BG, 10BG, 5B, 10B, or 5PB with value 4 or more in the matrix or along channels containing dead roots or no roots within 30 cm of the soil surface. The matrix has dominant chroma 2 or less.

F16 — High Plains Depressions. Soils in closed depressions subject to ponding with a layer at least 10 cm thick within the upper 35 cm (13.5 in.) of the mineral soil that has chroma 1 or less and:

a. 1% or more redox concentrations as nodules or concretions, or

b. redox concentrations as nodules or concretions with distinct or prominent corona.

---

* Selected hydric soil indicators have been approved for use in each land resource region (Hurt et al. 1998). See Table 8.6.

Published Soil Surveys and Hydric Soil Lists

The published soil survey is an excellent place to start offsite investigation before making onsite wetland determinations. Soil surveys have been completed for more than 90% of the private and non-federal lands in the continental United States. Most of these have been published at a scale of 1:12000 to 1:24000 at the local or county level. Hydric soil lists are also available at the local or county level. These lists contain soil survey map units that have a strong probability of being hydric. They were developed by comparing the estimated soil properties found in a published soil survey with the hydric soil criteria. Hydric soil lists must be used with caution because they have at least three limitations. First, the scale limitation must be considered. Most soil surveys do not show soil bodies that are less than about 1.2 hectares in size. Second, the presence of a soil on a hydric soil list does not mean that it is in fact hydric; this is only an interpretive rating and

Table 8.3   Test Hydric Soil Indicators of the United States*

TA. TEST INDICATORS FOR ALL SOILS REGARDLESS OF TEXTURE:

TA1 — Playa Rim Stratified Layers. Soils with several stratified layers starting within the upper 15 cm of the soil surface. At least one layer has value 3 or less and chroma 1 or it has value 2 or more and chroma 2 or less with 2% or more distinct or prominent redox concentrations as soft masses or pore linings. The upper 15 cm has dominant chroma 2 or less.

TA2 — Structureless Muck. Soils with a layer of muck 2 cm (0.75 in.) or more thick that has no soil structure starting within 15 cm (6 in.) of the soil surface on concave positions or in depressions.

TS. TEST INDICATORS FOR SOILS WITH SANDY SOIL MATERIAL:

TS1 — Iron Staining. Soils with a continuous zone, 3 cm or more thick, of iron staining with value 4 or more and chroma 6 or more within 15 cm of the soil surface. The zone is immediately below a horizon in which iron/manganese oxides have been removed from the matrix and exposed the primary base color of the silt and sand grains.

TS2 — Thick Sandy Dark Surface. Soils with a layer at least 15 cm thick with a depleted matrix that has 60% or more chroma 2 or less or a gleyed matrix starting below 30 cm of the soil surface. The layer(s) above the depleted or gleyed matrix have hue N and value 3 or less; or hue 10YR or yellower with value 2 or less and chroma 1 to a depth of 30 cm and chroma 1 or less in the remainder of the epipedon.

TS3 — Dark Surface 2. Soils with a layer 10 cm or more thick starting within 15 cm of the soil surface with matrix value 2 or less and chroma 1 or less. At least 70% of the soil materials are covered, coated, or masked with organic material. The matrix color of the layer immediately below the dark surface must have value 4 or more and chroma 2 or less.

TS4 — Sandy Neutral Surface. Soils with a layer at least 10 cm thick with a depleted matrix that has 60% or more chroma 2 or less or a gleyed matrix starting within 30 cm of the soil surface. The layer(s) above the depleted or gleyed matrix have hue N and value 3 or less.

TS5 — Chroma 3 Sandy Redox. Soils with a layer starting within 15 cm of the soil surface that is at least 10 cm (4 in.) thick and has a matrix chroma 3 or less with 2% or more distinct or prominent redox concentrations as soft masses and/or pore linings.

TF. TEST INDICATORS FOR SOILS WITH LOAMY AND CLAYEY SOIL MATERIAL:

TF1 — ? cm Mucky Peat or Peat. Soils with a layer of mucky peat or peat ? cm thick with value 4 or less · and chroma 3 or less starting within 15 cm of the soil surface.

TF2 — Red Parent Material. Soils formed in parent material with a hue of 7.5YR or redder that have a layer at least 10 cm thick with a matrix value 4 or more and chroma 4 or less and 2% or more redox depletions and/or redox concentrations as soft masses and/or pore linings. The layer is entirely within 30 cm of the soil surface. The minimum thickness requirement is 5 cm if the layer is the mineral surface layer.

TF3 — Alaska Concretions. Soils that have within 30 cm of the soil surface redox concentrations as nodules or concretions greater than 2 mm in diameter that occupy more than approximately 2% of the soil volume in a layer 10 cm or more thick with a matrix chroma 2 or less.

TF4 — 2.5Y/5Y Below Dark Surface. Soils with a layer at least 15 cm thick with 60% or more hue 2.5 Y or yellower, value 4 or more, and chroma 1; or hue 5Y or yellower, value 4 or more, and chroma 2 or less starting within 30 cm of the soil surface. The layer(s) above the 2.5Y/5Y layer have value 3 or less and chroma 2 or less.

TF5 — 2.5Y/5Y Below Thick Dark Surface. Soils with a layer at least 15 cm thick with 60% or more hue 2.5Y or yellower, value 4 or more, and chroma 1; or hue 5Y or yellower, value 4 or more, and chroma 2 or less starting below 30 cm of the soil surface. The layer(s) above the 2.5Y/5Y layer have hue N and value 3 or less; or have hue 10YR or yellower with value 2 or less and chroma 1 or less to a depth of 30 cm and value 3 or less and chroma 1 or less in the remaining epipedon.

TF6 — Calcic Dark Surface. Soils with a layer with an accumulation of calcium carbonate ($CaCO_3$), or calcium carbonate equivalent, occurs within 40 cm of the soil surface. It is overlain by a layer(s) with value 3 or less and chroma 1 or less. The layer of $CaCO_3$ accumulation is underlain by a layer within 75 cm of the surface 15 cm or more thick having 60% or more by volume one or more of the following:

a. depleted matrix, or

b. gleyed matrix, or

c. hue 2.5Y or yellower, value 4 and chroma 1.

TF7 — Thick Dark Surface 2/1. Soils with a layer at least 15 cm thick with a depleted matrix that has 60% or more chroma 2 or less (or a gleyed matrix) starting below 30 cm of the soil surface. The layer(s) above the depleted or gleyed matrix have hue 10YR or yellower, value 2.5 or less and chroma 1 or less to a depth of 30 cm and value 3 or less and chroma 1 or less in the remainder of the epipedon.

TF8 — Redox Spring Seeps. Soils with a layer that has value 5 or more and chroma 3 or less with 2% or more distinct or prominent redox concentrations as soft masses or pore linings. The layer is at least 5 cm thick and is within the upper 15 cm of the soil surface.

Table 8.3   Test Hydric Soil Indicators of the United States* *(continued)*

TF9 — Delta Ochric. Soils with a layer 10 cm or more thick that has 60% or more of the matrix with value 4 or more and chroma 2 or less with no redox concentrations. This layer occurs entirely within the upper 30 cm of the soil surface.

TF10 — Alluvial Depleted Matrix. Soils on frequently flooded floodplains that have a layer with a matrix that has 60% or more chroma 3 or less with 2% redox concentrations as soft iron masses, starting within 15 cm of the soil surface and extending to a depth of more than 30 cm.

* Selected hydric soil indicators have been approved for testing in each land resource region (Hurt et al. 1998). See Table 8.6.

must be verified in the field. Finally, most soil surveys were produced prior to development of the hydric soil concept. If any portion of the range of estimated properties for a soil is within any portion of any of the hydric soil criteria, that soil will appear on hydric soils lists. For example (Criteria 2B3), if a soil with a permeability of less than 15 cm/hour has an estimated water table of 30 to 60 cm during any portion of the growing season, that soil would appear on a hydric soil list, even though most of the range in estimated water table (> 30 cm) is outside the criteria. Just as with all interpretations based on information in published soil surveys, hydric soil interpretations are confirmed by onsite investigations.

National Wetland Inventory Maps

Also available for offsite examination are National Wetland Inventory (NWI) maps produced by the U.S. Fish and Wildlife Service. NWI maps contain wetland delineations as defined in "Classification of Wetlands and Deepwater Habitats of the United States" (Cowardin et al. 1979) at a scale of 1:24000. The NWI maps were produced by interpreting high-altitude photography, usually at a scale of 1:80000 to 1:40000. The NWI have three limitations for wetland delineation. First, the definition of wetlands used to produce the NWI maps is not the same as the definitions used to delineate jurisdictional wetlands. Jurisdictional wetlands are determined based on the three parameters of soils, hydrology, and vegetation, whereas NWI wetland maps may have delineations based on only one parameter and often fail to delineate cropped fields and borderline wetlands. Second, many NWI maps were produced from poor-quality aerial photography. Finally, scale limitations do not allow for delineation of areas less than about 1.6 hectares.

Topographic Maps

Another source of information is the topographic quadrangle series of maps produced by USGS. These maps contain topographic features including swamp and marsh symbols at a scale of 1:24000 and may be useful as a source of offsite wetland information. Limitations of these maps for wetland delineation include the following points. First, not all areas with marsh and swamp symbols are wetlands. Conversely, there are areas of wetlands that lack marsh and swamp symbols. Second, the quality of the topographic maps varies from quadrangle to quadrangle and within any given quadrangle; however, the degree of field verification is indicated on the legend for each map. Finally, the scale limitation is the same as for the NWI maps.

Federal Emergency Management Agency Maps

Another source of information is the topographic quadrangle series of maps produced by the Federal Emergency Management Agency (FEMA). These maps contain delineations of areas that FEMA has determined are flood prone at a scale of 1:24000. The limitations of FEMA maps for wetland delineation include the following. First, flood-prone areas delineated contain many areas of uplands flooded as rarely as once every 1 to 500 years. Although many areas of wetlands will be within areas delineated as flood-prone areas, there will also be many areas of uplands. Second, saturated wetlands and many depressional wetlands are not identified on these maps. Finally, the scale limitation is approximately the same as for the NWI maps and the USGS topography quadrangle maps.

Because of the limitations listed above, onsite investigation is recommended to decide if hydric soils occur and to determine the exact location and extent of hydric soils. However, valuable insight can be gained by reviewing these sources of information before attempting hydric soil delineations. Time needed to locate and delineate hydric soils will be lessened.

DETAILED EXAMINATION AND DELINEATION PROCEDURES

Landform Recognition

A landscape is the land surface that an eye can comprehend in a single view (Tuttle 1975, U.S. Department of Agriculture 1993a). Most frequently it is a collection of landforms. Landforms are physical, recognizable forms or features on the earth's surface that have characteristic shapes produced by natural processes. Hydric soils occur on landforms (U.S. Department of Agriculture 1993a) that include backswamps, bogs, depressions, estuaries, fens, interdunes, marshes, flats, floodplains, muskegs, oxbows, playas, pocosins, potholes, seep slopes, sloughs, and swamps (Figure 8.1). One of the most important factors in hydric soil determination and delineation is landform recognition.

Hydric soils develop because unoxygenated water saturates the soil or collects on the soil surface. A concave surface frequently augmented by slower percolating subsurface soil horizons allows this process to occur. Hydric soil indicators normally begin to appear at this concave slope break and continue throughout the extent of the wetland even though concavity may not exist throughout the wetland (see Figure 8.1). The concave slope break may be very subtle, but it will be present in almost all natural landscapes. Wetland delineators need to become very familiar with the landscapes and hydrology of their areas in order to recognize the often very subtle slope break. They need to anticipate where inundated or saturated soils are likely to occur. Water is the driving force behind the development of hydric soils (wetlands) and hydrology of the landscape must be understood prior to making hydric soil determinations and delineating wetlands.

Hydric Soil Indicators

Hydric soil indicators are formed predominantly by accumulation, loss, or transformation of iron, manganese, sulfur, or carbon compounds (Plates 15 through 18). The presence of H_2S (a rotten egg odor) is a strong indicator of a hydric soil, but this indicator is found in only the wettest sites containing sulfur. While indicators related to Fe/Mn depletions or concentrations are the most common, they cannot form in soils with parent materials that contain very low amounts of Fe/Mn. Soils formed in such materials may have low chroma colors (2 or less) that are not related to saturation and reduction. For these soils, features related to accumulations of organic carbon are most commonly used.

Field indicators of hydric soils are routinely used in conjunction with the definition to confirm the presence or absence of a hydric soil. The publication *Field Indicators of Hydric Soils in the United States* (Hurt et al. 1998) is the current guide that should be applied to identify and delineate hydric soils in the field. The National Technical Committee for Hydric Soils (NTCHS) is responsible for revising and maintaining the hydric soil indicators. Indicators currently approved for identifying and delineating hydric soils are given in Table 8.2; examples are provided in Plates 19 through 22.

The list of hydric soil indicators is not static. Changes are anticipated as new knowledge of morphological, physical, chemical, and mineralogical soil properties accumulates. Revisions and additions will continue as we gain a better understanding of the relationships between the devel-

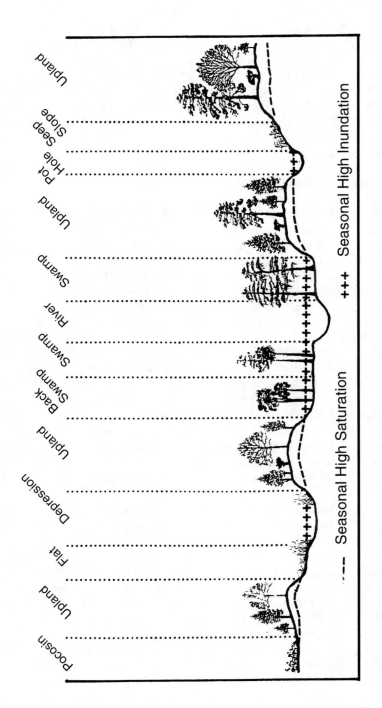

Figure 8.1 Idealized landscape depicting uplands and the hydric soil landforms pocosin, flat, depression, back swamp, swamp, pot hole, and seep slope. Note that each hydric soil area begins at a slightly concave slope break, although not all of each hydric soil area expresses concavity throughout the landform (seep slope). Vertical scale is exaggerated.

opment of recognizable soil properties and anaerobic soil conditions. Indicators that NTCHS has identified for testing are given in Table 8.3. Comments regarding the test indicators and field observations of hydric soil conditions that cannot be documented using the presently recognized hydric soil indicators are welcome; however, any modifications must be approved by NTCHS. Many of these test indicators are known to provide reliable guidelines for hydric soil delineation.

A minimal number of terms (Table 8.4) must be defined correctly to interpret Tables 8.2 and 8.3. To apply indicators properly, a basic knowledge of soil science, soil–landscape relationships, and soil survey procedures is also necessary. Many hydric soil indicators are landform specific. Professional soil or wetland scientists familiar with local conditions are best equipped to make an onsite hydric soil determination. Each Land Resource Region (LRR) and some Major Land Resource Regions (MLRA) have lists of indicators that have been approved by NTCHS for use and testing (Table 8.5). Geographic extent of LRRs (Figure 8.2) and MLRAs in the United States and Puerto Rico has been defined in USDA Ag. Handbook 296 (U.S. Department of Agriculture 1981).

Hydric Soil Indicators for Delineation and Identification

Table 8.6 differentiates those indicators used primarily for delineation and those used primarily for identification. Those identified as primarily identification hydric soil indicators usually occur in the wettest of wetlands and are normally saturated or inundated for much of most years, and those identified as primarily delineation hydric soil indicators occur at the much drier delineation boundary.

Indicators A1 (Histosols), A2 (Histic Epipedon), and A3 (Black Histic) are not normally used to identify the delineation boundary of hydric soils except possibly in Alaska (Land Resource Regions W, X, and Y). Other indicators with organic soil material (A8, A9, and A10) are used more often to delineate hydric soils. If indicator A1 is used to identify hydric soils, organic soil material and Histosol requirements contained in *Soil Taxonomy* must be met (U.S. Department of Agriculture, Soil Survey Staff, 1994, pp. 51–55, 58–59 and 305–323). If indicator A2 is used to identify hydric soils, all the requirements contained in *Soil Taxonomy* must be met (U.S. Department of Agriculture, Soil Survey Staff, 1994, pp. 4–5). Unlike indicators A1 and A2, no taxonomic requirements exist for A3. Indicator A3 identifies those Histic Epipedons that are always wet in natural conditions.

Indicators A4 (Hydrogen Sulfide), S4 (Sandy Gleyed Matrix), and F2 (Loamy Gleyed Matrix) are not normally used to identify the delineation boundary of hydric soils. Presence of the "rotten egg" odor for A4 and the gleying for S4 and F2 indicates the soils are very reduced for much of each year and would therefore identify only the wetlands saturated or inundated for very long periods. These three indicators normally occur inside the delineation line established by the delineation indicators.

Indicator A5 (Stratified Materials) is routinely used to delineate hydric soils on floodplains and some flats. Soils on the non-hydric side of delineations are stratified, but the chroma in one or more layers is 3 or higher.

Indicator A6 (Organic Bodies) is routinely used to delineate hydric soils dominantly on flats of the southern United States and Puerto Rico. Soils on the non-hydric side of delineations usually have organic accreted areas, but these bodies lack the required amount of organic carbon.

Indicators A7 (5 cm Mucky Mineral), A8 (Muck Presence), A9 (1 cm Muck), A10 (2 cm Muck), S1 (Sandy Mucky Mineral), S2 (3 cm Mucky Peat or Peat), S3 (5 cm Mucky Peat or Peat), and F1(Loamy Mucky Mineral) are routinely used to delineate hydric soils throughout various regions of the U.S. and Puerto Rico. Soils on the non-hydric side of delineations usually have surface layers that lack the required amount of organic carbon.

Indicators S5 (Sandy Redox), S6 (Stripped Matrix), and S7 (Dark Surface) are routinely used to delineate hydric soils throughout various regions of the U.S. and Puerto Rico. Soils on the non-hydric side of delineations usually lack chroma 2 or less within 6 inches of the surface (S5), have a layer that meets all the requirements of a stripped matrix except depth (S6), or the surface layer has a salt-and-pepper appearance (S7).

Table 8.4   Definition of Terms (These definitions are needed to understand certain terms used in Tables 8.2 and 8.3)

Abrupt Boundary — Used to describe redoximorphic features that grade sharply from one color to another. The color grade is commonly less than 0.5 mm wide. Clear and gradual are used to describe boundary color gradations intermediate between abrupt and diffuse.

Covered, Coated, Masked — These are terms used to describe all of the redoximorphic processes by which the colors of soil particles are hidden by organic material, silicate clay, iron, aluminum, or some combination of these.

Depleted Matrix — A depleted matrix refers to the volume of a soil horizon or subhorizon from which iron has been removed or transformed by processes of reduction and translocation to create colors of low chroma and high value. A, E, and calcic horizons may have low chromas and high values and may therefore be mistaken for a depleted matrix; however, they are excluded from the concept of depleted matrix unless common or many, distinct or prominent redox concentrations as soft masses or pore linings are present. In some places the depleted matrix may change color upon exposure to air (reduced Matrix); this phenomenon is included in the concept of depleted matrix. The following combinations of value and chroma identify a depleted matrix:

1. Matrix value 5 or more and chroma 1 or less with or without redox concentrations as soft masses and/or pore linings; or
2. Matrix value 6 or more and chroma 2 or less with or without redox concentrations as soft masses and/or pore linings; or
3. Matrix value 4 or 5 and chroma 2 and has 2% or more distinct or prominent redox concentrations as soft masses and/or pore linings; or
4. Matrix value 4 and chroma 1 and has 2% or more distinct or prominent redox concentrations as soft masses and/or pore linings.

Diffuse Boundary — Used to describe redoximorphic features that grade gradually from one color to another. The color grade is commonly more than 2 mm wide. *Clear* is used to describe boundary color gradations intermediate between *sharp* and *diffuse*.

Distinct — Readily seen but contrast only moderately with the color to which compared; a class of contrast intermediate between faint and prominent. In the same hue or a difference in hue of one color chart (e.g., 10YR to 7.5YR or 10YR to 2.5Y), a change of 2 or 3 units in chroma and/or a change of 3 units of value, or a change of 2 or 3 units of value and a change of 1 or 2 units of chroma, or a change of 1 unit of value and 2 units of chroma. With a change of 2 color charts of hues (e.g., 10YR to 5Y or 10YR to 5YR), a change of 0 to 2 units of value and/or a change of 0 to 2 units of chroma is distinct.

Faint — Evident only on close examination. In the same hue or 1 hue change (e.g., 10YR to 7.5YR or 10YR to 2.5Y) a change of 1 unit in chroma, or 1 to 2 units in value, or 1 unit of chroma and 1 unit of value.

Gilgai — A type of microrelief produced by expansion and contraction of soils that results in enclosed microbasins and microknolls.

Glauconitic — A mineral aggregate that contains micaceous mineral resulting in a characteristic green color, e.g., glauconitic shale or clay.

Gleyed Matrix — Soils with a gleyed matrix have the following combinations of hue, value, and chroma, and the soils are not glauconitic:

1. 10Y, 5GY, 10GY, 10G, 5BG, 10BG, 5B, 10B, or 5PB with value 4 or more and chroma is 1; or
2. 5G with value 4 or more and chroma is 1 or 2; or
3. N with value 4 or more; or
4. (For testing only) 5Y, value 4, and chroma 1.

In some places the gleyed matrix may change color upon exposure to air (reduced matrix). This phenomenon is included in the concept of gleyed matrix.

Hemic — See Mucky Peat.

Histic Epipedon — A thick (20 to 60 cm) organic soil horizon that is saturated with water at some period of the year unless artificially drained and is at or near the surface of a mineral soil.

Hydric Soil Definition (1994) — A soil that formed under conditions of saturation, flooding, or ponding long enough during the growing season to develop anaerobic conditions in the upper part.

Loamy and Clayey Soil Material — Refers to those soil materials with a USDA texture of loamy very fine sand and finer.

Muck — A sapric organic soil material in which virtually all of the organic material is decomposed not allowing for identification of plant forms. Bulk density is normally 0.2 g/cm³ or more. Muck has <1/6 fibers after rubbing, and sodium pyrophosphate solution extract color has lower value and chroma than 5/1, 6/2, and 7/3.

Mucky Modified Texture — A USDA soil texture modifier, e.g., mucky sand. Mucky modified mineral soil with 0% clay has between 5 and 12% organic carbon. Mucky modified mineral soil with 60% clay has between 11 and 18% organic carbon. Soils with an intermediate amount of clay have intermediate amounts of organic carbon.

Table 8.4   Definition of Terms (These definitions are needed to understand certain terms used in Tables 8.2 and 8.3) *(continued)*

Mucky Peat — A hemic organic material with decomposition intermediate between that of fibric and sapric organic material. Bulk density is normally between 0.1 and 0.2 g/cm^3. Mucky peat does not meet fiber content (after rubbing) or sodium pyrophosphate solution extract color requirements for either fibric or sapric soil material.

Organic Soil Material — Soil material that is saturated with water for long periods or artificially drained and, excluding live roots, has an organic carbon content of: 18% or more with 60% or more clay, or 12% or more organic carbon with 0% clay. Soils with an intermediate amount of clay have an intermediate amount of organic carbon. If the soil is never saturated for more than a few days, it contains 20% or more organic carbon. Organic soil material includes Muck, Mucky Peat, and Peat (Figure 8.2).

Peat — A fibric organic soil material with virtually all of the organic material allowing for identification of plant forms. Bulk density is normally <0.1 g/cm^3. Peat has 3/4 or more fibers after rubbing, or 2/5 or more fibers after rubbing and sodium pyrophosphate solution extract color of 7/1, 7/2, 8/2, or 8/3.

Prominent — Soils contrasting strongly with the color to which they are compared. In the same hue or a 1 hue change (e.g., 10YR to 2.5Y or 10YR to 7.5YR), a change of 4 units in chroma and/or 4 units in value. With a change of 2 hues (e.g., 10YR to 5Y or 10YR to 5YR), a change of 3 or more units of value and/or a change of 3 or more units of chroma is prominent.

Sandy Soil Material — Refers to those soil materials with a USDA texture of loamy fine sand and coarser.

Sapric — See Muck.

Sharp Boundary — Used to describe redoximorphic features that grade sharply from one color to another. The color grade is commonly less than 0.1 mm wide. *Clear* is used to describe boundary color gradations intermediate between *sharp* and *diffuse*.

Table 8.5   Hydric Soil Indicators by Land Resource Region (LRR) (The indicators approved for use or testing by the National Technical Committee for Hydric Soils are as follows)

LRR	Indicators
A	A1, A2, A3, A4, A10, S1, S4, S5, S6, F1 (except MLRA 1), F2, F3, F4, F5, F6, F7, F8, TF2, TF7.
B	A1, A2, A3, A4, A10, S1, S4, S5, S6, F1, F2, F3, F4, F5, F6, F7, F8, TF2, TF7.
C	A1, A2, A3, A4, A10, S1, S4, S5, S6, F1, F2, F3, F4, F5, F6, F7, F8, F9, TF2, TF7.
D	A1, A2, A3, A4, A9, S1, S4, S5, S6, F1, F2, F3, F4, F5, F6, F7, F8, F9, TA1, TF2, TF5, TF7, TF8.
E	A1, A2, A3, A4, A10, S1, S4, S5, S6, F1, F2, F3, F4, F5, F6, F7, F8, TF2, TF7.
F	A1, A2, A3, A4, A5, A9, S1, S3, S4, S5, S6, F1, F2, F3, F4, F5, F6, F7, F8, TS2, TS5, TF1, TF2, TF4, TF5, TF6, TF7.
G	A1, A2, A3, A4, A9, S1, S2, S4, S5, S6, F1, F2, F3, F4, F5, F6, F7, F8, TS3, TS5, TF1, TF2, TF6, TF7.
H	A1, A2, A3, A4, A9, S1, S2, S4, S5, S6, F1, F2, F3, F4, F5, F6, F7, F8, F16, TS5, TF1, TF2, TF7.
I	A1, A2, A3, A4, A9, S1, S4, S5, S6, F1, F2, F3, F4, F5, F6, F7, F8, TF2, TF7.
J	A1, A2, A3, A4, A9, S1, S4, S5, S6, F1, F2, F3, F4, F5, F6, F7, F8, TF2, TF7.
K	A1, A2, A3, A4, A5, A10, S1, S4, S5, S6, F1, F2, F3, F4, F5, F6, F7, F8, TS5, TF2, TF7.
L	A1, A2, A3, A4, A5, A10, S1, S4, S5, S6, F1, F2, F3, F4, F5, F6, F7, F8, TS5, TF2, TF7.
M	A1, A2, A3, A4, A5, A10, S1, S3, S4, S5, S6, F1, F2, F3, F4, F5, F6, F7, F8, F12, TS4, TS5, TF1, TF2, TF4, TF5, TF6, TF7, TF10.
N	A1, A2, A3, A4, A5, A10, S1, S4, S5, S6, S7, F2, F3, F4, F5, F6, F7, F8, F12, TF2, TF7, TF10.
O	A1, A2, A3, A4, A5, A9, S1, S4, S5, S6, F1, F2, F3, F4, F5, F6, F7, F8, F12, TF2, TF9.
P	A1, A2, A3, A4, A5, A6, A7, A9, S4, S5, S6, S7, F2, F3, F8, F12, F13.
R	A1, A2, A3, A4, A5, A10, S1, S3, S4, S5, S6, S7, S8, S9, F2, F3, F4, F5, F6, F7, F8, TA2, TF2, TF7.
S	A1, A2, A3, A4, A5, A10, S1, S4, S5, S6, S7, S8, S9, F2, F3, F4, F5, F6, F7, F8, TF2, TF4, TF10.
T	A1, A2, A3, A4, A5, A6, A7, A9, S4, S5, S6, S7, S8, S9, F2, F3, F6, F8, F11 (MLRA 151 only), F12, F13.
U	A1, A2, A3, A4, A5, A6, A7, A8, S4, S5, S6, S7, F2, F3, F5, F6, F10, F13.
V	A1, A2, A3, A4, A5, A8, S4, S5, S6, S7, F2, F3, F4, F5, F6, F7, F8, TF2, TF7.
W	A1, A2, A3, A4, A10, S10, F4, F5, F6, F7, F8, F14, F15, TS1, TF2, TF3, TF7.
X	A1, A2, A3, A4, A10, S10, F4, F5, F6, F7, F8, F14, F15, TS1, TF2, TF3, TF7.
Y	A1, A2, A3, A4, A10, S10, F4, F5, F6, F7, F8, F14, F15, TS1, TF2, TF3, TF7.
Z	A1, A2, A3, A4, A5, A6, A7, A8, S4, S5, S6, S7, F2, F3, F4, F5, F6, F7, F8, TF2, TF7.

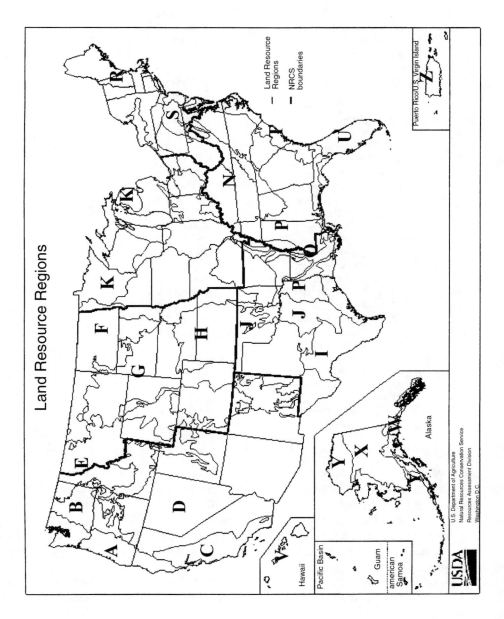

Figure 8.2   Land Resource Regions of the United States and Puerto Rico.

Table 8.6 Hydric Soil Indicators Used Primarily for Delineation and Identification

Hydric Soil Indicator		Type of Indicator
AI	Histosol	Identification
A2	Histic Epipedon	Identification
A3	Black Histic	Identification
A4	Hydrogen Sulfide	Identification
A5	Stratified Layers	Delineation
A6	Organic Bodies	Delineation
A7	5 cm Mucky Mineral	Delineation
A8	Muck Presence	Delineation
A9	1 cm Muck	Delineation
A10	2 cm Muck	Delineation
S1	Sandy Mucky Mineral	Delineation
S2	3 cm Mucky Peat or Peat	Delineation
S3	5 cm Mucky Peat or Peat	Delineation
S4	Sandy Gleyed Matrix	Identification
S5	Sandy Redox	Delineation
S6	Stripped Matrix	Delineation
S7	Dark Surface	Delineation
S8	Polyvalue Below Surface	Identification
S9	Thin Dark Surface	Identification
S10	Alaska Gleyed	Delineation
F1	Loamy Mucky Mineral	Delineation
F2	Loamy Gleyed Matrix	Identification
F3	Depleted Matrix	Delineation
F4	Depleted Below Dark Surface	Delineation
F5	Thick Dark Surface	Identification
F6	Redox Dark Surface	Delineation
F7	Depleted Dark Surface	Delineation
F8	Redox Depressions	Delineation
F9	Vernal Pools	Delineation
F10	Marl	Delineation
F11	Depleted Ochric	Delineation
F12	Iron/Manganese Masses	Delineation
F13	Umbric Surface	Delineation
F14	Alaska Redox Gleyed	Delineation
F15	Alaska Gleyed Pores	Delineation
F16	High Plains Depressions	Delineation

Indicators S8 (Polyvalue Below Surface) and S9 (Thin Dark Surface) are not normally used to identify the delineation boundary of hydric soils. These two indicators normally occur inside the delineation line established by indicators S5, S6, and/or S7.

Indicators S10 (Alaska Gleyed), F14 (Alaska Redox Gleyed), and F15 (Alaska Gleyed Pores) are routinely used to delineate hydric soils in Alaska. Soils on the non-hydric side of delineations usually lack chroma 2 or less within the required depths or lack the required amounts and kinds of redox features.

Indicators F3 (Depleted Matrix), F4 (Depleted Below Dark Surface), F6 (Redox Dark Surface), F7 (Depleted Dark Surface), and F13 (Umbric Surface) are routinely used to delineate hydric soils throughout various regions of the U.S. and Puerto Rico. Soils on the non-hydric side of delineations usually lack chroma 2 or less within the required depths or lack the required amounts and kinds of redox features (F3, F4, F6, and F8) or the surface layer is too thin or not dark enough (F13).

Indicator F5 (Thick Dark Surface) is not normally used to identify the delineation boundary of hydric soils. This indicator normally occurs inside the delineation line established by indicators F3, F4, F6, F7, and/or F13.

Indicators F8 (Redox Depressions), F9 (Vernal Pools), and F16 (High Plains Depressions) are used to delineate hydric soils that occur in closed depressions subject to ponding throughout various regions of the U.S. and Puerto Rico. Soils on the non-hydric side of delineations usually lack any redox features within the required depths.

Indicator F10 (Marl) is used to delineate hydric soils in southern Florida. Soils on the nonhydric side of delineations may meet all the requirements of marl, but the chroma is 2 or more or they are dry Histosols (Folists).

Indicators F11 (Depleted Ochric) and F12 (Iron/Manganese Masses) are used to delineate hydric soils that occur on floodplains that frequently flood dominantly in the southern U.S. Soils on the non-hydric side of delineations usually lack any redox features within the required depths.

Regional Hydric Soil Indicators

Why Regional Indicators?

During 1994 and 1995 a national team of soil scientists and other wetland scientists representing the U.S. Department of Agriculture Natural Resources Conservation Service (NRCS), the U.S. Environmental Protection Agency (EPA), the U.S. Department of Interior Fish and Wildlife Service (FWS), and the U.S. Army Corps of Engineers (COE), universities, and the private sector tested proposed indicators nationwide. Quickly, during the review process this team realized: (1) some indicators, such as A1 (Histosol), expressed a maximum of anaerobiosis and identified but did not delineate hydric soils nationwide, (2) some redox process driven indicators, such as S6 (Stripped Matrix) and F3 (Depleted Matrix), identified and delineated hydric soils virtually nationwide, and (3) some indicators, such as presence of muck, were good virtually nationwide but the required thickness varied by both latitude and longitude. For these reasons hydric soil indicators have been selected for use in each Land Resource Region (Table 8.5) of the United States and Puerto Rico. These are the only indicators approved by NTCHS for each specific LRR.

System for Regionalizing Indicators

Indicators were developed for very wet conditions nationwide by observing the centers of wetlands. In addition, by observing the delineation edge of ecological wetlands throughout the nation, indicators for edges were developed region by region. Other than the exceptions given above, few of the indicators can be used for delineation nationwide. The reasons for this is that an indicator is good if it separates uplands from wetlands in any area regardless of area size. More than 40 indicators have been developed for use and testing; however, rarely will more than a few indicators be used for delineation purposes in any specific region.

For example, 23 of the national indicators are identified for use or testing in LRR N, an area that includes all or parts of Arkansas, Missouri, Alabama, Georgia, North and South Carolina, Tennessee, Kentucky, Virginia, West Virginia, Pennsylvania, Maryland, Ohio, and Indiana. Of these 23, seven occur only in very wet areas (A1, A2, A3, A4, S4, F2, and F5) and are not needed to delineate wetlands. Of the remaining 16, four (F4, F6, F7, and TF7) are useful only for delineating wet Mollisols, three (A10, S1, and F1) are for delineating muck or mucky soils, and three others (S5, S6, S7) are for sandy soils only. Wet Mollisols, muck, mucky, and sandy soils are rare in LRR N. Therefore, to delineate most hydric soils in LRR N the number of indicators with which one must become proficient is six (A5, F3, F8, F12, TF2, and TF10).

Twenty indicators are for use or testing in California's LRR C. Of these, seven identify very wet conditions (A1, A2, A3, A4, S4, F2, and F5) and four (F4, F6, F7, and TF7) are for very dark Mollisols, which are rare in LRR C. Therefore the number of indicators with which one must become proficient is nine for all of LRR C. If one's area of interest is Sacramento and San Joaquin

Valleys, the number of indicators needed is only four (F3, F8, F9, and F12). Seventeen indicators are for use or test in LRR P. Of these, six (A1, A2, A3, A4, S4, and F2) are for identifying very wet conditions. The remaining 11 indicators are needed to delineate hydric soils in LRR P. It is not necessary to become familiar with all of the 42 different hydric soil indicators. It is only necessary to become familiar with the few indicators used to delineate hydric soils in any given area.

Additional investigations are needed to fine-tune and verify the validity of certain test hydric soil indicators. However, the present body of knowledge is sufficient to allow accurate hydric soil delineations throughout most of the United States. These delineations are integrated with additional spatial data for evaluating, assessing, designing, planning, managing, and regulating wetlands.

Hydric Soil Determination and Delineation

To document a hydric soil first remove all loose leaf matter, bark, and other easily identified plant parts (often called "duff") to expose the surface. Dig a hole approximately 20 to 30 cm in diameter and describe the soil profile to a depth of approximately 50 cm using the procedures outlined in the *Soil Survey Manual* (U.S. Department of Agriculture 1993b). Using the completed soil description, specify which of the hydric soil indicators (Tables 8.2 and 8.3) have been identified.

Deeper examination of soil may be required where hydric soil indicators are not easily seen within 50 cm of the surface. It is always recommended that soils be excavated and described as deep as necessary to make reliable interpretations. For example, examination to less than 50 cm may suffice in soils with surface horizons of organic material or mucky mineral material (Chapter 1) because these shallow organic accumulations only occur in hydric soils. Conversely, depth of excavation will often be greater than 50 cm in Mollisols because the upper horizons of these soils, due to the masking effect of organic material, often contain no visible redoximorphic features. At many sites, making exploratory observations to a meter or more is necessary. These observations should be made with the intent of documenting and understanding the variability in soil properties and hydrologic relationships on the site.

Depths used in making hydric soil determinations are measured from the muck or, if muck is not present, the mineral soil surface unless otherwise indicated. All colors refer to moist Munsell colors (Kollmorgen Instruments Corporation 1994). The Munsell value and chroma required in the indicators are whole numbers, but colors do occur between Munsell chips. Soil colors can match individual chips or, as in many cases, occur between two adjacent chips. For example, a soil matrix with a chroma between 2 and 3 should be listed as having a chroma of 2+. This soil material does not have a chroma 2 and would not meet any indicator that requires a chroma 2 or less.

The process of delineating hydric soil boundaries on undisturbed landscapes is really rather simple in concept but can be difficult in practice. Where the landscape is undisturbed, the upland boundary of hydric soils is at a landform change. That change is usually a convex/concave slope break. Hydric soils occur at the concave slope change, and soils that are non-hydric occur at the convex slope change. The slope break may be very subtle or hidden with vegetation, but it will be there. Often the boundary delineates a very intricate pattern of extremely small areas of hydric soils and soils that are non-hydric.

The easiest way to delineate hydric soils is to begin on the upland side of a wetland and traverse toward the wetland looking for concave slope breaks. Not all concave slope breaks delineate hydric soils; however, the hydric/non-hydric boundary of undisturbed soils will usually be at a concave slope break (see the section on Disturbed Soils for an explanation of how to delineate these soils). By traversing once or twice, the hydric soil boundary can frequently be located. Once the boundary is located, using vegetation is most expeditious (or, where vegetation is absent, the landform change convex to concave slope break) in completing the delineation. Most often, if vegetation is present, one or two species can be correlated to the hydric soil boundary and thereby provide the key to a correct delineation. For example, in the flatwoods and associated landform areas of the southeastern United States (LRRs T and U), the uplands have the shrub saw palmetto (*Serenoa repens* L.), which disappears near the hydric soil boundary to be replaced by other shrubs, such as gallberry (*Ilex glabra* L.), and fetterbush *(Lyonia lucida* L.), in LRR T or by herbaceous plants, such as blue maidencane (*Amphicarpum mulenbergianum* L.), in LRR U.

Understanding that the field indicators are indicators known to identify hydric soils is important. They were developed by observing soil pedons both inside and outside ecological wetlands. Pedons inside the line were described; descriptions of pedons outside the line were not deemed necessary. For example, S7 (Dark Surface) requires a layer 10 cm or more thick starting within the upper 15 cm of the soil surface with a matrix value 3 or less and chroma 1 or less. In this layer at least 70% of the visible soil particles must be covered, coated, or similarly masked with organic material, and the matrix color of the layer immediately below the dark layer must have chroma 2 or less. This does not mean that the pedons outside the hydric soil boundary had all requirements of this indicator except thickness of the dark surface. It means that, because of the concave slope break, pedons outside the line are normally very dissimilar to pedons inside the line. Normally, neighboring pedons outside the line have a surface layer that has a salt-and-pepper appearance and is more of a 50/50 mixture of soil material covered, coated, or masked with organic material (pepper) and soil material not covered, coated, or masked (salt).

Vertical and Horizontal Soil Variability

Soil variability occurs vertically within a soil and horizontally across the landscape. Vertical variability is related to depositional and soil-forming processes. Horizontal variability is related to vertical variability and to site-specific landscape expressions of geomorphic processes; therefore, in most soils that represent simple landforms, soil variability is relatively unimportant in making a hydric soil determination. Most soils identified to be hydric on a specific landform are hydric throughout the extent of that landform, and where an upland is encountered the soils are no longer hydric. Soils have a vertical sequence of horizons that have perceptible and predictable changes with depth; however, a significant portion of soils with high shrink/swell potential are different. In the section "Problem Hydric Soil Delineations," high shrink/swell soils and other difficult to delineate hydric soils will be discussed.

DISCHARGE VS. RECHARGE HYDRIC SOILS

Discharge hydric soils release groundwater to the land surface through springs, seeps, and other discharge zones. Recharge hydric soils transmit water to the groundwater/aquifer and to discharge hydric soils. Hydric soils in the humid southeastern and eastern United States generally are discharge hydric soils; however, they may function as season dependent recharge systems. Both recharge and discharge hydric soils exist in the sub-humid Midwest, Southwest, and West of the United States (see Chapter 3 for more discussion concerning this topic). The significance to hydric soils is that discharge systems generally have different morphological indicators than recharge systems. Classic discharge hydric soils have morphologies that reflect water moving to the soil surface. This water carries materials, such as reduced Fe, and these become part of the soil. Discharge hydric soils below a depth of about 0.5 m often lack evidence of saturation, most often because of additions of Fe and low available organic matter needed for microbial activity. The following are examples of discharge hydric soil indicators: A5 (Stratified Layers), TA1 (Playa Rim Stratified Layers), F3 (Depleted Matrix) where the depleted matrix is the surface layer, F8 (Redox Depressions, F9 (Vernal Pools), F12 (Iron/Manganese Masses), F16 (High Plains Depressions), TF2 (Red Parent Materials), and TF8 (Redox Spring Seeps).

Recharge hydric soils are wettest at the surface and remain wet there longer than discharge hydric soils. The amount of organic matter and microbial activity is very high, and these hydric soils have maximum expressions of anaerobiosis. Recharge activities often leach soils, creating acidity. The acidity may be reflected in plants that produce tannin. Tannins in turn create organic surfaces that aid in holding water for anaerobiosis. Classic recharge indicators include A1 (Histosols), A3 (Black Histic), S2 (Sandy Gleyed Matrix), F2 (Loamy Gleyed Matric), F3 (Depleted Matrix) where the depleted matric is not the surface layer and is continuous, F5 (Thick Dark Surface), and TF7 (Thick Dark Surface 2/1).

Indicators not specified as one of the discharge or recharge hydric soil indicators above have either discharge/recharge dependent morphologies or they are for hydric soils that function as season dependent discharge and recharge hydric soils. For example, indicator F6 (Redox Dark Surface) occurs in both discharge and recharge hydric soils. In recharge hydric soils, the layer immediately below the dark surface should have a depleted or gleyed matrix. In discharge hydric soils, the depleted/gleyed matrix may be absent below the dark surface. It is recommended that delineators evaluate the hydrologic source and examine soils accordingly.

PROBLEM HYDRIC SOIL DELINEATIONS

High Shrink/Swell Potential Soils

Most soils with high shrink/swell potential (Vertisols) have microvariability within a soil body (pedon). Vertical sequence (horizons) and horizontal variability vary greatly within a short distance from any point (Williams et al. 1996). For determining the hydric status of Vertisols, understanding this variability is important. As a result of the required slickensides or wedge-shaped aggregates, Vertisols have micro lows and micro highs that are approximately 2 to 5 m from the centers of the lows to the centers of the highs. A maximum expression of the subsurface highs and lows is gilgai (U.S. Department of Agriculture 1975). Micro lows have more organic carbon, less gypsum and carbonates, a higher coefficient of linear extensibility, and a higher probability of being hydric than micro highs.

Hydric high shrink/swell soils (Vertisols) are hydric because of surface inundation. Vertisols become hydric where water remains on the surface long enough for anaerobic bacteria to deplete the soil water of O_2. Rarely does the resulting anaerobiosis penetrate into the soil to a significant depth. Therefore, it is important to look at near-surface soil morphologies to determine the hydric status of Vertisols. Vertisols in depressions (both large scale and micro lows) and other concave landforms are most often hydric. Vertisols in micro highs are most often non-hydric.

The exact extent of hydric Vertisols in a particular area is highly variable. For example, hydric Vertisols of the Alabama, Mississippi, and Arkansas Blackland Prairies region of LRR P occur exclusively on depressional landforms of floodplains and lack significant gilgai relief. In these soils the hydric soil indicator F3 (Depleted Matrix) is used almost exclusively to delineate hydric Vertisols (Table 8.2, Table 8.4). The depleted matrix has a dominant value 4 or more and chroma 2 or less. It must begin within 25 cm of the soil surface and continue for at least 15 cm; the minimum thickness requirement is 5 cm if the depleted matrix is within 15 cm of the soil surface. Redox concentrations including iron/manganese soft masses and/or pore linings are required in soils with matrix values/chromas of 4/1, 4/2, and 5/2. Where the depleted matrix is within 25 cm of the mineral soil surface the soil is hydric. Vertisols that are non-hydric have a depleted matrix starting below 25 cm or have chroma 3 or more in a surface layer that is more than 15 cm thick. Approximately 5 to 25% of the Vertisols occur in depressional positions on the floodplains of the Alabama, Mississippi, and Arkansas Blackland Prairies (LRR P) and are hydric; the remaining 75 to 95% are non-hydric (estimated from personal observations).

Hydric Vertisols in the Mississippi River Delta of LRR O also occur on depressional landscape positions. Indicator F3 (Depleted Matrix) is also used to delineate these Vertisols. The depressional hydric Vertisols of the Mississippi River Delta are easy to delineate; most often they pond water for much of the year. However, Vertisols in this area that do not occur in depressions are difficult to delineate. Large areas may be either entirely hydric or entirely non-hydric, and large areas may have hydric and non-hydric soils so intricately mixed that separation of individual areas of hydric and non-hydric soils is extremely difficult. All of the Vertisols occurring in depressional landforms in the Mississippi Delta are hydric. Approximately 70% of the Vertisols that are not in depressional landforms are hydric. Rises, knolls, and micro highs in gilgai Vertisols normally lack indicator F3 (Depleted Matrix). Vertisols in a particular region may be almost entirely hydric or almost entirely non-hydric.

Playas

The term *playa*, is used to describe two different conditions. One occurs in the saline and alkaline flats of LRR D, extending from Oregon to New Mexico. The other occurs in the depressional landscapes of the High Plains in LRR H from Nebraska to Texas (Mitsch and Gosselink 1993). Saline and high plains playas are hydric due to wetness from surface water and not from below-ground saturation. Therefore the indicators used to delineate them are based on near-surface morphological characteristics.

Saline Playas

Playas in LRR D are either sparsely vegetated or they are not currently capable of producing vegetation because of high salinity and/or alkalinity. They range in size from less than a hectare to many thousands of hectares. These areas are characteristically lacking in any significant pedological development or morphology that results in the hydric soil characteristics. Although the non-vegetated areas are not considered soils by accepted USDA definition, two indicators have been developed to delineate the boundary of hydric conditions in saline playas. For sparsely vegetated saline playas of LRR· D, indicator TA1 (Playa Rim Stratified Layers) is used (Table 8.3). A dominant species on the sparsely vegetated playas is *Allenrolfea occidentalis* L. The areas that are hydric due to the occurrence of the indicator TA1 are the areas that lack vegetation (Figure 8.3). Usually the difference in elevation between the vegetated areas and non-vegetated areas is more than 8 to 15 cm. Areas where vegetation occurs and the difference in elevation between the vegetated and non-vegetated areas is less than 8 to 15 cm are non-hydric (indicator TA1 is not present). Indicator TF8 (Redox Spring Seeps) occurs in seeps and flow-through areas adjacent to springs and playas (Table 8.3).

Although the majority of saline playas lack a morphological indicator of hydric soils, indicators TA1 and TF8 allow for the proper delineation of hydric conditions at their upland boundary and thereby satisfy the main intent of the indicators (delineation of hydric soils). The vegetated rims of the saline playas have the indicators TA2 and TF8 and, although the lower non-vegetated areas, due to the lack of pedogenic processes, often lack hydric soil indicators, they are presumed to meet the intent of the hydric soil definition. Other nonsoil areas that may lack morphologies characteristic of hydric soils include beaches, riverwash, salt flats, slickens, and slickspots (U.S. Department of Agriculture 1993a)

High Plains Playas

The playas of LRR H are not true saline playas. They are vegetated depressions. Two indicators were developed to delineate the areas of hydric soils in these high plains playas. Indicator F8 (Redox Depressions) is restricted to use in closed depressions subject to ponding. Hydric soils are recognized where 5% or more distinct or prominent redox concentrations occur as soft masses or pore linings in a layer 5 cm or more thick entirely within the upper 15 cm of the soil surface. Indicator F16 (High Plains Depressions) is also restricted to use in closed depressions subject to ponding. This indicator is used to recognize hydric soils where a layer at least 10 cm thick within

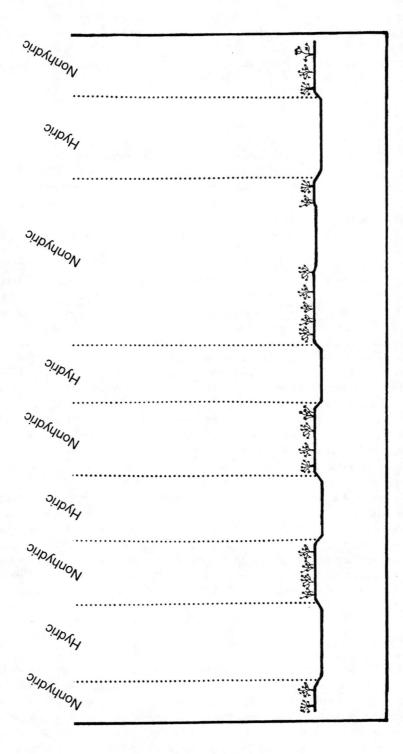

Figure 8.3 Saline playas rims have hydric soils in areas that lack vegetation and where the difference in elevation between the vegetated areas and nonvegetated areas is more than 8 to 15 cm. Areas where vegetation occurs and where the difference in elevation between the vegetated areas and nonvegetated areas is less that 8 to 15 cm are non-hydric. Individual areas of vegetated and nonvegetated areas are approximately 20 to 100 meters wide. Vertical scale is exaggerated.

the upper 35 cm of the mineral soil has a chroma 1 or less and 1% or more redox concentrations as nodules or concretions with distinct or prominent coronas. These two indicators (Table 8.2) along with indicators F3 and F6 are most often used to differentiate the hydric playas from non-hydric playas. Indicator F8 occurs most often in Texas and Oklahoma, and F16 most often in Kansas and Nebraska. F3 and F6 occur throughout the U.S. The high plains playas are depressional landforms that either have or lack indicators F3, F6, F8, and/or F16. Where present, the playas are hydric; where lacking, the playas are non-hydric.

Soils with Red Parent Material

Soils with red parent material often present a delineation challenge. These soils occur in areas such as the Triassic/Jurassic sediments in the Connecticut River valley, the Permian "Red Beds" in Kansas, clayey red till and associated lacustrine deposits around the Great Lakes, Jurassic sediments associated with "hogbacks" on the eastern edge of the Rocky Mountains, and river alluvium of rivers such as the Red, Congaree, Chattahoochee, and Tennessee. The indicator TF2 (Red Parent Material) was developed specifically for hydric soil delineation in areas with red parent material. Other indicators useful in delineating the hydric component of soils that formed in red parent material include F8 (Redox Depression), F9 (Vernal Pools), F12 (Iron/Manganese Masses), and F3 (Depleted Matrix), where the depleted matrix is within 15 cm of the soil surface.

Disturbed Soils

Hydric soil determinations and delineating hydric soils in areas that have been filled, dredged, land leveled, or otherwise disturbed can be a difficult and an extremely challenging assignment. In some instances of disturbance, the vegetation has been destroyed or removed; therefore, soils are the only onsite indicator of predisturbance hydrology and the only feasible means of identifying wetlands. Where upturned soil disturbance is recent, sufficient clods of various soil horizons may remain that will aid experienced soil scientists in verifying the original soil morphology. Predisturbance soil surveys should be consulted where available. Undisturbed areas in the vicinity may be investigated to provide information of predisturbance soil morphology. Small areas of unaltered soil may be found at the base of remaining trees; however, most frequently, the disturbance is more extreme. Fill materials spread on disturbed sites usually compound the difficulties of making hydric soil determinations. Guidelines have been established to determine the hydric status of disturbed soils after varying amounts of fill materials have been added. These guidelines are based on insights and observations and are not to be considered official guidance for CWA and FSA use.

Hydric soil requirements are the same for disturbed areas as they are for undisturbed areas. Most significantly, the hydric soil definition must be satisfied (*Federal Register,* July 13, 1994). This is normally exemplified by the presence of a hydric soil indicator (Hurt and Carlisle 1997). The amount of fill that can be placed on a hydric soil and still allow that soil to be considered hydric is directly related to the hydric soil indicator and the reason (inundation or saturation) it is hydric prior to filling.

Although the areas that meet criteria 1, 3, and 4 usually have a hydric soil indicator, this is not a requirement. According to the deliberations of NTCHS, areas that satisfy criteria 1, 3, or 4 are considered hydric if they are anaerobic in the upper part regardless of whether an indicator is present or not (*Federal Register,* February 24, 1995). It is important to remember that the criteria must be met based on actual data and not the estimated soil properties, such as those found in the National Cooperative Soil Survey publications.

For areas that meet the requirements for Histosols, except Folists (Criteria 1), fill can be placed on the soil surface to the extent that the soil, after the placement of the fill, still meets the taxonomic requirements of Histosols (U.S. Department of Agriculture 1996). Therefore, the maximum amount of fill material that can be added to a hydric Histosol and still have that soil retain its hydric status is 40 cm (60 cm if 3/4 or more of the organic soil material is moss fibers). This would apply to hydric Histosols that have organic soil material starting at the soil surface that is 40 or more cm thick (60 or more cm thick if 3/4 or more of the organic soil material is moss fibers). For Histosols with thinner organic layers to maintain their hydric status, the thickness would be less. The indicator that meets this indicator is A1 (Histosol). Table 8.2 provides additional information concerning this and all other indicators developed to help identify and delineate hydric soils and that are approved for such use by NTCHS.

For soils that are frequently ponded for long or very long duration during the growing season (Criteria 3), or soils that are frequently flooded for long or very long duration during the growing season (Criteria 4) to maintain their hydric status after filling, the added fill may be very thick or very thin. The thickness of the fill must be slightly less than the height of frequent ponding or flooding of long duration (more than 7 days). This height may be either measured or estimated. If estimated, professional judgment that the definition (anaerobiosis) is met must be carefully exercised. Although any of the indicators listed in Table 8.2 may occur on inundated landforms, indicators A4 (Hydrogen Sulfide), F8 (Redox Depressions), F9 (Vernal Pools), F12 Iron/Manganese Masses), F13 (Umbric Surface), and F16 (High Plains Depressions) are restricted to inundated landforms.

For other soils that are hydric due to saturation (Criteria 2) an indicator should be present (Hurt et al. 1998). The depth of fill that can be placed on these soil in order to maintain their hydric status is variable. The range is from slightly less than 15 cm to 0 cm in soils with sandy soil materials and the range is from slightly less than 30 cm to 0 cm in other soils. After fill materials are added, an indicator must be present in the original soil material within the prescribed depths in order for that soil to retain its hydric status. Table 8.7 can be used to determine the depth of fill material that would adversely affect the hydric status of a soil that is hydric due to saturation. This table is not to be used for indicator A1 and the indicators restricted to inundated landforms (F8, F9, F11, F12, and F16).

Soils with an indicator starting depth that is intermediate to those listed in Table 8.7 can have an intermediate amount of fill without changing the hydric status of the soil. For example, a soil with a stripped matrix starting at 10 cm can have up to 5 cm of any type of fill material placed on the surface without changing the hydric status of the soil. Conversely, more than 5 cm of fill would change the status of the soil to non-hydric.

Table 8.7　Hydric Status of Soils with Varying Amounts of Fill[1]

Depth to Indicator	Type of Indicator[2]	Thickness of Fill Material	Hydric Status
Surface	All, Sandy	Up to 15 cm	Hydric
Surface	All, Sandy	More than 15 cm	Non-hydric
Surface	Loamy or Clayey	Up to 30 cm[3]	Hydric
Surface	Loamy or Clayey	More than 30 cm[3]	Non-hydric
15 cm	All, Sandy	Zero	Hydric
15 cm[4]	All, Sandy	More than zero	Non-hydric
15 cm	Loamy or Clayey	Up to 15 cm[4]	Hydric
30 cm[3]	Loamy or Clayey	Zero	Hydric
30 cm[3]	Loamy of Clayey	More than zero	Non-hydric

[1] This table is used to determine the depth of fill material that would adversely affect the hydric status of a soil that is hydric due to saturation and is based on the presence or absence of an indicator; however, if an indicator is absent, a soil may well be hydric if, according to NTCHS guidance, the definition is met.

[2] See *Field Indicators of Hydric Soils in the United States* for additional information concerning use of All (A), Sandy (S), and Loamy and Clayey (F) indicators.

[3] Depths and thicknesses would be 10 inches if the indicator present is F3 (Depleted Matrix).

[4] Depth would be 4 inches if the indicator present is F3 (Depleted Matrix).

To determine the hydric status of land-leveled areas the same procedure as outlined above is used. Soils that are hydric due to criteria 1, 3, or 4 prior to land leveling are evaluated after land leveling to determine their hydric status. Soils that are hydric due to saturation prior to land leveling are evaluated by applying the guidelines outlined in Table 8.6 to determine their hydric status.

The presence of structures that provide increased drainage (ditches, tile drains, etc.) and protection from ponding and/or flooding (dikes, levees, etc.) does not alter the hydric status of a soil.

REFERENCES

Best, G.R., D. Segal, and C. Wolfe. 1990. Soil–vegetation correlations in selected wetlands and uplands of north-central Florida. U.S. Fish and Wildlife Service. Biol. Rep. 90(9), Washington, DC.

Christensen, N.L., R.B. Wilbur, and J.S. McLean. 1988. Soil–vegetation correlations in the pocosins of Croatan National Forest. U.S. Fish and Wildlife Service Biol. Rep. 88(28), Washington, DC.

Cowardin, L.M., V. Carter, F.C. Golet, and E.T. LaRoe. 1979. Classification of wetlands and deepwater habitats of the United States. U.S. Fish and Wildlife Service, FWS/OBS-79/31, Washington, DC.

Eicher, A.L. 1988. Soil–plant correlations in wetlands and adjacent uplands of the San Francisco Bay Estuary, California. U.S. Fish and Wildlife Service Biol. Rep. 88(21). Washington, DC.

Environmental Laboratory. 1987. *Corps of Engineers Wetland Delineation Manual.* Technical Report Y-87-1. U.S. Army Engineers Waterways Experiment Station, Vicksburg, MS.

Erickson, N.E. and D.M. Leslie, Jr. 1989. Soil–vegetation correlations in coastal Mississippi wetlands. U.S. Fish and Wildlife Service Biol. Rep. 89(3). Washington, DC.

Faulkner, S.P., W.H. Patrick, Jr., R.P. Gambrell, W.B. Parker, and B.J. Good. 1991. Characterization of soil processes in bottomland hardwood wetland–nonwetland transition zones in the lower Mississippi River Valley. U.S. Army Corps of Engineers, Waterways Experiment Station, WRP-91-1. Vicksburg, MS.

Federal Register. July 13, 1994. Changes in hydric soils of the United States. Washington, DC.

Federal Register. Feb. 24, 1995. Hydric soils of the United States. Washington, DC.

Florida Soil Survey Staff. 1992. Soil and water relationships of Florida's ecological communities. G.W. Hurt (Ed.). USDA, Soil Conservation Service, Gainesville, FL.

Hammer, D.A. 1992. Creating freshwater wetlands. pp. 20–37. Lewis Publishers, Boca Raton, FL.

Hurt, G.W. and R.B. Brown. 1995. Development and application of hydric soil indicators in Florida. *Wetlands* 15(1):74–81.

Hurt, G.W. and V.W. Carlisle. 1997. Proper use of hydric soil terminology. *Soil Survey Horizons* 38(4):98–101.

Hurt, G.W., P.M. Whited, and R.F. Pringle (Eds.). 1998. *Field Indicators of Hydric Soils in the United States.* Version 4.0. USDA, NRCS, Forth Worth, TX.

Kollmorgen Instruments Corporation. 1994. *Munsell Soil Color Charts.* Munsell Color, Baltimore, MD.

Mausbach, M.J. and J.L. Richardson. 1994. Biogeochemical processes in hydric soils. pp. 68–127. *In Wetland Biogeochemistry, Volume 1.* Wetlands Biogeochemistry Institute, Louisiana State University, Baton Rouge, LA.

Mitsch, W.J. and J.G. Gosselink. 1993. *Wetlands.* Van Nostrand Reinhold, New York.

Reed, P.B., Jr. 1988. National list of plant species that occur in wetlands: 1988 national summary. U.S. Fish and Wildlife Service, Biol. Rep. 88(24). Ft. Collins, CO.

Segal, S.D., S.W. Sprecher, and F.C. Watts. 1995. Relationships between hydric soil indicators and wetland hydrology for sand soils in Florida. U.S. Army Corps of Engineers, Waterways Experiment Station, WRP-DE-7. Vicksburg, MS.

Skaggs, R.W., D. Amatya, R.O. Evans, and J.E. Parsons. 1991. Methods of evaluating wetland hydrology. ASAE Paper No. 912590, American Society of Agricultural Engineering, St. Joseph, MI.

Skaggs, R.W., D. Amatya, R.O. Evans, and J.E. Parsons. 1994. Characterization and evaluation of proposed hydrology criteria for wetlands. *J. Soil and Water Conservation* 49(5):501–510.

Tiner, R.W. and D.G. Burke. 1995. Wetlands of Maryland. pp. 7. U.S. Fish and Wildlife Service, Hadley, MA and Maryland Department of Natural Resources, Annapolis, MD, Cooperative publication.

Tuttle, S.D. 1975. *Landforms and Landscapes.* Wm. C. Brown Company, Dubuque, IA.

U.S. Department of Agriculture, Soil Conservation Service. 1981. *Land Resource Regions and Major Land Resource Areas of the United States.* USDA–SCS Agricultural Handbook 296. U.S. Govt. Printing Office. Washington, DC.

U.S. Department of Agriculture, Soil Conservation Service. 1994. *National Food Security Act Manual.* 3rd edition. USDA, NRCS, Washington, DC.

U.S. Department of Agriculture, Soil Survey Staff. 1993a. *National Soil Survey Handbook.* USDA, Soil Conservation Service, U.S. Govt. Printing Office. Washington, DC.

U.S. Department of Agriculture, Soil Survey Staff. 1993b. *Soil Survey Manual.* USDA Agricultural Handbook 18. U.S. Govt. Printing Office. Washington, DC.

U.S. Department of Agriculture, Soil Survey Staff. 1975. *Soil Taxonomy: A Basic System of Soil Classification For Making and Interpreting Soil Surveys.* USDA Agricultural Handbook 436. U.S. Govt. Printing Office. Washington, DC.

U.S. Department of Agriculture, Soil Survey Staff. 1996. *Keys to Soil Taxonomy,* 7th edition. U.S. Govt. Printing Office. Washington, DC.

Venemann, P.L.M. and R.W. Tiner. 1990. Soil–vegetation correlations in the Connecticut River floodplain (sic.) of western Massachusetts. U.S. Fish and Wildlife Service Biol. Rep. 90(6). Washington, DC.

Vepraskas, M.J. 1994. *Redoximorphic Features for Identifying Aquic Conditions.* Tech. Bulletin 301. North Carolina Ag. Research Service, North Carolina State Univ., Raleigh, NC.

Williams, D., T. Cook, W. Lynn, and H. Eswaran. 1996. Estimating the field morphology of Vertisols. Soil Survey Horizons, 37(4):123–131.

PART II

Wetland Soil Landscapes

CHAPTER 9

Wetland Soils and the Hydrogeomorphic Classification of Wetlands

J. L. Richardson and Mark M. Brinson

INTRODUCTION

Purpose of Classification

The hydrogeomorphic (HGM) classification for wetlands was developed as a starting point for applying functional assessments used in the determination of the effects of impacts (Brinson 1993a). The main purpose of the classification was to aggregate wetlands with similar geomorphic settings so that altered or degraded conditions could be evaluated relative to unaltered states. In so doing, wetland assessments could be tailored to a much narrower range of natural variation than if a single assessment procedure was designed for all wetland classes. By controlling for the degree of natural variation through classification, alteration due to human activities could be detected more effectively (Smith et al. 1995). Although *Soil Taxonomy* (Soil Survey Staff 1975) was not taken into account in this classification, we recognize that many of the same factors associated with geomorphic setting are also related to soil. Geomorphic setting of wetlands is used in this chapter to examine the correspondence between wetland characteristics and soils.

The hydrogeomorphic classification considers two additional factors as critical to the functioning of wetlands: dominant source of water and hydrodynamics (Brinson 1993a). We recognize that these three factors are highly interdependent and autocorrelated, just as climate is highly influential on soil-forming processes in wetlands (Richardson 1997). Since the original publication, the classes have been modified (Brinson 1993a) into seven hydrogeomorphic groups: riverine, depressional, slope, mineral soil flats, organic soil flats, estuarine fringe, and lacustrine fringe (Table 9.1). These classes differ from but do not substitute for the Fish and Wildlife Service (FWS) classification (Cowardin et al. 1979). For example, National Wetland Inventory maps are useful for functional assessment in addition to their usefulness in other aspects of wetland resource management. A major difference in nomenclature, however, is that the HGM riverine class refers to the entire river and its floodplain, while the FWS classification encompasses only the channel, bank to bank.

1-56670-484-7/01/$0.00+$.50

Table 9.1 Hydrogeomorphic Classes of Wetlands Showing Associated Dominant Water Sources, Hydrodynamics, and Examples of Subclasses

Hydrogeomorphic Class	Dominant Water Source	Dominant Hydrodynamics	Examples of Subclasses	
			Eastern U.S.	Western U.S. & Alaska
Riverine	Over bank flow from channel	Unidirectional and horizontal	Bottomland hardwood forests	Riparian forested wetlands
Depressional	Return flow from groundwater & interflow	Vertical	Prairie pothole marshes	California vernal pools
Slope	Return flow from groundwater	Uni-directional, horizontal	Fens	Avalanche chutes
Mineral soil flats	Precipitation	Vertical	Wet pine flatwoods	Large playas
Organic soil flats	Precipitation	Vertical	Peat bogs; portions of Everglades	Peat bogs
Estuarine fringe	Over bank flow from estuary	Bidirectional, horizontal	Chesapeake Bay marshes	San Francisco Bay marshes
Lacustrine Fringe	Over bank flow from lake	Bidirectional, horizontal	Great Lakes marshes	Flathead Lake marshes

From Brinson, M. M., F. R. Hauer, L. C. Lee, W. L. Nutter, R. D. Rheinhardt, R. D. Smith, and D. Whigham. 1995. *Guidebook for Application of Hydrogeomorphic Assessments to Riverine Wetlands.* Technical Report TR-WRP-DE-11, Waterways Experiment Station, Army Corps of Engineers, Vicksburg, MS.

Origin and Driving Forces of Hydrologic and Geomorphic Classifications

Position and movement of water in landscapes explain the distribution of wetlands that result in the separation of most landscapes into drier uplands and more moist wetlands. While the location of the boundary between the two has initiated much debate (Committee on Characterization of Wetlands 1995), the boundary is really a part of a landscape continuum that is maintained by precipitation, which varies in both frequency and intensity. Once a landscape receives precipitation, the water is redistributed until it is exported via stream flow, groundwater flow, or evapotranspiration. In humid climates that support well-vegetated landscapes, the runoff factor is reduced to near zero. In warm, dry climates with low vegetation cover, evaporation becomes a more significant loss of water from the hydrologic cycle in contrast to more humid regions.

In the sections that follow, we will describe each of the geomorphic settings, consider the nature of their hydrology with illustrations, and cite at least one example of a soil hydrosequence for each of the settings. Our rationale is that the current hydrologic processes are similar to those that have taken place in the past and have been largely responsible for the soil-forming processes (Richardson 1997). Generally, soils reflect the long-term hydrology in wetlands, with the exception of relict soils. Because of extensive coverage in other publications, organic soil flats (peatlands) will not be discussed in any detail here.

DESCRIPTION OF CLASSES

Depressional Wetlands

Geomorphic Setting

Depressional wetlands occur in basins that lie below the surrounding topography. Figure 9.1 depicts a basin containing a wet meadow in the center that is surrounded by a low prairie. This depression is typical of thousands in the prairie pothole glacial terrain (Stewart and Kantrud 1971).

Depressional wetlands have restricted surface outflow due to closed topographic contours. Examples include interdunal areas in the sandhills of Nebraska, kettle depressions in till, and some

TEMPORARY WETLAND

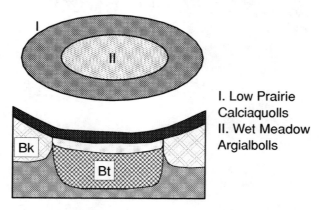

I. Low Prairie
Calciaquolls
II. Wet Meadow
Argialbolls

Figure 9.1 A recharge wetland or temporary pond with wet meadow vegetation. The cross-section illustrates the soil distribution. (Adapted from Knuteson, J. A., J. L. Richardson, D. D. Patterson, and L. Prunty. 1989. Pedogenic carbonates in a Calciaquoll associated with a recharge wetlands. *Soil Sci. Soc. Am. J.* 53:495–499.)

karst features such as sinkholes or dolines with high water tables. Some geographic regions are dominated by this class of wetland, such as the extensive glacial terrain in northern states, karst in the southeastern states, the various "playas" in Texas, and depressions of the intermountain region in the western U.S.

We use for illustration a system of till plain depressions in eastern North Dakota studied by Arndt and Richardson (1989) and Steinwand and Richardson (1989). These depressions are completely closed to surface outflow and are located in a subhumid, continental climate where they receive about 0.5 m annual precipitation and have a potential evapotranspiration of roughly 0.75 m. Salinity of the surface water varies greatly among wetlands as well as within a single wetland over time. The landscape is rolling ground moraine of relatively homogeneous till; the wetland density is high (often well over 80 depressions per square mile; and the surface is hummocky. Bedrock is 10 m or so below the lowest wetlands. The overlying till consists mainly of dead-ice facies of the Pleistocene Coleharbor Formation. The till contains dolomite and high sulfur marine shales (Arndt and Richardson 1989).

The distribution of some of the wetlands is illustrated in Figure 9.2a (Arndt and Richardson 1989). All of these wetlands have plant communities arranged in zones in the typical manner described by Stewart and Kantrud (1971). Wetland 22 is a temporary pond with wet meadow vegetation in the pond center (460 m above mean sea level); wetland 18 is a semipermanent pond with deep marsh species in the center (457 m elevation); and wetland 20 is a saline lake with open water in the center (456 m). The lake dried out in 1988 (and probably in many other very dry years), but it is the high salinity that prevents establishment of emergent vascular plants.

Hydrology

We are able to predict the general flow of water in these landscapes based on many observations from this and other landscapes in the area from the early 1980s through the drought years of 1988 to 1992 and into the pluvial years of 1993 to 1996. Figure 9.2b is a cross-section of three depressions in North Dakota and illustrates equipotential lines (points of equal hydraulic gradient or head) in a depressional landscape. The water table is shown at the surface in the three wetlands, reflecting Sloan's (1972) comment, that "wetlands are windows to the water table." Flow of water is perpendicular to the equipotential lines. Under the wetland marked recharge (wetland 22 in Figure

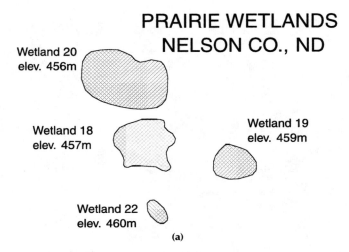

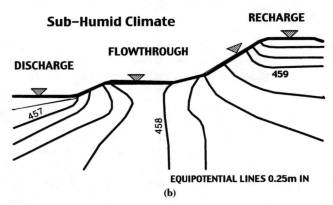

Figure 9.2   (a) The distribution of a few wetlands in the Arndt and Richardson (1989) study. (b) A flownet illustration of a general local flow based on a landscape in Nelson County, ND. (Adapted from Richardson, J. L., L. P. Wilding, and R. B. Daniels. 1992. Recharge and discharge of groundwater in aquic conditions illustrated with flownet analysis. *Geoderma* 53:65–78.)

9.2a), the equipotential lines are roughly parallel to the surface and are decreasing. Water moves down and away from that wetland. In the flowthrough wetland (number 18 in Figure 9.2a), water intersects the wetland at the upper side (discharge) and recharges on the lower side. The equipotential lines are perpendicular to the land so that flow is through the wetland. In the lower wetland, which is marked "Discharge" (wetland 20 in Figure 9.2a), the equipotential lines become more parallel and orient themselves with the surface. Because they are increasing upward, water is moving to the surface.

Postglacial exposure of these landscapes over the long term creates groundwater flow systems modified by fracturing, local changes in stratigraphy, and soil types. Because of the numerous depressions in many till landscapes, the water flow is often in isolated local flow water systems (Toth 1963). In a subhumid climate, water is transported slowly from the upper ponds as depression focused recharge. Depression focused recharge, as used by Lissey (1971), implies that recharge in the center of the depression is the dominant hydrological process, although Hayashi et al. (1998) noted that most recharge water actually moves laterally and is lost to evaporation from the wetland edge. If a wetland center has both recharge and discharge, the wetland is a flowthrough wetland in which water discharges on one side and recharges on the other. This type of wetland is illustrated in the middle wetland in Figure 9.2b. In highly fractured tills, fewer flowthrough ponds would be expected because the water is conducted downward away from the nearby wetlands (Rosenberry and Winter 1997). In ablation till landscapes with typically coarser textured materials, flowthrough ponds are more frequent because buried and surficial aquifers are common. The lower wetlands receive water from the entire landscape as groundwater by a phenomenon known as depression focused discharge.

Interpond high areas do not participate in the flow process in subhumid climates (Toth 1963). As water flows in long groundwater flow paths, wetlands in the low areas receive substantial amounts of dissolved ions and other solutes from the higher areas. The recharge wetlands release material, and the discharge wetlands accumulate the material.

In humid regions or in subhumid regions during pluvial or wet cycles (as in eastern North Dakota 1993 to 1996), however, the interpond high areas develop water tables that were the subdued replicas of the topography as commonly expressed in basic geology textbooks. The same three wetlands in Figure 9.2b under these humid conditions develop three more localized flow systems, thus becoming isolated from the larger flow groundwater systems (see Figure 3.34 for an example). Under localized flow, the interpond highs become recharge areas and each pond is a discharge pond. This situation is illustrated in Chapter 3 earlier in this book. The uplands become leached and the ponds become enriched with dissolved material in humid regions.

In semiarid climates such as at St. Denis, Saskatchewan, studied by Miller et al. (1985), nearly all ponds (15 of 16) were recharge ponds. Water mounded under the wetlands during the wet times and dropped quickly in the dry periods of the year. These observations are similar to the gaining or losing streams in the same climate. During drought cycles in subhumid areas, the drawdown of the water table, particularly at the edge of a wetland, creates a situation in which a pond may switch from being a discharge pond to being a recharge pond similar to the semiarid situation. Arndt and Richardson (1993) studied such a pond with a 20-year record. The pond had always been a discharge pond and had a substantial amount of salt. During the 1988 drought, the pond became a recharge pond and dried up most of the year. It stayed dominantly dry for the next 3 years, during which time the salt was leached from the pond.

Each pond can also have a much smaller local flow system focused at the edge of the wetland. In a wetland studied by Whittig and Janitzky (1963), water ponded in the depression was evaporated from the edge, creating an accumulation in labile minerals illustrated by differentiated chemistry and soil. Their system was an edge focused evaporative discharge type. Knuteson et al. (1989) observed a similar system (Figure 9.1) and measured the actual unsaturated or upward flow. From these estimates, they calculated that about 9000 years would be needed to form such a horizon. The flow envisioned by them is illustrated in Figure 9.3. Saturated flow in the pond interior moves materials down and away from the pond interior. This leaches the soil free of calcite and translocates the clay enough to form a Bt-horizon. The pond edges, however, remain an evaporative dry soil surface where wind and drying exert a tension on the wet soils below. This creates a water potential gradient such that matric tension moves the water from where the soil is wet to where it is drier. As the water evaporates, more water from the edge moves upward. The dissolved materials accumulate, and the edges become enriched with calcite, gypsum, and salt and as illustrated in Figure 9.4. The B horizon at the edges is a Bkyzg: k means that substantial calcite has accumulated; y means that much gypsum has accumulated; z means that dissolved salts occur in the pores of the soil; and g means the soil is wet for substantial periods of time.

Soil Hydrosequences

The edge of a wetland, as indicated above, is the focus of evaporative discharge (Whittig and Janitzky 1963, Arndt and Richardson 1989, Knuteson et al. 1989, Steinwand and Richardson 1989, Seelig et al. 1990). Figure 9.4 shows a wetland edge with two types of calcareous soils (Steinwand

LOCAL EDGE FOCUSED FLOW

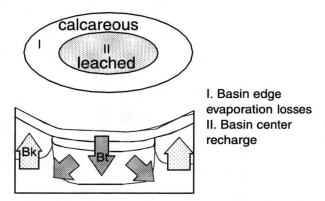

I. Basin edge
evaporation losses
II. Basin center
recharge

Figure 9.3 The flow in the pond center is saturated flow that recharges the water table and is dominantly downward and outward. Flow leaches the soils. The evapotranspiration on the edge of the pond creates an unsaturated upward flow with water loss. These soils will become enriched with calcite.

and Richardson 1989). This semipermanent flowthrough pond edge is similar to wetland 18 in Figure 9.2a. While the pond center has reducing conditions, the edge has strong evaporative discharge where a calcic (Bk) horizon forms due to the upward flow of water and a concentration of calcium. These two types of soils commonly occur in semipermanent ponds in the prairie areas: a wet soil that ranges from very poorly drained to poorly drained with a very gray Bk (Typic Calciaquoll) and a somewhat poorly drained soil with a khaki-colored Bk (Aeric Calciaquoll). The wetter soils frequently accumulate gypsum and dissolved salts (Steinwand and Richardson 1989), as indicated by the letters to designate the B horizon mentioned in the preceding section. The sequence here from driest to wettest is Udic Haploborloll, Aeric Calciaquoll, Typic Calciaquoll, and Cumulic Endoaquoll.

WET EDGE EFFECT
Evaporative Discharge

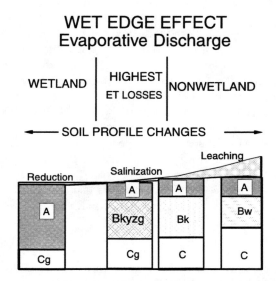

Figure 9.4 The wet soils at pond edges are subject to high evapotranspiration stresses because the water is near the surface and evaporation at the surface increases the matric tension and lifts more water to the surface. As the water is evaporated, the calcite increases, creating a Bk horizon.

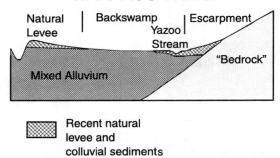

Figure 9.5   Cross-section of landforms for the Wakarusa River floodplain that includes the natural levee, back swamp, and the yazoo stream. These are bounded by the entrenched river channel and the escarpment.

Riverine Wetlands

Geomorphic Setting

Riverine wetlands are linear features of the landscape that normally consist of a flood plain and stream channel. Because they are incised into the landscape, their topographically low position often makes them discharge areas for groundwater. Exceptions to this are "losing streams" of arid regions where the groundwater table slopes away from the channel and the floodplain. Riverine wetlands of floodplains of larger streams also receive water from upstream via overbank flow from the channel during flood events. The importance of this water source, relative to groundwater discharge, becomes greater as one moves downstream and the valley widens (Brinson 1993b).

We use for illustration the Wakarusa River valley near Lawrence, Kansas. The Wakarusa entrenched its valley by cutting into an upland that is underlain by limestone bedrock. The escarpment that forms the boundary of the floodplain is a steep landform connecting the upland and floodplain (Figure 9.5). The floodplain has three landforms: a back swamp, a natural levee and entrenched channel of the Wakarusa River, and a minor channel of a "yazoo" stream. Back swamp, as used here, is a geomorphic term implying the area "in back of the natural levee" or away from the stream. This low area receives water from the upland via the escarpment and flood water from the trunk stream via the natural levee (Figure 9.5). A small stream usually drains the back swamp and flows parallel to the trunk stream, such as the Yazoo River flowing parallel to the Mississippi River in Mississippi; hence the name *yazoo*. The stream has been channeled into the typical straight drainageway and travels some distance parallel to the main stream. The drainageway then turns abruptly and enters the main channel, a pattern that is typical of a back swamp stream.

The back swamp supports a variety of hydrophytes in the natural area maintained by Baker University. The elevated natural levee at this site supports herbaceous species that are tall prairie grasses (not hydrophytes). Portions have been converted to farm land by drainage and are maintained for agriculture by both tile and ditch drainages.

Hydrology

We expect the following hydrology based on the soils and geomorphology of the area. Both groundwater and surface water are discharged from the upland into the back swamp where water becomes focused toward the yazoo. During most times of the year, the back swamp will release groundwater in the floodplain to the river by flowing laterally under the natural levee (Figure 9.6).

NORMAL GROUND WATER FLOW

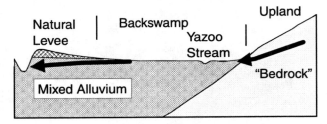

Figure 9.6 Flow reaches the river from the uplands by discharging on the floodplain near the escarpment. The yazoo stream carries away much of the water, but this stream is rather low gradient. Abundant water flows in the floodplain because of the permeable strata. Many floodplains are underlain by permeable strata.

During wetter times, such as during heavy rains, a groundwater and surface water divide develops very near the stream (Figure 9.7). The slow throughflow system of water to the yazoo from the levee should be reflected in a progression of wetter soils toward the yazoo. The soils nearer the upland escarpment should be rather wet and possibly calcareous, reflecting the discharge of waters from the limestone that dominates the region. The transition of wetness from the back swamp to the escarpment would be quite sharp.

Soil Hydrosequences

The soil survey identifies two landform mapping units for the Wakarusa valley: the natural levee and the back swamp, including the soils near the escarpment. The latter included the transitional soils that occurred on the lower levee (somewhat poorly drained and very poorly drained soils). Typically, older soil surveys did not differentiate poorly drained and very poorly drained soils in many areas, although the surveyors clearly were aware that these soils had the drainage inclusions. However, the distinction is now made due to the economic consequences of the Food Security Act, which requires more detailed wetland boundary identification. The soil mapping unit in the back swamp was divided into three units: a very poorly drained calcareous unit near the escarpment, a very poorly drained unit on both sides of the yazoo, and a somewhat poorly drained to poorly drained unit transitional from the levee to the yazoo drainageway. The last unit was not hydric. Field observations of the vegetation on the Baker University preservation site reflected a

WET SEASON WATER FLOW

Cross-Section of the Wakarusa Floodplain

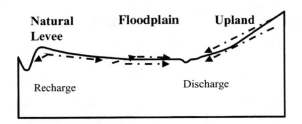

Figure 9.7 The natural levee acts as a drainage divide in wet periods, and the water table rises under this landform.

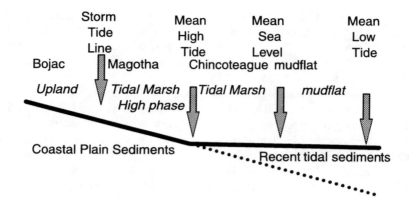

Figure 9.8 The general landscape for Accomack and Northampton Counties, VA for the salt marsh landforms. (Adapted from Edmonds, W. J., G. M. Silberhorn, P. R. Cobb, C. D. Peacock, Jr., N. A. McLoda, and D. W. Smith. 1985. Soil classifications and floral relationships of seaside salt marsh soils in Accomack and Northampton Counties, Virginia. Virginia Agric. Exp. Sta. Bull. 85-8.)

strong correspondence between hydric soil indicators and hydric vegetation (personal observations of the senior author and Kelly Kindscher of the Kansas Biological Survey, 1992).

Estuarine Fringe

Geomorphic Setting

Salt marshes are common coastal features found on nearly level landscapes behind barrier islands and spits, and along bays and lower tidal river shorelines (Edmonds et al. 1985). Portions of these areas are flooded daily by tidal waters that carry abundant salts. Edmonds et al. (1985) describe the salt marsh soil types they observed and studied in Accomack and Northampton Counties of Virginia's eastern shore (37° to 38°N, 75° to 76°W). Their study encompasses 32,780 ha (81,000 acres) of seaside salt marshes that lie between the barrier islands and the mainland. The islands protect the marshes from storms in the Atlantic Ocean.

The Edmonds et al. (1985) study provides an example of a hydrogeomorphic unit for tidal systems. They sequence their hydrogeomorphic unit into four soil landforms (Figure 9.8): (1) upland marine terraces with the Bojac series as a common representative which lies above the spring tide level; (2) salt meadow (Group 2 marsh plants of Silberhorn and Harris 1977) between storm tide line and the mean high tide represented by the Magotha series; (3) salt marsh cordgrass community (Group 1 marsh plants of Silberhorn and Harris 1977) between the mean high tide and mean sea level, represented by the Chincoteague series; and (4) tidal mudflats that lack vegetation of vascular plants. No soil series is used in the last landform because landforms that do not support vascular plants are not considered soils. The Magotha soils were former uplands located on the higher landscape positions in the salt marshes (Edmonds et al. 1985). In earlier soil surveys, landforms dominated by Magotha were a miscellaneous land class called Tidal Marsh, High Phase; the landforms occupied by Chincoteague soils were included in the miscellaneous land class Tidal Marsh.

Hydrology

The basic hydrologic feature of these landscapes is the daily or intermittent tidal inundation by brackish water containing abundant sodium and other dissolved ions from sea water. The lower two landforms (tidal marsh and mud flat) illustrated in Figure 9.8 flood and drain surficially, and thus remain saturated or have a peraquic moisture regime. The Chincoteague soils are dominated by salt marsh cordgrass (*Spartina alterniflora*), which is typically divided into tall, medium, and short grass growth forms. Tall forms tend to be restricted to creek bank environments, where flushing prevents accumulation of sulfides, and hypersaline conditions that tend to be associated with short growth forms (Delaune et al. 1983). Nitrogen supply may also be a factor (Broome et al. 1975). In the two lower landforms, the water table can be considered to be at or near the surface at low tide and above the surface at high tide.

The landform above the mean high tide and the spring tide line can be flooded or saturated to the surface for extensive periods of time but the source of water is mostly precipitation (Stasavich 1998). However, during storm and extreme tides, enough saltwater is transported to these sites to support the growth of marsh halophytes. Salt meadow hay (*Spartina patens*) is the dominant plant, along with admixtures of saltgrass (*Distichlis spicata*) where drainage is more restricted and more saline. Black needlerush (*Juncus roemerianus*) also occurs on this landform, although the study site is at the northern biogeographical distribution of the species. The water table is usually close to the surface of the soil in this landform, and precipitation is the dominant source of water.

Our interpretation of the hydrodynamics is based on the comments and data of Edmonds et al. (1985) and our own experience at the Virginia Coast Reserve (Hmieleski 1994, Brinson et al. 1995, Stasavich 1998). In the lower two landforms, the tidal water table changes are surficial and are only partly influenced by the soil itself. Typical weathering transformations that would cause profile development (other than reduction) do not take place because of the lack of infiltration and drawdown. Where coarser textured soils occur, such as on the barrier islands, groundwater discharge toward these landforms may poise the water table (Hayden et al. 1995). In fact, bioturbation by invertebrates, particularly fiddler crabs, and surficial sediment deposition are dominant factors in soil development. In the upper reaches of the Chincoteague landform, some drainage at low tide may allow for some oxidation and minor translocation to a very shallow depth (Harvey et al. 1987), but most transport processes occur above rather than within the soil. In some locations, salt pans can develop where infrequent flooding (spring tides only), combined with evaporation, creates hypersaline conditions too salty for plant growth (Hayden et al. 1995).

In the high salt marsh areas, flow reversals of two types can be envisioned. The first is the spring tide flood or storm event bringing in saline waters. These events discharge water into the soils when unsaturated. Recession of the water and the subsequent lowering of the water table by evapotranspiration then allows precipitation and possibly some surface runoff from the upland to infiltrate. The landform maintains a relatively high water table in spite of infrequent flooding from estuarine sources (Stasavich 1998). Groundwater here is saline in contrast to fresh conditions in upland soils. In the transition between high marsh and forest, microrelief plays a role in the redistribution of salts. Microrelief highs act as local recharge areas, and the lows act as discharge areas, which flushes the highs, resulting in lower soil salinity. Edmonds et al. (1985) mentioned that halophytes such as *Distichlis spicata* and *Salicornia* spp. were present in these areas, while trees are restricted to hummocks (*Pinus taeda* and *Juniperus virginiana*). The combination of high water table and evaporation, called "evaporative discharge" by Seelig et al. (1990), is common in other saline landforms.

The uplands are freshwater-dominated systems that have better drainage than the salt marsh systems below them. These are freshwater recharge areas. The freshwater ponds on the denser saltwater and protects the upland soils from encroachment of the saline water under the soils. In drier climates the saltwater table may move inland much farther because of the lower infiltration. The Bojac series, however, has a rapid infiltration and exists in an environment with annual precipitation that ranges from 25 to 60 inches (64 to 152 cm) (Edmonds et al. 1985).

Example Soil Sequence

The soil hydrosequence (catena) for the salt marsh landscape is illustrated in Figure 9.9. The better-drained Bojac soil (non-hydric) is a coarse-loamy mixed thermic Typic Hapludult that is well drained and leached free of salts. At the spring tide line to the mean high tide the Magotha

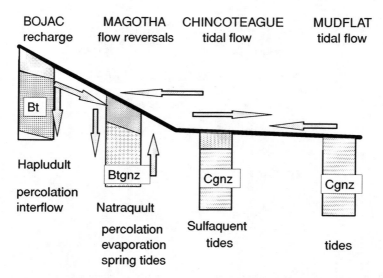

Figure 9.9 The soils schematic for salt marsh soils contrasted to upland and mudflat conditions. (Adapted from Edmonds, W. J., G. M. Silberhorn, P. R. Cobb, C. D. Peacock, Jr., N. A. McLoda, and D. W. Smith. 1985. Soil classifications and floral relationships of seaside salt marsh soils in Accomack and Northampton Counties, Virginia. Virginia Agric. Exp. Sta. Bull. 85-8.)

series classifies as a coarse-loamy mixed thermic Typic Natraqualf. This soil is saline and sodic throughout its solum. If drained, the high sodicity of this soil would create a sodic condition that would dry to brick hard consistency. The tidal salt marsh soil, Chincoteague, reflects little profile development and classifies as fine-silty, mixed, nonacid Thermic Typic Sulfaquent. If drained, it would not remain nonacid. When exposed to oxidizing conditions, the sulfide oxidizes sulfuric acid. The mudflat is sediment.

The uplands are leached soils with moisture regimes that are drier than the soils lower in the landscape (Figure 9.8). These soils create their own regional water table of freshwater. Soils in the salt marshes contain enough soluble salts that most are both saline and sodic (Ec >4.0 dS/m and SAR >13) in sharp contrast to the associated upland. Sodium increases clay dispersion and possibly its translocation. In the lower two landforms, environmental conditions include high sodium and magnesium ion contents, chronic wetness, reducing conditions, accumulation of sulfate and chloride ions, and slow weathering other than reduction. Accumulation of sulfide minerals in salt marsh soils results in acid sulfate soils with a drastic reduction of pH if these areas are drained and oxidized. For example, Edmonds et al. (1985) incubated Chincoteague soils in an oxidizing condition and measured a decrease from 7.0 to 3.0 in 24 days. The latter pH would significantly increase the solubility of aluminum, a plant toxin. The Magotha soil, however, did not significantly change in pH on incubation, which suggests it lacks sulfide accumulation.

Lacustrine Fringe

Figures 9.10a and b depict a lacustrine fringe wetland based on an area along the western side of Lake Erie. The barrier sands create a lagoon system that extends from open water to emergent marsh to wet meadow and then non-wetland. In this example, mineral soils dominate the wetlands but buried peat deposits occur in the area, illustrating that water level fluctuations created and later destroyed fringe wetlands. The sequence probably is first mineral wetland soils and later Histosol development. Currently fringe wetlands along Lake Erie are diked to create waterfowl impoundments. The dikes and causeways for roads and the canals in the wetlands and lagoons sever the original water connections with the lake.

LACUSTRINE FRINGE WETLANDS

Monroe County, MI and Lake Erie

ESTUARINE LAGOONAL FACIES   BEACH FACIES

DelRey   Lenawee   Lenawee ponded   Metea

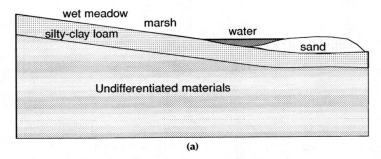

(a)

SOIL PROFILE COMPARISON

from an example of a FRINGE WETLAND

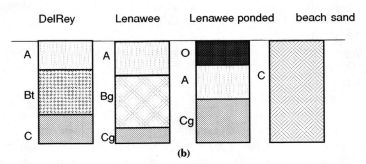

(b)

Figure 9.10   (a) Lacustrine fringe wetland based on a site along the western side of Lake Erie. (b) Chronose-quence of lacustrine fringe soils in Monroe County, Michigan Soil Survey. (Adapted from Bowman, W. 1981. *Soil Survey of Monroe County, Michigan,* U.S. Govt. Printing Office. Washington, DC.

Using the Woodtick Bay area as an example and data from the Monroe County Soil Survey (Bowman 1981), the sequence of soils and landforms illustrated in Figures 9.10a and b were developed. The sands on the outside yield to the shoreward finer textured, wetter soils and eventually to open water in the lagoon landform. As the water becomes shallower toward the upland, a marsh develops. The soil is mapped as Lenawee, a fine, mixed, nonacid, mesic Mollic Haplaquept. It probably is an Endoaquept in the revised classification. This high clay soil with a thin dark surface and neutral reaction is formed under conditions of "endo-saturation" or groundwater saturation. These soils are used for both the ponded marsh phase (mapping unit 10) and the wet meadow phase (mapping unit 21), which may mean that two distinct soil taxa exist but are not separated. Inclusions of Saprists in the ponded marsh phase are high and may dominate some areas. The wet meadow phase can be farmed with some land modification. Herdendorf et al. (1981) relates the hydrophytes of these two mapping units.

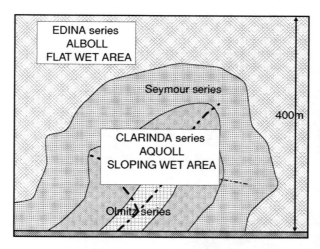

Figure 9.11 The distribution of soils on the landscape from the Wayne County, Iowa, Soil Survey. Note that the Edina upland is flat, and the other units have 2 to 7% slopes. (Adapted from Lockridge, L. D. 1971. *Soil Survey of Wayne County, Iowa.* USDA NRCS, U.S. Govt. Printing Office. Washington, DC.

The somewhat poorly drained Del Rey series completes the hydrosequence. This Aeric Ochraqualf is fine textured with profile development suggesting frequent drying as well as ponding phases. The presence of carbonates within 2 or 3 feet of the surface and an argillic horizon indicate greater soil development than for the Lenawee, which is an Inceptisol lacking horizon development. The Lenawee soil does not dry out enough to allow for the downward movement of clay necessary to create an argillic horizon.

Flats

Geomorphic Setting

"Planosols" were a clear concept from an older soil classification used to describe upland wet areas that developed high clay Bt horizons. The Bt horizon usually had over 40% smectitic (montmorillonitic) clay, which acts as an aquitard to downward movement of water. These soil–landform units were extensive in Illinois, Missouri, and Iowa in areas where interstream divides are essentially flat. Albaqualfs and Albolls are attempts by *Soil Taxonomy* to encompass these soil units. These soils may or may not be hydric, but if undrained they are certainly seasonally wet. Most have now been surface or tile drained. We have chosen to represent the "flat" HGM class by an area of "Planosols," and in particular the Edina series (fine, smectitic, mesic, Typic Argialboll) from Wayne County in southern Iowa, the type just south of the village of Harvard (Lockridge 1971). It is a flat upland summit covered with 3 m of loess. Below the loess is a paleosol developed in highly weathered till of exceedingly high clay content, apparently having been a planosol also. The map view of the landscape depicted in Figure 9.11 illustrates the dendritic stream dissection typical of this landscape and the flat upland.

Hydrology

During the spring thaw and rainy times that produce much water, the water on the landscape cannot run off easily because lateral flow is restricted by gradient rather than by texture. Downward movement is retarded by two restrictive barriers that act as severe aquitards: the modern Bt horizon and the buried underlying paleosol argillic horizon. The A and E horizons over the Bt horizon and the loess below the Bt horizon have relatively rapid permeability. The horizontal to downward

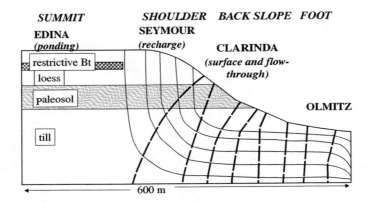

Figure 9.12   The flownet of the Edina landscape in the vicinity of Harvard, Wayne County, Iowa.

saturated conductivity based on the NRCS estimated data is about 30/1. The combination of flat landscapes with low hydric gradient and restricted downward flow creates a large wet area. We detail the stratigraphy and flow in a flownet modeled after the Wayne County type location for Edina series (Figure 9.12). The flat is shown without any flow at all, though some may occur laterally in the thin soil surface. The "perched" water on the Bt horizon of the Alboll (Edina series) may saturate the horizons below, but the flow is so slow that the flowlines are concentrated in the shoulder position. This is the recharge area for flats (Richardson et al. 1992). A significant amount of water flows on the paleosol and discharges on the slope. The area used for this model includes a cove or headslope area (Figures 9.11 and 9.13). The convergence of flow in these areas creates a sloping wetland.

The following points detail our conclusions: (1) the upland flat can get wet very fast and flows laterally slowly because of the low elevation gradient; (2) at the back-slope where the paleosol soil crops out, another wet area occurs; (3) recharge is concentrated at the shoulder; and (4) the upland releases little water to downward flow. The Aquoll area developed on the paleosol is especially expressed in the coves or swales because of the convergence of lateral flowing water. The stratigraphy here produces potentially two wetland types, an upland flat and a sloping wetland. Local farmers, of course, are well aware of these wet areas because crops do not do well and tractors may get mired in the slope. The local name for these areas is "blue clays," and they are not spoken of with much fondness.

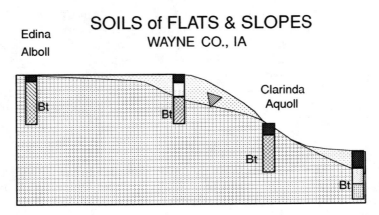

Figure 9.13   The cross-section with high water table and the soil types distributed on the landscape.

Example of a Soil Hydrosequence

The flat area of the landscape has a two soil system. The interior of the flat area is wet and has a "planosol" soil or Alboll. The edge of the summit area has a better drained non-hydric soil (Figure 9.12). Daniels and Gamble (1967) called this the *red edge* after the reddish-colored soils in North Carolina in similar landscapes. The soils of planosols are located high on the landscape and therefore dry out late in the season. They are subject to translocation of clay and leaching of soluble constituents and develop a distinct profile. These are some of the few soils developed under prairie vegetation that have E or eluvial horizons reflective of the wetting and drying aspect of the soil.

Slope Wetlands

We favor the idea that two types of slope wetlands can be differentiated in the field based on the slope and geological conditions. We call these stratigraphic slope wetlands and topographic slope wetlands. The first relates to a stratum that intersects the land surface and forces the water to discharge on the slope. The second relates to slopes that converge the water in coves or draws. In places, combinations of the two occur which amplifies discharge on slopes. The topographic slope wetland disappears in semiarid and arid regions, but the stratigraphic type can form in any climate.

Topographic Slope Wetlands

The topographic slope wetland occurs in concave convergent positions on landscapes, as illustrated in Figure 9.14, which shows the seasonally high water table position. Hack and Goodlett (1960) discussed the formation of these wet areas, which they called *hollows*, in the mountains of Virginia (other terms are *headslopes* and *coves*). The convergence of flows occurs in zones at the margins of incipient channels that receive water from more than one direction. Thick soil provides the capacity to store water for long periods so that sudden rainfall events are followed by infiltration and slow movement in the landscape. The accumulation of the water at slope bases was noticeable to Hack and Goodlett (1960) and others from many landscapes (Chorley 1978). The areas of substantial wetness are the heads of drainages that had short slopes and a flat convergent shape with deep soils. Throughflow water moving by gravity is greatly slowed by infiltrating and moving in the soil. Penetration to depth in forest soils is often constricted by the soil subsurface horizons,

Topographic Slope Wetland

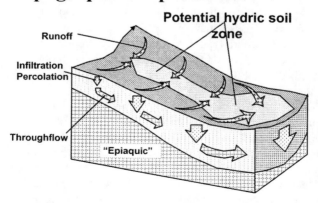

Figure 9.14 An illustration of a Topographic Slope Wetland with both runoff and throughflow water converging in the swale, creating an episaturated transient wetland.

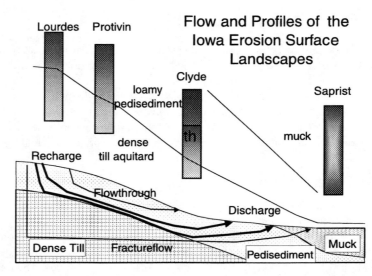

Figure 9.15 An area in Iowa with a topographic slope wetland that is tile drained. (Adapted from Kirkham, D. 1947. Studies of hillslope seepage in the Iowan drift area. *Soil Sci. Soc. Am. Proc.* 12:73–80; Buckner, R. and J. Highland. 1974. *Howard County Soil Survey Report.* USDA, NRCS, U.S. Govt. Printing Office. Washington, DC.

such as argillic horizons, or from lack of macropores in the C horizon. Flow within the soil is slow if contrasted to runoff. However, once the pores are water filled, the wet area in the convergent landform expands upslope in all directions. The wettest area is the lower and central part of the convergent landform. Usually all soils in these landscapes are recharge and are leached. The increase in upland soil features and decrease in hydric soil indicators occur from the center and lower part of the convergent landform. Eventually the hydric:non-hydric line is reached. The central soils may dry out. If they do, the strong wetting and drying contrast would aid in developing an argillic horizon.

These wet areas relate to the idea of "varying source area" of Hewlett and Nutter (1970). The wetlands that form would grow up the slope with additional wetness. Nutter (1973) observed during his studies in the forests of the southeastern U.S. that water fed to the water table during storm events came from water that had been infiltrated and not from overland flow. Second, the water came not just from above a point on the landscape but also laterally from upslope and converged on the lower segments of the slope. Effective storage in these portions of the landscape was reduced. At the beginning of the drainage cycle actual flow may have been downward, but the net flow was downslope. As drainage continued, the flow lines slowly oriented more parallel with the surface. The upper boundary is very diffuse, making it difficult to map for wetland delineation, especially if contrasted with the stratigraphic type of wetland. These wetlands tend to have mineral soils at the top. Histosols may occur downslope if the concavity is wet enough. In the Howard County, Iowa, situation described in the following section, the Histosol occurred in the flat out from the sloping portion of the wetland (Figure 9.15).

Kirkham (1947) conducted a wetness survey on areas that did not drain well despite having tile drains on the Iowan erosion surface in northeastern Iowa. These areas were foot slopes and usually had convergent water flow. On close inspection and measurement with piezometers, he determined that flow differed by landscape position. The flow was in the soil and little runoff occurred, even though some of the study area was cultivated. The upper areas were distinctly recharge areas with downward pressures. The side slopes had horizontal flow (parallel to the slope), and the lower slope areas had upward artesian pressures and discharge.

The Howard County Soil Survey Report (Buckner and Highland 1974) reveals that the soils used by Kirkham were strongly anisotropic, and the impact on water had been observed (Figure 9.15). The Lourdes mapping unit is described as occurring on convex ridges and was an acid-

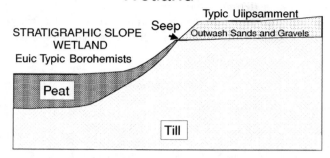

Figure 9.16   An illustration of a stratigraphic slope wetland that has developed into a fen with an organic soil; the area used to model this landscape is from the western part of North Dakota, suggesting that if such wetlands develop here, this is a universal process. (From Malterer, T. J., J. L. Richardson, and A. L. Dusbury. 1986. Peatland soils associated with the Souris River, McHenry County, North Dakota. *North Dakota Acad. Sci. Proc.* 40:103.)

leached soil. After heavy rains or extended wet periods, the water perches on the impermeable dense lower till and creates side hill seeps. Coupling the observations of Kirkham (1947) and the later analysis of Nutter (1973) and Chorley (1978), it seems that some deep water penetration occurs with abundant throughflow that discharges in the Clyde soil. The actual flow mechanism has created the downward flowing, well-drained Lourdes that suffers periodic wet periods with ponding. The water will flow laterally but is restricted by gradient laterally and by saturated hydraulic conductivity from flowing downward. The sloping Protivin soil is deeper to the dense restrictive till stratum and receives water from above. This soil is somewhat poorly drained and has strong lateral flow tendencies. It is leached in its upper part but has carbonates in places in the restrictive stratum. The Clyde at the concave area of the slope is poorly drained and receives water from above. The soil of the flat area extending out from the hillslope wetland has a muck surface. The muck surface becomes deep enough to be a Saprist. This sequence is rather typical of fens; in fact Kratz et al. (1981) describe piezometric data in mounded peats similar to the sequence here but almost entirely on Histosols.

Stratigraphic Slope Wetlands

Mausbach and Richardson (1994) described several aspects of fens, some of which are examples of stratigraphic slope wetlands. One example from Malterer et al. (1986) and Des Lauriers (1990) will be used here as an example. Stratigraphic slope wetlands occur because landscape geology creates exceptional anisotropic conditions that focus water flow to a point on the landscape where the water discharges. Stratigraphic slope wetlands have sharp, narrow upper boundaries when contrasted to topographic slope wetlands. The strata conducting the water create a narrow area, just above the wetland, while the diffuse nature of the topographic system has a broad continuum of ever-increasing wetness downslope. Figure 9.16 depicts a dense till with overlying sand and gravels of an outwash unit. The water moves freely in the gravels, but its downward movement is severely retarded in the till. The resulting point of discharge on the valley edge creates a calcareous fen with a 3% slope 15 m distance before starting to decrease to a nearly level contour. The soil types are Hemist and Saprists (Malterer et al. 1986). The organic layer is >4 m thick at the base of the slope. The hydrology is simply that water discharges at the spring or seep on the hillslope. As the vegetation develops, some organic matter develops on the surface. The water tends to flow below the organic layer and is protected from evapotranspiration. The organic accumulation starts to act as an aquitard and confines the water to flow below the layer. The water often flows under positive head or artesian pressures as can be noted by the fountain created when the surface peaty-muck is penetrated with an auger or peat sampler. The water that moves through the landscape picks up substantial dissolved ions. These ions are concentrated and precipitated at the surface in places, but the high organic matter also holds the ions as adsorbed or exchangeable ions. The fens of stratigraphic slope wetlands are nutrient rich contrasted to ombrotrophic bogs that only receive rainwater. Bogs would be considered in the HGM class of organic soil flats.

REFERENCES

Arndt, J. L. and J. L. Richardson. 1989. Geochemical development of hydric soil salinity in a North Dakota prairie-pothole wetland system. *Soil Sci. Soc. Am. J.* 53:848–855.

Arndt, J. L. and J. L. Richardson. 1993. Temporal variations in the salinity of shallow groundwater from the periphery of some North Dakota wetlands (USA). *J. Hydrology* 141:75–105.

Bowman, W. 1981. *Soil Survey of Monroe County, Michigan,* U.S. Govt. Printing Office. Washington, DC.

Brinson, M. M. 1993a. *A Hydrogeomorphic Classification for Wetlands.* Technical Report WRP-DE-4, Waterways Experiment Station, Army Corps of Engineers, Vicksburg, MS.

Brinson, M. M. 1993b. Gradients in the functioning of wetlands along environmental gradients. *Wetlands* 13:65–74.

Brinson, M. M., F. R. Hauer, L. C. Lee, W. L. Nutter, R. D. Rheinhardt, R. D. Smith, and D. Whigham. 1995. *Guidebook for Application of Hydrogeomorphic Assessments to Riverine Wetlands.* Technical Report TR-WRP-DE-11, Waterways Experiment Station, Army Corps of Engineers, Vicksburg, MS.

Broome, S. W., W. W. Woodhouse, Jr., and E. D. Seneca. 1975. The relationship of mineral nutrients to growth of *Spartina alterniflora* in North Carolina. II. The effect of N, P, and Fe fertilizers. *Soil Sci. Soc. Am. J.* 39:301–307.

Buckner, W. and J. Highland. 1974. *Howard County Soil Survey Report.* U.S. Govt. Printing Office. Washington, DC.

Chorley, R. J. 1978. The hillslope hydrological cycle. pp. 1–42. *In* M. J. Kirkby (Ed.) *Hillslope Hydrology,* John Wiley & Sons, New York.

Committee on Characterization of Wetlands. 1995. *Wetlands: Characteristics and Boundaries.* National Research Council, National Academy of Sciences, Washington, DC.

Cowardin, L. M., V. Carter, F. C. Golet, and E. T. LaRoe. 1979. *Classification of Wetland and Deepwater Habitats of the United States.* FWS/OBS-79/31. U.S. Fish and Wildlife Service, Washington, DC.

Daniels, R. B. and E. E. Gamble. 1967. The edge effect in some Ultisols in the North Carolina Coastal Plain. *Geoderma* 1:117–124.

Darmody, R. G. and J. E. Foss. 1978. *Tidal Marsh Soils of Maryland.* Md. Agric. Exp. Stra. Misc. Publ. 930.

DeLaune, R. D., C. J. Smith, and W. H. Patrick, Jr. 1983. Relation of marsh elevation, redox potential, and sulfide to *Spartina alterniflora* productivity. *Soil Sci. Soc. Am. J.* 47:930–935.

Des Lauriers, L. L. 1990. *Soil Survey of McHenry County, North Dakota.* USDA Soil Conservation Service, U.S. Govt. Printing Office. Washington, DC.

Edmonds, W. J., G. M. Silberhorn, P. R. Cobb, C. D. Peacock, Jr., N. A. McLoda, and D. W. Smith. 1985. *Soil Classifications and Floral Relationships of Seaside Salt Marsh Soils in Accomack and Northampton Counties, Virginia.* Virginia Agric. Exp. Sta. Bull. 85-8.

Hack, J. T. and J. G. Goodlett. 1960. *Geomorphology and Forest Ecology of a Mountain Region in the Central Appalachians.* US Geol. Surv. Prof. Pap. 347. U.S. Govt. Printing Office. Washington, DC.

Hayashi, M., G. van der Kamp, and D. L. Rudolph. 1998. Water and solute transfer between a prairie wetland and adjacent uplands, 1. Water balance. *J. Hydrology* 207:42–55.

Harvey, J. W., P. F. Germann, and W. E. Odum. 1987. Geomorphological control of subsurface hydrology in the creekbank zone of tidal marshes. *Estuarine, Coastal and Shelf Science* 25:677–691.

Hayden, B. P., M. C. Rabenhorst, F. V. Santos, G. Shao, and R. C. Kockel. 1995. Geomorphic controls on coastal vegetation at the Virginia Coast Reserve. *Geomorphology* 13:283–300.

Herdendorf, C. E., S. M. Hartley, and M. D. Barnes, (Eds.). 1981. *Fish and Wildlife Resources of the Great Lakes Coastal Wetlands within the United States. Volume One: Overview.* U.S. Fish and Wildlife Service, Washington, DC. FWS/OBS-81/02-v1.

Hewlett, J. D. and W. L. Nutter. 1970. The varying source area of streamflow from upland basins. pp. 65–83. *Proceedings of the Symposium on Interdisciplinary Aspects of Watershed Management.* Montana State Univ. Bozeman. Amer. Soc. Civil Engr. NY.

Hmieleski, J. I. 1994. High marsh-forest transitions in a brackish marsh: the effects of slope. Master's thesis, Biology Department, East Carolina University, Greenville, NC.

Kirkham, D. 1947. Studies of hillslope seepage in the Iowan drift area. *Soil Sci. Soc. Am. Proc.* 12:73–80.

Knuteson, J. A., J. L. Richardson, D. D. Patterson, and L. Prunty. 1989. Pedogenic carbonates in a Calciaquoll associated with a recharge wetland. *Soil Sci. Soc. Am. J.* 53:495–499.

Kratz, T. K. M. J. Winkler, and C. B. De Witt. 1981. Hydrology and chronology of a pear mound in Dane County southern Wisconsin. *Wisc. Acad. Sci. Arts and Letters* 69:37–45.

Lissey, A. 1971. *Depression-Focused Transient Groundwater Flow Patterns in Manitoba.* Geol. Assoc. Can. Spec. Pap. 9:333-341.

Lockridge, L. D. 1971. *Soil Survey of Wayne County, Iowa.* USDA NRCS, U.S. Govt. Printing Office. Washington, DC.

Malterer, T. J., J. L. Richardson, and A. L. Duxbury. 1986. Peatland soils associated with the Souris River, McHenry County, North Dakota. *North Dakota Acad. Sci. Proc.* 40:103.

Mausbach, M. J. and J. L. Richardson. 1994. Biogeochemistry processes in hydric soil formation. *In* W. H. Patrick, Jr. (Ed.) *Current Topics in Wetland Biogeochemistry.* 1:68–127.

Miller, J. J., D. F. Acton, and R. J. St. Arnaud. 1985. The effect of groundwater on soil formation in a morainal landscape in Saskatchewan. *Can. J. Soil Sci.* 65:293–307.

Mills, J. G. and M. Zwarich. 1986. Transient groundwater flow surrounding a recharge slough in a till plain. *Can. J. Soil Sci.* 66:121–134.

Nutter, W. L. 1973. The role of soil water in the hydrologic behavior of upland basins. pp. 181–193. *Field Soil Water Regime.* Soil Science Soc. Amer. Madison, WI.

Peacock, C. D., Jr. and W. J. Edmonds. 1992. *Supplemental Data for Soil Survey of Accomack County, Virginia.* Virginia Agric. Exp. Sta. Bull. 92-3.

Richardson, J. L. 1997. Soil development and morphology in relation to shallow ground water: an interpretation tool. pp. 229–233. *In* K. W. Watson and A. Zaporozec (Eds.) *Proceedings of the 4th Decade of Progress Symposium,* Tampa Bay, FL, American Institute of Hydrology, St. Paul, MN.

Richardson, J. L., L. P. Wilding, and R. B. Daniels. 1992. Recharge and discharge of groundwater in aquic conditions illustrated with flownet analysis. *Geoderma* 53:65–78.

Rosenberry, D. O. and T. C. Winter. 1997. Dynamics of water-table fluctuations in a upland between two prairie-pothole wetlands in North Dakota. *J. Hydrology* 191:266–289.

Seelig, B. D., J. L. Richardson, and W. T. Barker. 1990. Characteristics and taxonomy of sodic soils as a function of landform position. *Soil Sci. Soc. Am. J.* 54:1690–1697.

Silberhorn, G. M. and A. F. Harris. 1977. *Accomack County Tidal Marsh Inventory.* Spec. Rep. No. 138, applied Marine Science and Ocean Engineering. Virginia Institute Marine Science, Gloucester Point, VA.

Sloan, C. E. 1972. *Ground-water Hydrology of Prairie Potholes in North Dakota.* U.S. Geol. Survey Prof. Pap. 585-C. U.S. Govt. Printing Office. Washington, DC.

Smith, R. D., A. Ammann, C. Bartoldus, and M. M. Brinson. 1995. *An Approach for Assessing Wetland Functions Using Hydrogeomorphic Classification, Reference Wetlands and Functional Indices.* Technical Report TR-WRP-DE-9, Waterways Experiment Station, Army Corps of Engineers, Vicksburg, MS.

Soil Survey Staff. 1975. *Soil Taxonomy.* Soil Conservation Service USDA Agr. Handbook 436, U.S. Govt. Printing Office. Washington, DC.

Stasavich, L. E. 1998. Quantitatively defining hydroperiod with ecological significance to wetland functions. In progress. MS thesis, Biology Department, East Carolina University, Greenville, NC.

Steinwand, A. L. and J. L. Richardson. 1989. Gypsum occurrence in soils on the margin of semipermanent prairie pothole wetlands. *Soil Sci. Soc. Am. J.* 53:836–842.

Stewart, R. E. and H. C. Kantrud. 1971. *Classification of Natural Ponds and Lakes in the Glaciated Prairie Region.* U.S. Fish. Wild. Serv., Resour. Publ. 92. 57 pp.

Toth, J. 1963. A theoretical analysis of groundwater flow in small drainage basins. *Proc. Hydrol. Symp. Groundwater* 3:75–96. Queen's Printer, Ottawa, Canada.

Whittig, L. D. and P. Janitzky. 1963. Mechanisms of formation of sodium carbonate in soils. I. Manifestations of biological conversions. *J. Soil Sci.* 14:322–333.

CHAPTER 10

Use of Soil Information for Hydrogeomorphic Assessment

J. A. Montgomery, J. P. Tandarich, and P. M. Whited

INTRODUCTION

Wetlands perform numerous important functions, including water quality maintenance, flood protection, and habitat for threatened species of plants and wildlife (Mitsch and Gosselink 1986). The scientific community and public have become increasingly aware of the importance of wetlands in maintaining environmental quality (Soil and Water Conservation Society 1992). Such heightened awareness is reflected in increased financial support for wetland research, and the enactment of a patchwork of federal, state, and local laws regulating the environmental impacts to wetlands (Hauer 1995, Smith et al. 1995).

Impacts to wetlands at the national scale are regulated by the Clean Water Act (33 U.S.C. 1344). Section (§) 404 of the Act directs the U.S. Army Corps of Engineers, in cooperation with the U.S. Environmental Protection Agency, to administer a program regulating discharge of dredge and fill materials in U.S. waters, including wetlands. The main goal of §404 is to maintain and improve the chemical, physical, and biological integrity of the nation's waters (40 CFR, Part 230.1). Operators desiring to discharge fill and dredge materials must apply for a §404 permit. Applications must undergo a public interest review process whereby both the project-specific and cumulative impacts of the proposed action on wetland functions are assessed.

Functional assessment is a procedure used to estimate the level of wetland performance of hydrological, biochemical, and habitat maintenance processes. Assessment results help determine whether or not activities in wetlands result in gains (e.g., mitigation) or losses (e.g., impacts) in functioning. Paragraph §320.4(a)(1) of the U. S. Army Corps Regulatory Program Regulations (33 CFR Parts 320–330) and EPA paragraph §404(b)(1) Guidelines (40 CFR Part 230) summarize the sequence of steps for reviewing permit applications. Functional assessment is required at several steps in this sequence (Smith et al. 1995). The results of the functional assessment are but one factor considered in the permit decision.

Various methods have been developed for assessing wetland functions, many of which are reviewed by Lonard et al. (1981). None of these methods, however, has totally met the technical and programmatic requirements of §404. As a result of these shortcomings, the Wetlands Research

Program at the U.S. Army Corps of Engineers Waterways Experiment Station was charged with developing a rapid functional assessment procedure that would satisfy these technical and programmatic guidelines, and at the same time be simple, efficient, accurate, and precise. The resulting *Hydrogeomorphic Approach (HGM)* to functional assessment of wetlands meets the technical and programmatic requirements of §404 through hydrogeomorphic classification, functional indices, and reference wetlands (Brinson 1995, Smith et al. 1995).

Given the preceding discussion, the objectives of this review article are to present: (i) an overview of the HGM approach to wetland functional assessment, (ii) a rationale for including soil–landscape information in the HGM development process, and (iii) case studies of how soil information can be and has been used in developing the HGM approach.

OVERVIEW OF THE HGM APPROACH

The HGM approach to wetland functional assessment consists of a development and application (assessment procedure) phase. An interdisciplinary team (A-Team) of individuals carries out the development phase. The A-Team should have expertise in wetland ecology, soil science, geomorphology, hydrology, geochemistry, wildlife biology, and plant ecology. Regulators, wetland managers, consultants, and other end-users of the HGM approach conduct the application phase.

In the development phase, the A-Team groups wetlands into *hydrogeomorphic classes* based on geomorphic setting, dominant source of water, and hydrodynamics. These criteria are believed to control most functions in wetlands. Seven hydrogeomorphic classes of wetlands have been recognized to date. Wetlands in a geographic region are then classified into *subclasses* based on hydrogeomorphic characteristics and other ecosystem and/or landscape characteristics that influence how wetlands function in the region (Smith et al. 1995). Classification into subclasses is necessary to achieve the degree of detail required for functional assessment (Brinson 1995). The number of regional wetland subclasses may depend on the diversity of wetlands in a region and regional assessment objectives. The A-Team then prioritizes regional subclasses for the purpose of developing HGM models and functional assessment guidebooks. The priority subclass may be the most common subclass in a particular geographic region (cf. depressional wetlands with temporary and seasonal hydroperiods *in* Lee et al. 1997), or it may be the subclass for which the most §404 permits have been granted.

In the HGM approach to functional assessment, gains or losses in functioning are quantified in terms of *functional capacity*. Functional capacity is the degree to which a wetland performs a particular function, and it depends on characteristics of the wetland and surrounding landscape, including plant composition, water source, and soil type. Functional capacity can be measured quantitatively or estimated qualitatively. In either case, the resulting metric, defined as the *functional capacity index (FCI),* is a measure of the capacity of a wetland to perform a particular function relative to other wetlands in the regional subclass. Determining the FCI thus requires that standards of comparison, or *reference standards*, be developed for the various functions performed by a particular regional subclass. Reference standards are determined for each subclass and are measured in the field on wetland sites that are self-sustaining and representative of the highest level of functioning. Examples of reference standards include average depth of flooding, level of sediment removal, and the number of trees per acre.

Reference standards are developed from *reference wetlands*. Reference wetlands are sites judged by the A-Team and other wetland professionals to encompass the known variation of the subclass due to natural processes and anthropogenic disturbances. They are used to establish ranges in wetland functions. Reference wetlands are selected from the *reference domain*, the geographic area that includes all or part of the area in which the wetland subclass occurs. The HGM reference system (e.g., reference wetlands, reference standards) is thus designed to incorporate all of the

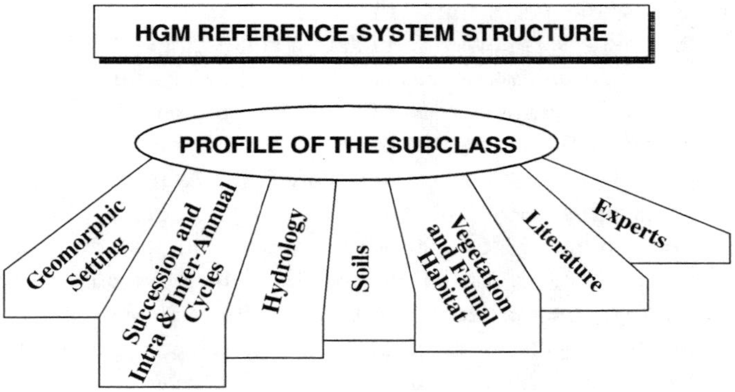

HGM REFERENCE SYSTEM STRUCTURE

PROFILE OF THE SUBCLASS

Geomorphic Setting

Succession and Intra & Inter-Annual Cycles

Hydrology

Soils

Vegetation and Faunal Habitat

Literature

Experts

Figure 10.1 HGM reference system structure. (From Lee, L.C., Brinson, M.M., Kleindl, W.J., Whited, P.M., Gilbert, M., Nutter, W.L., Whigham, D.F., and DeWalk, D. 1997. *Operational Draft Guidebook for the Hydrogeomorphic Assessment of Temporary and Seasonal Prairie Pothole Wetlands.* Seattle, WA.)

conditions that affect functions performed by a particular subclass. Use of a reference system allows end-users of the HGM approach to use the same standard of comparison (Lee et al. 1997).

Data collected during sampling of reference wetlands can be used to develop a *functional profile* of the priority subclass (Figure 10.1). The functional profile describes the physical, chemical, and biological characteristics of the priority subclass, the functions it is most likely to perform, and the variety of ecosystem and landscape attributes that control these functions (Brinson 1993). The functional profile of the regional subclass can be used to develop an HGM assessment model to detect net changes in functional capacity in the priority subclass, as a template for restoration, as a basis for developing a monitoring program, and as the basis for identifying contingency measures (Figure 10.2).

After the functional profile has been developed, the A-Team must define the *variables* of those functions. Variables are attributes and processes of the wetland ecosystem and surrounding landscape that influence the capacity of a wetland to perform a function (Smith et al. 1995). Examples of variables include soil organic matter, wetland land use, and depth of flooding. Variables can be selected using literature sources, available data from reference wetlands, and the best professional judgment of A-Team members and regional experts. Model variables should be directly measured

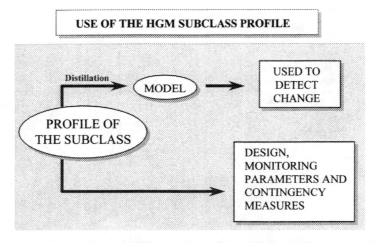

USE OF THE HGM SUBCLASS PROFILE

Distillation

MODEL

USED TO DETECT CHANGE

PROFILE OF THE SUBCLASS

DESIGN, MONITORING PARAMETERS AND CONTINGENCY MEASURES

Figure 10.2 Use of the HGM subclass profile. (From Lee, L.C., Brinson, M.M., Kleindl, W.J., Whited, P.M., Gilbert, M., Nutter, W.L., Whigham, D.F., and DeWalk, D. 1997. *Operational Draft Guidebook for the Hydrogeomorphic Assessment of Temporary and Seasonal Prairie Pothole Wetlands.* Seattle, WA.)

or estimated whenever possible. For example, the variable "flood frequency" can be measured using stream gauge data. In some cases, however, model variables cannot be directly measured. In this case it is necessary to define *indicators*, easily observed or measured characteristics that can be used to estimate variables. For example, flood frequency could be estimated using indicators such as aerial photographs or drift lines.

Once the variables and indicators have been defined, the A-Team then develops a conceptual *assessment model* representing the relationship between measurable variables of the particular wetland ecosystem function and the capacity of the wetland to perform a function (e.g., surface water storage). Assessment models consist of several variables that are aggregated into a simple algorithm to produce a functional capacity index (FCI). For example, the model for the function "Maintenance of static surface water storage," developed for temporary and seasonal prairie pothole wetland ecosystems (Lee et al. 1997), can be expressed by the variables (V):

$$FCI = [V_{out} \times (V_{source} + V_{upuse})/2 + (V_{wetuse} + V_{sed} + V_{pore} + V_{subout})/4)/2]^{0.5}$$

These variables are defined in Table 10.4. An HGM model thus consists of functions, variables, and indicators (Figure 10.3), and the relationship among these model components is based on data collected from reference wetlands.

Because model variables have different units and measurement scales, they must be transformed to a ratio scale prior to aggregation in the model. Each variable in the model algorithm is assigned a subindex value ranging from 0.0 to 1.0 based on the relationship between the variable and the functional capacity. Subindices are assigned based on data collected from reference wetlands, the literature, and the best professional judgment of the A-Team and other regional experts. A subindex of 1.0 is assigned to a variable if it is similar to the reference standard assigned for that variable. As the condition of a variable deviates from the reference standard, it is assigned a lower subindex value, reflecting a decrease in functional capacity.

HGM models can be used in the §404 permitting process to determine the least damaging alternative for the proposed project, describe the potential impacts of the proposed action, determine mitigation requirements, guide restoration design, and compare wetland management alternatives or results. HGM is a rapid assessment method that depends on using the reference system and on the assumption that wetland ecosystem functions can be inferred from ecosystem structure. HGM is not a "one size fits all" approach to functional assessment. Indeed, one of the strengths of the

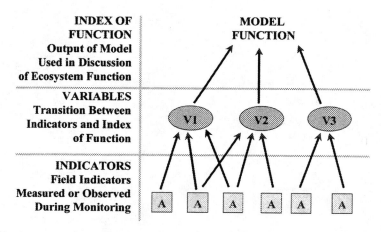

Figure 10.3 Structure of the HGM approach. (From Lee, L.C., Brinson, M.M., Kleindl, W.J., Whited, P.M., Gilbert, M., Nutter, W.L., Whigham, D.F., and DeWalk, D. 1997. *Operational Draft Guidebook for the Hydrogeomorphic Assessment of Temporary and Seasonal Prairie Pothole Wetlands.* Seattle, WA.)

HGM approach is its flexibility, allowing for the integration of additional data and a mix of other assessment methodologies.

The output obtained from applying any model depends both on whether or not the model variables constitute the best suite of variables to accurately describe the function, and on whether or not measurements collected for each variable are accurate and precise. Soil scientists and geomorphologists are concerned that variables describing fundamental soil biological, chemical, and physical processes and soil–geomorphic relationships are not being fully considered or used in the development of HGM assessment models. In the following section we present our rationale for including soil information in HGM assessment models.

The Need for Soil Information in HGM Assessment Models

Soil is critical to living organisms, including humans. It constitutes a major structural component of terrestrial and transitional ecosystems, including wetlands, and it has several important functions and values within these ecosystems (Brady and Weil 1996). First, soil is a medium for plant growth. It provides structural support for higher plants, and it supplies essential nutrients to the entire plant. Soil biological, chemical, and physical properties also influence the structure and function of plant ecosystems. Second, soil properties control the fate of water in the hydrologic cycle. The soil acts as a system for water supply and purification. Third, soil provides habitat for living organisms. Many of these organisms feed on waste products and body parts of other living organisms, releasing their constituent elements back into the soil for uptake by plants. The soil thus acts as a recycling system for nutrients and organic wastes. Finally, soil acts as an engineering medium, providing important building materials and foundations for anthropogenic structures.

Knowledge and understanding of these various soil functions is important in building wetland functional profiles (Figure 10.1), developing HGM models of wetland functions, delineating the reference domain, selecting reference wetland sites, and defining reference standards (e.g., the reference system). The type(s) of soil information used in these endeavors depends in part on the assessment objectives established by the A-Team, and on the suite of functions that they deem most likely to be performed by the subclass. This suite of functions in turn reflects both the structural characteristics of the wetland ecosystem and the nature of the surrounding landscape.

Table 10.1 shows the phases and associated steps in developing HGM model guidebooks. Phases I to III were discussed in the preceding section ("Overview of the HGM Approach"). In the discussion that follows, we will describe how various types and scales of soil information can be used in the HGM Development Phase, specifically, to help identify regional wetland assessment needs (Phase II) and develop functional profiles and HGM models (Phase III).

Use of Soil Information in Phase II of Draft Guidebook Development

The objective of Phase II is to identify regional wetland assessment needs, prioritize regional wetland subclasses, delineate the reference domain, and review pertinent literature pertaining to all aspects of the wetland subclasses. The A-Team also may identify potential reference wetland sites and establish working definitions of the subclasses to be sampled during Phase III development.

Identifying regional wetland assessment needs requires analysis of various types and scales of data, including topographic, geologic, soil, land use and NWI maps, aerial photographs, and a review of the literature pertaining to regional climate, and plant and animal species. Geographic information system (GIS) technology may also be useful in identifying regional wetland assessment needs. A geographic information system is a type of information system that is designed to work with data referenced by spatial or geographic coordinates. A GIS is both a database system with specific capabilities for spatially referenced data, as well as a set of operations for working with the data. A GIS can be thought of as a higher-order map. Just as there are maps designed for specific tasks

Table 10.1   Steps in Development of Model Guidebook

Phase I: Organization of Regional Assessment Team
  A. Identify A-Team members
  B. Train members in HGM classification and assessment
Phase II: Identification of Regional Wetland Assessment Needs
  A. Identify regional wetland subclasses
  B. Prioritize regional wetland subclasses
  C. Define reference domains
  D. Initiate literature review
Phase III: Draft Model Development
  A. Review existing models of wetland functions
  B. Identify reference wetland sites
  C. Identify functions for each subclass
  D. Identify variables and measures
  E. Develop functional indices
Phase IV: Draft Regional Wetland Model Review
  A. Obtain peer-review of draft model
  B. Conduct interagency and interdisciplinary workshop to critique model
  C. Revise model to reflect recommendations from peer-review and workshop
  D. Obtain second peer-review of draft model
Part V: Model Calibration
  A. Collect data from reference wetland sites
  B. Calibrate functional indices using reference wetland data
  C. Field test accuracy and sensitivity of functional indices
Phase VI: Draft Model Guidebook Publication
  A. Develop draft model guidebook
  B. Obtain peer-review of *Draft Guidebook*
  C. Publish as an Operational Draft of the Regional Wetland Subclass
  D. HGM Functional Assessment Guidebook to be used in the field
Phase VII: Implement Draft Model Guidebook
  A. Identify users of HGM functional assessment
  B. Train users in HGM classification and evaluation
  C. Provide assistance to users
Phase VIII: Review and Revise Draft Model Guidebook

From *Federal Register,* August 16, 1996. v. 61m, no. 160.

(e.g., *thematic maps,* such as topographic, geologic, NWI maps), GIS software can also be customized for specific users (soil scientists, geologists, geographers, etc.; Star and Estes 1990).

With respect to Phase II draft guidebook development, geologic, topographic, soil, land use, and other spatially referenced natural resource data could be imported into GIS software and superimposed to produce thematic maps at different spatial scales. Thematic maps could assist the A-Team in identifying and prioritizing wetland subclasses and in delineating the reference domain. Digital soil map databases prepared by the U.S. Department of Agriculture–Natural Resources Conservation Service (USDA–NRCS) can be used with GIS software to address planning and management initiatives at site-specific, regional, watershed, and statewide scales. For small-scale planning problems, the State Soil Geographic (STATSGO) database soil maps are quick, efficient, and cost-effective tools. Soil maps for the STATSGO database are prepared by generalizing the detailed county soil survey data. The base map used is the U.S. Geological Survey 1:250,000 topographic quadrangle. The minimum area mapped is approximately 1500 acres. Each STATSGO map is linked to a Soil Interpretation Record (SIR) attribute database. This database gives the proportional extent of the component soils and the properties for each map unit. The STATSGO map units consist of 1 to 21 components each. The SIR database includes over 25 physical and chemical soil properties, interpretations, and productivity. Examples of information that can be queried from the database include available water capacity, soil reaction, salinity, water table, and flooding.

For site-specific, large-scale planning and management initiatives, soil maps in the Soil Survey Geographic (SSURGO) database provide detailed soil resource information at scales ranging from 1:12,000 to 1:63,360. SSURGO is the most detailed level of soil mapping done by the NRCS. SSURGO mapping bases are either orthophotoquads or 7.5-minute topographic quadrangles. SSURGO data are collected and archived in 7.5-minute quadrangles and distributed as complete coverage for a soil survey area. SSURGO is linked to a Map Unit Interpretation Record (MUIR) attribute database. This database gives the proportionate extent of the component soils and their properties for each map unit. The MUIR contains over 25 physical and chemical soil properties. Examples of properties that can be accessed from the database include soil reaction, available water capacity, salinity, water table, and bedrock. The following case study illustrates the use of soil map databases and soil survey information in Phase II model guidebook development.

Case Study: Use of STATSGO, SSURGO, and GIS Technology to Determine Pre-European Settlement Wetland Acreage — Applications to Phase II Model Guidebook Development

Tandarich and Elledge (1996) used STATSGO and SSURGO soil maps to estimate the percentage cover of hydric soil and pre-European settlement wetlands in three southeastern Wisconsin watersheds (Figures 10.4 and 10.5). They assumed that currently mapped hydric soils are a direct reflection of the pre-European settlement wetland conditions that produced them. Acreage estimates of hydric soils in a watershed should be a fair estimate of pre-European settlement wetlands (SAST and FMRC 1994). With respect to Phase II model guidebook development, pre-European settlement wetland maps could be imported into GIS software and combined with topographic, vegetation,

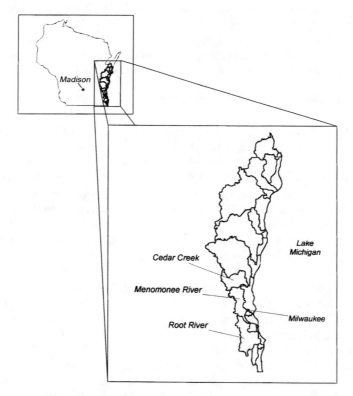

Figure 10.4   Location of the Cedar Creek Watershed. (Tandarich, J.P. and Elledge, A.L. 1996. *Determining the Extent of Presettlement Wetlands from Hydric Soil Acreages: A Comparison of SSURGO and STATSGO Estimates.* Hey & Associates, Inc. Chicago, IL. With permission.)

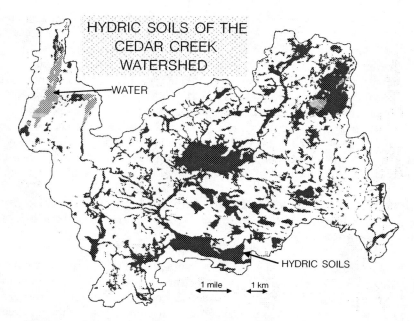

Figure 10.5 Hydric Soils of the Cedar Creek watershed. (From Tandarich, J.P. and Elledge, A.L. 1996. *Determining the Extent of Presettlement Wetlands from Hydric Soil Acreages: A Comparison of SSURGO and STATSGO Estimates.* Hey & Associates, Inc. Chicago, IL. With permission.)

and land use data to produce a variety of thematic maps. Examples of such thematic maps include: (i) the acreage and types of pre-European settlement wetland subclasses that have been lost through anthropogenic impacts in a region (i.e., the reference domain), (ii) the acreage and type of pre-European settlement wetland subclasses that remain in the reference domain, and (iii) the relationship between vegetation community types and soil taxa (Tandarich and Mosca 1990).

Use of Soil Information in Phase III of Draft Guidebook Development

In Phase III the A-Team develops a draft assessment model of wetland functions. Model development requires a literature review of existing models of wetland functions, identification of reference wetland sites and functions, identification of variables and indicators of wetland functions, and development of functional capacity indices (FCI). The A-Team conducts site visits of each regional wetland subclass to refine their assessment needs, select the priority wetland subclass, collect data to build the functional profile of the priority subclass, and identify a gradient of reference wetland sites with different land uses in the reference domain of the priority subclass (Lee et al. 1997).

One critical component in Phase III development is the identification of variables and indicators of wetland functions. A variable is defined as an attribute of a wetland ecosystem or the surrounding landscape that influences the capacity of a wetland to perform a function (Smith et al. 1995). Implicit in this definition is that a variable is an ecosystem attribute that can be quantified either in the field or in the laboratory. Calibrating and scaling HGM model variables should use quantitative data whenever possible. While this may require a greater expenditure of resources (e.g., time, money, etc.) by the A-Team during the reference wetland-sampling phase, we feel that such expenditures will lead to the development of a more robust HGM model. However, we are also cognizant of the fact that time constraints encountered in developing and performing a rapid functional assessment often preclude such an investment of resources. In this case, it is often more practical to use indirect indicators or qualitative measures of model variables. There are numerous soil properties that reflect and/or affect wetland functions. Many of these properties are easily measured in the field or laboratory and should be incorporated into an HGM assessment model

(Appendix 1). Appendix 2 lists several examples of HGM soil variables, their primary and secondary indicators, and how these variables and indicators might be scaled for use in HGM functional assessment. The following case study illustrates the use of soil information in the development of a draft HGM model guidebook.

Case Study: Hydrogeomorphic Assessment of Functions in Temporary and Seasonal Prairie Pothole Wetland Ecosystems — Use of Soil Information in Phase III Model Guidebook Development

Data Collection

The NRCS is mandated to assist federal, state, and local agencies in meeting the provisions of the Clean Water Act, in particular, ... "to restore the physical, chemical, and biological integrity of the Nations' waters" (33 U.S.C. 1344). As part of this mandate, NRCS is often called upon to assess the impacts of agricultural activities on wetland functions. The *Operational Draft Guidebook to Hydrogeomorphic Assessment of Functions in Temporary and Seasonal Prairie Pothole Wetland Ecosystems* (Lee et al. 1997), was developed by NRCS to satisfy the mandate of the "National Action Plan to Develop the Hydrogeomorphic Approach for Assessing Wetland Functions," and in response to NRCS's need for a "... consistent and scientifically based assessment procedure for assessment of functions of wetlands in the Northern Prairie Region" (Lee et al. 1997).

The A-Team and associated wetland professionals selected 25 reference wetland sites in the reference domain and collected data during concomitant field reconnaissance. Data collection and analysis occurred at four scales: (i) landscape, (ii) catchment area, (iii) site, and (iv) "within site" (Figure 10.6). *Landscape scale analysis* was performed within a 1-mile radius from the centroid

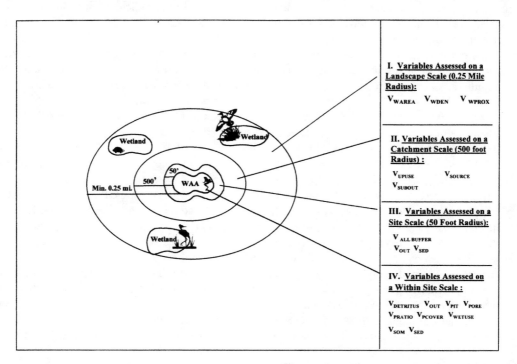

Figure 10.6 Observation Areas of Scale-Dependent Variables. (From Lee, L.C., Brinson, M.M., Kleindl, W.J., Whited, P.M., Gilbert, M., Nutter, W.L., Whigham, D.F., and DeWalk, D. 1997. *Operational Draft Guidebook for the Hydrogeomorphic Assessment of Temporary and Seasonal Prairie Pothole Wetlands.* Seattle, WA. With permission.)

of each pothole. GIS software was used in combination with digital NWI data and NRCS soil and land use data to evaluate wetland complexity and faunal characteristics. Wetland complexity was assessed by two metrics, *wetland area* and *wetland density*. Wetland area is the ratio of the total area of temporary and seasonal wetlands to the total area of semi-permanent and permanent wetlands, within a 1-mile radius from the center of the wetland. Wetland density is the absolute density of wetlands in a given water regime within a 1-mile radius from the center of the wetland. Landscape scale analysis also involved classifying soil map units into slope range classes (0 to 3%, 3 to 9%, >9%), and consolidating land use categories into distinct cover classes. *Catchment-scale analysis* was conducted within a 500-foot radius from the perimeter of each pothole complex. Data were collected on the dominant land use of the upland watershed that contributes to the wetland, subsurface flow from the wetland, and the area surrounding the wetland that defines the catchment or watershed of the wetland. Acreage estimates were made of wetland structural components, soil slope classes, land use cover classes, and linear coverages (e.g., transportation data). *Site-level analysis* was conducted within a 50-foot radius of the perimeter of each pothole complex. Data were collected on the extent of sediment delivery to the pothole complex from anthropogenic sources, the width of grassland buffer zones surrounding the outermost edge of the pothole complex (i.e., 50 feet), the continuity of the grassland buffer within 50 feet of the outermost edge of the complex, and the dominant land use condition within 50 feet of the outermost edge of the complex. *Within-site analysis* involved measuring plant species abundance and characterizing the soil resource, including making pedon descriptions and taxonomic classifications. Pedon descriptions included measurements of the thickness and degree of decomposition of litter, thickness of the A-horizon, quantity and continuity of soil pores, moist consistence, and soil structure. Litter thickness measurements were made in the temporary and seasonal zones and served as an indicator of the detrital pool. A-horizon thickness was used as an indicator of sediment delivery to the pothole complex from anthropogenic sources, including agriculture. Other indicators of sediment delivery were the presence of a lighter-colored A-horizon overlying a darker-colored A-horizon and/or the presence of calcareous "overwash."

Soil morphological features, such as pores, consistence, and structure, influence water and air movement through the soil. Anthropogenic activities can disrupt and destroy these features, resulting in significant changes in soil porosity and permeability and, hence, water and air movement (Bouma and Hole 1971). The A-Team designed metrics to describe the *quantity and continuity of pores* as well as consistence and structure. The quantity and continuity of pores partly control saturated hydraulic conductivity. Reduced hydraulic conductivity results in decreased recharge of the water table. The quantity and continuity of pores received a score of 1 through 3. Many "very fine" and "fine" pores in the A-horizon received a score of 3. Common pores received a score of 2, and few pores received a score of 1. *Consistence* is defined as the combination of soil properties that determine its resistance to crushing and its ability to be molded or changed in shape. Consistence is often used as an indicator of compaction. Increased compaction results in increased bulk density, reduced porosity and permeability, reduced hydraulic conductivity, and, therefore, reduced recharge of the water table. "Very friable" and "friable" consistence received a score of 3. "Firm" consistence received a score of 2; and "very firm" and harder received a score of 1. *Soil structure* is defined as the arrangement of soil particles into secondary units called *peds*. Structure that was "moderate" or "weak prismatic" parting to "moderate" and "strong subangular blocky," or parting to "moderate granular" in the A-horizon received a score of 3. Moderate to weak grades of "subangular blocky" and "granular" structure in the A-horizon received a score of 2. "Massive" structure, "strong coarse" and "very coarse subangular blocky" structure, and evidence of a plowpan received a score of 1.

Model Structure

The A-Team identified 11 important functions (Table 10.2) performed by temporary and seasonal depressional wetlands in the Northern Prairie Pothole Region. These functions were grouped into three functional classes: (a) physical/hydrological; (b) biogeochemical, and (c) biotic/habitat. FCI model algorithms were developed to describe the response of these various functions to anthropogenic activities, particularly agricultural practices. Each model algorithm consists of a group of variables that represents a particular ecosystem attribute that is sensitive to anthropogenic impacts (Table 10.3). The Prairie Pothole Region HGM draft model contains fifteen variables (Table 10.4). Variables may be used in one or several functions (Tables 10.5).

Table 10.2   Definitions of Functions for Temporary and Seasonal Northern Prairie Wetlands

Physical/Hydrologic Functions

Maintenance of Static Surface Water Storage. The capacity of a wetland to collect and retain inflowing surface water, direct precipitation, and discharging groundwater as standing water above the soil surface, pore water in the saturated zone, and/or soil moisture in the unsaturated zone.

Maintenance of Dynamic Surface Water Storage. The capacity of the wetland to detain surface water above the wetland surface as it flows through the wetland to be discharged via groundwater recharge and/or surface outlet.

Retention of Particulates. Deposition and retention of inorganic and organic particulates (>0.45 µm) from the water column, primarily through physical processes.

Biogeochemical Functions

Elemental Cycling. Short- and long-term cycling of elements and compounds on site through the abiotic and biotic processes that convert elements (e.g., nutrients and metals) from one form to another; primarily recycling processes.

Removal of Imported Elements and Compounds. Nutrients, contaminants, and other elements and compounds imported to the wetland are removed from cycling processes.

Biotic and Habitat Functions

Maintenance of Characteristic Plant Community. Characteristic plant communities are not dominated by non-native or nuisance species. Vegetation is maintained by mechanisms such as seed dispersal, seed banks, and vegetative propagation, which respond to variations in hydrology and disturbances such as fire and herbivores. The emphasis is on the temporal dynamics and structure of the plant community as revealed by species composition and abundance.

Maintenance of Habitat Structure Within Wetland. Soil, vegetation, and other aspects of ecosystem structure within a wetland are required by animals for feeding, cover, and reproduction.

Maintenance of Food Webs Within Wetland. The production of organic matter of sufficient quantity and quality to support energy requirements of characteristic food webs within a wetland.

Maintenance of Habitat Interspersion and Connectivity Among Wetlands. The spatial distribution of an individual wetland in reference to adjacent wetlands within the complex.

Maintenance of Taxa Richness of Invertebrates. The capacity of a wetland to maintain characteristic taxa richness of aquatic and terrestrial invertebrates.

Maintenance of Distribution and Abundance of Vertebrates. The capacity of a wetland to maintain characteristic density and spatial distribution of vertebrates (aquatic, semiaquatic, and terrestrial) that utilize wetlands for food, cover, and reproduction.

From Lee, L.C., Brinson, M.M., Kleindl, W.J., Whited, P.M., Gilbert, M., Nutter, W.L., Whigham, D.F., and DeWald, D. 1997. *Operational Draft Guidebook for the Hydrogeomorphic Assessment of Temporary and Seasonal Prairie Pothole Wetlands.* Seattle, WA.

Table 10.3   Indices of Functions for Temporary and Seasonal Northern Prairie Wetlands

Function 1. Maintenance of Static Surface Water Storage

$$\text{Index} = (V_{OUT} \times ((V_{SOURCE} + V_{UPUSE})/2 + (V_{WETUSE} + V_{SED} + V_{PORE} + V_{SUBOUT})/4)/2)1/2$$

Function 2. Maintenance of Dynamic Surface Water Storage
If V_{OUT} is less than .75, then function index is 0.0. Otherwise use:

$$\text{Index} = (V_{OUT} + (V_{SOURCE} + V_{UPUSE})/2 + (V_{PORE} + V_{WETUSE})/2)/3$$

Function 3. Elemental Cycling

$$\text{Index} = ((V_{SOURCE} + V_{OUT})/2 + (V_{UPUSE} + V_{WETUSE} + V_{SED})/3 + (V_{PCOVER} + V_{DETRITUS})/2 + V_{PORE})/4$$

Function 4. Removal of Imported Elements and Compounds

$$\text{Index} = ((V_{SOURCE} + V_{OUT} + V_{SUBOUT})/3 + (V_{UPUSE} + V_{WETUSE} + V_{SED})/3 + (V_{PCOVER} + V_{DETRITUS})/2 + V_{PORE})/4$$

Function 5. Retention of Particulates

$$V_{OUT} \leq 0.5, \text{ use: } (V_{UPUSE} + V_{WETUSE} + V_{SED} + V_{OUT})/4.$$

If $V_{OUT} > 0.5$ use: $(V_{UPUSE} + V_{SED})/2$

Function 6. Maintenance of Characteristic Plant Community

$$\text{Index} = (V_{WETUSE} + V_{SED} + V_{OUT} + V_{PRATIO} + V_{PCOVER} + V_{DETRITUS})/6$$

Function 7. Maintenance of Habitat Structure Within Wetland

$$\text{Index} = (V_{UPUSE} + V_{WETUSE} + V_{SED} + (V_{PRATIO} + V_{PCOVER})/2 + V_{DETRITUS} + V_{OUT} + (V_{BWIDTH} + V_{BCONTINUITY} + V_{BCONDITION})/3)/7$$

Function 8. Maintenance of Food Webs Within Wetland

$$\text{Index} = (V_{WETUSE} + V_{SED} + V_{PRATIO} + V_{PCOVER} + V_{DETRITUS} + V_{OUT} + (V_{BWIDTH} + V_{BCONTINUITY} + V_{BCONDITION})/3)/7$$

Function 9. Maintenance of Habitat Interspersion and Connectivity Among Wetlands

$$\text{Index} = (((V_{UPUSE} + V_{WETUSE} + V_{OUT})/3) \times ((V_{DEN} + V_{WAREA})/2))1/2$$

Or use number of breeding pairs of ducks

Function 10. Maintenance of Taxa Richness of Invertebrates
Note: Due to complexities of rapid assessment of invertebrates in the field, no index currently applies to this function.

Function 11. Maintenance of Distribution and Abundance of Vertebrates
Note: Due to complexities of rapid assessment of vertebrates in the field, no index currently applies to this function.

From Lee, L.C., Brinson, M.M., Kleindl, W.J., Whited, P.M., Gilbert, M., Nutter, W.L., Whigham, D.F., and DeWald, D. 1997. *Operational Draft Guidebook for the Hydrogeomorphic Assessment of Temporary and Seasonal Prairie Pothole Wetlands.* Seattle, WA.

Table 10.4   Definitions of Variables for Temporary and Seasonal Northern Prairie Wetlands

$V_{BCONDITION}$	**Grassland Buffer Condition.** Dominant land use condition within 50 feet of the outermost edge of the wetland.
$V_{BCONTINUITY}$	**Grassland Buffer Continuity.** Continuity of grassland buffer within 50 feet of the outermost edge of the wetland.
V_{BWIDTH}	**Grassland Buffer Width.** Width of grassland buffer surrounding outermost wetland edge ($\leq$50 feet from wetland edge).
$V_{DETRITUS}$	**Detritus.** The presence of litter in several stages of decomposition (e.g., litter).
V_{OUT}	**Wetland Outlet.** The presence of a low elevation (threshold elevation) over which water could flow from the wetland. Change in outlet invert elevation modifies wetland water surface elevation.
V_{PCOVER}	**Plant Density.** The abundance of woody and herbaceous plants in all vegetation zones within the wetland.
V_{PORE}	**Soil Pores.** The physical integrity of the soil above the Bt horizon. This includes the number and continuity of pores and the type, grade, and size of soil structure.
V_{PRATIO}	**Ratio of Native to Non-Native Plant Species.** The ratio of native to non-native plant species present in wetland zones as indicated by the top 4 dominants or by a more extensive species survey. Dominants are the most abundant species that immediately exceed 50% of the total dominance for a given stratum when the species are ranked in descending order of abundance and cumulatively totaled. Dominants also include any additional species comprising 20% or more of the total.
V_{SED}	**Sediment Delivery to Wetland.** Extent of sediment delivery to wetland from anthropogenic sources including agriculture.
V_{SOURCE}	**Source Area of Flow Interception by the Wetland.** The area surrounding a wetland that defines the catchment or watershed of that wetland.
V_{SUBOUT}	**Subsurface Outlet.** Presence of a subsurface flow from the wetland. Subsurface or surface drain and distance from the wetland impacts groundwater surface elevation.
V_{UPUSE}	**Upland Land Use.** Dominant land use and condition of upland watershed that contributes to the wetland. When possible, an assessment of the entire watershed is recommended. When this is not possible, an assessment of 500 foot perimeter from the outer temporary edge is recommended.
V_{WAREA}	**Wetland Area in the Landscape.** The ratio of total area of temporary and seasonal wetlands to the total area of semipermanent and permanent wetlands within a 1-mile radius of the assessment site.
V_{WDEN}	**Density of Water Regime in the Landscape.** The absolute density of wetlands in a given water regime within a 1-mile radius from the center of the wetland.
V_{WETUSE}	**Wetland Land Use.** Dominant land use and condition of wetland.

Table 10.5   Relationship of Variables to Wetland Functions for Temporary and Seasonal Northern Prairie Wetlands

Functions Variables	Static	Dynamic	Cycling	Removal	Retention	Plant	Structure	Food	Habitat
$V_{BCONDITION}$							X	X	
$V_{BCONTINUITY}$							X	X	
V_{BWIDTH}							X	X	
$V_{DETRITUS}$			X	X		X	X	X	
V_{OUT}	X	X	X	X	X	X	X	X	X
V_{PCOVER}			X	X		X	X	X	
V_{PORE}	X	X	X	X					
V_{PRATIO}						X	X	X	
V_{SED}	X	X	X	X	X	X	X	X	
V_{SOURCE}	X	X	X	X					
V_{SUBOUT}	X			X					
V_{UPUSE}	X	X	X	X	X		X		X
V_{WAREA}									X
V_{WDEN}									X
V_{WETUSE}	X	X	X	X	X	X	X	X	X

Note: Due to complexities of rapid assessment of vertebrates and invertebrates, no variables currently apply to these related functions.

KEY
Functions

Static	Maintenance of static surface water storage
Dynamic	Maintenance of dynamic surface water storage
Cycling	Elemental cycling
Removal	Removal of imported elements and compounds
Retention	Retention of particulates
Plant	Maintenance of characteristic plant community
Structure	Maintenance of habitat structure within wetland
Food	Maintenance of food webs within wetland
Habitat	Maintenance of habitat interspersion and connectivity among wetlands
Vertebrate	Maintenance of distribution and abundance of vertebrates
Invertebrate	Maintenance of taxa richness of invertebrates

Variables

$V_{BCONDITION}$	Buffer condition
$V_{BCONTINUITY}$	Buffer continuity
V_{BWIDTH}	Buffer width
$V_{DETRITUS}$	Detritus
V_{OUT}	Wetland outlet
V_{PCOVER}	Plant density
V_{PORE}	Soil pores
V_{PRATIO}	Ratio of native to non-native plant species
V_{SED}	Sediment delivery to wetland
V_{SOURCE}	Source area of flow interception by wetland
V_{SUBOUT}	Constructed subsurface/surface outlet
V_{UPUSE}	Upland land use
V_{WAREA}	Wetland area in the landscape
V_{WDEN}	Density of water regime in the landscape
V_{WETUSE}	Wetland land use

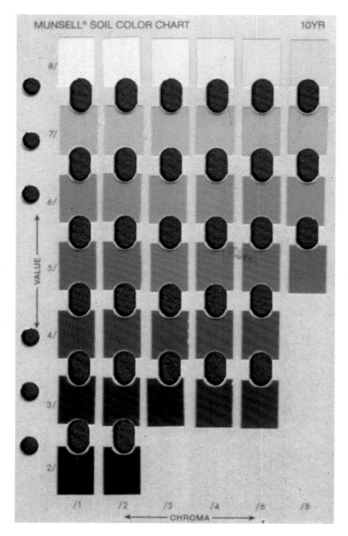

Plate 1. The 10YR soil color chart. The 10YR 4/2 chip is located in the third row from the bottom and the second column from the left on the 10YR chart. Only color quality of original charts is adequate for color determinations. (Courtesy of GretagMacbeth, Munsell®, Corp., New Windsor, NY.)

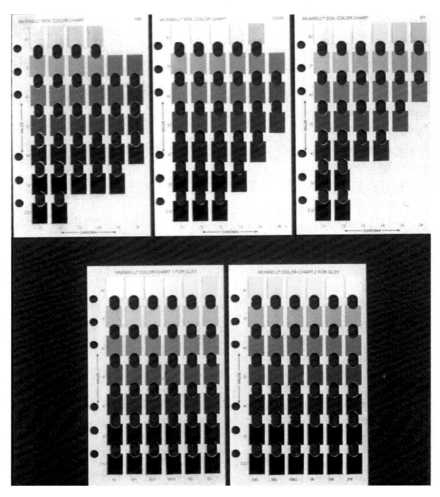

Plate 2. Top row shows the 10R, 7.5YR, and 5Y Munsell® Soil Color Charts, which span the range of hues normally found in soils in the 48 contiguous states. Bottom row is a photograph of the two gley charts of the Munsell® Soil Color Charts. Only color quality of original charts is adequate for color determinations. (Courtesy of GretagMacbeth, Munsell®, Corp., New Windsor, NY.)

Plate 3. Photograph of the Muskego soil, a Histosol, in Minnesota. The thickness of the organic materials in this soil is 100 cm. The underlying materials are limnic (coprogenous earth). (From Domeier, M.J. 1997. Soil Survey of Sibley County, Minnesota. USDA–NRCS, in cooperation with the Minn. Ag. Exp. Sta., U.S. Govt. Printing Office, Washington, DC.)

Plate 4. Soils formed in organic materials along a tidal creek in New Hampshire. These soils are Sulfihemists and have a high content of sulfidic materials above a depth of 100 cm. (From Kelsea, R.J. and J.P. Gove. 1994. Soil Survey of Rockingham County, New Hampshire. USDA–SCS in cooperation with the New Hampshire Ag. Exp. Sta., U.S. Govt. Printing Office, Washington, DC.)

Plate 5. Water lines on the trees in this swamp in Florida show the fluctuations in the water levels that occur in these areas during the dry and wet seasons. The soils in this depression have a thin organic surface layer (histic epipedon).

Plate 6. Photograph showing the amount of organic materials that have subsided in the Florida Everglades Agricultural Area. The top of the column was at the soil surface in 1924. Marks are in feet.

Plate 7. Iron masses (approximately 5 cm wide) with diffuse boundaries (arrow).

Plate 8. Pore linings (arrow) along channel. Blocks are 5 cm wide.

Plate 9. Pore linings (red coatings) along cracks. Roots range from 1 to 2 mm in diameter. Matrix of the soil is gray.

Plate 10. Iron/manganese nodules that have been removed and washed from a Bt horizon. Nodules range from approximately 2 to 10 mm in diameter.

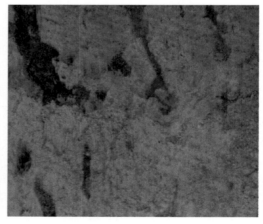

Plate 11. Iron depletions (gray colored soil) in matrix with Fe masses (yellow colors). Black areas are root channels filled with A horizon material. Field is 6 cm across.

Plate 12. Iron depletions (gray colors) along channels. Channels are approximately 10 mm wide.

Plate 13. Iron depletions (gray spots) in matrix of A horizon. Largest depletions are approximately 20 mm wide.

Plate 14. Iron depletions (gray colors identified by arrow) in sand. Partially decomposed root material appears as black dots (1 to 2 mm diameter) with the gray depletions.

Plate 15. This soil shows accumulation of carbon in the upper part and reduction, translocation, and reoxidization of iron in the lower part. Scale is feet.

Plate 16. Some soils are low in iron, and the only evidence of anaerobiosis is the accumulation of organic matter. Scale is inches.

Plate 17. In hydric soils with loamy and clayey material, a common redoximorphic feature is redox concentrations around stable macropores such as this diffuse pore lining.

Plate 18. In sandy materials macropores are not as stable as in loamy and clayey materials; a common redoximorphic feature is redox depletions within the matrix, which appear as oval splotches.

Plate 19. Indicator F3 (Depleted Matrix). The matrix is chroma 2 with redox concentrations as pore linings. Scale is inches.

Plate 20. Indicator S7 (Dark Surface). The dark surface is about 15 cm thick. Indicator S8 (Polyvalue Below Surface) occurs below 15 cm. Scale is inches.

Plate 21. Indicator S5 (Sandy Redox). The redox masses occur throughout this soil in a matrix with 60% or more chroma 2 or less. Scale is inches.

Plate 22. Indicator F13 (Umbric Surface). The umbric surface is about 30 cm thick and the chroma is 2 or less below the umbric surface. Scale is inches.

REFERENCES

Amoozegar, A. 1989. A compact constant-head permeameter for measuring saturated hydraulic conductivity of the vadose zone. *Soil Sci. Am. J.* 53:1356–1361.

Blake, G.R. and Hartge, K.H. 1986. Bulk density, pp. 363–375. *In* Klute, A. (Ed.) *Methods of Soil Analysis, Part I: Physical and Mineralogical.* Am. Soc. Agron., Madison, WI.

Bouma, J. and Hole, F.D. 1971. Soil structure and hydraulic conductivity of adjacent virgin and cultivated pedons at two sites: a Typic Argiudoll and a Typic Eutrochrept. *Soil Sci. Soc. Am. Proc.* 35:316–319.

Brady, N.C. and Weil, R.R. 1996. *The Nature and Property of Soils,* 11th ed. Prentice-Hall, Englewood Cliffs, NJ, 740 p.

Brinson, M.M. 1993. *A Hydrogeomorphic Classification for Wetlands.* Technical Report WRP-DE-4, U.S. Army Engineer Waterways Experiment Station, Vicksburg, MS.

Brinson, M.M. 1995. The HGM approach explained. *National Wetlands Newsletter.* November–December, 17:7–13.

Brinson, M.M., Hauer, F.R., Lee, L.C., Nutter, W.L., Smith, R.D., and Whigham, D.F. 1994. Developing an approach for assessing the functions of wetlands, *In* W.J. Mitsch and R.E. Turner, (Eds.) *Wetlands of the World: Biogeochemistry, Ecological Engineering, Modelling and Management.* Elsevier Publishers, Amsterdam.

Federal Register, August 16, 1996. v. 61m, no. 160.

Hauer, F.R. 1995. *The Hydrogeomorphic Functional Assessment of Wetlands: The Characterization of Reference Wetlands and Development of a Regional Assessment Guidebook in the Northern Rocky Mountain Region.* Flathead Biological Station, Univ. Montana, Polson, MT.

Jenny, H. 1941. *Factors of Soil Formation.* McGraw-Hill. New York.

Larson, W.E. and Pierce, F.J. 1991. Conservation and enhancement of soil quality, p. 175–203. *In Evaluation of Sustainable Management in the Developing World. Vol. 2* Tech. Papers. IBSRAM Proc. 12(2). Intl. Board of Soil Res. and Manage., Bangkok, Thailand.

Lee, L.C., Brinson, M.M., Kleindl, W.J., Whited, P.M., Gilbert, M., Nutter, W.L., Whigham, D.F., and DeWald, D. 1997. *Operational Draft Guidebook for the Hydrogeomorphic Assessment of Temporary and Seasonal Prairie Pothole Wetlands.* Seattle, WA.

Lonard, R.I., Clairain, E.J., Jr., Huffman, R.T., Hardy, J.W., Brown, C.D., Ballard, P.E., and Watts, J.W. 1981. *Analysis of Methodologies Used for Assessment of Wetland Values.* Final Report. U.S. Army Engineer Waterways Experiment Station, U.S. Water Resources Council, Washington, DC.

Lowery, B., Hickey, W.J., Arshad, M.A., and Lal, R. 1996. Soil water parameters and soil quality. *In* Doran, J.W. and Jones, A.J. (Eds.) *Methods for Assessing Soil Quality.* SSSA Spec. Pub. No. 49. Soil Sci., Soc. Am., Madison, WI.

Mitsch, W.J. and Gosselink, J.G. 1993. *Wetlands,* 2nd ed., Van Nostrand Reinhold, New York.

Rhoads, J.D. 1993. Electrical conductivity methods for measuring and mapping soil salinity. *Adv. Agron.* 49:201–251.

SAST (Scientific Assessment and Strategy Team) and FMRC (Interagency Floodplain Management Review Committee). 1994. *Science for Floodplain Management Into the 21st Century. A Blueprint for Change/Part V.* Report to FMRC to the Administration Floodplain Management Task Force. Washington, DC.

Smith, R.D., Ammann, A., Bartoldus, D., and Brinson, M.M. 1995. *An Approach for Assessing Wetland Functions Using Hydrogeomorphic Classification, Reference Wetlands, and Functional Indices.* Technical Report WRP-DE-9, U.S. Army Engineer Waterways Experiment Station, Vicksburg, MS.

Soil and Water Conservation Society. 1992. SWCS adopts wetland policy statement. *J. Soil and Water Cons.* Nov–Dec.

Star, J. and Estes, J. 1990. *Geographic Information Systems: An Introduction.* Prentice-Hall, Englewood Cliffs, NJ.

Stewart, R.E. and Kantrud, H.A. 1972. Vegetation of prairie potholes, North Dakota, in relation to quality of water and other environmental factors. U.S. Geol. Surv. Prof. Paper 585-D.

Tandarich, J.P. and Elledge, A.L. 1996. *Determining the Extent of Presettlement Wetlands from Hydric Soil Acreages: A Comparison of SSURGO and STATSGO Estimates.* Hey & Associates, Inc. Chicago, IL.

Tandarich, J.P. and Mosca, V. 1990. Soil maps and natural area data: useful tools in restoration planning. Note 147. *Restoration and Management Notes.* 8:65.

Vepraskas, M.J. 1995. *Redoximorphic Features for Identifying Aquic Conditions.* Tech. Bull. 301, North Carolina Ag. Res. Ctr., North Carolina State Univ., 33 p.

West, L.T., Chiang, S.C., and Norton, L.D. 1992. The morphology of surface crusts. *In* Sumner, M.D. and Stewart, B.A. (Eds.) *Soil Crusting, Chemical and Physical Processes.* Advances in Soil Science. Lewis Publishers, Chelsea, MI.

APPENDIX I:
Fundamental Soil Variables and HGM Variables

Infiltration — $V_{sinfilt}$

Definition — the downward entry of water into the soil through the soil surface. Infiltration flux (or rate) is the volume of water entering a specified cross-sectional area per unit time.

Measurement Techniques — infiltrometer, numerous methods include ponded double ring sprinkler methods. For soil quality evaluation, Lowery et al. (1996) have recommended a simplified "coffee can" method.

Units of Measure — length per time (m/s, in/hr)

Qualitative Indicators — arrangement, continuity, and size distribution of pores, soil structure, structural and sedimentary crusts (West et al. 1992), compaction (bulk density), surface sealing by sediment, consistence, soil tilth, root quantity.

Variability — infiltration is a temporally and spatially variable property. Anthropogenic activities, however, can significantly impact infiltration rate. Mechanical activities, including tillage, plant removal, traffic patterns of vehicles and livestock, typically affect infiltration. Deposition by water, orientation, and/or packing of a thin layer of fine soil particles on the surface of the soil (soil sealing) can also greatly reduce infiltration.

Importance to Wetland Function — infiltration is important for maintaining plant growth, preventing erosion, carrying solutes into the soil biological "filter," maintaining anaerobic conditions, and contributing to groundwater recharge. Reduced infiltration on upland areas surrounding wetlands can increase sediment and toxicant delivery.

Importance to HGM Functions — Maintenance of plant community; conversion, removal, and cycling of elements and compounds; groundwater recharge; maintenance of characteristic hydrologic regime.

Saturated Hydrologic Conductivity (Ksat) — V_{shcond}

Definition — the amount of water that would move downward through a unit area of saturated in-place soil in unit time under unit hydraulic gradient.

Measurement Techniques — numerous techniques are available. In recent years, the Amoozemeter (Amoozegar 1989) has been used by the National Cooperative Soil Survey Program as an *in situ* field method. For soil quality evaluation, Lowery et al. (1996) suggest a simplified falling head permeameter technique.

Units of Measure — length per unit time (m/s, in/hr)

Qualitative Indicators — *Ksat* indicators include the arrangement, continuity, and size distribution of visible pores (e.g., worm holes, root channels, animal burrows [krotovina], grade and size of structural aggregates, relative strength and vertical axes of aggregates, compaction (i.e., bulk density), consistence, root quantity, rooting depth, presence of plow pans and other mechanically produced structural features (e.g., coarse, platy structure). Textural discontinuities, such as occur in filled and created wetlands, can greatly reduce hydraulic conductivity.

Variability — *Ksat* is a naturally temporal and spatially variable property that can vary both within and among soil horizons. Any comparison of field-measured *Ksat* to an HGM reference *Ksat* standard must be made on the same soil horizons (e.g., A to A, Bt to Bt, etc.). Anthropogenic activities such as mechanical activities, associated with agriculture and urbanization, typically reduce *Ksat*. Changes in *Ksat* are often more evident in surface horizons, although there are exceptions. One notable exception is the deep ripping of hardpan soil horizons to increase permeability and hydraulic conductivity

Importance to Wetland Functions — water moving through the soil is important for maintaining plant growth, preventing erosion, carrying solutes into the soil biological "filter," maintaining soil water storage capability, and contributing to groundwater recharge. Large average *Ksat* values for similar soils under different management types may be indicative of soils that have improved aggregation and greater macroporosity, both of which may be related to greater biological activity (Lowery et al. 1996). In soils with perched water tables, soil horizons with low *Ksat* influence the maintenance of saturated conditions in horizons above the perched zone.

Importance to HGM Functions — Maintenance of plant community; conversion, removal, and cycling of elements and compounds; groundwater recharge; maintenance of characteristic hydrology (e.g., perched water tables).

Bulk Density (p_b) — V_{sbd}

Definition — the mass (weight) of a unit volume of dry soil. The volume includes both solids and pores. Bulk volume. The bulk volume is determined before drying at 105°C to a constant weight.

Measurement Techniques — techniques include the core, excavation, clod, and radiation methods (Blake and Hartge 1986).

Units of Measure — the SI unit is kilograms per cubic meter (kg/m^3). Other common units derived from the SI unit include Megagrams per cubic meter (Mg/m^3) and grams per cubic centimeter (g/cm^3).

Qualitative Indicators — One commonly described morphologic indicator is *dry consistence*. Consistence is the combination of properties of soil material that determines its resistance to crushing and its ability to be molded or changed in shape (Brady and Weil 1996). Anthropogenic indicators would include anything that indicates compaction.

Variability — bulk density is not an invariant quantity for a given soil. It varies with structural conditions, soil texture, packing, clay mineralogy, water content, and system of land management. Increases in bulk density generally indicate a poorer environment for root growth and undesirable changes in hydrologic functions. Anthropogenic activities, including removal of forest trees by clear cutting and mechanical activities such as vehicle and animal traffic, can lead to increased bulk densities.

Importance to Wetland Function — increased bulk density reduces porosity, infiltration, and hydraulic conductivity and may contribute to increased overland (surface) flow, erosion, and sedimentation.

Importance to HGM Functions — maintenance of hydrologic functions; maintenance of microbial habitat.

Organic Matter — V_{som}

Definition — the organic fraction of a soil, including living organisms (biomass), carbonaceous remains of soil organisms, and organic compounds produced by current and past metabolism in the soil (Brady and Weil 1996).

Measurement Techniques — direct determination of organic matter can be measured by loss on ignition; however, measurement of organic carbon is often used as an indirect indicator of organic matter. Organic carbon is commonly measured using the Walkley–Black wet oxidation method or by use of an automated carbon analyzer (e.g., Leco, Perkin-Elmer, Fisons). Organic matter can be quantified from organic carbon measurements using the equation: *Organic matter = organic carbon × 1.724.*

Units of Measure — %

Qualitative Indicators — common soil morphologic indicators include soil color and texture. Other indicators include plant and root abundance, historic land use, and drainage.

Variability — soil organic matter/carbon is a natural spatially and temporally variable property. Organic matter content can be altered by anthropogenic activities. Larson and Pierce (1991) describe organic matter as the most important property for assessment of soil quality. In addition to considering the amount of organic matter, the thickness of organic soil layers should be evaluated in some wetland systems. An example would be the harvesting of organic material for horticultural peat.

Importance to Wetland Functions — the influence of organic matter on soil properties and plant growth is tremendous. Organic matter binds soil particles together into granular soil structure, thus aiding aeration, infiltration, and water-holding capacity. It is a major source of the plant nutrients phosphorous and sulfur, and it is the main source of nitrogen. Organic matter is the main food that supplies carbon and energy to heterotrophic soil organisms.

Importance to HGM Functions — maintenance of plant community; maintenance of food webs; retention, conversion, and cycling of elements and compounds; organic carbon retention and/or release.

Oxidation–Reduction (Redox) Potential (Eh) — V_{redox}

Definition — a measure of the oxidation–reduction potential status of a soil. Redox potential is the electrical potential (measured in volts or millivolts) of a system due to the tendency of the substances in it to give up or acquire electrons (Brady and Weil 1996). The potential is generated between an oxidation or reduction half-reaction and the hydrogen electrode in the standard state.

Measurement Techniques — soil redox potential is typically measured using platinum (Pt) electrodes, a mercury–chloride (HgCl or *calomel*) or silver chloride (AgCl) reference electrode, and a portable voltmeter. A minimum of three Pt electrodes should be placed at each depth in the soil profile, and readings should be taken every 1 to 2 weeks

Units of Measure — millivolts (mV), adjusted for pH. pH should be measured concurrently.

Qualitative Indicators — soil redox conditions can be manifested in distinguishing morphologic (redoximorphic) features, including iron masses, oxidized rhizospheres, and reduced matrices (Vepraskas 1995), and by the presence or absence of drainage, hydrophytic plant communities, reaction to α,α'-dipyridyl, and water table data.

Variability — soil redox potential is a natural spatially and temporally variable property. Long-term monitoring is required to assess this variability. Redox potential varies with soil aeration and pH.

Importance to Wetland Functions — soil redox controls most of the important chemical biogeochemical reactions in wetland soils, particularly the availability of essential plant nutrients (e.g., NO_3).

Importance to HGM Functions — all biogeochemical functions; maintenance of characteristic plant communities.

Electrical Conductivity (EC_e) — V_{sec}

Definition — the electrolytic conductivity of an extract from saturated soil. EC is one of the three primary properties, including exchangeable sodium percentage (ESP) and sodium adsorption ratio (SAR), that is used to characterize salt-affected soils.

Measurement Techniques — EC_e is measured by both laboratory and field methods. The *saturation paste extract method* is the most commonly used laboratory procedure. A soil sample is saturated with distilled water to a paste-like consistency, allowed to stand overnight to dissolve the salts, and the electrical conductivity of the water extracted is measured. A variant of this method involves the EC of the solution extracted from a 1:2 soil–water mixture after 0.5 hours of shaking (Brady and Weil 1996). Field methods include the use of sensors to measure bulk soil conductivity that is in turn related to soil salinity. A more rapid field method involves electromagnetic induction of electrical current in the soil. Electrical current is related to conductivity and soil salinity (Rhoads 1993).

Units of measure — SI units are siemens per meter (S m^{-1}) at 25°C, and the tesla (T). Non-SI units include millimhos per centimeter (mmho cm^{-1}) and the gauss (G).

Qualitative Indicators — because EC_e is related to salt content in the soil, indicators include the presence of salts on the soil surface (e.g., white alkali) and throughout the soil profile, and the presence of salt-tolerant plant communities.

Variability — EC_e is a temporally and spatially variable property that can be significantly affected by anthropogenic activities, particularly irrigation practices and groundwater extraction.

Importance to Wetland Functions — excess salts detrimentally affect plant and microbial communities. High pH and low concentrations of essential plant micronutrients such as iron, manganese, and zinc, characterize alkaline soils. High soluble salt concentrations affect osmotic potentials in plants and thus retard their growth. A change in EC of as little as 1 s m^{-1} can cause significant shifts in microbial activity (J. Doran, personal communication).

Importance to HGM Functions — maintenance of characteristic plant communities; conversion, retention, and cycling of elements and compounds.

pH — V_{sph}

Definition — the negative logarithm of the hydrogen ion activity of a soil. The degree of acidity or alkalinity of a soil.

Measurement Techniques — typically measured using glass, quinhydrone, or other suitable electrodes, colorimetric indicators, or paper strips.

Units of Measure — expressed in terms of the pH scale (0 to 14).

Qualitative Indicators — some plant communities, parent materials, and climates are indicative of soil pH conditions.

Variability — pH is a function of the five soil-forming factors (Jenny 1941); however, it can be influenced by anthropogenic factors. pH values of 6.0 to 7.5 typically do not directly affect plant roots or soil microbes. Therefore, "small" deviations in pH from an HGM reference standard are not ecologically significant. pH may be an ecologically significant variable in wetlands that have been affected by acid mine drainage or which are developed in acid sulfate soils.

APPENDIX II:
Using Soil Morphological Descriptions as Indicators of Wetland Function

The following pedon descriptions are from two reference wetlands in Benson County, MN.

Site 1 — Native prairie, never tilled, reference standard community, pedon description from temporarily inundated Wet Meadow Zone (Stewart and Kantrud 1972).

Oa, 1 — 0"   Undecomposed organic matter.

A1, 0 — 6"   N2/(black) loam. Weak, medium subangular blocky structure parting to moderate fine and medium granular. Friable. Common very fine tubular pores with moderate vertical continuity, few fine prominent 10YR 4/6 (dark yellowish brown) redoximorphic concentrations in root channels. Many very fine and common fine roots. EC < 1. Few worms.

A2, 6 — 8"   10YR 2/1 (black) loam. Weak medium prismatic structure parting to moderate medium subangular block. Very friable. Many fine tubular pores with moderate vertical continuity. Common very fine and few fine roots.

Bt, 8 — 15"   10YR 2/1 (black) clay loam. Moderate medium prismatic structure parting to moderate medium subangular blocky. Friable. Common fine tubular pores with moderate vertical continuity. Common very fine and few fine roots.

Site 2 — Farmed wetland, frequently cropped, pedon description from historic temporarily inundated Wet Meadow Zone (Stewart and Kantrud 1972). Soybeans this year. Site is partially drained.

Ap, 0 — 6"   10YR 2/1+ (black) silt loam. Moderate medium subangular blocky structure. Firm. Few very fine tubular pores with low vertical continuity, few very fine roots. EC < 1. Very slight effervescence (dilute HCl).

A2, 6 — 12"   10YR 2/1 (black) loam. Moderate medium subangular blocky structure parting to weak medium platy (mechanical structure?). Friable. Common very fine tubular pores with low vertical continuity. Few very fine roots. Few worms.

Bt, 12 — 19"   10YR 2/1 (black) clay loam. Moderate medium prismatic structure parting to moderate medium subangular blocky. Friable. Common fine tubular pores with low vertical continuity. Few very fine roots.

The pedon descriptions allow us to make some inferences concerning wetland function. Site 2 is characterized by greater sediment influx than Site 1. Soil morphological evidence to support this conclusion includes the calcareous overwash in the upper horizons of pedon #2, its silt loam texture and lighter color (2/1+ vs. N/2), and the greater depth to the Bt horizon.

APPENDIX III:
Examples of HGM Variables, Indicators, and Corresponding Subindices

Direct Measure	Primary Indicator	Secondary Indicator	Subindex
Variable — Soil Organic Matter Content			
6–8%	Soil color of A horizon is N2 or N3	Lightly to moderately grazed pasture; abundant roots; no evidence of tillage; detritus 1–2 in. thick.	1.0
4–6%	Color value and chroma of A horizon is 2/1, 2/2, or 3/1 and no evidence of tillage; sedimentary overwash or fill.	Heavily grazed pasture of hayed; no evidence of tillage; abundant roots; detritus < 1 in. thick buy present throughout site.	0.25–0.75
< 4%	Color value and chroma of A horizon is 2/1, 3/1, or 3/2; evidence of sedimentary overwash and fill.	Frequently tilled; few roots; evidence of erosion on surrounding landscape; some detritus but lacking throughout site.	0.1–0.25
0%	Non-soil material.	Parking lot.	0.0
Variable — Soil Infiltration			
75–125% of reference standard	Many continuous pores in A horizon; very friable consistence; compound soil structure.	Many roots; undisturbed "natural" vegetation; no evidence of historic mechanical disruption of soil surface.	1.0
25–75% of reference standard	Common discontinuous pores or many discontinuous, pores; friable to firm consistence; structure somewhat degraded compared to reference standard.	Common to many roots; vegetated; heavily grazed with evidence of trampling; evidence of rutting from machinery; historic tillage.	0.25–0.75
1–25% of reference standard	Few pores; firm or very firm consistence; massive structure; evidence of plow pan or other "mechanical" structure; surface sealing due to sediment or fill.	Common to many roots; vegetated; heavily grazed with evidence of trampling; evidence of rutting from machinery; historic tillage.	0.1–0.25
No infiltration	Nonporous surface	Parking lot	0.0
Variable — Permeability			
$Ksat$ = 75–125% of reference standard	Many continuous pores; compound structure, i.e., weak/moderate prismatic parting to moderate subangular blocky parting to moderate granular; friable or very friable consistence.	Many roots; no evidence of historic mechanical disruption.	1.0
$Ksat$ = 25–75% of reference standard	Common continuous and discontinuous pores; structure weaker compared to reference standard, i.e., subangular blocky parting to granular; firm consistence.	Common to many roots; evidence of historic mechanical disruption; rutting from machinery; some roots "plastered" to ped faces.	0.25–0.75
$Ksat$ = 1–25% of reference standard	Few discontinuous pores; massive or coarse subangular blocky structure; plow pan present; firm to very firm consistence.	Few roots; frequently tilled; roots growing horizontally across plow pan.	0.1–0.25
$Ksat$ = 0	Substrate is a nonporous medium.	Parking lot.	0.0

APPENDIX III: *(continued)*
Examples of HGM Variables, Indicators, and Corresponding Subindices

Direct Measure	Primary Indicator	Secondary Indicator	Subindex
Variable — Soil Redox Potential			
Monitoring of redox potential on the site compared to monitoring data from reference standard sites.	Oxidized rhizospheres present; hydric indicators present; no evidence of drainage; soil organic matter is comparable to reference standard.	Hydrophytic plants common, present, not "removed" by harvesting; ratio of hydric soil to hydrophytic plant community area matches reference standard (both aerial and indicator status[1]); in agricultural areas the site is termed a "wetland farmed under natural conditions."	1.0
It may be possible to substitute water table data for redox data, however this is not recommended.	Hydric soil indicators are present; some drainage is evident, or water is prevented from reaching the sites (e.g., levees); soil organic matter is less than reference standard.	Plant community to hydric soil ratio deviates from the reference condition; plant community is removed. In agricultural areas, the site is termed a "farmed wetland."	0.25–0.75
	No hydric indicators; site is "effectively" drained or protected from flooding.	Nonhydrophytic plant community; in agricultural areas, the site is "prior converted" if a playa, pothole, or pocosin.[2]	
	Site is completely drained; no soil organic matter; site is filled.	Parking lot.	0.0

[1] One type of field data that can be easily collected is an aerial ratio of hydrophytic plant communities to hydric soils. One could also assess the indicator status of the plant community using a method such as the Prevalence Index. It may be possible for an area to have a spatial ratio of hydrophytic plant communities to hydric soil that equals the reference standard, however, the plant community may reflect a drier indicator status than the standard. This could be used as an indication of reduced soil redox potential. The field data may be valid as an indicator of several variables in the HGM model.

[2] The use of the Food Security Act (FSA) designation of "Prior Converted' as an indicator of redox potential is not appropriate in many parts of the U.S., especially areas that use the 15-day surface water criteria to separate "Prior Converted" from "Farmed Wetland."

Wetland Soils of Basins and Depressions of Glacial Terrains

C. V. Evans and J. A. Freeland

INTRODUCTION

In closed depressions subject to ponding, hydric soil morphology is indicated simply by the presence of 5% or more distinct or prominent redox concentrations as soft masses or pore linings in a layer 5 cm or more thick within the upper 15 cm (Hurt et al. 1996). In these "redox depressions," soils are determined to be hydric primarily on the basis of landscape position and documentation of at least seasonal ponding. There is no fixed requirement for Munsell value or chroma in the soil matrix. The accompanying notes state, "Most often soils pond water because of two reasons: they occur in landscape positions that collect water and/or they have a restrictive layer(s) that prevent water from moving downward through the soil" (Hurt et al. 1996). Such flat or depressional landscapes may be created by a variety of geological processes. Examples of depressional features include glacial kettles, vernal pools, playas, till plain swales, and potholes. Water can be received directly as rain, from throughflow, overland flow, or from groundwater discharge (Mausbach and Richardson 1994). Most simply, inflow exceeds the capacity of the system to remove the water, at least for a significant period of time in most years.

The most direct relationship between soil water table maxima and landscape position is described in basic terms of gravitational potential, whereby water seeks the lowest potential energy level — usually the lowest point in the landscape. In many landscapes, however, soils are formed in anisotropic materials. By the nature of the formation of horizons, soils are anisotropic as well. In these landscapes stratigraphy combines with topography to influence soil moisture regime by controlling movement of water across and through the landscape (Zaslavsky and Rogowski 1969). Stratigraphic control of water potential is based on hydraulic conductivity, which is a function of soil bulk density, structure, and texture (King and Franzmeier 1981). Several studies (Daniels et al. 1971, Vepraskas and Wilding 1983, Evans and Franzmeier 1986, Steinwand and Fenton 1995) have described water tables affected at least partially by differential conductivity of stratigraphic layers or soil horizons.

Movement of water in such landscapes is often by overland flow or by saturated subsurface flow. Overland flow is most important when precipitation rates exceed infiltration rates (Hortonian overland flow), or when precipitation and run-on exceed the hydraulic conductivity of the most limiting layer (reflow). If infiltration rates are sufficient, and the soil becomes saturated, drainage may continue along subsurface gradients, often along the surface of the limiting layer, if one exists. Such restrictions may lead to the presence of a perched water table that may be considerably above stream level and separated from "true" groundwater by an unsaturated layer. Saturated conditions involving these apparent water tables are referred to as "episaturation" (Soil Survey Staff 1994). Many depressional soils are characterized by episaturation.

Closed depressions lack stream outlets; thus, all water, sediment, and other materials from the surrounding slopes are trapped and must either evaporate, be transpired by vegetation, or recharge groundwater and flow through to groundwater. In these conditions finer sediments tend to accumulate and organic matter may be preserved in the depression center. Both of these conditions differ from soil-forming processes in better-drained soils, in which fine materials are often translocated downward, and additions and losses of organic matter reach a steady state. Additionally, stagnation of the ponded water usually results in anaerobic conditions in these soils (Mausbach and Richardson 1994). Thus, hydric soils develop in these depressions through a variety of conditions. Often, the spatial transition is abrupt from depressional hydric soils to non-hydric soils in associated landscape positions.

Geomorphology and stratigraphy combine with regional climate to distribute water in the landscape and determine the maximum height of the saturated zone, as well as the duration of saturation. Interactions among precipitation, evapotranspiration, geomorphometry, and hydraulic conductivity create — at least seasonally — a positive water balance in these wetland soils of basins and depressions. Thus, both geomorphology and stratigraphy, by their influence on hydrologic properties, must be viewed as important factors in the development of soil moisture regimes, soil drainage classes, and hydric soil properties.

These interactions may occur in a variety of climates, and two of those are exemplified in this chapter. Vernal pools, another type of depressional wetland, are considered in Chapter 10B.

WETLAND SOILS OF PRAIRIE POTHOLES

Landscape and Geomorphic Features

The Prairie Pothole Region (PPR) of central North America (Figure 11a.1) is a geologically young landscape generally ranging in age from 13,000 to 9000 years old. Continental ice sheets

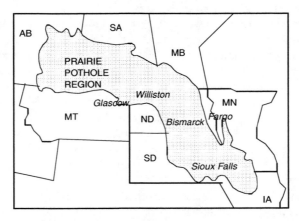

Figure 11a.1 Prairie pothole region of North America. (Adapted from Mann, G. E. 1974. The Prairie Pothole Region — a zone of environmental opportunity. *Naturalist* 25(4):2–7.)

melted northward at the end of the Pleistocene Epoch, creating this complex mosaic of gentle swells and swales, rugged kettle and kame topography, moraines, outwash plains, and glacial lake basins. Maximum relief in the PPR is over 100 m, but typically, relief is on the order of meters to tens of meters.

Surficial sediments, which are the soil parent materials, consist mostly of glacial till, outwash, and lacustrine muds derived from the glacial erosion of Mesozoic and Cenozoic sedimentary rocks (Winter 1989). Additional glacial sediment was derived from Paleozoic carbonates and sandstones found in the northern and western portions of the PPR, and Precambrian gneisses, greenstones, and granites found north and east of the PPR (Teller and Blumele 1983). Depth to bedrock typically ranges from 60 to 120 m under stagnant ice moraines, to usually less than 30 m beneath lake plains and ground moraines (Bluemle 1971, Winter 1989). Soil parent materials of the PPR, for the most part, tend to be silty, clayey, calcareous marine deposits. The relative youth of the landscape, together with geomorphic and climatic factors, accounts for the absence of well-developed integrated drainage systems and, alternatively, the existence of relatively small, prairie pothole lakes and wetlands (Bluemle 1991). Charles Froebel (1870, quoted by Kantrud et al. 1989) summarized the landscape of the region by writing, "The entire face of the country is covered with these shallow lakes, ponds and puddles, many of which are, however, dry or undergoing a process of gradual drying out." One could say the same today, realizing that the processes associated with the flooding and drying out of these wetlands are what produce the suite of wetland soils found in the PPR.

Climatic Conditions

Annual precipitation decreases from east to west across the PPR and is highly variable from year to year. The western half of the PPR is usually under the influence of dry continental air masses descending the eastern slope of the Rocky Mountains. In the eastern PPR, atmospheric low pressure cells frequently draw relatively moist air northward from the Gulf of Mexico. These air masses are capable of releasing large amounts of precipitation when they meet colder, drier continental air masses. Strong, isolated convective storms are common, causing heavy precipitation over short-range land areas. Over several years, portions of the PPR vacillate between arid and humid conditions (Table 11a.1). Winters tend to be relatively cold and dry, with most of the annual precipitation occurring between April and September (Abel et al. 1995, Wood 1996).

Annual temperatures also fluctuate widely in the PPR. Without a large body of water to moderate warm and cold temperatures, or mountains to block the flow of arctic air masses, the PPR generates surface temperatures, generally, between −40°C during winter to 40°C in the summer (Winter 1989). Awareness of the cyclic, though largely unpredictable, shifts in climatic conditions is requisite to an understanding of prairie pothole soils.

Table 11a.1   1964–93 Precipitation Data from Cities of the PPR

City	Annual Precipitation (cm)		Mean
	Minimum (Yr)	Maximum (Yr)	
Sioux Falls, SD	29.01 (1976)	91.71 (1993)	63.93
Fargo, ND	22.45 (1976)	81.99 (1977)	52.68
Aberdeen, SD	20.04 (1976)	71.45 (1993)	48.44
Bismarck, ND	25.83 (1988)	68.55 (1993)	40.91
Williston, ND	23.27 (1976)	55.47 (1986)	36.19
Glasgow, MT	17.12 (1984)	41.33 (1993)	29.18

Data from Wood, R.A. (Ed.). 1996. *Weather of U.S. Cities,* 5th ed., Gale Research, Inc., Detroit, MI.

Table 11a.2   Permeability of Wetland Soils from 0–150 cm Deep

Soil Series	Permeability (cm/hr)	
Southam	0.15–1.52	(slow)
Parnell	0.15–0.51	(slow)
Vallers	0.51–1.52	(moderately slow)
Hamerly	1.52–5.08	(moderately slow)

Data from Abel, P. L., A. Gulsvig, D. L. Johnson, and J. Seaholm. 1995. *Soil Survey of Stutsman County, North Dakota*. USDA–NRCS. U.S. Govt. Printing Office. Washington, DC.

Hydrologic Properties

Seasonal saturation of wetlands is variable, but, usually, spring runoff raises water levels in prairie pothole wetlands (Winter 1989). Shjeflo's work (1968) showed that, although snow only accounted for 25% of the region's precipitation, it accounted for 50% of the precipitation reaching the wetlands. The east–west precipitation gradient, and the spatial and temporal variability of precipitation throughout the region, however, complicates hydrologic conditions at specific wetland sites. Wetlands may be sites of either groundwater recharge or discharge, and through the course of a year, may do both (Meyboom 1966, Arndt and Richardson 1989a,b, Winter and Rosenberry 1996). On an annual basis, potential evapotranspiration is usually greater than precipitation in most of the PPR, especially in the central and western areas (Geraghty et al. 1973). Groundwater recharge usually occurs in the spring, when evapotranspiration rates are still low (Winter and Rosenberry 1996). Lissey (1971), working in the western PPR, noted numerous recharge sites he called "depression focused recharge wetlands." Essentially, wetlands filled during spring runoff, and water leached the soil profiles in the interior of the wetlands, resulting in profiles that were low in soluble salts and calcium carbonate. In the western, more arid ranges of the PPR, groundwater tends to mound beneath wetlands as groundwater is recharged, whereas in the eastern, more humid areas of the PPR, groundwater tends to follow the surface topography along subdued contours. In the eastern PPR, then, wetlands are topographic lows that usually discharge groundwater (Richardson et al. 1991, Richardson et al. 1992).

Hydraulic conductivity of soils and substrate is generally slow, due to the fine texture of the glacial till (Table 11a.2). However, hydraulic conductivity is often inconsistent due to the presence of fractures and the shrink–swell behavior of many wetland soils. Prairie pothole soils generally contain high concentrations of smectite, a clay mineral with high shrink–swell potential. During periods of drought, deep vertical cracks develop, creating high hydraulic conductivity rates through soil macropores. When soils are moist, cracks close, macropores narrow, and hydraulic conductivity becomes slower. Hence, the hydraulic conductivity within a particular wetland soil will depend, in part, on the soil texture, the clay mineralogy, and the antecedent moisture conditions.

Soil Morphology, Classification, and Genesis

The PPR is an extensive area found in two countries that have different soil classification systems. No attempt is made here to present a detailed and comprehensive discussion of all wetland soil types from Iowa to Alberta, but rather, to focus on four soils of North Dakota, which display the wetland soil morphologies and concepts associated with soil-forming processes widespread throughout the PPR.

The Hamerly–Parnell complex consists of a calcareous wetland edge soil, the Hamerly, and a leached interior soil, the Parnell (Figure 11a.2). The genesis of the soils in the Hamerly–Parnell complex of seasonal wetlands requires water to flow dominantly in opposite directions. The Parnell

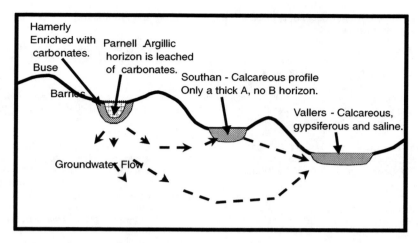

Figure 11a.2 Landscape profile of the Hamerly–Parnell complex.

has to have strong downward flow and the Hamerly needs upward evapotranspiration loss of water. In the spring, the water fills the shallow marsh and then infiltrates into the soil and leaches downward, removing the carbonates and translocating the clay. An argillic horizon greatly enriched in clay forms that restricts downward movement and creates lateral water flow. The edges, or wet meadow portion of the wetland, receive much water. Since the matric potential of the relatively dry pond edge is high, water and dissolved solutes are drawn away from the pond center and toward the edges, and the water evaporates and leaves the carbonates behind. The result is the calcic horizon in the Hamerly soil. The centers of the ponds tend to dry out for at least part of the year. The smectitic clays shrink upon desiccation, forming deep vertical cracks. When wetlands fill, typically during spring, water infiltrates through the cracks, carrying dissolved minerals and clays. In this fashion, then, the Parnell series is formed with its characteristic argillic horizon. Dissolved minerals move out of the Parnell both by gravity flow under saturated conditions, and by matric flow under unsaturated conditions. During the growing season, matric potentials in the Hamerly are kept high by evapotranspiration. The Hamerly soil is classified as a Calciaquoll based on the presence of a mollic epipedon and a calcareous (Bk) horizon within 40 cm of the soil surface. High-chroma matrix colors in the Bk, however, place the Hamerly in the Aeric subgroup. The Parnell classifies as a Typic Argiaquoll subgroup because of its mollic epipedon, argillic horizon, and its low-chroma matrix colors. Typifying pedons for the Hamerly and Parnell soils are shown in Tables 11a.3 and 11a.4 (Abel et al. 1995).

Table 11a.3   Typical Description of a Pedon in the Hamerly Series: Fine-Loamy, Frigid, Aeric Calciaquolls

Ap — 0 to 23 cm; black (10YR 2/1) loam, dark gray (10YR 4/1) dry; moderate medium subangular blocky structure parting to moderate medium granular; slightly hard and friable; slightly sticky and slightly plastic; common fine roots; about 3% gravel; common fine rounded soft masses of lime; strong effervescence; moderately alkaline; abrupt smooth boundary.

Bk — 23 to 70 cm; light olive brown (2.5Y 5/4) loam, light gray (2.5Y 7/2) dry; moderate medium subangular blocky structure; slightly hard and friable; slightly sticky and slightly plastic; few fine roots; about 3% gravel; common medium rounded soft masses of lime; violent effervescence; moderately alkaline; gradual wavy boundary.

C — 70 to 150 cm; olive brown (2.5Y 4/4) loam, light yellowish brown (2.5Y 6/4) dry; few fine prominent red (2.5Y 4/8) and common medium prominent light gray (N 7/0) mottles; massive; slightly hard and friable; slightly sticky and slightly plastic, about 3% gravel; strong effervescence; moderately alkaline.

From Abel, P. L., A. Gulsvig, D. L. Johnson, and J. Seaholm. 1995. *Soil Survey of Stutsman County, North Dakota.* USDA–NRCS. U.S. Govt. Printing Office. Washington, DC.

Table 11a.4   Typical Description of a Pedon in the Parnell Series: Fine, Montmorillonitic, Frigid, Frigid, Typic Argiaquolls

A1 — 0 to 20 cm; black (10YR2/1) silty clay loam, dark gray (10YR 4/1) dry; moderate fine subangular blocky structure parting to moderate medium granular; slightly hard and friable; slightly sticky and slightly plastic; many fine and medium roots; neutral; clear smooth boundary.

A2 — 20 to 40 cm; very dark gray (10YR 3/1) silty clay loam, gray (10YR 5/1) dry; weak fine subangular blocky structure parting to weak fine platy; slightly hard and friable; slightly sticky and slightly plastic; many fine and medium roots; neutral; clear smooth boundary.

Bt1 — 40 to 70 cm; very dark gray (10YR 3/1) silty clay, dark gray (10YR 4/1) dry; moderate coarse prismatic structure parting to strong medium angular blocky; hard and firm; sticky and plastic; common very fine and fine roots; common faint black (10YR 2/1) clay films on faces of peds; neutral; clear wavy boundary.

Bt2 — 70 to 90 cm; very dark grayish brown (10YR 3/2) silty clay, grayish brown (10YR 5/2) dry; weak medium prismatic structure parting to strong medium angular blocky; hard and firm; sticky and plastic; common very fine roots; few distinct black (10YR 2/1) clay films on faces of peds; neutral; gradual wavy boundary.

Cg — 90 to 150 cm; olive gray (5Y 5/2) loam, light olive gray (5Y 6/2) dry; common fine prominent strong brown (7.5YR 5/6) and few fine prominent dark red (2.5YR 3/6) mottles; massive; slightly hard and friable; slightly sticky and slightly plastic; few very fine roots; few fine rounded iron concretions of manganese oxide; about 2% gravel; neutral.

From Abel, P. L., A. Gulsvig, D. L. Johnson, and J. Seaholm. 1995. *Soil Survey of Stutsman County, North Dakota.* USDA–NRCS. U.S. Govt. Printing Office. Washington, DC.

The Southam series occurs in semipermanently and permanently ponded wetlands compared to the seasonally ponded wetlands with Hamerly, Parnell, and Vallers soils. The Southam has a cumulic (thick) mollic epipedon, and is calcareous throughout the mineral soil profile, indicating that little vertical leaching occurs in the soil. Horizonation and soil structure are not well developed in the soil because of the lack of wetting and drying cycles. The Southam soil receives precipitation and runoff, which are relatively low in dissolved minerals, as well as groundwater that is relatively enriched with dissolved minerals (Figure 11a.2). The Southam is found in "flowthrough" wetlands (Richardson et al. 1992, Richardson et al. 1994). Such wetlands are situated along an essentially horizontal hydraulic potential gradient, whereby water and dissolved solutes can enter and exit, i.e., flow through, the wetland. A typifying pedon from Stutsman County, ND (Abel et al. 1995), is given in Table 11a.5.

Discharge wetlands with a large influx of groundwater accumulate abundant carbonate, gypsum, and more labile or saline materials. A saline Vallers is often classified for these conditions. The Vallers soil is, for the most part, saturated from the bottom up by groundwater of relatively high ionic concentration. A relatively small proportion of the water entering these soils comes directly from precipitation or runoff. Evapotranspiration enable salts and carbonates to precipitate in the soil profile. Vallers soils are found in relatively low positions in the landscape (Figure 11a.2). A typifying pedon from Stutsman County, ND, is in Table 11a.6.

Characteristic Vegetation

Potential natural vegetation, i.e., the vegetation under which the soils of the PPR were formed, follow the precipitation gradient from east to west. Vegetation zones include the bluestem (Andropogon–Panicum–Sorghastrum) prairie in the eastern PPR, wheatgrass–bluestem–needlegrass (Andropyron–Andropogon–Stipa) prairie in the central PPR, and the wheatgrass–needlegrass (Agropyron–Stipa) prairie in the more arid, western PPR (Kuchler 1964). The natural prairie vegetation replaced spruce forest about 6000 YBP. For the past 100 years, however, most of the land has been placed into cultivation to grow a variety of grains including wheat, barley, flax, and sunflower in the northwestern and central portions of the PPR, as well as corn and soybeans in the southeastern area.

Table 11a.5   Typical Description of a Pedon in the Southam Series: Fine, Montmorillonitic (Calcareous), Frigid, Cumulic Endoaquoll

Oe — 5 cm to 0; black (5Y 2/1) peat, very dark grayish brown (2.5Y 3/2) dry; neutral; clear wavy boundary.

Ag1 — 0 to 15 cm; black (5Y 2/1) silty clay loam, dark gray (5Y 4/1) dry; massive; hard and firm; sticky and plastic; few coarse and many medium and fine roots; slight effervescence; slightly alkaline; gradual wavy boundary.

Ag2 — 15 to 45 cm; black (5Y 2/1) silty clay loam, dark gray (5Y 4/1) dry; massive; hard and firm; sticky and plastic; few fine roots; few fine snail shells, strong effervescence; moderately alkaline; gradual wavy boundary.

Ag3 — 45 to 69 cm; black (5Y 2/1) clay loam, dark gray (5Y 4/1) dry; massive; hard and firm; sticky and plastic; few fine roots; few fine

A1 — 0 to 20 cm; black (10YR2/1) silty clay loam, dark gray (10YR 4/1) dry; moderate fine subangular blocky structure parting to moderate medium granular; slightly hard and friable; slightly sticky and slightly plastic; many fine and medium roots; neutral; clear smooth boundary.

Cg1 — 69 to 104 cm; dark greenish-gray (5GY 4/1) silty clay, gray (5Y 5/1) dry; massive; hard and firm; sticky and plastic; few fine roots; common fine snail shells, strong effervescence; moderately alkaline; gradual wavy boundary.

Cg2 — 104 to 150 cm; dark gray (5Y 4/1) silty clay, light gray (5Y 6/1) dry; massive; hard and firm; sticky and plastic; few fine snail shells; violent effervescence; moderately alkaline.

From Abel, P. L., A. Gulsvig, D. L. Johnson, and J. Seaholm. 1995. *Soil Survey of Stutsman County, North Dakota.* USDA–NRCS. U.S. Govt. Printing Office. Washington, DC.

Table 11a.6   Typical Description of a Pedon in the Vallers Series: Fine-Loamy, Frigid, Typic Calciaquolls

Apz — 0 to 18 cm; black (10YR 2/1) silty clay loam, very dark gray (10YR 3/1) dry; weak fine granular structure; slightly hard and firm; slightly sticky and slightly plastic; few fine roots; common nests of salts; violent effervescence; moderately alkaline; abrupt smooth boundary.

Bkzg — 18 to 33 cm; gray (5Y 6/1) silty clay loam, light gray (5Y 7/1) dry; weak medium prismatic structure; slightly hard and firm; slightly sticky and slightly plastic; tongues of very dark grayish brown (10YR 3/2) A horizon material; common nests of salts.

Bkyg1 — 33 to 55 cm; olive gray (5Y 5/2) clay loam, light olive gray (5Y 6/2) dry; few coarse prominent yellowish brown (10YR 5/8) mottles; weak medium prismatic structure; slightly hard and friable; slightly sticky and slightly plastic; common nests of gypsum crystals; common fine rounded soft masses of lime; violent effervescence; moderately alkaline; clear smooth boundary.

Bkyg2 — 55 to 75 cm; olive gray (5Y 5/2) clay loam, light olive gray (5Y 6/2) dry; few fine prominent yellowish brown (10YR 5/8) mottles; weak medium prismatic structure; slightly hard and friable; slightly sticky and slightly plastic; few fine nests of gypsum; common fine rounded soft masses of lime; violent effervescence; moderately alkaline; clear smooth boundary.

Cg — 75 to 150 cm; gray (5Y 5/1) clay loam, light gray (5Y 6/1) dry; common medium prominent yellowish brown (10YR 5/8) and few medium prominent dark brown (7.5YR 3/4 mottles; massive; slightly hard and friable; slightly sticky and slightly plastic; few fine nests of gypsum crystals; few fine rounded soft masses of lime; violent effervescence; moderately alkaline.

From Abel, P. L., A. Gulsvig, D. L. Johnson, and J. Seaholm. 1995. *Soil Survey of Stutsman County, North Dakota.* USDA–NRCS. U.S. Govt. Printing Office. Washington, DC.

Importance Within the Geographic Region

Prairie wetlands in the United States have been threatened with drainage since European settlers arrived in the area. The U.S. Swamp Lands Acts of 1849, 1850, and 1860 encouraged the drainage of American wetlands for what was believed to be sound agricultural and public health policy. Government-sponsored drainage continued into the early 1970s when it encountered serious opposition from environmental interests (Leitch 1989). The PPR is now recognized as a valuable international resource, supporting biologically rich communities of plants and animals (Kantrud et al. 1989). The waterfowl that depend on prairie wetlands for nesting cover, feeding, and habitat support a multimillion-dollar hunting industry in North and South Dakota. Agricultural growers are often bothered by having to drive large, awkward farm implements around small wetlands. Chronically wet soils are not suitable for growing most commercial crops, so growers have increased production and eliminated wetlands by adding drain tiles and ditches to their fields (Leitch 1989). As farmers, environmentalists, commercial interests, and wildlife managers battle over the fate of the prairie pothole wetlands, science needs to communicate its best information about what role wetlands play in the local, regional, and global ecosystems. Since water is so critical in the productivity of natural or managed ecosystems, and because soils act as a kind of "Rosetta Stone" that can be used to interpret the history of water and chemical movement in the landscape, soil scientists need to play a prominent role in future decision-making processes affecting the management of prairie pothole wetlands.

WOODLAND SWALES

Landscape and Geomorphic Properties

Tippecanoe County, in the western part of north central Indiana, is within a region that typifies the Tipton Till Plain (Schneider 1966). The till plain was flattened and scraped by repeated glacial advances and retreats, then dissected by glacial melt-waters on their way to the Wabash and Ohio rivers. Within the Tipton Till Plain, which comprises approximately the northern one third of the state, glacial and eolian deposits are Wisconsin age and strongly influenced by the underlying sedimentary rocks — chiefly limestone and shale — over which the glacier rode. As a result, the till is loamy and calcareous. It is also characteristically very compact, although lenses of water-worked material occur locally.

A loess cap overlies the glacial till, and most of the soils here are formed in varying depths of silty loess and in the underlying loamy glacial till (Figure 11a.3). The dominant soil type in the area is Fincastle silt loam (fine-silty, mixed, mesic, Aeric Epiaqualf), a somewhat poorly drained soil found on the weak relief of till plain swells. Fincastle and Crosby soils (fine-loamy, mixed, mesic, Aeric Epiaqualf) occupy similar landscape positions, but Fincastle soils have a thicker loess cap. Moderately well-drained Celina soils (fine-loamy, mixed, mesic, Aquic Hapludalf) also occur in small areas where the loess is thinner, and well-drained Russell (fine-silty, mixed, mesic, Typic Hapludalf) and Strawn (fine-loamy, mixed, mesic, Typic Hapludalf) soils are at the dissected upland edges. Poorly drained Treaty soils (fine-silty, mixed, mesic, Typic Argiaquoll) occupy shallow drainageways. The depressional soil in this landscape is the Montgomery series (fine, mixed, mesic, Typic Endoaquoll). The Montgomery soil is formed in water-lain silts and clays above the compact till. Landscape relief is slight between major drainageways — most slopes are less than 6%, and many are less than 3%. Dissected edges of the upland have steeper slopes, however, frequently greater than 15%. At the Soldiers' Home Woods site presented here, elevation differences are slight, and slope rarely exceeds 2%, except at drainage edges (Figure 11a.4). The maximum elevation difference is about 4 m per 75 m. The soil surface is about 33 m above stream level.

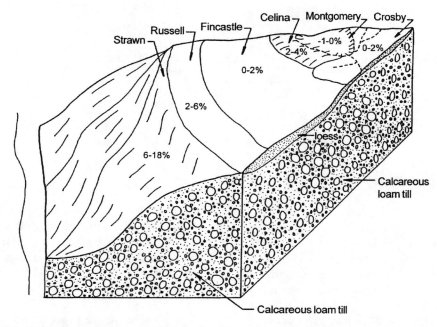

Figure 11a.3 Block diagram with representative soils of the western Tipton Till Plain. (Adapted from Schneider, A. F. 1966. Physiography. *In* A. A. Lindsey (Ed.) *Natural Features of Indiana.* Indiana Academy of Science, Indianapolis, IN.)

Climatic Conditions

Soil temperature regime in this region is mesic, and the regional moisture regime is udic (Soil Survey Staff 1994), although soils in lower lying depressions and drainageways often have aquic moisture regimes. The average annual precipitation is about 910 mm, and potential evapotranspiration is about 720 mm. The wettest months are April through July, and maximum potential evapotranspiration occurs between June and August. The mean minimum temperature in January is about –6°C, and the mean maximum temperature in July is about 31°C (Schaal 1966). There is typically a plentiful moisture supply for plant growth, since surplus groundwater accumulates prior to the growing season, and there is normally no deficit during the summer months.

Soldiers' Home Woods

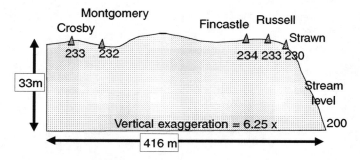

Figure 11a.4 Landscape profile of the Soldiers' Home Woods site.

Table 11a.7   Geometric Mean Values for Saturated Hydraulic Conductivity (K_{sat})

Pedon	0.6 m	1.1 m	1.6 m
		mm s^{-1}	
Montgomery	0.28	0.08	0.96
Crosby	1.38	0.72	0.03
Fincastle	1.18	1.85	0.23
Russell	ND	1.45	3.69
Strawn	3.25	0.08	1.04

Hydrologic Properties

The plentiful moisture supply is due not only to sufficient rainfall, but also to the high available water-holding capacity of the regional soils. Subsoil textures are usually silty clay loams or clay loams, which provide ample storage capacity for plant-available water. Thus, soil–water balances have a pronounced seasonality. Coincidence of fall precipitation and biological dormancy signal the beginning of surplus water accumulation, and water tables are further elevated by early spring rains. When biological activity resumes and plants require moisture for spring growth, the water tables fall.

The compact till is much less permeable than the Bt horizons above it, regardless of whether the Bt horizons are developed in loess or till (Harlan and Franzmeier 1974, King and Franzmeier 1981). These general relationships were supported by saturated hydraulic conductivity (K_{sat}) data at Soldiers' Home Woods (Table 11a.7). In these landscapes, overland flow is important only during major storms, as infiltration rates and water-holding capacities are adequate for most precipitation events. Due to the low K_{sat} values in the basal till, landscape drainage relies heavily on saturated subsurface flow. The result is that absolute elevation differences do not necessarily correspond to drainage differences. For example, the Montgomery soil, in the closed depression, is actually higher in elevation than the Strawn soil (232 m vs. 230 m). The Montgomery soil, however, is surrounded on all sides by sloping soils. Thus, run-on and flow-through accumulate rapidly, while drainage from the ponded soil is very slow because of the extremely low K_{sat} values. Lateral flow away from the depression is probably nonexistent, and water losses occur almost exclusively from evapotranspiration.

Seasonal distribution of saturation patterns (Figure 11a.5) confirms this. The water table in the Montgomery soil is at or near the surface most of the time from late fall to mid-spring. Fluctuations

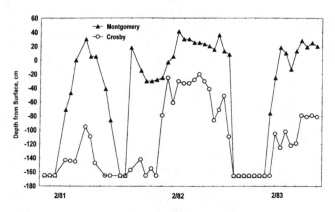

Figure 11a.5 Water table data for Montgomery and Crosby pedons.

Table 11a.8   Pedon Description for Montgomery Silty Clay Loam (Fine, Mixed, Mesic Typic Endoaquoll) at Soldiers' Home Woods

A1 — 0 to 25 cm; very dark gray (10YR 3/1) silty clay loam; weak, fine, angular blocky structure; very friable; few, fine, distinct yellowish brown (10YR 5/6 and 10YR 5/8) redox concentrations; clear, smooth, boundary.

A2 — 25 to 51 cm; very dark gray (10YR 3/1) silty clay; weak, medium angular and subangular blocky structure; friable; common, fine, distinct yellowish brown (10YR 5/6 and 10YR 5/8) redox concentrations; thin, discontinuous black (10YR 2/1) organic coats on faces of peds; clear, smooth boundary.

Bg1 — 51 to 76 cm; dark gray (2.5Y 4/0) silty clay; moderate, medium prismatic structure; firm; few, fine, distinct yellowish brown (10YR 5/8) redox concentrations; patchy black (10YR 2/1) and very dark gray (10YR 3/1) organic coats on ped faces; clear, wavy boundary.

Bg2 — 76 to 91 cm; gray (2.5Y 6/0) light silty clay; weak, fine, platy and angular blocky structure; friable; few, fine, distinct olive yellow (2.5Y 6/8) redox concentrations; patch dark gray (10YR 4/1) and very dark gray (10YR 3/1) organic coats on faces of peds; clear, wavy boundary.

Cg1 — 91 to 96 cm; gray (2.5Y 6/0) stratified silts; massive structure; very friable; few, fine distinct olive yellow (2.5Y 6/8) and yellowish brown (10YR 5/6) redox concentrations; thin, patchy dark gray (10YR 4/1) organic coats on ped faces; gradual, wavy boundary.

Cg2 — 96 to 150 cm; gray (2.5Y 6/0 and 2.5Y 5/0) stratified silt, clay, and very fine sand; massive; friable; few, fine, distinct light olive brown (2.5Y 5/6) redox concentrations.

---

of the water table in the adjacent Crosby soil closely parallel those of the Montgomery soil, suggesting that water losses from Crosby are at least partially controlled by the hydrology of the Montgomery site. Water table levels in the somewhat poorly drained Crosby soil are never as high as those in the very poorly drained Montgomery soil, however. The Fincastle pedon, which is also somewhat poorly drained, had a different hydrologic pattern than the Crosby pedon, presumably because the Fincastle's position in the landscape made it more independent of the depressional hydrology (Evans and Franzmeier 1986).

In general, water tables were higher, and saturation persisted longer, at landscape positions where subsurface lateral flow was likely to be suppressed by lack of potential gradient, reduced hydraulic conductivity, or both (Evans and Franzmeier 1986). Water tables showed strong relationships to hillslope position and substratum permeability, but, despite the disparity in K_{sat} values, "perched" water tables were not observed within the soil profiles. Lower horizons — including compact till horizons — were always saturated more frequently and for a longer duration than upper horizons (Evans and Franzmeier 1986). Furthermore, all soils were considerably above stream level, so none could be saturated by "true" groundwater. Instead, these apparent water tables were temporary saturation caused by impeded throughflow.

Soil Morphology, Genesis, and Classification

The Montgomery pedon (Table 11a.8) has a thick (51 cm), dark (10YR 3/1) epipedon over a subsoil horizon with a gleyed matrix. Redox concentrations are apparent in the epipedon. The epipedon is nearly thick enough to classify as cumulic (Soil Survey Staff 1994). The surface horizon genesis can be partially attributed to accumulations of fine organic matter and/or organic matter bound with silt and clay particles that move into the depression from adjacent soils. Saturation and the associated reducing conditions tend to preserve the organic matter. Thus, this mollic epipedon has a very different genetic history than those in the Prairie Pothole Region above. Redox potentials (Figure 11a.6) and dissolved oxygen levels (Evans and Franzmeier 1986) were also consistent with aquic conditions (Soil Survey Staff 1994). Although subsoil textures are fine, there is no evidence of clay illuviation. This is somewhat remarkable in a landscape dominated by Alfisols. Two factors may provide an explanation, however. First, the extremely low K_{sat} value (0.08 mm s-1) in the Cg2

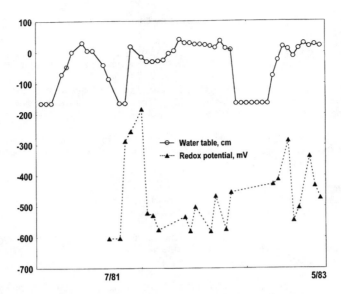

Figure 11a.6 Redox potentials and water table levels from Montgomery pedon.

horizon would restrict percolation of water carrying suspended clay from upper horizons. Second, the extreme length of saturation duration in this pedon most likely fails to permit sufficient wetting and drying cycles for effective translocation (Fanning and Fanning 1989). Other studies (e.g., Smeck et al. 1981) have also documented the absence of argillic horizons in similar poorly and very poorly drained soils.

Immediately adjacent to the Montgomery depression, the Crosby soil has high chroma matrices in the upper portion of the profile, although redox depletions and concentrations are common below 18 cm (Table 11a.9). Crosby soils reflect their hydrologic differences from Montgomery soils in other ways, as well. First, the Crosby soils have ochric epipedons that are too thin to be a mollic epipedon. This is presumably because the Crosby soil does not receive as much run-on as the Montgomery soil, and thus does not accumulate organically enriched material. Second, the Crosby soil has a very well-developed argillic horizon with abundant, distinct clay films. As shown in Figure 11a.5, the Crosby pedon experiences more frequent wetting and drying cycles than the Montgomery. In addition, the Crosby BC horizon has a mean K_{sat} value that is an order of magnitude greater than that of the Montgomery Cg2 at a comparable depth (Table 11a.7). Morphology of the Fincastle soil (Table 11a.10) is similar to that of the Crosby pedon. Subsoil matrices have high chroma, but redox depletions and concentrations are not present above 33 cm. Both the Bt1 horizon, developed in loess, and the 2Bt3 horizon, developed in glacial till, have comparable K_{sat} values, and both are substantially greater than the mean K_{sat} value of the 2Cd horizon — the compact glacial till.

Both the Crosby and Fincastle series are classified as Aeric Epiaqualfs. The aeric subgroup is due to the presence of high chroma matrices throughout most of the subsoil. Both soils are Aqualfs because redox depletions and concentrations are present in the upper argillic horizon. The Fincastle pedon (Table 11a.10) lacks redox features within 25 cm, however, and has a generally "better-drained" appearance than the Crosby soil. As noted above, the Fincastle soil was saturated less frequently than the Crosby pedon, so the color differences between the two pedons correspond to saturation and aeration regimes (Evans and Franzmeier 1988).

Assignment to the Epiaqualf great group is due to the assumption that the compact glacial till restricts downward water flow in these soils. As noted above, however, C horizons were actually saturated more frequently and for longer duration than the B horizons above them. Although the C horizons were very slowly permeable, they were not unsaturated during the periods when

Table 11a.9   Pedon Description for Crosby Silt Loam (Fine-Loamy, Mixed, Mesic, Aeric Epiaqualf) at Soldiers' Home Woods

A — 0 to 8 cm; very dark grayish brown (10YR 3/2) silt loam; moderate, medium granular structure; very friable; clear, wavy boundary.

E — 8 to 18 cm; pale brown (10YR 6/3) silt loam; weak, fine, platy structure; very friable; clear, wavy boundary.

BE — 18 to 30 cm; yellowish brown (10YR 5/4) silt loam; weak, medium angular blocky structure; friable; common, fine, distinct gray (10YR 5/1) redox depletions and common, fine, faint yellowish brown (10YR 5/6) redox concentrations; patch, discontinuous very dark gray (10YR 3/1) and black (10YR 2/1) stains on ped faces and in channels; gradual, smooth boundary.

Bt1 — 30 to 41 cm; yellowish brown (10YR 5/4) silty clay loam; moderate medium subangular blocky structure; friable; common, medium distinct dark grayish brown (10YR 4/2) redox depletions and few, fine, faint yellowish brown (10YR 5/6) redox concentrations; continuous gray (10YR 5/1) clay films on faces of peds and in channels; gradual, wavy boundary.

2Bt2 — 41 to 70 cm; yellowish brown (10YR 5/6) clay loam; weak, coarse prismatic structure parting to moderate, medium subangular blocky; firm; common, fine, distinct dark grayish brown (10YR 4/2) redox depletions and few, fine, faint yellowish brown (10YR 5/8) redox concentrations; thick, continuous gray (10YR 5/1) clay films on faces of peds and in channels.

2BC — 70 to 113 cm; gray (10YR 5/1) loam; moderate, coarse angular blocky structure; firm; common, coarse, distinct yellowish brown (10YR 5/6) and few, fine, distinct yellowish brown (10YR 5/8) redox concentrations; gradual wavy boundary.

2Cd — 113 to 150 cm; gray (10YR 5/1) and yellowish brown (10YR 5/6) loam; coarse platy and angular blocky structure; firm; light gray (10YR 7/1) carbonate coats in cracks and on ped faces; effervescent.

Table 11a.10 Pedon Description for Fincastle Silt Loam (Fine-Silty, Mixed, Mesic, Aeric Epiaqualf) at Soldiers' Home Woods

A — 0 to 8 cm; very dark grayish brown (10YR 3/2) silt loam; weak, fine, subangular blocky structure parting to moderate, medium granular; very friable; clear, wavy boundary.

E1 — 8 to 18 cm; pale brown (10YR 6/3) silt loam; weak, fine subangular blocky structure; very friable; gradual, smooth boundary.

E2 — 18 to 33 cm; light yellowish brown (10YR 6/4) silt loam; weak, fine platy structure; friable; clear, smooth boundary.

BE — 33 to 51 cm; yellowish brown (10YR 5/4) heavy silt loam; moderate, medium subangular blocky structure; friable; few, fine, faint yellowish brown (10YR 5/6) redox concentrations; patchy, discontinuous light brownish gray (10YR 6/2) and light yellowish brown (10YR 6/4) clay films and silt coats on faces of peds; gradual, smooth boundary.

Bt1 — 51 to 71 cm; yellowish brown (10YR 5/4) silty clay loam; moderate, medium subangular blocky structure; friable; few, fine, faint yellowish brown (10YR 5/6) redox concentrations; common, continuous light brownish gray (10YR 6/2) clay films on faces of peds and in channels; gradual, smooth boundary.

2Bt2 — 71 to 107 cm; yellowish brown (10YR 5/6) clay loam; moderate, medium subangular blocky structure, firm; common, medium, faint strong brown (7.5YR 5/6) redox concentrations; common, fine, distinct black (7.5YR 2/0) Mn concentrations in channels; continuous grayish brown (10YR 5/2) clay films on ped faces and in channels and voids; gradual, smooth boundary.

2Bt3 — 107 to 119 cm; yellowish brown (10YR 5/6) clay loam; moderate, medium angular blocky and subangular blocky structure; firm; common, medium, faint strong brown (7.5YR 5/6) redox concentrations; continuous dark grayish brown (10YR 4/2) clay films on faces of peds; common, black (7.5YR 2/0) Mn stains in root channels; gradual, smooth boundary.

2CB — 119 to 150 cm; yellowish brown (10YR 5/6) loam; moderate, medium and coarse angular blocky structure; firm; common, fine, faint strong brown (7.5YR 5/8) redox concentrations; thin, patchy white (10YR 8/1) and light gray (10YR 7/1) carbonate coats in cracks and channels; common, fine pebbles; strong effervescence.

saturation occurred in the sola. While it is true that the zone of saturation is perched on top of a relatively impermeable layer (Soil Survey Staff 1994), it is also true that the requisite unsaturated layers are often below the soil profile (i.e., >200 cm). Nonetheless, the concept of episaturation (Soil Survey Staff 1994) is the most appropriate concept to apply to these soils. Episaturation, we believe, is more appropriate than endosaturation because, as noted above, water tables in these soils are several meters above the true groundwater table. Furthermore, even though the entire soil profile is saturated, saturation is not continuous to the actual groundwater table. Evidence of this comes from the presence of free carbonates in the C horizons of most of these soils (Tables 11a.9 and 11a.10). If true groundwater were fluctuating into these pedons, it is not likely that carbonate accumulations would remain so exclusively associated with C horizons. (*Note:* Compare with the Calciaquoll pedons in Tables 11a.3 and 11a.6).

Characteristic Vegetation

The native vegetation is deciduous forest. At the study area, oak and hickory were dominant. This is noteworthy for two reasons. First, the evapotranspirative demand of the trees is an important factor in the seasonality of hydrologic patterns in this soil landscape. Water begins to accumulate in the fall at about the time that deciduous trees lessen their demands for moisture, due to their impending leaf drop and dormancy. When leaf-out begins in the spring, the demand on stored soil water resumes. As temperatures rise and leaves mature, evapotranspirative demands increase through the summer months. Near the end of the summer, and just before leaf drop, the water table in the Montgomery soil briefly falls below the soil surface.

The second reason that native vegetation is noteworthy is that the Montgomery series is a Mollisol. In some sense, however, it is not a "natural" Mollisol even though it has a thick, dark epipedon and sufficiently high base saturation. As noted above, the epipedon is nearly thick enough to be classified in a cumulic subgroup. Other soils in the landscape are Alfisols, however, and genesis of the mollic epipedon does not follow the classic prescription of development under native tall grass prairie (Fanning and Fanning 1989). Although the Alfisols here are also relatively base-rich, due to calcareous parent material, they lack the mollic epipedon, as do most forest soils. Clearly, the reason for the mollic epipedon in the Montgomery soil is that organic matter and fines wash into the depression from higher landscape positions. The long duration of ponding and saturation preserve the organic matter from oxidation; in some locations, Montgomery soils may have a thin, mucky Oa horizon at the surface. Thus, the Montgomery soil is not only an Aquoll, it is also a hydrologic Mollisol because the mollic features result from the hydrologic regime of this depressional pedon.

Importance Within the Geographic Region

Montgomery soils in wooded swales are no longer common landscape features in northern Indiana because most of the area has been cleared for farming. These areas remain wooded, in fact, because they were deemed too difficult to clear and/or too unprofitable to drain. Many were cleared initially, but abandoned to woods when maintenance of drainage made them unsustainable as crop land. Most of these woods have served as woodlots or livestock browsing areas. Recent wetlands protection acts now render them preserved areas.

REFERENCES

Abel, P. L., A. Gulsvig, D. L. Johnson, and J. Seaholm. 1995. *Soil Survey of Stutsman County, North Dakota.* USDA–NRCS. U.S. Govt. Printing Office. Washington, DC.

Arndt, J. L. and J. L. Richardson. 1989a. A comparison of soils to wetland classification types. pp. 76–90. *In Proc. 32nd Annual Manitoba Soil Sci. Soc.*, Dep. Soil Science, Univ. Manitoba, Winnipeg.

Arndt, J. L. and J. L. Richardson. 1989b. Geochemical development of hydric soil salinity in a North Dakota prairie pothole wetland system. *Soil Sci. Soc. Am. J.* 53:848–855.

Bluemle, J. P. 1971. Depth to bedrock in North Dakota. N. D. Geol. Surv., Misc. Map 13. Bismarck, ND.

Bluemle, J. P. 1991. *The Face of North Dakota*, revised edition, Educational Series 21. NDGS, Bismarck, ND.

Daniels, R. B., E. E. Gamble, and L. A. Nelson. 1971. Relations between soil morphology and water-table levels on a dissected North Carolina coastal plain surface. *Soil Sci. Soc. Am. Proc.* 35:781–784.

Evans, C. V. and D. P. Franzmeier. 1986. Saturation, aeration and color patterns in a toposequence of soils in north-central Indiana. *Soil Sci. Soc. Am. J.* 50:975–980.

Evans, C. V. and D. P. Franzmeier. 1988. Color index values to represent wetness and aeration in some Indiana soils. *Geoderma* 41:353–368.

Fanning, D. S. and M. C. B. Fanning. 1989. *Soil Morphology, Genesis, and Classification.* John Wiley & Sons, New York.

Froebel, C. 1870. Notes of some observations made in Dakota, during two expeditions under command of General Alfred Sully against the hostile Sioux, in the years 1864 and 1865. *Proc. Lyc. Nat. Hist.* New York 1:64–73.

Geraghty, J. J., D. W. Miller, F. van der Leeden, and F. L. Troise. 1973. *Water Atlas of the United States.* Water Information Center, Inc., Port Washington, NY.

Harlan, P. W. and D. P. Franzmeier. 1974. Soil-water regimes in Brookston and Crosby soils. *Soil Sci. Soc. Am. Proc.* 36:638–643.

Hurt, G. W., P. M. Whited, and R. F. Pringle. (Eds.). 1996. *Field Indicators of Hydric Soils in the United States.* USDA, NRCS, Fort Worth, TX.

Kantrud, H. A., G. L. Krapu, and G. A. Swanson. 1989. Prairie basin wetlands of the Dakotas: a community profile. U.S. Fish Wild. Svc. Biol. Rep. 85(7.28). 116 pp. U.S. Govt. Printing Office. Washington, DC.

King, J. J. and D. P. Franzmeier. 1981. Estimation of saturated hydraulic conductivity from soil morphological and genetic information. *Soil Sci. Soc. Am. J.* 45:1153–1156.

Kuchler, A. W. 1964. *The Potential Natural Vegetation of the Conterminous United States.* American Geographical Society Special Publ. No. 36, American Geographical Society, New York.

Leitch, J. A. 1989. Politico-economic overview of prairie potholes. pp. 2–14. *In* A. van der Valk (Ed.). *Northern Prairie Wetlands.* Iowa State University Press, Ames, IA.

Lissey, A. 1971. Depression-focused transient groundwater flow patterns in Manitoba. *Geol. Assoc. Can. Spec. Pap.* 9:333–341.

Mann, G. E. 1974. The Prairie Pothole Region — a zone of environmental opportunity. *Naturalist* 25(4):2–7.

Mausbach, M. J. and J. L. Richardson. 1994. Biogeochemical processes in hydric soil formation. *In Current Topics in Wetland Biogeochemistry.* Vol. 1, pp. 68–127. Wetland Biogeochemistry Institute. Louisiana State University, Baton Rouge.

Meyboom, P. 1966. Unsteady groundwater flow near a willow ring in hummocky moraine. *J. Hydrol.* 4:32–62.

Richardson, J. L., J. L. Arndt, and R. G. Eilers. 1991. Soils in three prairie pothole wetland systems. Pap. 34th Annual Manitoba Soc. Soil Sci. Meet., pp. 15–30. Manitoba Soil Sci. Soc., Dep. Soil Science, Univ. Manitoba, Winnipeg.

Richardson J. L., J. L. Arndt, and J. Freeland. 1994. Wetland soils of the prairie potholes. pp. 121–171. *In* D. L. Sparks (Ed.) *Advances in Agronomy.* Vol. 52. Academic Press, San Diego, CA.

Richardson, J. L., L. P. Wilding, and R. B. Daniels. 1992. Recharge and discharge of groundwater in aquic conditions illustrated with flownet analysis. *Geoderma* 53:63–78.

Schaal, L. 1966. Climate. *In* A. A. Lindsey (Ed.) *Natural Features of Indiana.* Indiana Academy of Science, Indianapolis, IN.

Schneider, A. F. 1966. Physiography. *In* A. A. Lindsey (Ed.) *Natural Features of Indiana.* Indiana Academy of Science, Indianapolis, IN.

Shjeflo, J. B. 1968. Evapotranspiration and the water budget of prairie potholes in North Dakota. *Hydrology of Prairie Potholes.* U.S. Geological Survey Prof. Pap. 585-B. 49 p. U.S. Govt. Printing Office. Washington, DC.

Smeck, N. E., A. Ritchie, L. P. Wilding, and L. R. Drees. 1981. Clay accumulation in sola of poorly drained soils of western Ohio. *Soil Sci. Soc. Am. J.* 45:95–102.

Soil Survey Staff. 1994. *Keys to Soil Taxonomy.* 6th ed. U.S. Govt. Printing Office. Washington, DC.

Steinwand, A. L. and T. E. Fenton. 1995. Landscape evolution and shallow groundwater hydrology of a till landscape in central Iowa. *Soil Sci. Soc. Am. J.* 59:1370–1377.

Teller, J. T. and J. P. Bluemle, 1983. Geological setting of the Lake Agassiz Region. pp. 7–20. *In* J. T. Teller and Lee Clayton (Eds.) *Glacial Lake Agassiz.* Geological Association of Canada Special Paper 26. Geologic Association of Canada, St. Johns, Newfoundland.

Vepraskas, M. J. and L. P. Wilding. 1983. Aquic moisture regimes in soils with and without low chroma colors. *Soil Sci. Soc. Am. J.* 47:280–285.

Winter, T. C. 1989. Hydrologic studies of wetlands in the northern prairie. *In* van der Valk (Ed.) *Northern Prairie Wetlands.* Iowa State University Press, Ames, Iowa.

Winter, T. C. and D. O. Rosenberry. 1996. The interaction of ground water with prairie pothole wetlands in the Cottonwood Lake area, east-central North Dakota, 1979–1990. *Wetlands* 15(3):193–211.

Wood, R. A. (Ed.). 1996. *Weather of U.S. Cities,* 5th ed., Gale Research, Inc., Detroit, MI.

Zaslavsky, D. and A. S. Rogowski. 1969. Hydrologic and morphologic implications of anisotropy and infiltration in soil profile development. *Soil Sci. Soc. Am. Proc.* 33:594–599.

Wetland Soils of Basins and Depressions: Case Studies of Vernal Pools

W. A. Hobson and R. A. Dahlgren

LANDSCAPE AND GEOMORPHIC PROPERTIES

Vernal pools and swales are episaturated, seasonal, freshwater wetlands found in the western United States, Mexico, and other Mediterranean-type climates of the world (Reifner and Pryor 1996). The boundary between grassland and vernal pools is sharply demarcated, with vegetation composition often changing completely in less than a meter (Holland and Jain 1977). The abundant grassland pools and swales are highly variable in size, and they typically occur in groups separated by tens or hundreds of meters (Holland and Jain 1981). Less commonly, they are found on coastal terraces and basalt mesas (Zedler 1990, Stone 1990, Weitkamp et al. 1996), on lava plateaus and scablands (Crowe et al. 1994), and in woodlands scattered throughout the landscape (Stone 1990, Heise et al. 1996).

These wetlands typically range in size from 50 to 5000 m^2 (Mitsch and Gosselink 1993), with some functioning pools being as small as 30 m^2 (Hobson and Dahlgren 1998a). Vernal pools usually have maximum water depths that range from 0.3 to 1.0 m. The drainageways, commonly referred to as swales or vernal marshes, occupy greater areas but lack the deep standing water (Broyles 1987). Their locations are characterized by poorly drained areas of level, gently undulating topography called "hogwallows" or mima mounds (Nikiforoff 1941, Broyles 1987, Stone 1990), with the majority of pools found on slopes $< 8\%$ (Smith and Verrill 1998). Vernal pools and swales are commonly found at elevations of 30 to 200 m on intermediate river terraces, alluvial fans, and coastal terraces (Holland and Jain 1977, Moran 1984, Zedler 1987). With lower frequency, pools also occur at elevations up to 1800 m in the valleys, plateaus, foothills, and lower montane environments throughout the western United States and Mexico (Holland and Jain 1977, Zedler 1987, Stone 1990, Crowe et al. 1994).

Geomorphic ages of pool-bearing landforms frequently range from early to late Pleistocene, 0.1 to 1.0 M.y.a. (Stone 1990, Crowe et al. 1994), although other pools occur on late Pliocene formations, 1.5 to 2.0 M.y.a. (Jokerst 1990, Stone 1990, Hobson and Dahlgren 1998a). Pools occur on a wide variety of geologic materials which include: the Pleistocene alluvium of the Great Central Valley of California and its associated older terraces (Holland and Jain 1981, Stone 1990); Pleis-

VERNAL POOL LANDFORMS

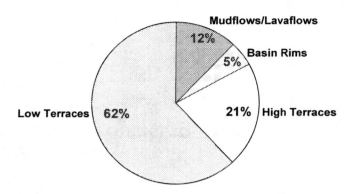

Figure 11b.1 Landform types and relative amounts of vernal pools found in California's Great Central Valley. (Data from Smith, D.W. and W.L. Verrill. 1998. Vernal pool landforms and soils of the Central Valley, California. In *The Conference on the Ecology, Conservation, and Management of Vernal Pool Ecosystems.* Sacramento, California. June 19–21.)

tocene coastal terraces near San Diego, California (Zedler 1987) and adjacent Mexico (Moran 1984); Pleistocene-age loess over Miocene Columbia River Basalts in eastern Washington (Crowe et al. 1994); Pliocene-age olivine basalt on the Santa Rosa Plateau, California (Weitkamp et al. 1996); Pliocene lahars (mudflows) and associated alluvium in northeastern California (Jokerst 1990, Hobson and Dahlgren 1998a); and Pleistocene basalts of the Modoc Plateau, California, and southeastern Oregon (Stone 1990).

The Great Central Valley of California contains numerous vernal pools on a variety of alluvial deposits. Some of these alluvial deposits have been in place for over 600,000 years (Arkley 1964). Landforms in the Great Central Valley on which vernal pools occur are low terraces, high terraces, volcanic mudflows and lava flows, and basin rims (Smith and Verrill 1998) (Figure 11b.1). As this geosynclinal basin filled with sediments from surrounding mountain ranges, the bottom of the sediments subsided, allowing large rivers to maintain grade, and meander across the valley. Soil development continued on these older valley-filling terraces (Holland and Jain 1981). Pedogenic processes have created indurated layers, claypans, and duripans (a silica cemented horizon) that perch the water table (episaturation) and form vernal pools. Vernal pools are abundant on the more developed soil profiles usually found on older terraces (Holland and Jain 1981, Stone 1990).

The microtopographic areas with vernal pools are characteristically hummocky, with low mounds (mima mounds) separated by closed to partially closed depressions (hogwallows) (Broyles 1987). These pools appear to indicate former drainages that once had increased gradients and were affected by large-scale alluvial processes. Today, the decreased gradients and nearly level conditions indicate a dominance of micro alluvial and eolian processes in these landscapes, as well as in former river terraces and coastal terraces. The pools on soils overlying lithic contacts are also dominated by micro alluvial and eolian processes. These processes occur because ephemeral drainage courses, cracks in the lava or mudflow, and existing illuviated clay layers restrict water flow.

CLIMATIC CONDITIONS

Vernal pools and swales are found in grassland and woodland ecosystems where Mediterranean-type rainfall patterns prevail (Holland and Jain 1981, Crowe et al. 1994). This xeric soil moisture regime exhibits moist/cool winters and warm/dry summers (Soil Survey Staff 1996). The winter moisture arrives when potential evapotranspiration is minimal, thus creating an effective soil-leaching environment (Soil Survey Staff 1996).

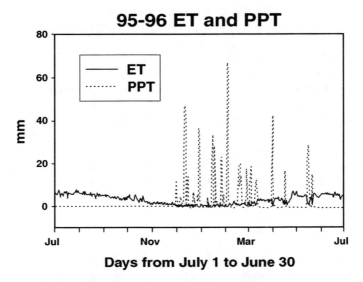

Figure 11b.2 Daily potential ET and PPT for Chico, California, July 1, 1995, to June 30, 1996. Totals for the year were: ET = 1238 mm, PPT = 712 mm. (From National Weather Service. 1995, 1996. Chico Weather Station, Butte County, CA. U.S. Department of Commerce. NOAA.)

The pools fill with winter and spring rains, or snow melt in colder climates. They gradually lose ponded water by late spring or early summer due to evapotranspiration. As the pools desiccate, vegetation in concentric rings grows around the shrinking pool. The vegetation zonation indicates a strong linkage between the preferred habitat of a given species and the combined hydrologic and pedogenic regimes.

The majority of remaining vernal pools and swales are found in the Great Central Valley of California, where the soil temperature regime is thermic (mean annual soil temperature (MAST): 15°C ≥ MAST > 22°C, with mean summer and mean winter soil temperatures varying by > 5°C). Other locations, such as the Modoc Plateau of northeastern California or the Channeled Scablands of eastern Washington, have colder climates where snow melt contributes to pool hydrology. These areas have mesic soil temperature regimes (mean annual soil temperature 8°C ≥ MAST > 15°C, with mean summer and mean winter soil temperatures varying by > 5°C).

The xeric soil moisture regime and the thermic or mesic soil temperature regimes contribute to the ephemeral nature of these wetland ecosystems. Water stands in the pools through most of the rainy winter season, or snow melt winter–spring season. As the rainy season ends, temperatures increase, and evapotranspiration dominates, eventually leaving the pool beds baked hard and dry (Figure 11b.2).

HYDROLOGIC PROPERTIES

Vernal pools are unique among wetlands because they function as wetlands for 4 to 5 months during a typical year before desiccating to conditions drier than permanent wilting point or soil water potentials less than about −1.5 Mpa. Zedler (1987) refers to vernal pools as intermittently flooded wet meadows. Yet in spite of their seasonal nature, they display all the hydrologic, soil, and vegetation characteristics needed to be classified as jurisdictional wetlands. The unique assemblage of flora and fauna has adapted to a seasonal regime of inundation followed by desiccation, which is attributed to many combinations of geologic, soil, and climatic factors. The dominant hydrologic factors that control pool water levels can be significantly different, depending on these factors (Hanes et al. 1990). In areas of abundant precipitation, direct precipitation may account for the majority of pool water regardless of the pool watershed topography. However, in more arid locations where direct precipitation is insufficient to offset evapotranspirative losses, overland or near-surface flow may contribute significantly to pool water depth.

Vernal pools fill by collecting precipitation (Holland and Jain 1981, Hanes et al. 1990), snow melt in colder climates (Crowe et al. 1994), and water that has infiltrated and moved to the pools by interflow. Once the pools have filled to capacity, water moves from pool to pool by interconnecting channels and swales, interpool reflow, and by shallow groundwater flow. These characteristically flow-through wetlands contribute water downslope to intermediate swales, other flow-through pools, and in some cases to seasonal streams.

Overland flow does not appear to be a dominant hydrologic pathway in soils overlying a claypan or duripan greater than 30 cm thick (Hanes et al. 1990). In extreme cases where the claypan or duripan is less than 30 cm, heavy precipitation events may quickly exceed the soil water-holding capacity, resulting in overland flow (reflow in this case). Reflow is water that flows on the soil surface because the underlying soils have become saturated. However, with the gentle slopes and dense vegetative cover, the infiltration rate of most vernal pool soils commonly exceeds the incident rainfall, preventing downslope surface flow.

Losses of water are dominantly attributed to evapotranspiration, and they are subordinately attributed to seepage into or through the pool bottom, outflow to a channel, or movement into the adjacent upland (Hanes et al. 1990, Crowe et al. 1994). Decreased levels of evapotranspiration during the winter, combined with abundant precipitation leads to the filling of pools. During late winter and early spring in the xeric moisture regime, temperatures warm, precipitation decreases, plants begin to grow, and evapotranspiration increases. Evapotranspiration continues to increase as temperatures rise, and rains diminish and become insignificant by April or May. Pools commonly reach near dry down levels in spring (mid-March to April in Figure 11b.3), then refill with the frequent heavy spring rains before completely desiccating (early May in Figure 11b.3).

The characteristic seasonality of these freshwater wetlands is not only attributed to the Mediterranean climate (xeric SMR, thermic or mesic STR) in which they are found, but also to their episaturated nature. They are underlain by an impervious layer, such as a hardpan (e.g., duripan, indurated layer) (Holland and Jain 1977), a dense clay layer (Schlising and Sanders 1982), a mudflow or lahar (Jokerst 1990), or a lithic contact (Weitkamp et al. 1996). These layers, or aquitards, perch

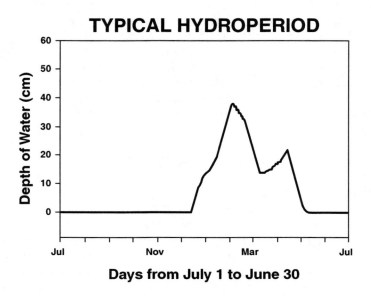

Figure 11b.3 Typical hydroperiod for a vernal pool in California's Great Central Valley.

the water table, allowing evapotranspiration to dominate water losses as the unique flora utilize the water-holding capacity of the pools, swales, and soils, while temperatures and daylight hours increase during spring. Seepage into or through the pool or swale bottom via cracks in lithic or paralithic contacts (duripans, or some indurated horizons) contributes to complete desiccation.

These wetlands may have discharge–recharge interchanges with surrounding areas, depending on local topography. A discharge pool or swale results when groundwater or surface waters are higher than the pool/swale, thus discharging into it. A flow-through pool or swale can have both inflows and outflows of groundwater or surface water. A recharge pool or swale occurs when these wetlands are higher than the surrounding episaturated water table, and groundwater or surface water flows from the pool to downslope areas or even into seasonal streams.

Hydraulic conductivity of soils and substrate is controlled by the relatively impervious, underlying layers such as a claypan, a duripan, an indurated layer, or lithic contact, and by the texture of the overlying soil. These provide effective barriers to downward movement of water, which results in a perched water table. Frequently, a well-developed clay enriched B horizon overlies the pedogenic hardpans, duripans, or lithic contacts (Holland 1978, Holland and Jain 1981, Jokerst 1990, Weitkamp et al. 1996). The low hydraulic conductivity and high water-holding capacity of the clay enriched B horizon may initially perch the water table above the impervious layer. One study revealed that an upward hydrologic gradient also exists, as average soil matric potential was −56 MPa at 2 to 10 cm depth, −27 MPa at 10 to 30 cm depth, and −2 MPa at 30 to 60 cm depth (Crowe et al. 1994). This indicates the upward movement of water due to evapotranspiration as pools desiccate.

SOIL MORPHOLOGY, GENESIS, AND CLASSIFICATION

The seasonality and microtopography of these freshwater wetlands create a catena or drainage toposequence as the shallow basins retain more water than the surrounding rim and upland geomorphic positions (Figure 11b.4). Typically, the properties of upland–rim–basin soils (and vegetation) differ laterally toward the basin as well as vertically down to the impervious layer (Lathrop and Thorne 1976, Bauder 1987, Crowe et al. 1994, Weitkamp et al. 1996). Pedogenic processes have created this three-dimensional biogeochemical environment, which dramatically affects hydrology and nutrient cycling (Hobson and Dahlgren 1998a). The dominant pedogenic processes are ferrolysis, organic matter accumulation, clay formation and translocation, and duripan formation (Hobson and Dahlgren 1998b).

The seasonal nature of vernal pool wetlands creates annual and shorter-term (e.g., weekly to monthly) cycles of anaerobic and aerobic conditions within the soil profile. These conditions allow the cyclic reduction and oxidation of Fe, termed *ferrolysis* (Brinkman 1970). Redox potential in wetland soils can be used to quantify the tendency of the soil to oxidize or reduce substances (Faulkner and Richardson 1989). Organic matter is oxidized in the soil under aerobic conditions between +600 and +400 mV. After aerobic organisms consume the available O_2, facultative and obligate anaerobes proliferate. Then a sequence of anaerobic conditions occurs at progressively lower Eh levels: disappearance of O_2 below +400 mV, disappearance of NO_3^- at +250 mV, appearance of Mn^{+2} at +225 mV, appearance of Fe^{+2} at +120 mV, disappearance of SO_4^{-2} at −75 to −150 mV, and the appearance of CH_4 at −250 to −350 mV. Organic matter is consumed in anaerobic, waterlogged soils in the above sequence at about pH 7 (Mitsch and Gosselink 1993). These redox potentials are not exact limits, because they are subject to the effects of temperature, pH, available organic matter, organic acids, saturation conditions, and the availability of reducible substrates. The addition of mineral nitrogen from atmospheric deposition, oxygen produced by photosynthetic aquatic plants within the vernal pools, and the abundance of manganese within a system (such as andesitic alluvium) tend to poise (buffer the Eh) the system. These limit the reduction of iron until

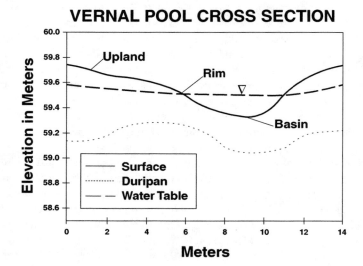

Figure 11b.4 Cross-section of vernal pool showing surface, duripan, and maximum height of pool water. Upland, rim, and basin positions indicate locations of *in situ* platinum reference electrodes, and approximate locations of pedons used in Tables 11b.1, 11b.2, and 11b.3. (From Hobson, W.A. and R.A. Dahlgren. 1998b. A quantitative study of pedogenesis in California vernal pool wetlands. pp. 107–128. *In* M.C. Rabenhorst, J.C. Bell, and P.A. McDaniel (Eds.) *Quantifying Soil Hydromorphology.* SSSA Spec. Publ. No. 51. Soil Sci. Soc. Am., Madison, WI. With permission.)

these substrates are consumed. Under the more acidic conditions (pH 5.5 to 6.5) that commonly occur above the duripan (pH $\geq$ 7.0), the reduction of Fe and Mn will occur at somewhat higher redox values (Ponnamperuma et al. 1967, 1969, Collins and Buol 1970).

Redox values measured *in situ* were sufficient to reduce nitrate, manganese, and sometimes iron (Hobson and Dahlgren 1998b) (Figure 11b.5). Consistent with the ferrolysis process is the inverse relationship of Eh and pH values (compare Figure 11b.5 to Figure 11b.6). Reduction reactions consume protons, increasing pH, while oxidation reactions generate protons, resulting in lower pH values (Brinkman 1970, van Breeman et al. 1984). Effects of ferrolysis include the release of bases, metal cations, and silicic acid into the soil solution for plant uptake, leaching, and accumulation of soluble constituents downward and toward the basin, and upon dry-down and oxidation of the soil, the creation of redoximorphic features (Soil Survey Staff 1996).

Redoximorphic features were most abundant in the basin and rim soils (Table 11b.1) corresponding to the lowest redox potentials (Hobson and Dahlgren 1998b). Depletions are zones of low chroma ($\leq$2) where Fe–Mn oxides, with or without clay, have been removed (Soil Survey Staff 1996). Depletions were abundant in the basin and rim positions above the duripan and common in the adjacent upland soil, as noted in Table 11b.1. Oxidation of Fe and Mn creates the redox concentrations of high chroma Fe mottles, and neutral Mn stains, concentrations, and masses. Manganese stains, concentrations, and masses are distributed more deeply within the soil profiles than are Fe mottles, because Mn^{+2} is more mobile in the soil solution than Fe^{+2} (McDaniel and Buol 1991) (Table 11b.1).

The Mn features are not diagnostic for hydric soil determinations. However, the Fe redoximorphic features, depletions, and low chroma matrix are diagnostic for identifying the rim and basin vernal pool soils as hydric soils (Vepraskas 1994, Hurt et al. 1996). The dominance of 3 chroma in the matrix of the upland soil (Table 11b.2) in the upper 30 cm makes the upland soil non-hydric (Hurt et al. 1996). Soils farther away from vernal pools lacking redoximorphic features are clearly non-hydric. Wetlands require wetland hydrology, hydric vegetation, and hydric soils to meet the requirements for a wetland (Environmental Laboratory 1987); therefore, only the rim and basin areas can qualify as wetlands.

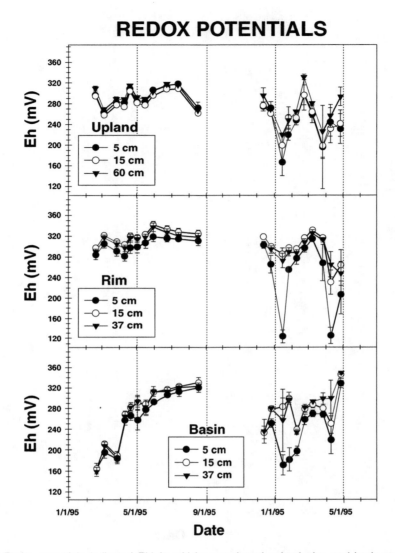

Figure 11b.5 Redox potentials (adjusted Eh) in mV for vernal pool upland, rim, and basin positions for the 1994–95 and 1995–96 seasons. Error bars represent standard deviations. Depths of *in situ* platinum reference electrodes are 5 cm, 15 cm, and at the respective duripans. Note the lower redox values at all 5 cm depths and the lower overall redox values in the basin and rim positions. (From Hobson, W.A. and R.A. Dahlgren. 1998b. A quantitative study of pedogenesis in California vernal pool wetlands. pp. 107–128. *In* M.C. Rabenhorst, J.C. Bell, and P.A. McDaniel (Eds.) *Quantifying Soil Hydromorphology.* SSSA Spec. Publ. No. 51. Soil Sci. Soc. Am., Madison, WI. With permission.)

Soil organic matter accumulates primarily in a thin upper layer of the mineral soil in vernal pools (Figure 11b.7). During the summer, when pools are dry, there is limited availability of water in the upper soil horizons for microbial activity and organic matter decomposition. Additionally, the seasonally anaerobic conditions during the winter and spring further inhibit organic matter decomposition. Soil organic matter distribution is also influenced by the high bulk density of the subsoil horizons, often exceeding 2 Mg m^{-3} (Hobson and Dahlgren 1998a), which limits the depth of penetration by roots into the subsoil (Table 11b.2). Beneath the A horizons, root growth is primarily restricted to ped faces. Significant inputs of atmospheric N, via precipitation and partic- ulate deposition, are quickly assimilated by biota, thus further increasing organic matter inputs to

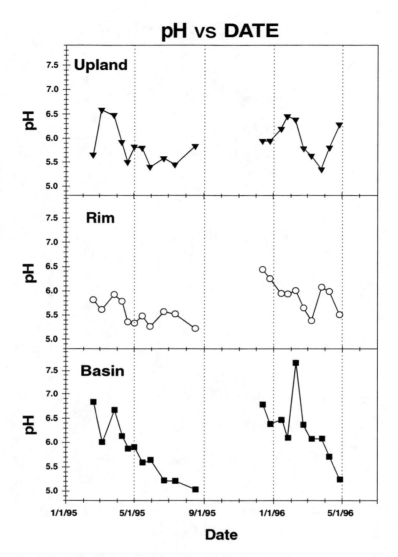

Figure 11b.6 *In situ* pH (0–5 cm depth) for vernal pool upland, rim, and basin for the 1994–1995 and 1995–1996 seasons. Compare the pH with the Eh in Figure 11b.5. Note the inverse relationship consistent with the ferrolysis process. (From Hobson, W.A. and R.A. Dahlgren. 1998b. A quantitative study of pedogenesis in California vernal pool wetlands. pp. 107–128. *In* M.C. Rabenhorst, J.C. Bell, and P.A. McDaniel (Eds.) *Quantifying Soil Hydromorphology.* SSSA Spec. Publ. No. 51. Soil Sci. Soc. Am., Madison, WI. With permission.)

the soil surface (Hobson and Dahlgren 1998a, 1998b) (Figure 11b.8). These environmental conditions result in relatively slow decomposition rates and the accumulation of organic matter primarily in the surface layers (Schlesinger 1991, Hobson and Dahlgren 1998a, 1998b) (Figure 11b.7).

The formation of silicate clays through alteration of existing primary or secondary minerals or from precipitation of oversaturated soil solutions is accelerated by ferrolysis (Brinkman 1970). The translocation of silicate clays from an overlying horizon (eluviation) into a lower horizon results in accumulation of silicate clays (illuviation) (Soil Survey Staff 1975, 1996). Clay content increases with depth to the duripan or lithic contact, and soils are frequently more strongly developed in the upland compared to the rim and basin positions for pools ≤ 100 m^2 (Holland and Jain 1981, Jokerst 1990; Hobson and Dahlgren 1998a). A large clay enrichment occurs immediately above the duripan

Table 11b.1 Redoximorphic Features and Selected Soil Chemistry in Vernal Pool Soils

Horizons	Depletions[a]	Fe Accumulations[a]	Mn Accumulations[a]	Fe_d[b]	Fe_o	Mn_d	Mn_o	Fe_o/Fe_d	Mn_o/Mn_d
						g kg^{-1}			
Upland: fine, smectitic, thermic Aquic Durixerert									
Ap	c2f(20%)7.5YR6/2	c1f&d mottles7.5YR6/8	f1f stains N 3/0	7.38	2.66	1.19	0.91	0.36	0.76
A	c2f(20%)7.5YR6/2	c1f&d mottles7.5YR6/8	f1f stains N 3/0	9.13	1.95	1.27	1.23	0.21	0.97
Btss1	c2f(20%)7.5YR6/2	c1f&d mottles7.5YR6/8	f1f stains N 3/0	9.68	1.69	1.16	0.91	0.17	0.78
Btss2	none	c1f&d mottles7.5YR6/8	f1f stains N 3/0	9.30	1.51	1.33	1.02	0.16	0.77
Bkqm	none	m2&3d mottles7.5YR7/8	m2p stains, m2r nod.N 3/0	7.30	0.34	0.45	0.31	0.05	0.69
BC1	none	c2d mottles7.5YR7/8	c1d stains, nod. c1r N 3/0	6.01	0.42	0.15	0.14	0.07	0.89
BC2	none	c2p mottles7.5YR6/8	none	6.47	0.48	0.40	0.33	0.07	0.83
Rim: clayey, mixed, superactive, thermic, shallow Vertic Duraquoll									
Ap	m3f(30%)7.5YR6/2	m1f&d mottles7.5YR6/8	f1f stains N 3/0	10.00	4.72	1.28	1.21	0.47	0.94
A	m3f(30%)7.5YR6/2	m1f&d mottles7.5YR6/8	c1d stains N3/0	8.99	2.76	1.26	1.14	0.31	0.90
Btss	c3f(20%)7.5YR6/2	c1f&d mottles7.5YR6/8	c1d stains N3/0	8.64	1.87	1.22	1.08	0.22	0.89
Btkqml	none	c1d mottles 7.5YR6/8	m3p masses (70%) N 3/0	8.24	0.90	1.75	1.40	0.11	0.80
Btkqm2	none	m2&3d mottles7.5YR7/8	m2p stains, m2rnod. N 3/0	7.54	0.37	0.36	0.34	0.05	0.94
Bkqm	none	c1&2f mottles7.5YR7/8	c1d stains, c1r nod. N 3/0	6.84	0.50	0.35	0.29	0.07	0.82
BC1	none	c1f mottles7.5YR7/8	c1f stains, c1r nod. N 3/0	6.20	0.51	0.40	0.34	0.08	0.84
Basin: clayey, mixed, superactive, thermic, shallow Vertic Duraquoll									
Ap	m3f(30%)7.5YR6/2	c2f&d mottles7.5YR6/8	c1d stains N 4/0	11.00	4.77	1.46	1.26	0.43	0.86
A	m3f(30%)7.5YR6/2	m1f&d mottles7.5YR6/8	c2f stains N 4/0	9.87	3.37	1.37	1.21	0.34	0.88
Btss	c3f(20%)7.5YR6/2	m1f&d mottles7.5YR6/8	c2d stains N 4/0	8.59	2.60	1.52	1.24	0.30	0.81
Btkqml	none	c1d mottles7.5YR6/8	m3p masses (70%) N 3/0	8.21	0.82	1.72	1.45	0.10	0.85
Btkqm2	none	m2d mottles5YR6/8	c1d stains & masses N 3/0	9.01	0.44	0.48	0.38	0.05	0.80
Bqm	none	c1f mottles7.5YR6/8	f1d stains, f1r nod. N 3/0	9.56	0.73	0.48	0.37	0.08	0.76
BC1	none	c1&2d mottles5YR5/8	f1d stains, f1r nod. N 3/0	8.53	0.60	0.45	0.45	0.07	0.99

[a] f = few (<2%), c = common (2–20%), m = many (>20%), 1 = fine (<5 mm), 2 = medium (5–15 mm), 3 = large (>15 mm), f = faint, d = distinct, p = prominent, r = rounded, nod. = nodules, all colors are dry.
[b] Fe_d and Mn_d are dithionite–citrate extractable Fe and Mn, respectively; Fe_o and Mn_o are acid oxalate extractable Fe and Mn, respectively.

From Hobson, W.A. and R.A. Dahlgren. 1998b. A quantitative study of pedogenesis in California vernal pool wetlands. pp. 107–128. In M.C. Rabenhorst, J.C. Bell, and P.A. McDaniel (Eds.) Quantifying Soil Hydromorphology. SSSA Spec. Publ. No. 51. Soil Sci. Soc. Am., Madison, WI. With permission.

Table 11b.2 Physical and Morphological Properties of Vernal Pool Soils

Horizon	Depth (cm)	Color Dry	Color Moist	Sand (%)	Silt (%)	Clay (%)	Texture[a]	Structure[b]	Bulk Density (Mg m⁻³)	Roots[c]	Clay Films[d]
Upland: fine, smectitic, thermic Aquic Durixerert											
Ap	0–6	10YR5/3	10YR3/3	39.9	38.0	22.1	l	3mpl	1.87	3vf&1f	—
A	6–16	10YR3/3	10YR3/3	29.9	38.2	31.9	cl	3mabk	2.32	2vf&1f	—
Btss1	16–30	10YR4/3	10YR3/3	26.6	27.5	45.9	c	3cpr	2.36	2vf&1f	1nco
Btss2	30–60	10YR4/2	10YR3/3	29.3	14.2	56.5	c	3vcpr	2.47	2vf	2npf
Bkqm	60–68	7.5YR4/4	7.5YR3/4	74.2	16.3	9.7	cosl	3vcpr	2.04	1vf	—
BC1	68–78	10YR7/3	10YR6/4	83.9	8.0	8.1	lcos	m	2.15	—	—
BC2	78–88	7.5YR5/2	7.5YR4/3	87.5	5.5	7.0	lcos	m	2.03	—	—
Rim: clayey, mixed, superactive, thermic, shallow Vertic Duraquoll											
Ap	0–7	10YR5/2	10YR4/2	25.5	45.6	28.9	cl	2cabk	1.94	3vf&1f	—
A	7–18	10YR5/2	10YR3/2	26.8	39.9	33.3	cl	2c&vcabk	2.47	2vf&1f	1nco
Btss	18–36	10YR5/2	10YR3/3	21.4	35.2	43.4	c	3vcpr	2.52	2vf	1nco
Btkqml	36–38	10YR6/4	7.5YR4/4	57.8	19.4	22.9	sl	3mpl	2.13	—	1npf
Btkqm2	38–58	10YR6/4	7.5YR4/4	58.7	21.7	19.6	scl-sl	3cpl	2.06	—	1npf
Bkqm	58–80	10YR6/4	10YR4/6	67.7	19.4	12.9	scl	m	2.23	—	—
BC1	80–118	10YR6/3	10YR4/4	88.1	4.3	7.6	cosl	m	2.44	—	—
Basin: clayey, mixed, superactive, thermic, shallow Vertic Duraquoll											
Ap	0–6	10YR5/2	10YR3/3	21.9	46.8	31.3	cl	3mpl	1.88	2vf&1f	—
A	6–19	10YR4/2	10YR3/3	25.0	33.6	41.4	c	3cabk	2.20	2vf	1nco
Btss	19–35	10YR4/2	10YR3/3	30.0	27.9	42.1	c	3vcpr	2.32	2vf	vinco
Btkqml	35–37	10YR6/4	7.5YR4/4	80.5	6.9	12.6	cosl	3mpl	2.14	—	vinco
Btkqm2	37–58	10YR5/6	10YR3/4	67.0	8.3	24.7	cosl	3cpl	2.27	—	1nco
Bqm	58–90	10YR6/6	10YR3/4	69.1	10.3	20.6	cosl	m	2.20	—	—
BC1	90–107	10YR5/6	10YR3/4	72.7	9.4	17.9	cosl	m	2.39	—	—

[a] l = loam, cl = clay loam, c = clay, cosl = coarse sandy loam, lcos = loamy coarse sand, scl = sandy clay loam, sl = sandy loam, m = massive.

[b] 1 = weak, 2 = moderate, 3 = strong, f = fine, m = medium, c = coarse, vc = very coarse, abk = angular blocky, pr = prismatic, pl = platy, m = massive.

[c] 1 = few, 2 = common, 3 = many, vf = very fine, f = fine.

[d] vi = very few (<5%), 1 = few (5–25%), 2 = common (25–50%), n = thin, pf = ped faces, po = lining pores, pf = ped faces, co = colloid stains on mineral grains.

From Hobson, W.A. and R.A. Dahlgren. 1998b. A quantitative study of pedogenesis in California vernal pool wetlands. pp. 107–128. *In* M.C. Rabenhorst, J.C. Bell, and P.A. McDaniel (Eds.) *Quantifying Soil Hydromorphology*. SSSA Spec. Publ. No. 51. Soil Sci. Soc. Am., Madison, WI. With permission.

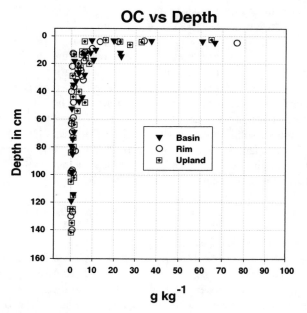

Figure 11b.7 Grams of organic carbon per kilogram of dry soil vs. depth in cm. Highest accumulations are in the surface horizons above 20 cm. These are average values from duplicate samples for five upland–rim–basin transects, 15 pedons, and 91 horizons. (From Hobson, W.A. and R.A. Dahlgren. 1998b. A quantitative study of pedogenesis in California vernal pool wetlands. pp. 107–128. *In* M.C. Rabenhorst, J.C. Bell, and P.A. McDaniel (Eds.) *Quantifying Soil Hydromorphology.* SSSA Spec. Publ. No. 51. Soil Sci. Soc. Am., Madison, WI. With permission.)

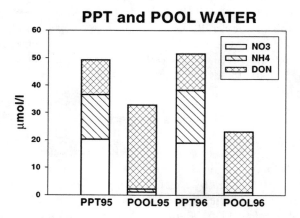

Figure 11b.8 Precipitation and water pool chemistry for the 1994–95 and 1995–96 seasons. Ammonium, nitrate, and dissolved organic nitrogen (DON) are shown. Totals are the sum of components. (From Hobson, W.A. and R.A. Dahlgren. 1998b. A quantitative study of pedogenesis in California vernal pool wetlands. pp. 107–128. *In* M.C. Rabenhorst, J.C. Bell, and P.A. McDaniel (Eds.) *Quantifying Soil Hydromorphology.* SSSA Spec. Publ. No. 51. Soil Sci. Soc. Am., Madison, WI. With permission.)

or lithic contact, due to their impervious nature, which effectively prevents transport of soluble constituents. A lateral gradient also exists as clay percentages increase closer to the soil surface along an upland to basin transect. Clay skins are formed when the soil solution transports the suspended clays along with any organic matter and iron compounds to the depth of leaching (Buol and Hole 1961). The absence of clay films in the surface horizons and the presence of colloidal stains and thin clay films beneath indicate a degrading zone above an accumulating zone (Bullock et al. 1974). An example of this distribution of clay films is seen in Table 11b.2.

The seasonal nature of vernal pools creates repetitive cycles of anaerobic and aerobic conditions. During the reduction phase, soluble manganese and iron (Mn^{+2} and Fe^{+2}) are formed that displace base cations from the exchange capacity. These base cations are leached from the soil with bicarbonate formed by biotic respiration and concentrated at the depth of leaching (e.g., $CaCO_3$ accumulation). Reduction reactions consume protons (H^+), resulting in an increase in soil pH during the reduction period. Subsequent oxidation of exchangeable Mn^{+2} and Fe^{+2} produces protons which acidify the soil and contributes to the lower base saturation in the more strongly reduced surface horizons (Table 11b.3). The low pH associated with the oxidation creates a harsh weathering environment that accelerates mineral weathering (Brinkman 1970, van Breeman et al. 1984).

A duripan is a diagnostic subsurface horizon that is cemented by silica so that less than 50% of the volume of air dry fragments slake in water or acid (HCl) after prolonged soaking (Soil Survey Staff 1996). Duripans underlying vernal pools vary in their degree of silica cementation, and they frequently contain accessory cementing agents, primarily iron oxides and calcium carbonate (Flach et al. 1969, Soil Survey Staff 1996). These pedogenic by-products accumulate at the depth of leaching within the profile (Arkley 1963, Flach et al. 1969) to form the duripan upon desiccation. Therefore, duripans, as well as other restrictive layers, comprise an integral component of vernal pool ecosystems due to their influence in episaturation, restriction of downward water flow and root growth, accumulation of weathering products, and reduction of soil volume in which pedogenesis and plant growth can effectively take place.

A classic soil for vernal pool landscapes is the San Joaquin soil series, a Typic Durixeralf or an Abruptic Durixeralf (claypan above duripan). It is the most extensive soil series mapped in natural vernal pool areas (Smith and Verrill 1998). The most abundant soil series and landforms on which vernal pools occur include: San Joaquin, a Durixeralf on low terraces, Redding, a Durixeralf on high terraces, Toomes, a Xerochrept on mudflows/lavaflows, and Pescadero, a Natrixeralf found in basins (Smith and Verrill 1998). Vernal pool upland–rim–basin soils are also mapped as Duraquolls, Durixerolls, Haploxerolls, Durochrepts, Xerochrepts, Xerorthents, Durixererts, Haploxererts, and Palexeralfs (Smith and Verrill 1998, Hobson and Dahlgren 1998a, 1998b). Frequently, less developed soils are found in the rim and/or basin positions compared to summit soils (Hobson and Dahlgren 1998a, 1998b) (Tables 11b.1 and 11b.2).

FLORA AND FAUNA

Vernal pools and swales represent small yet complete ecosystems that constitute "islands" in the surrounding grassland vegetation (Stebbins 1976). They are highly variable in size, and they typically occur in clusters separated by tens or hundreds of meters (Holland and Jain 1981). The majority of vernal pool plant species are annuals, and a large portion are endemic to California vernal pools (Holland and Jain 1977). Rarity and endemism exemplify these ecosystems as over 100 plants are typical vernal pool species and 69 are restricted to only vernal pools (Holland and Jain 1977).

These ecosystems provide habitat for a variety of local and migratory fauna which includes: feeding grounds for waterfowl on the Pacific Flyway (Baker et al. 1992); local burrowing owls, hawks, gophers (Hunt 1991), field mice, and a host of invertebrates (which provide avian food) including fairy shrimp and tadpole shrimp (Alexander and Syrdahl 1992, Gallagher 1996).

Table 11b.3 Soil Chemical Properties of Vernal Pool Soils

Horizons	pH	CEC (cmol$_c$ kg^{-1})	Base Sat. (%)	Organic C	Organic N	C/N	CaCO$_3$
				g kg^{-1}			g kg^{-1}
Upland: fine, smectitic, thermic Aquic Durixerert							
Ap	5.78	25.0	47	16.4	0.9	18.2	0.6
A	5.95	23.8	73	4.8	0.4	12.0	<0.1
Btss1	6.17	26.5	85	3.4	0.4	8.5	0.9
Btss2	7.00	29.5	89	2.8	0.3	9.3	0.9
Bkqm	8.03	35.9	100	1.3	<0.1	—	12.0
BC1	7.94	28.9	100	0.7	<0.1	—	2.4
BC2	8.09	25.7	96	0.6	<0.1	—	1.1
Rim: clayey, mixed, superactive, thermic, shallow Vertic Duraquoll							
Ap	5.70	26.0	75	12.9	0.8	16.1	0.9
A	5.52	26.6	83	4.6	0.4	11.5	<0.1
Btss	6.02	30.8	88	2.8	0.3	9.3	<0.1
Btkqml	6.50	36.9	94	0.9	<0.1	—	1.0
Btkqm2	7.26	45.7	100	0.4	<0.1	—	3.2
Bkqm	7.41	31.8	100	0.7	<0.1	—	1.0
BC1	7.59	27.7	78	0.8	<0.1	—	<0.1
Basin: clayey, mixed, superactive, thermic, shallow Vertic Duraquoll							
Ap	6.03	25.9	75	9.4	0.7	13.4	<0.1
A	5.65	26.3	72	5.3	0.5	10.6	<0.1
Btss	5.90	28.6	82	3.1	0.4	7.8	<0.1
Btkqml	6.27	43.4	90	0.7	<0.1	—	1.6
Btkqm2	7.76	40.5	93	2.3	<0.1	—	4.2
Bqm	7.86	28.9	98	0.7	<0.1	—	0.4
BC1	7.23	25.0	100	0.6	<0.1	—	0.7

From Hobson, W.A. and R.A. Dahlgren. 1998b. A quantitative study of pedogenesis in California vernal pool wetlands. pp. 107–128. *In* M.C. Rabenhorst, J.C. Bell, and P.A. McDaniel (Eds.) *Quantifying Soil Hydromorphology.* SSSA Spec. Publ. No. 51. Soil Sci. Soc. Am., Madison, WI. With permission.

EXTENT AND IMPORTANCE WITHIN GEOGRAPHIC REGION

Once a common feature in the Great Central Valley of California (Hoover 1937), many vernal pools have been degraded or destroyed by agricultural development and urbanization (Holland 1978, Zedler 1987). The majority of the remaining vernal pools and swales are located in California's Great Central Valley, and they are frequently located on Pleistocene-age alluvium on river terraces and on Pliocene-age mudflows/lavaflows (Holland and Jain 1981). Nearly one third of the Great Central Valley, or approximately 1.7 M hectares, may have supported vernal pools before development (Holland 1978).

Vernal pools provide numerous beneficial functions in the environment, which include feeding grounds for migrating birds on the Pacific Flyway, reservoirs of biodiversity for flora (Stone 1990) and fauna (Alexander and Syrdahl 1992, Baker et al. 1992, Gallagher 1996), and water-holding capacity of pools and soils that buffer against regional flooding, groundwater recharge, and transpirational needs for vegetation as the pools dry down (Hanes et al. 1990). Perhaps the most important, and frequently overlooked, aspect of vernal pool ecosystems is their ability to cycle nutrients. As in other freshwater wetland ecosystems, nitrogen and phosphorus inputs are assimilated by biota, and the turnover of nutrients is controlled through biotic and redox interactions (Schlesinger 1991, Mitsch and Gosselink 1993). Thus, vernal pools act as biological filters for watersheds by performing functions similar to other freshwater wetlands.

ACKNOWLEDGMENTS

The authors thank Randal J. Southard and Michael J. Singer for their technical assistance; David Kelley, Kelley & Associates Environmental Sciences, for his support, insight, and cooperation in the use of Wurlitzer Ranch Preserve and Doe Mill Preserve; Lisa Stallings for her assistance in the field; Jason Barnes, Rebecca Sutton, and Zeng-Shou Yu for their help with laboratory analysis.

REFERENCES

Alexander, D.G. and R. Syrdahl. 1992. Invertebrate biodiversity in vernal pools. *NW Environ. J.* 8(1):161–163.

Arkley, R.J. 1963. Calculation of carbonate and water movement in soil from climatic data. *Soil Sci.* 96:239–248.

Arkley, R.J. 1964. Soil survey of the eastern Stanislaus area, California. USDA, *Soil Conservation Service* 1957(20):1–160.

Baker, W.S., F.E. Hayes, and E.W. Lathrop. 1992. Avian use of vernal pools at the Santa Rosa Plateau Preserve, Santa Ana Mountains, California. *Southwestern Naturalist* 73(4):392–403.

Bauder, E.T. 1987. *Species Assortment Along a Small Scale Gradient in San Diego Vernal Pools.* Dissertation. University of California, Davis, California, and San Diego State University, San Diego, California.

Brinkman, R. 1970. Ferrolysis, a hydromorphic soil forming process. *Geoderma* 3:199–206.

Broyles, P. 1987. A flora of Vina Plains Preserve, Tehema County, California. *Madrono* 34(3):209–227.

Buol, S.W. and F.D. Hole. 1961. Clay skin genesis in Wisconsin soils. *Soil Sci. Soc. Proc.* 25:377–379.

Bullock, P., M.H. Milford, and M.G. Cline. 1974. Degradation of Argillic horizons in Udalf soils of New York State. *Soil Sci. Soc. Am. Proc.* 38:621–628.

Collins, J.F. and S.W. Buol. 1970. Effects of fluctuations in the Eh–pH environment on iron and/or manganese equilibria. *Soil Sci.* 110(2):111–118.

Crowe, E.A., A.J. Busacca, J.P. Reganold, and B.A. Zamora. 1994. Vegetation zones and soil characteristics in vernal pools in the channeled scablands of eastern Washington. *Great Basin Naturalist* 54(3):234–247.

Environmental Laboratory. 1987. *Corps of Engineers Wetland Delineation Manual.* Tech. Rpt. Y-87-1. U.S. Army Engineer Waterways Experiment Station, Vicksburg, MS.

Faulkner, S.P. and C.J. Richardson. 1989. Physical and chemical characteristics of freshwater wetland soils. pp. 41–72. *In* D.A. Hammer (Ed.) *Constructed Wetlands for Wastewater Treatment.* Lewis Publ., Chelsea, MI.

Flach, K.W., W.D. Nettleton, L.H. Gile, and J.G. Cady. 1969. Pedocementation: formation of indurated soil horizons by silica, calcium carbonate, and sesquioxides. *Soil Sci.* 107:442–453.

Gallagher, S.P. 1996. Seasonal occurrence and habitat characteristics of some vernal pool Branchiopoda in northern California, U.S.A. *J. Crustacean Biol.* 16(2):323–329.

Hanes, W.T., B. Hecht, and L.P. Stomberg. 1990. Water relationships of vernal pools in the Sacramento region, California. pp. 49–60. *In* D.H. Ikeda and R.A. Schlising (Eds.) *Vernal Pool Plants Their Habitat and Biology.* Studies from the Herbarium, CA State Univ., Chico, No. 8. Chico, CA.

Heise, K., A. Merenlender, and G.A. Giusti. 1996. Vernal pools in oak woodlands: puddles or unique habitats? pp. 1–2. In *Oaks 'n' Folks,* September, 11(2), University of California, Berkeley.

Hobson, W.A. and R.A. Dahlgren. 1998a. Soil forming processes in vernal pools of northern California, Chico area. In *The Conference on the Ecology, Conservation, and Management of Vernal Pool Ecosystems.* Sacramento, California. June 9–21. CA Native Plant Society, Sacramento.

Hobson, W.A. and R.A. Dahlgren. 1998b. A quantitative study of pedogenesis in California vernal pool wetlands. pp. 107–128. *In* M.C. Rabenhorst, J.C. Bell, and P.A. McDaniel (Eds.) *Quantifying Soil Hydromorphology.* SSSA Spec. Publ. No. 51. Soil Sci. Soc. Am., Madison, WI.

Holland, R.F. 1978. The geographic and edaphic distribution of vernal pools in the Great Valley, California. Special Publication No. 4, CA Native Plant Soc., Berkeley, CA.

Holland, R.F. and S.K. Jain. 1977. Vernal pools. pp. 515–533. *In* M.G. Barbour and J. Major (Eds.) *Terrestrial Vegetation of California.* Wiley-Interscience, New York.

Holland, R.F. and S.K. Jain. 1981. Insular biogeography of vernal pools in the Central Valley of California. *Am. Nat.* 117:24–37.

Hoover, R.F. 1937. *Endemism in the Flora of the Great Valley of California.* Ph.D thesis, University of California, Berkeley.

Hunt, J. 1991. Feeding ecology of Valley Pocket Gophers (*Thomomys bottae sanctidiegi*) on a California coastal grassland. *Am. Midl. Nat.* 127:41–51.

Hurt, G.W., P.M. Whited, and R.F. Pringle. (Eds.) 1996. *Field Indicators of Hydric Soils in the United States.* USDA, Natural Resources Conservation Service, Fort Worth, TX.

Jokerst, J.D. 1990. Floristic analysis of volcanic mudflow vernal pools. pp. 1–26. *In* D.H. Ikeda and R.A. Schlising (Eds.) *Vernal Pool Plants: Their Habitat and Biology.* Studies from the Herbarium, CA State Univ., Chico No. 8. Chico, CA.

Lathrop, E.W. and R.F. Thorne. 1976. The vernal pools on De Burro of the Santa Rosa Plateau, Riverside County, California. *Aliso* 8(4):433–445.

McDaniel, P.A. and S.W. Buol. 1991. Manganese distribution in acid soils of the North Carolina Piedmont. *Soil Sci. Soc. Am. J.* 55:152–158.

Mitsch, W.J. and J.G. Gosselink. 1993. *Wetlands.* Van Nostrand Reinhold, New York.

Moran, R. 1984. Vernal pools in northwest Baja California, Mexico. pp. 173–184. *In* S. Jain and P. Moyle (Eds.), *Vernal and Intermittent Streams.* Institute of Ecology Publ. 28, University of California, Davis.

National Weather Service. 1995, 1996. Chico Weather Station, Butte County, CA. U.S. Department of Commerce. NOAA.

Nikiforoff, C.C. 1941. Hardpan and microrelief in certain soil complexes of California. Tech. Bull. No. 745. U.S.D.A., U.S. Govt. Printing Office. Washington, DC.

Ponnamperuma, F.N., E.M. Tianco, and T. Loy. 1967. Redox equilibria in flooded soils: I. the iron hydroxide systems. *Soil Sci.* 103(6):371–382.

Ponnamperuma, F. N., E.M. Tianco, and T. Loy. 1969. Redox equilibria in flooded soils: II. the manganese oxide systems. *Soil Sci.* 108(1):48–57.

Riefner, R.E. and D.R. Pryor. 1996. New locations and interpretations of vernal pools in Southern California. *Phytologia* 80(4):296–327.

Schlising R.A. and E.L. Sanders. 1982. Quantitative analysis of vegetation at the Richvale vernal pools, California. *Am. J. Bot.* 69(5):734–742.

Schlesinger, William H. 1991. *Biogeochemistry: An Analysis of Global Change.* Academic Press, New York.

Smith, D.W. and W.L. Verrill. 1998. Vernal pool landforms and soils of the Central Valley, California. In *The Conference on the Ecology, Conservation, and Management of Vernal Pool Ecosystems.* Sacramento, California. June 19–21.

Soil Survey Staff. 1975. *Soil Taxonomy: A Basic System of Soil Classification for Making and Interpreting Soil Surveys.* USDA–CSC Agr. Handbook 436. U.S. Govt. Printing Office. Washington, DC.

Soil Survey Staff. 1996. *Keys to Soil Taxonomy.* 7th ed. U.S. Govt. Printing Office. Washington, DC.

Stebbins, G.L. 1976. Ecological islands and vernal pools of California. *In* S. Jain (Ed.) *Vernal Pools: Their Ecology and Conservation,* Vol. 1–4. Institute of Ecology Publication 9, University of California, Davis.

Stone, D.R. 1990. California's endemic vernal pool plants: some factors influencing their rarity and endangerment. pp. 89–107. *In* D.H. Ikeda and R.A. Schlising (Eds.) *Vernal Pool Plants: Their Habitat and Biology.* Studies from the Herbarium, CA State Univ., Chico No. 8. Chico, CA.

Van Breeman, N., J. Mulder, and C.T. Driscoll. 1984. Acidification and alkalinization of soils. *Plant and Soil* 75:283–308.

Vepraskas, M.J. 1994. *Redoximorphic Features for Identifying Aquic Conditions.* Tech. Bulletin 301. North Carolina Ag. Research Service, North Carolina State Univ., Raleigh, NC.

Weitkamp, W.A., R.C. Graham, M.A. Anderson, and C. Amrhein. 1996. Pedogenesis of a vernal pool Entisol–Alfisol–Vertisol catena in southern California. *Soil Sci. Soc. Am. J.* 60:316–323.

Zedler, P.H. 1987. The ecology of southern California vernal pools: a community profile. U.S. Fish and Widl. Serv. Biol. Rep. 85(7.11). U.S. Govt. Printing Office. Washington, DC.

Zedler, P.H. 1990. Life histories of vernal pool vascular plants. pp. 123–146. *In* D.H. Ikeda and R.A. Schlising (Eds.) *Vernal Pool Plants: Their Habitat and Biology.* Studies from the Herbarium, CA State Univ., Chico No. 8. Chico, CA.

CHAPTER 12

Hydric Soils and Wetlands in Riverine Systems

David L. Lindbo and J. L. Richardson

INTRODUCTION

The term *riverine system* as used here refers to a river or stream valley measured from the stream channel to the valley edge, including floodplain or terraces that can be inundated or flooded frequently. An active riverine system is one that lacks upstream dams, has not been channelized or has constructed levies or has been entrenched to a degree that flooding no longer occurs. They occur throughout the world in virtually every climate. In the following discussion we concentrate our comments on riverine systems that are related to meandering rivers. Meandering rivers create floodplains that are associated with extensive wetlands. Although rivers themselves do not account for a large percentage of the Earth's surface, their influence is nonetheless of paramount importance. Riverine systems contain some of the world's most fertile agricultural and silvacultural lands. Additionally, these land areas are home to numerous large cities. As a result, land within these systems is under increasing pressure from development and is exposed to major environmental hazards.

The riverine system is formed and constantly modified by fluvial (channel stream flow) and other hydrologic processes. These processes influence wetland occurrence and extent. Because of the fluvial and groundwater interaction with landform, soil, and vegetation, our discussion will start with these dynamic processes. The discussion of features within the riverine system focuses on the floodplain, which is their most common and often defining feature. Wetlands and hydric soils also occur at the interface of uplands and river valley terraces, at the headwaters of the riverine system as groundwater seeps (see Chapters 3 and 9), and they are associated with oxbows and related features.

THE RIVERINE SYSTEM

We believe that John Playfair (1802) best expressed the idea of streams and their valleys nearly 200 years ago:

"Every river appears to consist of a main trunk, fed from a variety of branches, each running in a valley proportioned to its size, and all of them together form a system of vallies (*sic*), communical with one another, and having such a nice adjustment of their declivities, that none of them join the principal valley, either on too high or too low a level, a circumstance which would be infinitely improbable if each of these vallies were not the work of the stream that flows in it."

We assume that the stream, or at least a precursor of that stream, formed the valley in which it flows. This includes misfit streams (streams in valleys larger or smaller than are suggested by the current stream), although the size of the stream may be greatly altered. The riverine system is restricted to the lower portions of the valley. The riverine system may be as small as a meter or so across or as extensive as the Mississippi Valley and extend for hundreds or thousands of kilometers. Regardless of the size of the area, these systems are the result of a common set of fluvial processes. Riverine systems respond to and have resulted from hydrologic input from all parts of their upstream drainage basin and to a degree from their downstream basin. The hydrologic input coupled with the kinetic energy resulting from landscape relief creates the geomorphic features. At any given point in a system, the valley cross-section that results reflects the upstream and downstream conditions over time. Some features are quite transitory, such as those on the lowest floodplains with the youngest soils. Higher terraces are progressively more stable and have older soils (Daniels et al. 1971).

Floodplains are often the most obvious geomorphic feature of the riverine system and are a direct result of fluvial processes (Ritter 1979). The same processes also contribute to other associated landforms, including levees, oxbows, meander scars, bars, sloughs, and backswamps. (See Leopold et al. 1964 and Allen 1970 for more detail on these landforms.)

FLUVIAL PROCESSES

Fluvial processes are driven by the kinetic energy of flowing water. Kinetic energy is derived from the elevation and gradient of the streams within their watershed. The watershed energy depends on overall relief from the highest portion of the watershed to the outlet. Within any given watershed, the fluvial processes depend on local relief factors and on the history of the stream itself; thus, the type and magnitude of the processes that occur in riverine systems vary depending on the location within the system. The basic fluvial activities in watersheds include runoff, landslides, channel erosion and deposition, and flood basin erosion and deposition. The material processes consist primarily of sediment entrainment, sediment transport, erosion, and deposition.

We generalize the fluvial processes within a watershed that form the valley and the landscape above the floodplain by using a hypothetical landform sequence based on the backwearing erosional model (Ruhe 1975). The first stage is a youthful stage without an integrated drainage network. This stage has high water tables and usually numerous wetlands, such as the prairie potholes or other closed depressions in young glacial terrain (Figure 12.1a). The headward erosion of a transgressing channel has potential energy both from gradient and the captured water volume. Once it taps into the wetlands and natural stream or channel drainage is initiated, the combination of increased relief and extra volume of water creates additional kinetic energy, and downcutting of the channel results (Figures 12.1a, b and c) (Nash 1996). Wetlands tend to disappear and the water table level is lowered. Eventually, some base level of downcutting is reached. Base level is a point of severe resistance to downcutting due to a water table (sea level in some cases), indurated rock, or similar phenomena that resists denudation. The edges of the valley can slowly be altered by erosion of footslopes over long periods of time and by the backwearing of the valley edge (Figures 12.1d and e).

The factors resulting in the actual riverine system (a stream forming landforms in its own alluvium) can easily be divided into two basic processes: channel process and floodplain process. Each forms its own type of floodplain; accretion floodplain from channel process and over-the-bank floodplain from floodplain process.

Closed Summit Depression

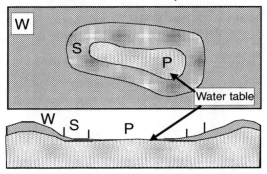

Figure 12.1a Closed summit depression illustrating a high water table. No channel or floodplain development has yet occurred. W is well drained; S is somewhat poorly drained; and P is poorly or very poorly drained.

Headward Erosion & Channel Development

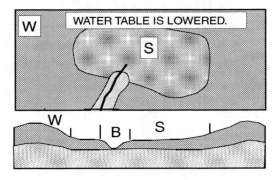

Figure 12.1b With the invasion of an either artificial or natural channel, the water table is lowered and only somewhat poorly drained soils (S) in this case remain in the original wetland. An incipient or immature backslope is started (B).

Down-cutting and dominance of summit (W) soils

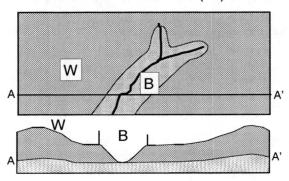

Figure 12.1c The channel is well established and has down-cut to a degree that the water table is only effective near the base of the backslope (B) with possible transient seeps above. The summit soils (W) are dominant and nearly all well drained.

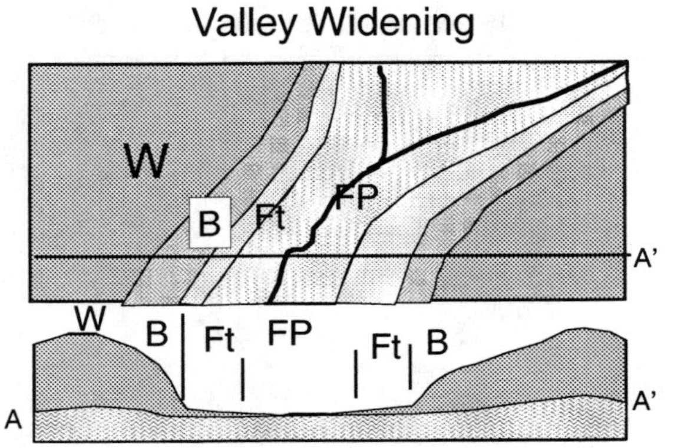

Valley Widening

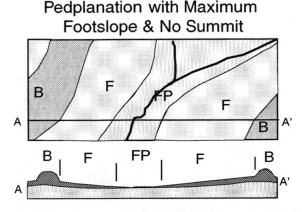

Figure 12.1d Backwearing or retreat of the backslope (B) into the upland summit (W) continues and the valley down-cuts to a water table or other restriction (base-level). The valley widens (FP) based on the size of the stream. An erosional footslope or pediment starts (Ft).

Pedplanation with Maximum Footslope & No Summit

Figure 12.1e Footslope (F) continues to develop at base-level created by the floodplain (FP). The backslope (B) retreats into the summit, and often the summit is just coalesced backslopes. The water table is high in the toeslope or floodplain and high in the lower footslope.

Valley widening and initial floodplain formation is a direct result of the flowing water contained in the river channel (Langbein 1964, Allen 1970, p. 128). Water in the channel flows at greatly varying velocities. Where the water is moving slowly, deposition of sediment occurs, often forming a point bar. On the opposite bank where the water is moving more rapidly, erosion occurs, resulting in the deepening of the channel, referred to as pools. A line through the deepest sections of the channel is referred to as the thalweg (Figure 12.2). Typically, the thalweg does not remain in the center of the channel; instead it migrates from side to side giving rise to a lateral component of the stream channel (Leopold et al. 1964). The shifting of the thalweg and subsequent differential erosion/deposition leads to the formation of meanders. The formation of meanders is related to the dynamic energy of the flowing water, the slope of the channel, and its sediment load (Figure 12.3). It is sufficient to say that this complicated process is common in most riverine systems.

Rivers meander, yet the channel will stay within the confines of its valley. The erosional and depositional processes result in cut and fill, deposition and erosion, or accretion that will cross the valley from edge to edge. An overall scroll-like pattern of sediments (Figure 12.3) across a flat plain results (Leopold et al. 1964, Allen 1970, Hickin 1974). Generally with time, the accretion system migrates laterally across the valley several times, creating a seasonal floodplain (Leopold et al. 1964).

STREAM CHARACTERISTICS

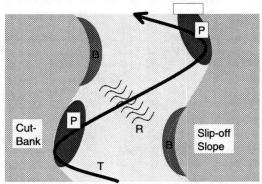

Figure 12.2   Essential parts of a stream or river channel. T = thalweg, P = pool, B = point bar, R = riffle.

The accretion floodplain is entrenched in a larger system that is flooded less frequently but with larger flood events (floodplain processes rather than channel processes). Periodic inundation of the valley occurs during times when the volume of water in the stream exceeds the capacity of the channel (e.g., flooding during rainfall runoff and snow melt). This results in floods where erosional and depositional processes combine (over-the-bank flooding). These events deposit coarse sediments at the edge of the stream and create a local topographic high called a *natural levee*. Behind the natural levee, away from the stream, a landform called a *backswamp* develops (Figure 12.4).

Meanders

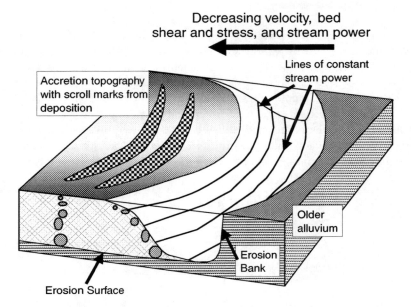

Figure 12.3   The thalweg creates strong water velocities on the outside that create a cutbank and fill on the inside (point bar). The point bar has coarser sediments deep, and fine upwards, reflecting the decreasing energy with shallow water. Minor depressional wetlands with crescent shapes occur in accretion floodplains. (Adapted from Allen, J. R. L. 1970. *Physical Processes of Sedimentation.* American Elsevier Publishing Company, New York.)

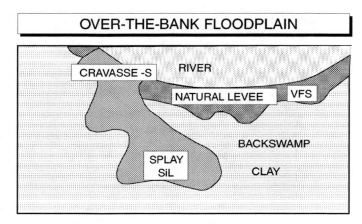

Figure 12.4 Map view of a small portion of a floodplain with over-the-bank landforms of various textures and elevations. The natural levee has the highest elevation, and backswamp is the lowest of these features.

Backswamps are the largest and most extensive of riverine wetlands. The combinations of these features are referred to as the over-the-bank floodplain (Figure 12.4). The water leaving the channel during flood stage often downcuts through the levee, creating a crevasse or a cut in the levee. The water then flows onto the floodplain from the levee, depositing coarser materials (splay deposits) (Figure 12.4). On the lower Mississippi, these distinct landforms are often named for the year they were first noted; cutoffs that produce oxbow lakes are similarly named.

HYDROLOGIC PROCESSES

Gaining and Losing Streams

In general, channels in riverine systems receive water that is discharging from the groundwater in the floodplain and are termed *gaining streams* (Figure 12.5a) (Todd 1980). Permanent streams are gaining streams. At high water, however, the channel yields water to the floodplain (recharge water) or releases flood water (surface water) (Figure 12.5b). In semiarid areas or in the upper reaches of streams in more humid areas, the stream may be a "losing" stream or a stream that gathers water above and recharges the water to groundwater (Figure 12.5c). Soils associated with losing streams are often unmodified sediments that classify as Fluvents. Fluvents are Entisols or recently formed soils that lack much horizon development. More soil development and features associated with hydric soils may be expected in soils associated with gaining streams.

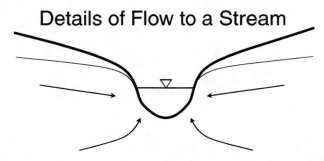

Figure 12.5a Groundwater flow directions in a gaining stream. Note that the water level in the stream (inverted triangle) is relatively lower than the adjacent water table and that flow direction is toward the stream.

Local Reversals of Flow at High Water

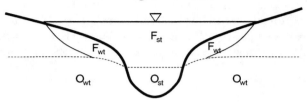

Figure 12.5b Groundwater flow directions in a flooding stream. The water level in the flooded stream (F_{st}) is higher than the original stream level (O_{st}). During the early stages of flooding, the zone of saturated soil beneath the flood waters (F_{wt}) may not extend completely down to the original water table (O_{wt}). This may result in areas of entrapped air and unsaturated soil during a flood event.

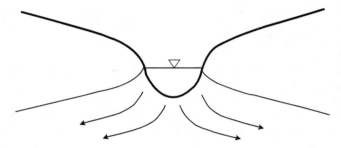

Figure 12.5c Groundwater flow directions in a losing stream. Note that the water level in the stream (inverted triangle) is relatively higher than the adjacent water table.

Surface and Throughflow Water

Overall stream flow is derived from two sources: surface runoff and base flow (Figure 12.6). Surface water flow is highly variable and travels rapidly to the stream. Surface runoff occurs during periods of high precipitation or snow melt. Water in excess of the soil's storage capacity results in overland flow to streams. In flood conditions, the stream channel's capacity to carry water is exceeded, and the stream channel overflows onto the floodplain. Wetland drainage and loss of

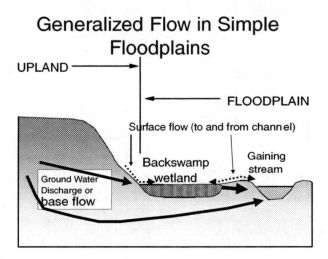

Figure 12.6 Surface and base flow into streams.

natural vegetation often increase overland flow and flooding (Leopold et al. 1964). Excess precipitation or rapid snow melt in one part of a basin frequently results in flooding downstream. Water from flooding is stored in floodplain depressions, such as oxbows, and in the backswamp landforms, creating wetlands. Evapotranspiration and slow groundwater release to the main stream occurs over time. An intermediate condition related to throughflow on slopes results in reflow or saturation of the soil and release or discharge of water at the base of slopes. The edge of floodplains often grades from a slope wetland to a riverine wetland. Slope wetlands and conditions are covered in Chapter 9.

Base Flow (Groundwater)

Base flow is that portion of the stream flow that is the result of groundwater discharge. Permanent streams are "gaining" streams that receive a steady influx of groundwater. Typically, base flow does not vary dramatically annually. Much of the groundwater is first discharged onto floodplain soils, however, before being forwarded to the stream in its channel either by some groundwater movement or via yazoo streams on the floodplain. Many backswamp landforms develop streams that flow parallel to the trunk stream for long distances. These are called *yazoo streams* after the Yazoo River in Mississippi, which flows in this manner before taking a sharp turn and discharging into the Mississippi River.

The general trend of both regional and local groundwater, however, is from the uplands to the river valley (Gonthier 1996) (Figure 12.6). These waters discharge in the floodplain (backswamp focused discharge) or more frequently at the valley edge (valley edge focused discharge). These discharge areas are termed *seeps* (Figures 12.7 and 12.8). Many seeps are actually slope wetlands with organic soils (fens), as illustrated in Figure 12.7 (Brinson 1993). Water discharging at the valley edge may be consistent enough for the formation of an organic soil (Histosol). The water at such discharge points is mineraltrophic (mineral rich) and has all the ingredients to form a fen or a graminoid-dominated, high-base wetland with organic soils. Many of these fens are slope wetlands or, as illustrated in Figure 12.7, combination wetlands of slope and riverine wetlands. The wetlands often have slopes steeper than 4%. Malterer et al. (1986) describe a fen slope–riverine wetland combination. The fen–slope wetland had an organic thickness greater than 1 m and a riverine wetland with over 4 m of muck.

Groundwater at discharge points often has a sudden alteration of its chemical regime. For instance, warming the water at the surface or evapotranspiration of water results in carbon dioxide

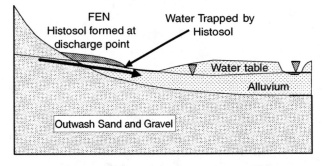

Figure 12.7 Discharge at the valley edge of the Des Plaines River in northern Illinois, creating a slope and riverine combination wetland.

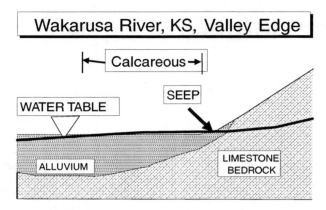

Figure 12.8 A detailed illustration of an edge seep with high amounts of Ca bicarbonate discharged into the soils. The limestone bedrock is slowly dissolved and added to the groundwater as calcium bicarbonate. At discharge into the soils, the temperature warms and the ability of the groundwater to hold carbon dioxide is reduced and calcite is precipitated.

(CO_2) being expelled from the solution. Loss of CO_2 removes the bicarbonate ions that are responsible for keeping soluble calcium (Ca) from precipitating as $CaCO_3$ (calcite). Some seeps have abundant calcite formed in this manner (Arndt and Richardson 1992, Almendinger and Leete 1998) as illustrated in Figure 12.8.

Deposits of ferric iron (Fe^{3+}) often form by oxidation of the groundwater as it is exposed to air at points of discharge. The mobile ferrous (Fe^{2+}) form of iron is reduced and moves with the groundwater. When exposed to the air, the Fe^{2+}-laden groundwater can often be observed as a plume of rusty colored water seeping out of a bank (Figure 12.9) (Rhoton et al. 1993). Deposits of Fe large enough to mine, called bog iron, were common along streams near the Atlantic Coast. Many were exploited as an iron ore during the colonial period. Sediment and soil beneath these oxidized surfaces remain reduced.

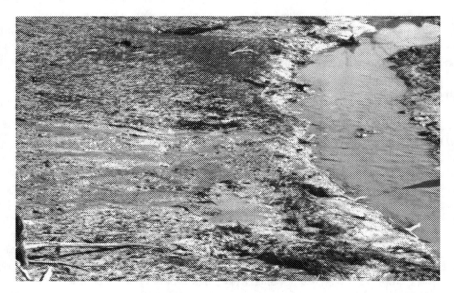

Figure 12.9 Example of oxidized iron entering a stream through a seep at valley edge. (Photo from Lafayette County, MS.)

Landforms of Simple Valley

High Energy ———►Low Energy

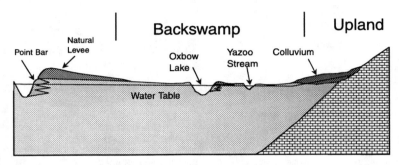

Figure 12.10 Cross-section of floodplain features indicating relative energy of deposition. Older, no longer accreting point, bar and natural levee occur adjacent to oxbow lake. Within the floodplain alluvium depositional facies are present but are not shown for the sake of clarity.

GEOMORPHIC FEATURES OF FLOODPLAINS

The current or modern floodplain associated with a river usually has accretion topography or is entrenching and has high slopes above the outside of its meanders and low "slip-off slopes" on the inside of the meanders (Figure 12.3). In nonentrenched streams, an over-the-bank floodplain usually has formed as a result of periodic inundation (Ritter 1979). The size and shape of the floodplain is the result of variable stream flow and sediment load throughout the drainage basin. The floodplain, and the riverine system itself, is directly related to a sum of factors including climate, age, topography, and geology of the entire drainage basin. The sediments found in the system are composed of alluvium transported and deposited by the stream or river. These sediments may have been transported from the headwaters of the system or derived locally from channel erosion and subsequent deposition. At the edge of the floodplain, colluvial sediment may be present, derived from the uplands at the river valley edge. The modern floodplain mitigates the effects of the flood by acting as storage for both water and sediment. As such, the floodplain has both form and function within the riverine system (Wolman and Leopold 1957).

Although floodplains appear to be topographically simple, they are composed of a variety of features (Figure 12.10). At the valley edge where the uplands and valley floor meet, it is common to observe colluvial material that has been deposited by mass movement off the valley sides. Colluvial material will grade into alluvial material toward the valley center. Numerous zones are common within the alluvial sediments. These include: coarse textured lag deposits in which the finer sediment has been selectively removed, poorly sorted fill resulting from bank and channel collapse, coarse to medium textured near-bank and point bar sediments deposited as stream energy decreases, and finer textured overwash sediments deposited as flood waters flow out over the floodplain. The sequence of textures in Figure 12.11 is based on soils from the Meherrin River floodplain in Virginia (Richardson and Edmonds 1987). The textures represent an energy of deposition geosequence, from the coarsest left by the strongest currents to the backswamp clays deposited in the most quiet conditions.

The overall topography of the floodplain becomes less flat and smooth in the area adjacent to the active river channel because of the active fluvial process of channel erosion called *accretion topography* (point bar area in Figure 12.11). Within the channel, point bars are likely to be seen in areas where flow is slower. The point bars are likely to be coarser textured than the surrounding

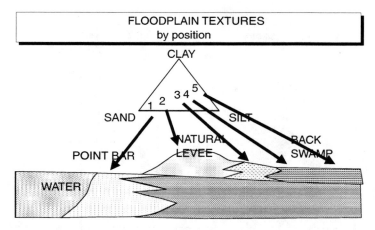

Figure 12.11 Texture of soils along the Meherrin River floodplain in Greenville County, Virginia, that form a geosequence based on decreasing energy of deposition. (Data from Richardson, J. L. and W. J. Edmonds. 1987. Linear regression estimation of Jenny's relative effectiveness of state factors equation. *Soil Sci.* 144:203–208.)

features (Figure 12.11). Over time as the river meanders, the point bars will be reworked into low ridges and troughs referred to as *meander scrolls*. The low ridges are rapidly vegetated and in some instances may act as a channel bank. It is common for a series of these scrolls to occur across the floodplain. The trough between the ridges, sometimes referred to as a slough or chute, may eventually fill with fine-grained material, but it is usually distinctly wetter than the surrounding meander scrolls. Such a sequence may be stable for a number of years and have a scroll-like appearance when viewed from above, hence the term *meander scrolls*. An example is shown in Figures 12.12 and 12.13. As noted by Allen (1970), the point bar deposits are coarser at depth and fine upwards.

Natural levees are, in some regards, similar to meander scrolls, because they appear as low ridges adjacent to the channel. Unlike the meander scrolls, however, natural levees are formed during flood events as the flood water flow decreases after the water leaves the channel and coarser suspended sediment is deposited. These ridges grade into the low flat areas of the floodplain referred to as the backswamp (Figure 12.12). The Wakarusa River near Lawrence, Kansas, has a distinct

LANDFORMS on an OVER-THE-BANK-
FLOODPLAIN
WAKARUSA RIVER, KANSAS

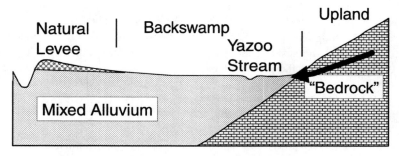

Figure 12.12 Cross-section of a valley with a distinct over-the-bank floodplain and an entrenched stream. The Wakarusa River near Lawrance, Kansas, has a distinct natural levee well above the river channel, which grades into the backswamp landform with a yazoo stream.

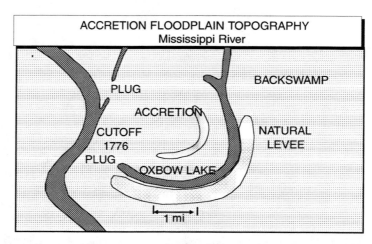

Figure 12.13 Map of an accretion topography with an oxbow wetland and lake created during 1776 on the Mississippi River.

natural levee well above the river channel. This river is deeply entrenched into its channel such that low flow water levels are well below the floodplain. Despite this deep entrenchment, the natural levee gradually grades into the floodplain and the backswamp landform. The backswamp is composed of fine-grained sediment. Periodically, a levee is breached during large floods. A channel is cut through the levee and forms a splay deposit, which consists of sediments that are finer than the levee but coarser than those of the backswamp (Figure 12.4). Most of the backswamp qualifies as having hydric soils.

As the river channel migrates laterally through the floodplain, sections of the channel may be cut off from the main river (Figure 12.13). These cutoffs form oxbows or oxbow lakes if they remain filled with water. Over time the oxbows will fill with fine-grained sediments. They may also be areas where organic matter accumulates, particularly if the oxbow remains water filled with little or no turnover of the water. Organic matter (leaf litter, woody vegetation, etc.) will decompose slowly in the anaerobic water. These features, even after filled in, remain visible for many years.

SOIL DISTRIBUTION

The complex nature of the riverine system, with backswamps, meander scar sloughs, and oxbows, results in a patchwork of environments based on topographic position, distance from the channel, relation to flood stage, and texture (Daniels and Hammer 1992, Daniels et al. 1999). The variety of sedimentary deposits gives rise to a complex distribution of soils across the floodplain (Figure 12.14). The nature of both lateral and horizontal changes in sediment deposition and texture result in soil series with a large degree of variability (Leab 1990). This combination makes standard soil survey maps of floodplains difficult to interpret, because they cannot adequately delineate all the details of the area at the scale of mapping used. Therefore, detailed investigations are often required to fully characterize these soils; detailed investigations are definitely needed for hydric soil delineation.

Wetlands and Hydric Soils

Any discussion of hydric soils and wetlands in the riverine systems must start with the definition of a wetland and the criteria of a hydric soil (Hurt et al. 1996). A wetland is defined as having wetland hydrology, a predominance of hydrophytic vegetation, and the presence of hydric soils. A

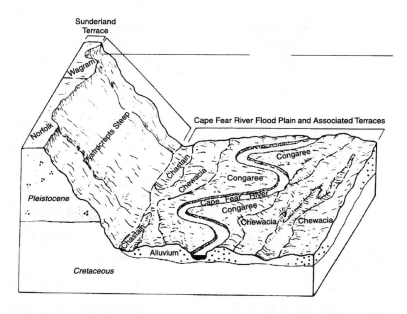

Figure 12.14 Soil distribution within the Cape Fear floodplain and associated terraces. (From Leab, R. J. 1990. *Soil Survey Report of Bladen Co., NC.* USDA, SCS. U.S. Govt. Printing Office. Washington, DC.)

hydric soil is identified based on its morphology or on the extent of flooding or ponding the soil is subject to over a given time span (Hurt et al. 1996). Some soils may be morphologically non-hydric, yet be hydric by virtue of their receiving enough water to be ponded or flooded.

Areas that are periodically inundated with floodwaters for a significant period during the growing season are considered to be wetlands containing hydric soils (*Federal Register,* July 13, 1994). The current 14 continuous days, 1 year in 2 years is used to define "significant period." In a broad sense this covers much of the active floodplain, within which can be found a range of geomorphic and hydrologic zones. These differences result in a gradation of soil morphologies and hydric soil locations across the floodplain which can be divided into two groups: those related to the channel and/or flood events, and those associated with seeps at the valley edge (Figure 12.15).

Wetlands and hydric soil distribution across the floodplain are directly related to its geomorphic features. In general, the higher areas, such as natural levees and larger scroll ridges, contain soils that are morphologically non-hydric, whereas the sloughs and backswamp areas, and features within the channel contain soils that are morphologically hydric. Farther away and higher in elevation than the active floodplain are the older terraces. Like the active floodplain, only the soils in low-lying areas are likely to be morphologically hydric. Oxbows are likely to contain morphologically hydric soils as well. Any of the natural levees or other ridges remaining near the oxbow are likely to be morphologically non-hydric. In addition to the fluvial and channel-related hydric soils are those associated with groundwater seeps at the valley edge. These occur when the underlying stratigraphy allows for lateral groundwater flow which eventually surfaces at the valley edge (Figure 12.6, 12.7 and 12.8).

Overall, the riverine system may alternate between hydric and non-hydric soils, depending on relative elevation and the presence or absence of groundwater discharge. Despite abundant soil variability, most active floodplains contain hydrophytic vegetation and are likely to be inundated frequently and, therefore, are wetlands for the most part (Veneman and Tiner 1990).

Hydric Soil Indicators

The distribution, delineation, and formation of morphologically hydric soils (as opposed to those that meet the criteria based only on flooding) are of particular interest. Several field indicators

Distribution of Soil Internal Drainage

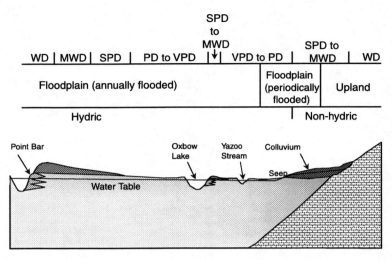

Figure 12.15 Cross-section showing hydric soil relations on a floodplain and at valley edge. Within the active (annual) floodplain well-drained (WD), moderately well-drained (MWD), and somewhat poorly drained (SPD) soils occur, yet due to annual inundation, these are considered hydric soils along with the poorly drained (PD) and very poorly drained soils (VPD). At the valley edge, a seep in colluvium a PD or VPD soil may occur and qualify as a hydric soil.

of hydric soils deal directly with hydric soils on floodplains (Hurt et al. 1996). It is probable that nearly all indicators occur in hydric soils of riverine systems, but we are focusing on particular indicators that we know to be important. These include: A5. Stratified Layers; and F12. Iron/Manganese Masses. In addition several test field indicators may be used in riverine systems; these include: TF2. Red Parent Material; TF8. Redox Spring Seeps; TF9. Delta Ochric; and TF10. Alluvial Depleted Matrix. The field indicators specific to floodplains (F12 and TF10, in particular) recognize some of their unique characteristics.

In particular, F12 stipulates that only "40% or more of the matrix" needs to have a chroma of 2 or less. Essentially, this recognizes that riverine wetlands in active floodplains frequently do not get the amount of iron depletion (reduction) that other wetland landscapes are subjected to, or at least it is not visible. This can be because the soils are younger, contain little to no leaf litter accumulation, have low carbon contents, or receive extra iron via groundwater discharge that is precipitated as bog iron. The young soils may be formed from oxidized sediments and are deposited quickly before reduction occurs. In indicator F12, the thickness requirement for the zone or horizon is waived if the indicator is found in the mineral surface layer. This recognizes that addition of sediment during a flood may result in the shallow or solely surface development of the indicator, because insufficient time has passed for deeper pedogenesis. Additionally because organic carbon (OC) is critical to soil reduction and the formation of low chroma colors, both reduction and low chroma colors will be absent if sufficient OC is not present. The extra iron in groundwater that discharges may reconstitute that high chroma coloration in these soils.

Complicating Factors

Several studies have shown that difficulties exist in identifying hydric soils on floodplains and riverine systems. The most common complicating factors include duration of relationship between oxygenated floodwaters and groundwater, reduction vs. oxidation, lithochromic colors inherited from the parent material, low carbon content, and relict morphologies, especially in soils located on terraces. In fact, terrace soils may have formed in saturated conditions possessing redoximorphic features, yet may currently exist in well-drained conditions (relict) because of entrenching. The actual creation of the terrace by entrenching streams leaves a portion of the terrace better drained than it was originally; these are the areas with relict wetness conditions.

Flood Relationships and Redox Status

Rainwater is generally oxygenated, and rivers swollen by rainwater may be oxygenated as well. During a flood, the water being added to the floodplain must become depleted of oxygen before the soil becomes reduced and begins to form redoximorphic features and hydric soil indicators. In order for oxygen to be depleted, the water must stagnate. This occurs more often in low-energy environments where the water is not flowing (Veneman and Tiner 1990, Faulkner et al. 1991). This relationship has been observed in a study of bottomland hardwood forests in the lower Mississippi River Valley. Soils in areas with higher hydrologic energy (active floodplains) developed anaerobiosis for shorter periods than soils in quieter, backwater areas (ponded conditions). Similar hydric soil morphologies (redoximorphic features) were observed in soils that were reduced and saturated for nearly 100% of the growing season and in those that were reduced and saturated for as little as 10% of the growing season. It seems plausible that hydric soil morphologies reflect a critical duration of anaerobiosis and change slowly after that period has been attained (Faulkner et al. 1991).

Another aspect of landscape position influencing the hydric status of the soil is the type of microtopography (Veneman and Tiner 1990). Closed drainage areas (depressions without surface flow outlets) on the upper floodplain, or terraces, tend to have hydric soils, while areas with open drainage, even on the active floodplain, do not. Thus, if aerated floodwaters are exchanged in the active floodplain, the soil water could remain aerobic even when flooded. Hydrophytic vegetation was observed throughout the active (young) floodplain despite some areas having soils that lacked hydric soil morphology.

Air may become entrapped as floodwaters saturate soil from the surface down. Such an occurrence has been observed on the floodplain of the Connecticut River (Chase-Dunn 1991). Some of the soils are assumed to remain aerobic as suggested by escaping air bubbles and the rapid fall of the water table after inundation. Further indication of aerobic conditions comes from *in situ* measurements indicating that reducing conditions do not always occur after flooding. Aerated floodwater flowing down macropores (such as root or worm channels) is the most probable explanation for the aerobic conditions observed in the soil. These macropores allow for the rapid exchange of aerated water or air into the soils. Measurement of conditions in a macropore would not reflect the true conditions in the soil matrix (Mukhtar et al. 1996). Additional research is needed in this area.

Lithochromic and Relict Colors

Two of the test field indicators for hydric soils on floodplains (TF2. Red Parent Material, and TF9. Delta Ochric) deal with the problem of soil colors inherited from the parent material, also referred to as *lithochromic colors*. In instances where the sediment accumulating on the floodplain comes from an area where the bedrock and/or soils are red (7.5YR or redder), then it too will have a red coloration. This color, likely due to hematite coatings, is more resistant to changes due to redox status, and persists longer in the soil. The result is a high chroma soil in a reducing environment. Such a situation was observed in a transect across the Red River floodplain in Louisiana (Faulkner et al. 1991). The alluvium in this floodplain was derived from Permian red bed parent material and was observed to be resistant to color change. For this reason it was not possible to compare soil morphologies to redox conditions in the soils.

Alluvium derived from gray or low chroma parent materials presents a contrasting dilemma. Such is the case in the Connecticut River Valley where soils with similar morphologies have different degrees of saturation and reduction (Veneman and Tiner 1990, Chase-Dunn 1991). In this situation two profiles of the Limerick series (coarse-silty, mixed, nonacid, mesic, Aeric Fluvaquent), both having similar morphologies, were monitored for saturation and reduction. The results showed one to have hydric soil conditions, while the other did not. Furthermore, both soils had similar OC and free iron (Fe) contents. Both of these soils were young, frequently flooded soils (possibly less than 20 years old in the upper 45 cm), and lacked the redoximorphic features that would help identify them as hydric soils. There was a similarity in color between the C horizons (or mineral strata) and the upper profile (0 to 30 cm). The lithochromic influence with matrix chromas less than 2 may suggest that a soil is saturated and reduced when it is not. As a result, these studies suggest monitoring may be necessary to confirm wetland and hydric soil status.

Organic Matter and Temperature Relationships

Reducing conditions are crucial for the formation of redoximorphic features common in iden-tifying hydric soils. In order for reduction to occur, sufficient organic matter (OM) must be present in the soil. One study of a constructed floodplain suggests that redoximorphic features are formed during short periods of inundation (after 1 event) when soil OM is > 30 g kg^{-1}, and are not found in soils with an OM concentration of < 15 g kg^{-1} (Vepraskas et al. 1995). The first features formed were small and difficult to see with the naked eye, but with time their abundance and size increased.

A study of some alluvial soils in the Puget Lowlands, WA, indicated that aerobic conditions persisted whether the soil was saturated or not (Cogger and Kennedy 1992, Cogger et al. 1992). It was further concluded that, despite the overall aerobic conditions, some microsite reducing conditions did occur. The study concluded that approximately 10% of the observed field variation was due to the inherent variability of the electrodes, while the remainder was attributed to microsites in the soil. This conclusion was based on several field and controlled laboratory investigations. The lack of overall reducing conditions was attributed to two factors: first, low levels of available carbon present; second, low temperatures during period of saturation (6°C in the surface). It was demon-strated that reducing conditions occurred only after 3 to 6 months of saturation in soils with low OM and at low temperatures. The combination of low OM and low temperatures inhibits or delays reducing conditions and redoximorphic feature formation.

Relict Features

Changes in the overall drainage due to variation in stream drainage (either natural or altered by humans) may result in the current morphology reflecting historic rather than current conditions. This can be observed on upper terraces associated with active floodplains. Some of these soils will retain their hydric morphology if insufficient time has passed to allow for pedogenic processes to reflect current conditions. This is influenced by OM and temperature, as indicated above, as well as by available Fe/Mn. If the soil were depleted in Fe, it is not likely to redden rapidly.

SUMMARY

Hydric soils and wetlands are influenced by the same processes that combine to form the features common in riverine systems. In general, all these processes are related to water as it moves through the system either as surface or groundwater flow. The dynamic features of the floodplain landscape contain several challenges to hydric soil identification. Most notable are the influences of aerated flood waters, low organic carbon, lithochromic colors, young sediments, and the possibility of relict features. Many of these problems can be overcome through detailed site evaluation and full understanding of the hydrology of the riverine system.

REFERENCES

Allen, J. R. L. 1970. *Physical Processes of Sedimentation.* American Elsevier Publishing Company, New York.

Almendinger, J. E. and J. H. Leete. 1998. Regional and local hydrology of calcareous fens in the Minnesota River Basin, USA. *Wetlands* 18:184–202.

Arndt, J. L. and J. L. Richardson. 1992. Carbonate and gypsum chemistry in saturated, neutral pH soil environments. pp. 179–187. *In* R. D. Robarts and M. L. Bothwell (Eds.) *Aquatic Ecosystems in Semi-Arid Regions.* Natl. Hydrol. Res. Symp. Se. 7. Environment Canada, Ottawa Ontario.

Brinson, M. M. 1993. *A Hydrogeomorphic Classification for Wetlands.* Tech. Rep. TR-WRP-DE-4. U.S. Army Corps of Engineers, Waterways Experiment Station, Vicksburg, MS.

Chase-Dunn, C. K. 1991. *Soil Mottling as an Indicator of Seasonal High Water Table in Massachusetts Floodplain Soils.* M.S. thesis, University of Massachusetts, Amherst, MA. 125p. (unpublished).

Cogger, C. G. and P. E. Kennedy. 1992. Seasonally saturated soils in the Puget Lowland. I. Saturation, reduction and color patterns. *Soil Sci.* 153:421–433.

Cogger, C. G., P. E. Kennedy, and D. Carlson. 1992. Seasonally saturated soils in the Puget Lowland. II. Measuring and interpreting redox potentials. *Soil Sci.* 154:50–58.

Daniels, R. B., S. W. Buol, H. J. Kleiss, and C. A. Ditzler. 1999. *Soil Systems of North Carolina.* North Carolina State University, Raleigh, NC.

Daniels. R. B., E. E. Gamble, and J. G. Cady. 1971. The relation between geomorphology and soil morphology and genesis. *Adv. Agr.* 23:51–88.

Daniels, R. B. and R. D. Hammer. 1992. *Soil Geomorphology.* John Wiley & Sons. New York.

Faulkner, S. P., W. H. Patrick, Jr., R. P. Gambrell, W. B. Parker, and B. J. Good. 1991. Characterization of soil processes on bottomland hardwood wetland–nonwetland transition zones in the lower Mississippi River Valley. Contract Report WRP-91-1. U.S. Army Corps of Engineers Waterways Experiment Station, Vicksburg, MS.

Federal Register. July 13, 1994. Changes in hydric soils of the United States. Washington, DC.

Gonthier, G. J. 1996. Groundwater flow conditions within a bottomland hardwood wetland, Eastern Arkansas. *Wetlands* 16:334–346.

Hickin, E. J. 1974. The development of meanders in natural river channels. *Am. J. Sci.* 274:412–442.

Hurt, G. W., P. M. Whited, and R. Pringle (Eds.). 1996. *Field Indicators of Hydric Soils of the United States.* USDA Natural Resources Conservation Service, U. S. Govt. Printing Office. Washington, DC.

Langbein, W. B. 1964. Geometry of river channels. *J. Hydraulics Div. ASAE.* 90:301–312.

Leab, R. J. 1990. *Soil Survey Report of Bladen Co., NC.* USDA, SCS. U.S. Govt. Printing Office. Washington, DC.

Leopold, L. B., M. G. Wolman, and J. P. Miller. 1964. *Fluvial Processes in Geomorphology.* W. H. Freeman, San Francisco.

Malterer, T. J., J. L. Richardson, and A. L. Duxbury. 1986. Peatland soils associated with the Souris River, McHenry County, North Dakota. *Proc. N.D. Acad. Sci.* 40:103.

Mukhtar, S., J. L. Baker, and R. S. Kanwar. 1996. Effect of short-term flooding and drainage on soil oxygenation. *Trans. ASAE.* 39:915–920.

Nash, D. J. 1996. Groundwater sapping and valley development in the Hackness Hills, North Yorkshire, England. *Earth Surface Processes and Landforms* 21:781–795.

Playfair, J. 1802. *Illustrations of the Huttonion Theory of the Earth.* William Creech, Edinburgh.

Richardson, J. L. and W. J. Edmonds. 1987. Linear regression estimation of Jenny's relative effectiveness of state factors equation. *Soil Sci.* 144:203–208.

Ritter, D. F. 1979. *Process Geomorphology.* W. C. Brown, Dubuque, IA.

Rhoton, F. E., J. M. Bigham, and D. L. Lindbo. 1993. Properties of iron-oxides in streams draining the loess uplands of Mississippi. *Agron. Abst.* pp. 345.

Ruhe, R. V. 1975. *Geomorphology.* Houghton Mifflin Company, Boston.

Todd, D. K. 1980. *Groundwater Hydrology.* John Wiley & Sons. New York.

Veneman, P. L. M. and R. W. Tiner. 1990. Soil–vegetation correlations in the Connecticut River floodplain of western Massachusetts. U.S. Fish. Wildl. Serv., Biol. Rep. 90(6). 51 pp. U.S. Govt. Printing Office. Washington, DC.

Vepraskas, M. J., S. J. Teets, J. L. Richardson, and J. P. Tandarich. 1995. *Development of redoximorphic features in constructed wetlands.* Technical Paper No. 5. 12p. Wetlands Research Inc., Chicago.

Wolman, M. G. and L. B. Leopold. 1957. *River floodplains: some observations on their formation.* USGS Prof. Paper. 282-C., U.S. Govt. Printing Office. Washington, DC.

CHAPTER 13

Soils of Tidal and Fringing Wetlands

M.C. Rabenhorst

INTRODUCTION

Within the hydrogeomorphic framework for classifying wetlands and understanding their functional processes, wetlands are described and grouped according to their geomorphology, water source, and hydrodynamics. By geomorphology, we mean the larger landscape and watershed setting within which the wetland occurs. The geomorphology of wetlands is largely responsible for the focusing of surface water or groundwater so as to maintain saturation, flooding, or ponding for significant periods of time. Typical examples might be upland depressions or flood plains along riverine systems. Alternatively, in the case of tidal or fringing wetlands, the geomorphological setting places the wetlands at an elevation and location in close proximity to a significant water body, such as an estuary or lake.

The source of water in a wetland can have a number of dramatic ramifications on soil–water processes, including water chemistry and energy vectors associated with water movement. Direct infall of precipitation can be important in all wetlands, but may be particularly so in humid regions. Generally, rainfall or snow melt will be lower in solutes than most surface or groundwater, and will move directly into the soil toward the groundwater, unless slow infiltration causes it to move laterally over the soil surface. Groundwater discharge to wetlands is perhaps the dominant source of water for many depressional systems, but can also be an important water source within smaller discharge stream systems. While there are a few instances where surface waters may dominate depressional wetland systems, such as the surface-focused recharge wetlands in the prairie pothole region, surface water is the main source of water entering wetlands associated with fluvial or tidal flooding conditions. This may occur only occasionally or seasonally within riverine systems, or it may occur on a more frequent basis, as is the case in tidal systems.

Hydrodynamics refers to the motion of water and the capacity of the water to accomplish work, such as the transport of sediments, the flushing of hypersaline water, or the transport of nutrients to roots, etc. There is both an energy and a direction associated with water entering and moving through wetlands. The kinetic energy of the water is related to its velocity and may be reflected in the particle size distribution of the suspended load. The direction of surface water has sometimes been classified as vertical (often corresponding to depressional systems), unidirectional (often corresponding to riverine systems), and bidirectional (often corresponding to fringe or tidal systems.)

GEOMORPHOLOGY OF TIDAL WETLANDS

Geomorphic Models

Based on studies in the Chesapeake Bay estuary, Darmody and Foss (1979) described three basic geomorphic types of tidal marshes. *Estuarine* marshes form in alluvial sediments deposited along tidally influenced rivers and streams (Figure 13.1a). The sediments generally have been eroded from within higher portions of the watershed during storm events and transported downstream to the tidal portion of the stream. During periods of especially high tides, the sediment-borne waters move beyond the channel and over the marsh where the velocity is decreased by the marsh vegetation. The carrying capacity consequently decreases, resulting in sediment deposition. Owing to the relatively low velocity of estuarine streams and rivers, the mineral component in these soils is mostly silts and clays. The mineral content of the marsh soils is dependent on the balance between the magnitude of erosion and deposition of mineral soil from upstream, and the rate of organic matter production within the marsh. For instance, if a stream provides a significant sediment source in close proximity to the marsh, these soils typically have a higher mineral content and are classified as Entisols. If extended periods occur without significant mineral deposition, organic lenses can be found stratified within the C horizons, or if these periods are especially prolonged (over decades), organic (O) horizons may form. Because these sediments have accumulated under water and have little opportunity for consolidation through dewatering, drying, or compaction, they typically have a high water content and low bulk density. Therefore, these soils tend to have a low bearing strength and have an *n*-value of >1 (Soil Survey Staff 1998).

In submerging (transgressive) coastal landscapes, Darmody and Foss (1979) described *coastal* type soils forming in marshes behind barrier island systems (Figure 13.1b). These marshes form from organic and mineral sediments within a protected lagoonal setting behind a barrier island. Initially, unvegetated intertidal flats may become colonized by marsh plants, sometimes aided by the growth of algae (Steers 1977). The growth of the plants provides organic materials directly to the soil and also aids in the trapping of suspended sediment from the tidal waters. Where these marshes occur directly adjacent to a barrier island, the main source of mineral sediment to these soils may be the sandy sediments of the island itself. Therefore, sandy lenses are common within the O horizons, and occasionally the marsh surface may become buried by a significant deposit of sand during violent storms. Because marshes in these locations are not typically in close proximity to sediment-laden estuarine streams, there is less opportunity for the accumulation of finer-textured mineral components. However, on the landward side of the bay or lagoon, the mineral sediments may be finer in texture, resulting in silty or clayey lenses within organic horizons, or even the formation of fine-textured mineral soils.

The third geomorphic setting for marsh soils in estuaries like Chesapeake Bay has been called *submerged uplands* (Darmody and Foss 1979) (Figure 13.1c). The essentially continuous (although punctuated) rise in sea level over the last several thousand years has caused the formation of marsh soils overlying what were once better-drained upland soils on very gently sloping to nearly level landscapes. The O horizons are thinnest at the upland margin of the marsh and usually thicken toward the estuary. Slow rates of sea level rise of less than a few mm/yr (actually an apparent sea level rise caused by both rising sea level and coastal subsidence) have permitted the vertical accretion of O horizons at rates that keep pace with the sea level rise (Rabenhorst 1997.) It has been suggested that the accelerated rates of sea level rise documented during the last one or two centuries are more rapid than the vertical accretion which can be maintained by the marshes, and thus have been implicated in causing marsh decline and loss (Stevenson et al. 1985). The properties of submerged upland type soils include both those inherited from the former upland soils (such as Bt horizons and high bulk densities) and also those acquired during the formation of organic horizons. Near the margins where the O horizons are thin, the soil classification is strongly affected by the old upland mineral soils. Where the O horizons are thick enough (>40 cm) the soils are

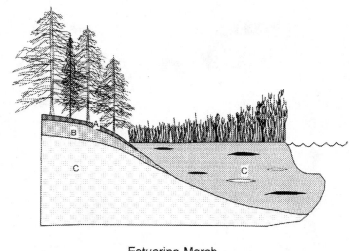

a) Estuarine Marsh

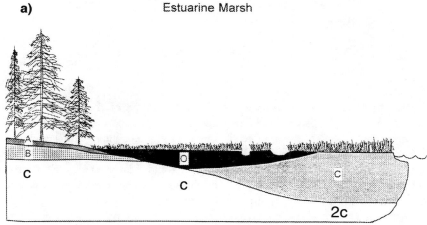

b) Submerging Coastal Marsh

c) Submerged Upland Marsh

Figure 13.1 Tidal marsh types classified according to geomorphologic settings. a) estuarine; b) submerging coastal; c) submerged upland; d) emerging coastal; e) floating. (Adapted from Darmody, R. G. and J. E. Foss, 1979. Soil–landscape relationships of the tidal marshes of Maryland. *Soil. Sci. Soc. Amer. J.* 43:534–541; Stevenson, J. C., M. S. Kearney, and E. C. Pendleton. 1985. Sedimentation and erosion in a Chesapeake Bay brackish marsh system. *Marine Geol.* 67:213–235.)

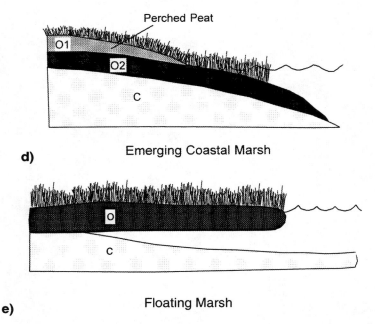

d) Emerging Coastal Marsh

e) Floating Marsh

Figure 13.1 *Continued.*

classified as Histosols. Also, the salinity and base saturation of the mineral portions of the old submerged soils are typically elevated due to the influence of more saline estuarine waters. Within these settings, there is also a pronounced, although gradual, change in vegetation, with marsh species pioneering as an under story below trees dying from excessive wetness and salinity.

In describing the geomorphic settings of fringing and tidal wetlands, Stevenson et al. (1986) have added two cases to those of Darmody and Foss (1979) in order to address two less common situations encountered in estuaries. In certain regions (such as along the California coastline), tectonic activity has been causing an emergence of the coast relative to sea level (regression). Under these circumstances, organic soil horizons that formed at or near sea level have been raised to higher elevations and effectively perched above their zone of formation. These have been termed *emerging coastal* types (Figure 13.1d) by Stevenson et al. (1986), which they differentiate from the *submerging coastal* type of Darmody and Foss (1979). Occasionally in tidal systems, but more common in lacustrine settings, densely interwoven organic horizons may be underlain by water. Since the marsh soil is effectively buoyed up by entrapped air and the low density of organic matter, these types have been termed *floating marshes* (Figure 13.1e).

Geomorphic Processes

Marshes have been differentiated into zones based on elevation and the resulting frequency of tidal inundation, and described as low, middle, or high marshes (Redfield 1972) (Figure 13.2). Low marsh areas are inundated frequently and have also been termed submergence marshes, while high marsh areas are inundated less frequently and have been termed by some as emergence marshes (Ranwell 1972, Adam 1990.) In England (Long and Mason 1983) and in the northeastern U.S. (Redfield 1972), where tidal ranges are moderate (1 to 3 m), particular types of vegetation have been reported to be associated with these elevational zones. In the middle portion of Chesapeake Bay estuary where the tidal range is lower (<1 m), the vegetational zonation is less pronounced, although some zonation may still be observed, where for example, at the highest elevations within some marshes, *Spartina patens* tends to dominate (Darmody 1975). More detailed investigations

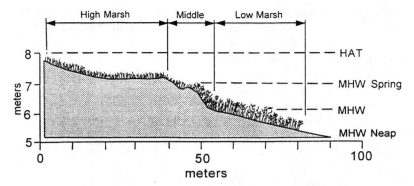

Figure 13.2 Classification of marsh zones into low, middle, and high marsh areas based on elevation and frequency of tidal inundation. The mean high water (MHW) for spring and neap tides are shown, as well as the high astronomical tide (HAT). (Adapted from Long, S. P. and C. F. Mason. 1983. *Saltmarsh Ecology.* Blackie Pub. Glasgow and London. 160 pp.)

have suggested, however, that even if vegetational–elevational associations are observed within an estuary the patterns may not be applicable to other estuaries (Adam 1990).

In general, both the nature and the geographic proximity of sediment sources have large effects on the mineral components of marsh soils, but the mechanics of physical transport tie the accumulation of mineral sediments in marsh soils both to geomorphology and to hydrodynamics. Because sand grains are larger and require greater transport energy than silt and clay, sandy sediments are usually transported only relatively short distances in an estuary and are added to marsh soils that are in close proximity to the source. Coastal marshes behind barrier islands commonly receive sandy sediments that are either blown or washed in during storm events. Finer-textured sediments (silt and clay) are more easily transported within the estuary and may be deposited in marshes at greater distances from their origin.

Where marshes occur adjacent to streams and rivers that carry a significant load of eroded sediment from further up the watershed, they will generally receive greater mineral additions and will most commonly be mineral soils (Hydraquents or Sulfaquents). Where the marshes are located farther away from the source of mineral sediment (such as in expansive submerged upland marshes), the relative input of mineral sediments will be less than from those sites nearer the source. These soils will be composed of a greater proportion of organic sediment derived from the marsh vegetation itself, and the soils in these marshes will more likely be organic (Histosols).

Where the mineral load to marshes is significant (that is, where marsh growth and accretion are strongly associated with mineral deposition [low OM soils]), the elevation of the marsh surface may have an important effect on sedimentation rates (Steers 1977). Because areas of lower elevation are submerged more frequently, they have greater potential for receiving sediment than those at higher elevations that are submerged less frequently. However, portions of the marsh directly adjacent to the sediment-laden stream may be rapidly accreting levees at slightly higher elevation because they receive greater sediment deposition as they decrease the velocity of the flooding water, lowering its energy and its transport capacity. Because the density and structure of marsh plants affect the movement of floodwater across the marsh, plant type and vigor can also affect the accumulation of mineral sediments.

Factors Affecting Marsh Vegetational Succession

A great deal has been written describing vegetational succession in tidal marshes, and the breadth and extent of the topic are beyond the scope of this paper. Nevertheless, a cursory sketch of the factors that govern the development, succession, and distribution of marsh vegetation is warranted. For a given marsh, the combination of elevation and tidal characteristics defines the frequency and duration of tidal inundation. Marsh plants are variably adapted to survival under submerged conditions, which helps to determine their distribution along elevational gradients. Also contributing to the distribution of plant populations is the salinity of estuarine and marsh soil pore waters. Salinity gradients are commonly observed within tidal estuaries between fresher portions toward the headwaters and more saline portions nearer the ocean. Thus on a broad scale, the general vegetational distribution within an estuary will in part be related to the salinity in that portion of the estuary. Locally, salinity tolerance may affect plant distribution within a marsh, while elevation and frequency of tidal inundation may also affect salinity of marsh soils. As a contributing factor, the physical nature of the mineral substrate is probably less influential on marsh vegetation than salinity or the tidal regime, but it can vary greatly between highly sandy sediments with very little fines to sediments dominated by silts and clays and essentially no sand.

WATER SOURCE IN TIDAL WETLANDS

Waters that enter tidal wetlands can come from any of three possible sources, including precipitation, surface water, or groundwater, although in comparison to the other two, groundwater inputs to tidal marshes are probably small. Surface waters represent the dominant water source for most tidal wetlands. Both astronomical and weather-related (storm) tides cause waters to flood and submerge the marsh surface at varying frequencies, depending on such factors as elevation and location. Of particular significance to tidal wetlands is the nature of these tidal floodwaters. The chemical composition of tidal waters ranges widely depending on their proximity to ocean water. Generally, coastal and estuarine marshes have salinities and levels of ionic solutes significantly above those of freshwater, and in some cases they may be orders of magnitude greater, eventually approaching levels found in seawater. Some tidally influenced marshes located along the upper reaches of estuarine rivers or in the interiors of coastal deltas may, however, have levels of soluble constituents similar to those of freshwater systems (Baxter 1973, Miehlich 1986). Elevated levels of solutes in tidal waters include many required plant nutrients, such as K, Ca, Mg, and S. Soluble N and P levels, however, tend to be fairly low in ocean waters relative to fresh waters (Long and Mason 1983). Therefore, much of the N and P that enters tidal marsh systems comes from either soluble or absorbed nutrients transported from upland sources. In addition, some N-fixing micro-organisms (free living or symbiotic) occur in marsh soils and contribute to plant available N.

By comparison, meteoric waters that enter tidal marshes contain very low levels of solutes. The significance of meteoric water entering a tidal marsh is dependent on the frequency with which the marsh is inundated with tidal water. In lower portions of the marsh that are flooded frequently by semidiurnal tides, the dilution affected by meteoric waters may be insignificant. Under those conditions, the marsh pore water is dominated by the chemistry of the tidal water. However, at higher elevations within the marsh, which may undergo extended periods without flooding by tidal water, the intrusion of fresh meteoric water may significantly affect short-term changes in the marsh pore water chemistry.

HYDRODYNAMICS OF TIDAL WETLANDS

Tidal Frequency and Range

In most of the coastal marshes of the world, astronomical conditions cause two tidal cycles per day with a period of approximately 12.5 h. The range in elevation between high and low tides varies widely depending largely on geography. Along the European Atlantic coast, the tidal range

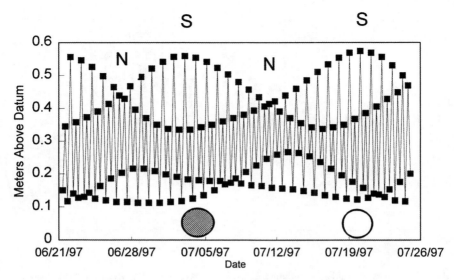

Figure 13.3   Simulated high and low tides at Solomon's Island (Chesapeake Bay) during a 5-week period. Superimposed upon the diurnal tides are lunar cycles causing unusually high, high tides (spring tides — S) (July 3 and July 20) near the time of full and new moons, and unusually low, high tides (neap tides — N) which occur approximately midway between spring tides (June 26 and July 11). Note new moon on July 4 and full moon on July 20.

is approximately 3 m, and along the North American Atlantic coast it ranges from 1 to 3 m. The Bay of Fundy is notorious for having the largest tidal range in the world — about 18 m (Steers 1977). The coastal bays of Delaware and Maryland (Sinepuxent, Assawoman) and the Chesapeake Bay estuary typically have smaller tidal ranges of <1 m, as do those along the gulf coast of Louisiana (Chabreck 1972). Perhaps the lowest tidal range occurs along the Baltic coast of Sweden, where it is 0.3 m (Ranwell 1972).

Superimposed upon the normal ranges of the twice-daily cycles of high and low tides are lunar cycles causing unusually high (spring) high tides when the sun, moon and earth are aligned (approximately 1.5 days after full and new moons) and unusually low (neap) high tides, which occur 7 days after spring tides (Figure 13.3). While particular patterns may vary from location to location, the magnitude of the tidal range is also related to annual cycles as illustrated in Figure 13.4. In addition to the astronomical conditions affecting heights of high and low tides, which can be calculated and predicted, are unpredictable meteorological conditions such as barometric pressure and prevailing winds, which can also affect tides. These are usually translated into infrequent and irregular storm tides with varying return frequencies. Analysis of a 50-year record of tides from Solomon's Island, MD, indicated that the frequency of occurrence of above normal high tides was described reasonably well by a log function (Rabenhorst 1997) (Figure 13.5).

Other Factors Affecting Hydrodynamics

While the daily tidal range may average 1 m or more in many areas, the fluctuation in water tables within most marsh soils is much less. The work of Haering (1986, Haering et al. 1989) indicated that the rise and fall of marsh soil water tables are only significant in the vicinity nearest the tidal creeks, with measurable effects occurring only within 20 m of the tidal creek. This is apparently due to the periodicity of tides (12.5-h cycle) and the modest rates of hydraulic conductivity in marsh soils. Thus, one should avoid embracing the simplistic caricature of water tables in marsh soils rising and falling with the daily tidal cycles.

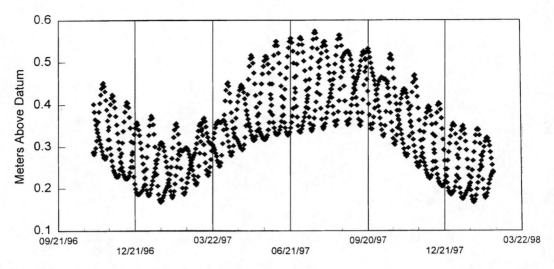

Figure 13.4   Simulated high tides at Solomon's Island (Chesapeake Bay) from 10/96 through 2/98. Note the annual cycle for this particular site, that has higher spring tides during the summer months and lower spring tides during the winter months.

At the lowest elevations within the marsh, the soils are generally flooded twice each day by tidal waters, and the water movement is predominantly bidirectional. The result is that the soil salinity is rather constant and very near that of the flooding water. In contrast, at the highest elevations within the marsh, the soils are flooded only by extreme tides and storm tides. Depending on the balance between precipitation and evapotranspiration, the water tables in these soils may drop to varying depths below the surface. During periods of high evapotranspiration, salts may accumulate in the marsh soil, causing highly saline conditions. During periods of rainfall, the salinity of the soils may be lowered considerably by leaching with meteoric water.

Stream Flow

The hydrodynamics in tidal wetlands can be affected by marsh accretion and succession, leading to the development of creeks or streams. Initially, on unvegetated sand or mud flats, tidal waters will basically have a uniform ebb and flow across the area. As vegetation becomes established,

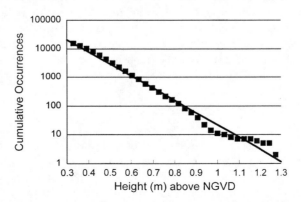

Figure 13.5   Analysis of a 50-year record of tidal measurements from Solomon's Island, MD, indicate that the cumulative frequency of occurrence of above-normal high tides is described well by a log function. Height of tides is given in relation to the national geodetic vertical datum of 1929 (NGVD) (formerly the sea level datum of 1929).

flow is restricted and channelized. Eventually most of the water movement becomes restricted to channelized flow within tidal creeks. Generally speaking, the presence of vegetation promotes accretion in those areas, while at the same time, tidal scour maintains or deepens creeks which can produce a steep-sided channel (Long and Mason 1983).

Most tidal marshes are interlaced with a network of tidal streams, often following a dendritic, but sometimes a trellis or rectangular, pattern (Steers 1977). Most tidal streams reach bank full roughly 360 times per year in the mid-marsh section, which stands in contrast to most alluvial systems, which reach bank full only occasionally (a few times per year at most). The migration of stream channels within the marsh can be very dynamic. Relative to other alluvial systems, the meanders move rapidly at rates reported to be up to 100 m per century (Long and Mason 1983). Most of the water in tidal marshes enters and leaves via the system of tidal creeks. Therefore, the marsh soil pore water near the creek banks is most similar to the tidal water itself, but its chemical composition changes dramatically with distance from the creek bank. Usually only during spring tides or storm tides will the water overflow the banks and cause the marsh to become fully submerged.

The flow regimes of streams in tidal marshes are distinctly bidirectional, with water moving against the bed gradient during flow regimes and moving along the bed gradient during ebb conditions. Some have suggested that the velocity and energy of water in tidal streams is lower during flow and greater during ebb conditions because of the effects of gravity (Long and Mason 1983). That there may be some variation in flow and ebb conditions is unquestionable, although the nature of this hysteresis may not be predictable. The relative velocity and energy of water flowing in tidal creeks during periods of flow and ebb play important roles in determining the relative balance between accretion/sedimentation and erosion in marsh development.

Although the water flow in tidal marshes is mainly bidirectional, there may be particular conditions when vertical and unidirectional flow may also occur. At higher elevations in the marsh which are only occasionally flooded by tidal waters, the reception and infiltration of meteoric water results in vertical movement and a downward leaching vector in the uppermost soil horizons above the free water surface. In the lower portions of the marsh, which are frequently inundated by tidal waters, meteoric infall would not have any noticeable effect. Also, storm events in the watershed supplying flow to an estuarine stream may have the effect of causing flood conditions in parts of the estuarine marsh that may resemble the unidirectional flow typical of other flooded alluvial systems.

GEOCHEMISTRY OF TIDAL WETLANDS

Effects of Peraquic Conditions

While the water tables in some of the higher portions of tidal wetlands may occasionally drop during periods of high evapotranspiration between spring tides, the water source in these wetlands is more or less constant, and the wetland hydrology is generally maintained. Because the hydrology is not particularly dependent on seasonal conditions and variations, the high water tables in fringing and especially tidal wetlands are essentially permanent, leading to what has been described as "peraquic" conditions in the soil (Soil Survey Staff 1975). Also for a variety of reasons (including dispersion and low permeability of estuarine sediments, and low hydrostatic head within tidal wetlands) the rate of water movement through wetland sediments is slow. Together, these factors result in wetland soils which not only have very low electrochemical redox potentials, but maintain these low redox potentials throughout much of the year. Because the development of reducing conditions in these wetland soils is microbially mediated, and because microbial activity is temperature dependent, there can still be significant seasonal trends in the soil redox conditions which may be related to soil temperature and to availability of labile C sources during plant senescence (Feijtel et al. 1988, Oenema 1988, Krairapanond et al. 1991).

Sulfidization and Methanogenesis

As heterotrophic bacteria decompose organic materials, they utilize various compounds and ions as electron acceptors under various Eh regimes and proceed to lower the redox potential as one acceptor is depleted and another is utilized. Generally, they are utilized in the order of O_2, NO_3^-, Mn^{+4}, Fe^{+3}, SO_4^{-2}, and CO_2. The peraquic conditions of tidal marsh soils create two important conditions that influence soil development. First, organic matter accumulates from the primary marsh vegetation that provides an energy source for microbes. Second, diffusion of atmospheric oxygen into the saturated soils is inhibited. Therefore, dissolved oxygen is quickly depleted, and the microbes move on to alternate electron acceptors. In tidal marshes, it is common for soils to become sufficiently reducing that microbes utilize sulfate as a primary electron acceptor. Therefore, tidal marsh soils often present a combination of conditions which are optimal for sulfidization, including organic matter as a microbial energy source, low redox potentials, the presence of sulfate as an electron acceptor, and sulfate-reducing bacteria (Rabenhorst and James 1992). While Gold-haber and Kaplan (1982) have indicated that rates of sulfate reduction are independent of sulfate concentrations when >10 mM, the work of Haering (1986) suggests that sulfur accumulation in tidal marsh soils may begin to be limited by sulfate concentrations only when levels in the estuarine water drop below 1 mM. This would suggest that most estuarine waters with salinities greater than 1 or 2 ppt generally will have adequate sulfate for sulfate reduction during microbial oxidation of organic carbon (Equation 1).

$$SO_4^{-2} + 2CH_2O \xrightarrow{\;\;SO_4^{-2}\text{ reducing bacteria}\;\;} H_2S + 2HCO_3^- \qquad \text{(Equation 1)}$$

If a source of reactive Fe is present in the marsh soil where sulfate reduction is occurring, iron sulfide minerals will form (Rabenhorst and James 1992). Both monosulfide (Equation 2) and disulfide (Equations 3 and 4) species can form (Rabenhorst 1990), although the disulfide (pyrite — FeS_2) is the thermodynamically favored phase. Both individual crystals of pyrite and framboids will form in marsh soils. Both Griffin and Rabenhorst (1989) and Rabenhorst and James (1992) have discussed the ways in which various of the factors necessary for sulfidization could affect or limit the formation of sulfides in tidal marsh systems. Any of a number of heavy and transition metals could substitute in small quantities for Fe in the pyrite structures. Thus, one of the environmental implications of sulfidization in tidal marsh soils is that heavy metals may accumulate as sulfides and other phases in the soils and sediments of tidal marshes and help ameliorate contaminated estuaries (Lindau and Hossner 1982, Griffin et al. 1989).

$$Fe^{2+} + S^{2-} \rightarrow FeS \qquad \text{(Equation 2)}$$

$$FeS + S^\circ \rightarrow FeS_{2\text{ (pyrite)}} \text{ (simplified)} \qquad \text{(Equation 3)}$$

or

$$Fe^{2+} + S_X^{Y-} \rightarrow FeS_2 + (X-2)S^{(Y-2)-} \qquad \text{(Equation 4)}$$

Because metal sulfides can accumulate in tidal marsh soils, they possess the potential for acid sulfate weathering if disturbed through dredging or excavating operations. When soil materials containing sulfide minerals are exposed to aerobic conditions, the sulfides can be oxidized through microbial activity, generating acidity in the form of sulfuric acid (Equation 5). If the neutralizing potential (through $CaCO_3$) or the buffering capacity of the soil is not adequate to counteract the acid generated, the soil itself can become extremely acid. In this way, soil materials from some tidal marshes, that under natural (peraquic) conditions usually have pH values that are neutral or slightly alkaline, can develop pH values as low as 3 or less when disturbed and oxidized.

$$FeS_2 + 5/2H_2O \rightarrow FeOOH + 2H_2SO_4 \qquad \text{(Equation 5)}$$

In the fresher reaches of some estuaries, the sulfate concentrations may be low enough that they limit sulfate reduction. When sulfate levels drop below some minimum threshold, an alternative methanogenic pathway will be utilized by the microorganisms, wherein CO_2 is reduced to CH_4. So long as sulfate is present, however, it has been shown that sulfate reduction is favored over methanogenesis (Widdell 1988). The low redox potentials found in tidal marsh soils should render nitrate highly unstable such that it would either be denitrified and lost as N_2 or reduced to ammonium.

MORPHOLOGY AND CLASSIFICATION OF TIDAL MARSH SOILS

Morphology and Horizonation

The dominant soil horizons described in marsh soils are O, A, and C. Only occasionally are B horizons described, and usually their properties are inherited from some prior episode of pedogenesis. Whether a marsh soil horizon is designated O, A, or C depends largely on the relative proportion of mineral and organic components. Where marsh vegetation is actively growing in areas removed from a sediment source, organic (O) horizons will form, whereas if conditions favor rapid accumulation of mineral sediments relative to organic materials, C or A horizons will form. It is also common during marsh soil genesis for the balance between these conditions to change so that occasional organic horizons or lenses may be interspersed within a dominantly mineral soil and vice versa.

Most organic soil horizons are dark in color with Munsell value/chroma of 3/2 or darker. They are differentiated by the degree of decomposition of the organic materials, which is based mainly on the quantity of recognizable plant materials remaining after rubbing (as Oi, Oe, and Oa, as the degree of decomposition increases).

The mineral soil horizons are almost always gleyed or gray in color with a Munsell chroma of 2 or less, and often 1 or less, regardless of whether they are sandy, loamy, or clayey. These gray colors can be attributed to the colors of the mineral grains, which lack coatings of iron oxides more typical of upland soils. When the mineral sediments are sandy, they tend to have higher density and lower n-value. The mineral horizons in tidal marshes which are loamy or clayey, however, tend to have a lower density, high water content, and consequently a high n-value (n > 1), which is one of the common characteristics of tidal marsh soils. The notable exception to this is when loamy subsoils of submerged upland marshes underlie more recently accumulated organic or mineral horizons. In these cases, the subsoil horizons are B rather than C horizons, and have properties inherited from upland pedogenic processes. These horizons have bulk densities typical for upland soils (1.3 to 1.7), have a low n-value (n < 0.7), and may even have such inherited features as illuvial clay films. It is also common for these soil B horizons to contain soft masses and concentrations of Fe (redox concentrations) within a gleyed or depleted matrix, which are generally absent from tidal marsh soils. Adaptations of some marsh plants enable them to oxygenate the environment immediately adjacent to their roots, forming an oxidized rhizosphere. If soluble ferrous Fe is present, the Fe may precipitate as oxyhydroxides, forming redox concentrations in the form of soil pore linings or coatings on roots.

Classification

If as much as 40 of the upper 80 cm of a tidal marsh soil is comprised of organic soil materials, it is classified as a Histosol. Most other tidal marsh soils are classified as Entisols. Histosols are divided into the suborders Fibrists, Hemists, and Saprists, depending upon the degree of decomposition in the subsurface organic horizons. In general, organic horizons forming in tidal marshes of cooler regions will be less decomposed than those forming in warmer regions. It is for this reason that Fibrists are mapped mainly in the tidal marshes of cold areas like Alaska. While both Hemists and Saprists can be found in tidal marshes all along the Atlantic, Gulf, and Pacific coasts, Hemists are dominant in New England and the Pacific Northwest, while Saprists dominate in the southeast Atlantic and Gulf coastal regions. At another level of classification, Histosols are differentiated between Terric and Typic subgroups based on whether they have a mineral layer 30 cm or more thick that has its upper boundary between 40 and 130 cm. A simpler, though not quite so accurate, way to describe this is to say that those soils where the base of the organic horizons is shallower than 130 cm are Terric, while those which have deeper organic materials are Typic. Some organic soils in tidal marshes of northern New England (Breeding et al. 1974) overlie rock at a shallow depth, which causes them to be classified in Lithic subgroups. Similarly, Histosols in Florida overlying coral and limestone are also classified in Lithic subgroups.

A significant characteristic used in classifying both organic and mineral tidal marsh soils (at the great group level) is the presence or absence of sulfidic materials within 100 cm of the soil surface. Sulfidic materials must contain a sufficient quantity of sulfide minerals (such as pyrite) so that when incubated under moist and ambient conditions, they will "show a drop in pH of 0.5 or more units to a pH value of 4.0 or less." Thus the sulfide is sufficient for evidencing the potential for acid sulfate weathering (Soil Survey Staff 1998).

Tidal marsh soils, which are dominated by mineral soil materials, are classified in the suborder of Aquents (Entisols). They are further differentiated based on the presence or absence of sulfidic materials and the nature and characteristics of the mineral horizons. If the soils contain sulfidic materials, they are classified as Sulfaquents. However, if sulfidic materials are absent and the soils are loamy and have a high n-value, they are classified as Hydraquents, while those that are predominantly sandy are classified as Psammaquents.

In submerged upland tidal marshes, if the recently accreted organic sediments are less than 40 cm thick, the soils will be classified based on the nature of the submerged soil. These submerged soils contain properties both inherited from the previous pedogenic environment and properties acquired from the present marsh environment. Gardner et al. (1992) have reported marshes forming over Spodosols on the South Carolina coastal plain, and soils with argillic (Bt) horizons have been reported under marshes in Chesapeake Bay (Stolt and Rabenhorst 1991, Rabenhorst 1997). In Dorchester County, MD, the Sunken soil series was established to accommodate tidal marsh soils with thin organic horizons (<20 cm) overlying soils which were previously Aqults (Brewer et al. 1998). The effect of brackish water diffusing into these soils following tidal submergence has resulted in their being changed into Alfisols, and they are classified as Endoaqualfs (Figure 13.6).

Because of the alluvial nature of tidal marsh soils, it is not uncommon for organic horizons or lenses to be interspersed within a mineral soil. In some classes of Aquents, where a buried organic horizon at least 20 cm thick beginning within the upper meter of the soil is recognized, it is accommodated in a Thapto–Histic subgroup.

Identification of Hydric Soils in Tidal Wetlands

The identification of hydric soils in fringing and especially tidal wetlands is probably among the easiest determinations to make for several reasons. First, they are geomorphologically constrained to locations essentially at sea level. Second, because of the nature of the tidal hydrology, the water table is basically permanent. Thus, unlike many hydric soils which have seasonally high water tables which drop significantly during certain times of the year, soil water tables in tidal wetlands can be readily observed at or near the soil surface any time of the year. Because these soils have peraquic moisture regimes, and because tidal water is often brackish or saline, there is commonly a distinctive vegetative community of obligate hydrophytes or halophytes which occupy

Figure 13.6 Profile of a soil in the *Honga* series, (loamy, mixed, euic, mesic Terric Sulfihemists) from Dorchester County, MD. This soil was probably a Typic Endoaquult before a gradual rise in sea level caused the accumulation of organic materials at the surface, which now are greater than 40 cm in thickness. The mineral subsoil contains a relict argillic (Btg) horizon that formed under a different pedogenic regime.

the hydric soils of tidal wetlands. While none of these characteristics is a soil morphological characteristic, they are nevertheless diagnostic for wetland identification. In addition, however, there are numerous soil morphological features that indicate the presence of hydric soils.

As was discussed above, many tidal marsh soils are Histosols, which alone is a diagnostic field indicator (field indicator A1; Hurt et al. 1998) in all land resource regions (LRRs). In all states but Alaska, the presence of a histic epipedon alone is an accepted field indicator that the soil is hydric (field indicator A2). There are also several field indicators tied to the occurrence of relatively thin layers of muck at the soil surface (field indicators A8, A9, A10, S2, S3). For any brackish or saline tidal marsh, the presence of the aroma of hydrogen sulfide gas not only indicates that the necessary conditions for sulfidization were met, but also that the soil is considered to be hydric (field indicator A4). In the case of Hydraquents or Psammaquents, in tidal marshes which lack histic epipedons or sulfidic materials, they will essentially all meet one or more of the indicators related to low chroma matrix colors (gleyed or depleted) in the upper portion of the soil. These would include field indicators S4, F2, and F3. The soils of tidal marshes are generally so clearly hydric that a given soil will often demonstrate numerous field indicators.

In some areas with very gently sloping landscapes, such as in areas of submerged upland marshes, there may be transitional zones grading from tidal wetlands to nontidal fringing wetlands. There may be areas of nontidal hydric soils within a meter or so of mean high water which are only occasionally inundated by storm tides (Rabenhorst 1997). These soils would need to be identified based on the field indicators used in the general vicinity, which are applicable to nontidal wetlands.

REFERENCES

Adam, P. 1990. *Saltmarsh Ecology.* Cambridge Univ. Press., Cambridge. 461 pp.

Baxter, J. 1973. Morphological, physical, chemical, and mineralogical characteristics of some tidal marsh soils in the Patuxent Estuary. M.S. thesis. Dept. of Agronomy, University of Maryland, College Park.

Breeding, C. H. J., F. D. Richardson, and S. A. L. Pilgrim. 1974. *Soil Survey of New Hampshire Tidal Marshes.* NH Agric. Exp. Sta. and Univ. of NH, and USDA–SCS. Research Report No. 40. Durham, NH.

Brewer, J. E., G. P. Demas, and D. Holbrook. 1998. *Soil Survey of Dorchester County, Maryland.* USDA–NRCS. In Coop. with MD. Agric. Exp. Sta., MD Dept. Agric., and Dorchester Soil Conserv. Dist., U.S. Govt. Printing Office. Washington, DC. 178 pp. 31 maps.

Chabreck, R. H. 1972. *Vegetation, Water and Soil Characteristics of the Louisiana Coastal Region.* LA State Univ. and Agric. Exp. Sta. Bull. 664.

Darmody, R. G. 1975. Reconnaissance survey of the tidal marsh soils of Maryland. M.S. thesis, Dept. of Agronomy, University of Maryland, College Park.

Darmody R. G. and J. E. Foss, 1979. Soil–landscape relationships of the tidal marshes of Maryland. *Soil. Sci. Soc. Amer. J.* 43:534–541.

Feijtel, T. C., R. D. DeLaune, and W. H. Patrick, Jr. 1988. Seasonal pore-water dynamics in marshes of Barataria Basin, Louisiana. *Soil Sci. Soc. Am. J.* 52:59–67.

Gardner, L. R., B. R. Smith, and W. K. Michener. 1992. Soil evolution along a forest–salt marsh transect under a regime of slowly rising sea level, southeastern United States. *Geoderma* 55:141–157.

Goldhaber, M. B. and I. R. Kaplan. 1982. Controls and consequences of sulfate reduction rates in recent marine sediments. pp. 19–36. *In* J. A. Kittrick, D. S. Fanning, and L. R. Hossner (Eds.) *Acid Sulfate Weathering.* SSSA Spec. Pub. 10. Amer. Soc. Agron., Madison, WI.

Griffin, T. M. and M. C. Rabenhorst. 1989. Processes and rates of pedogenesis in some Maryland tidal marsh soils. *Soil Sci. Soc. Am. J.* 53:862–870.

Griffin, T. M., M. C. Rabenhorst, and D. S. Fanning. 1989. Iron and trace metals in tidal marsh soils of the Chesapeake Bay. *Soil Sci. Soc. Am. J.* 53:1010–1019.

Haering, K. C. 1986. Sulfur distribution and partitionment in Chesapeake Bay tidal marsh soils. MS thesis, Univ. of Maryland, College Park, 172pp.

Haering, K. C., M. C. Rabenhorst, and D. S. Fanning. 1989. Sulfur speciation in Chesapeake Bay Tidal Marsh Soils. *Soil Sci. Soc. Am. J.* 53:500–505.

Hurt, G. W., P. M. Whited, and R. F. Pringle (Eds.). 1998. *Field Indicators of Hydric Soils in the United States: A Guide for Identifying and Delineating Hydric Soils,* Version 4.0, USDA, NRCS, Fort Worth, TX.

Krairapanond, N., R. D. DeLaune, and W. H. Patrick, Jr. 1991. Seasonal distribution of sulfur fractions in Louisiana salt marsh soil. *Estuaries* 14(1):17–28.

Lindau, C. W. and L. R. Hossner. 1982. Sediment fractionation of Cu, Ni, Zn, Cr, Mn, and Fe in one experimental and three natural marshes. *J. Environ. Qual.* 11:540–545.

Long, S. P. and C. F. Mason. 1983. *Saltmarsh Ecology.* Blackie Pub. Glasgow and London. 160 pp.

Miehlich, G. 1986. Freshwater-marsh of the Elbe river. pp. 99–128. In *Guidebook for a Tour of Landscapes, Soils and Land Use in the Federal Republic of Germany.* 13th Congress International Soc. of Soil Science, Hamburg, Germany. Mittielungen der Deutschen Bodenkundlichen Gesellschaft. Band 51. ISSN-0343-107X.

Oenema, O. 1988. Early diagenesis in recent fine-grained sediments in the Eastern Scheldt. Ph.D. dissertation. Instituut voor Aardwetenschappen, Rijksuniversiteit Utrecht, Budapestlaan 4, 3508 TA Utrecht, Nederland. 222 pp.

Rabenhorst, M. C. 1990. Micromorphology of induced iron sulfide formation in a Chesapeake Bay (USA) Tidal Marsh. pp. 303–310. *In* L. A. Douglas (Ed.) *Micromorphology: A Basic and Applied Science.* Proceedings of the 8th Int. Working Meeting on Soil Micromorphology, July 10–15, 1988, San Antonio, TX. Elsevier, Amsterdam.

Rabenhorst, M. C. 1997. The chrono-continuum: an approach to modeling pedogenesis in marsh soils along transgressive coastlines. *Soil Science* 167:2–9.

Rabenhorst, M. C. and B. R. James. 1992. Iron sulfidization in tidal marsh soils, pp. 203–217. *In* R. W. Fitzpatrick and H. C. W. Skinner (Eds.) *Biomineralization Processes of Iron and Manganese in Modern and Ancient Environments.* Catena Supplement No. 21, West Germany.

Ranwell, D. S. 1972. *Ecology of Salt Marshes and Sand Dunes.* Chapman and Hall Pub. London. 258 pp.

Redfield, A. C. 1972. Development of a New England salt marsh. *Ecol. Monogr.* 42:201–237.

Soil Survey Staff. 1975. *Soil Taxonomy: A Basic System of Soil Classification for Making and Interpreting Soil Surveys.* USDA Agric. Handbook No. 436, U.S. Govt. Printing Office. Washington, DC.

Soil Survey Staff. 1998. *Keys to Soil Taxonomy.* 8th Ed. USDA–NRCS. U.S. Govt. Printing Office. Washington, DC. 326 pp.

Steers, J. A. 1977. Physiography. pp. 31–60 *In* V. J. Chapman (Ed.) *Ecosystems of the World. 1. Wet Coastal Ecosystems.* Elsevier Pub. Amsterdam. 428 pp.

Stevenson, J. C., M. S. Kearney, and E. C. Pendleton. 1985. Sedimentation and erosion in a Chesapeake Bay brackish marsh system. *Marine Geol.* 67:213–235.

Stevenson, J. C., L. G. Ward, and M. S. Kearney. 1986. Vertical accretion rates in marshes with varying rates of sea-level rise. pp. 241–259. *In* D. A. Wolfe (Ed.) *Estuarine Variability.* Academic Press, New York.

Stolt, M. H. and M. C. Rabenhorst. 1991. Micromorphology of argillic horizons in an upland/tidal marsh catena. *Soil Sci. Soc. Am. J.* 55:443–450.

Widdell, F. 1988. Microbiology and ecology of sulfate- and sulfur-reducing bacteria. pp. 469–585. *In* A.J.B. Zehnder (Ed.) *Biology of Anaerobic Microorganisms.* John Wiley & Sons, New York.

CHAPTER 14

Flatwoods and Associated Landforms of the South Atlantic and Gulf Coastal Lowlands

Frank C. Watts, V.W. Carlisle, and G.W. Hurt

INTRODUCTION

A landscape is a collection of related natural landforms. It is usually the land surface that the eye can comprehend in a single view. Most landscapes contain many unique and readily identifiable landforms. A landform is a physical, recognizable form or feature on the earth's surface that has a characteristic shape and is produced by natural causes (Tuttle 1975, Soil Survey Staff 1996a). This chapter will discuss the soils, hydrology, and related features of flatwoods and other landforms as they occur on the south Atlantic and Gulf Coastal Lowlands landscape (Figure 14.1). Other areas of flatwoods occur in the United States, most notably in southwestern Louisiana and areas parallel to and south and west of the Mississippi and Alabama Blackland Prairies; these will not be discussed.

The term "flatwoods" was coined by Europeans who settled in the southeastern United States to designate the flat areas that support forests of pine (Ober 1954, Abrahamson and Hartnett 1990). Flatwoods has long been used to designate a landform and a landscape (Caldwell et at. 1958, Watts et al. 1996). This chapter will refer to flatwoods as landforms as they occur on the south Atlantic and Gulf Coastal Lowland landscapes where they are interspersed with other landforms, such as depressions, flood plains, flats, and rises and knolls (Watts and Carlisle 1997). Flatwoods and associated landforms occur in approximately 50% of Florida (Edmisten 1963, Davis 1967, Abrahamson and Hartnett 1990) and in southeastern Georgia. These landforms comprise relatively smaller portions of the landscapes as they extend northward to Virginia and westward to Louisiana. Subtle differences in local relief and somewhat impervious, geologic strata have primarily influenced the evolution of these different landforms. Flatwoods and associated landforms generally have elevations that are less than 100 m above sea level. The climate is characterized by long, humid, warm summers and mild winters with annual rainfall of about 1000 to 1650 mm and average annual temperatures of 13 to 25°C (Soil Conservation Service 1981).

SOILS

The soils of the area classify in seven orders and 11 suborders. Below is a discussion of each order and suborder. This discussion is abbreviated from *Soil Taxonomy: A Basic System of Soil*

Figure 14.1   Major extent of flatwoods and associated landforms of the south Atlantic and Gulf coastal lowlands landscape.

Classification for Making and Interpreting Soils Surveys (Soil Survey Staff 1999) and *Keys to Soil Taxonomy: Seventh Edition* (Soil Survey Staff 1996b).

Histosols

Histosols are soils that consist of organic materials (more than 12 to 18% organic carbon) in at least two thirds of the thickness above bedrock and mineral soil layers are less than 10 cm thick *or* are saturated for most of the year and half of the upper 80 cm of soil is organic material. Histosols, as they occur in these landscapes, have organic soil material over mineral soil material or bedrock at varying depths. Sapric is the type of organic soil material that occurs in the area. Sapric soil material (muck texture) is organic material in which most of the plant remains have decomposed such that plant forms cannot be identified (<1/6 fibers after rubbing). Saprists are the Histosols (*ist* is the formative word element from Histosol that appears in the suborder name) that have more sapric material than other, less decomposed organic (hemic and fibric) material. These soils are wet most of the year unless artificially drained.

Spodosols

Spodosols are mineral soils that contain a spodic horizon 10 cm or more thick. A spodic horizon is a subsurface horizon, usually black to dark reddish brown, in which organic material has accumulated in combination with aluminum and iron due to downward translocation. Depths to spodic horizons vary from <25 cm to 2 m, and thicknesses vary from 10 cm to >1 m; some Spodosols have more than one spodic horizon (Figure 14.2). Bedrock or an argillic horizon (a zone of clay accumulation) may occur at varying depths beneath the spodic horizon (Figure 14.3). Aquods are the Spodosols (*od* is the formative word element from Spodosol that appears in the suborder name) that are wet for extended periods of most years unless they have been artificially drained to reduce the duration of saturation. Orthods are Spodosols that are drier than Aquods.

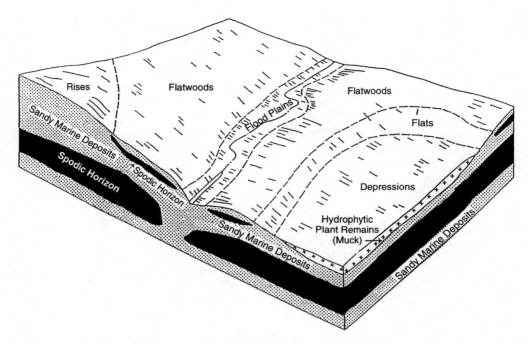

Figure 14.2 Common landscape of flatwoods and associated landforms in the northern and western ranges of their occurrence where the flatwoods landform is dominant. This landform pattern is repetitive across the landscape with individual landforms of varying extensiveness. Some landforms may not be present in all landscapes.

Ultisols

Ultisols are other mineral soils that have an argillic or kandic horizon and a base saturation of less than 35%. Argillic and kandic horizons are subsurface zones of clay accumulations from the horizon(s) above. Base saturation is the ratio of chemical exchange sites on soil particles occupied by base cations (Ca^{2+}, Mg^{2+}, K^+, Na^+) to acid cations (H_3O^{2+}, Al^{3+}) times 100. Depths to argillic or kandic horizons vary from <25 cm to 2 m, and thicknesses vary from <20 cm to >2 m. Bedrock or other material may occur at varying depths beneath the argillic horizon. Base saturation is less than 35%. Ultisols have a lower base saturation status because enough time has passed to make them more highly weathered than Alfisols (see below). Aquults are the Ultisols (*ult* is the formative element) that are wet for extended periods of most years unless they have been artificially drained to reduce the duration of saturation. Udults are the Ultisols that are drier than Aquults; however, the ability of Udults to retain plant-available water varies widely with depth to and thickness of the argillic or kandic horizon.

Mollisols

Mollisols are other mineral soils that have a dark (usually black to very dark gray) mineral surface horizon that is more than 25 cm thick (10 cm is underlain by bedrock) and that has a base saturation of 50% or more. Limestone bedrock and argillic horizons may be present or absent in these soils. Aquolls are the Mollisols (*oll* is the formative element) that are wet for extended periods of most years unless they have been artificially drained to reduce the duration of saturation.

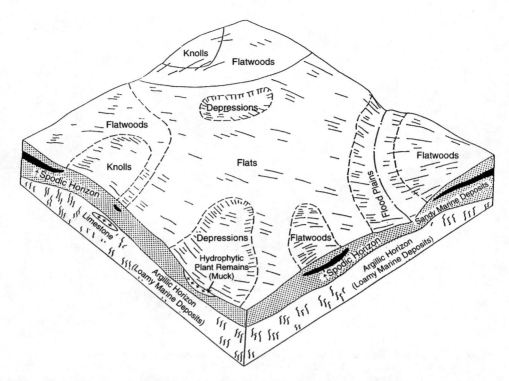

Figure 14.3   Common landscape of flatwoods and associated landforms in the southern range of their occurrence where the flats landform is dominant. Similar to the northern and western ranges of occurrence the landform pattern is repetitive across the landscape with individual landforms of varying extensiveness. Some landforms may not be present in all landscapes.

Alfisols

Alfisols are other mineral soils that have an argillic or kandic horizon and a base saturation of 35% or more. Argillic and kandic horizons are subsurface zones of clay accumulations from horizon(s) above. Depths to argillic or kandic horizons vary from <25 cm to 2 m, and thicknesses vary from <20 cm to >2 m. Bedrock or other material may occur at varying depths beneath the argillic horizon. The higher base saturation status of Alfisols is due to the fact that not enough time has passed to make them as highly weathered as Ultisols. Aqualfs are the Alfisols (*alf* is the formative element) that are wet for extended periods of most years unless they have been artificially drained to reduce the duration of saturation. Udalfs are the Alfisols that are drier than Aqualfs; however, the ability of Udalfs to retain plant-available water varies widely with depth to and thickness of the argillic or kandic horizon.

Inceptisols

Inceptisols are other mineral soils that have horizon development exemplified by color and/or structure. They have had some horizon development and parent material differentiation but not enough to class as Spodosols, Ultisols, or other soil orders already described. The limited expression of profile development is most often expressed by differentiating color and structure changes. Inceptisols are most often found on flood plains and marine terraces. Aquepts are the Inceptisols (*ept* is the formative element) that are wet for extended periods of most years unless they have been artificially drained to reduce the duration of saturation.

Entisols

Entisols are all other mineral soils. These soils lack significant profile development to classify in any of the soil orders described above. Entisols are essentially unaltered parent material. Aquents are the Entisols (*ent* is the formative element) that are wet for extended periods of most years unless they have been artificially drained to reduce the duration of saturation. Psamments are the sandy Entisols. These are the driest soils in the area. They generally have sandy layers to 2 m or more. Both mineral soils and the limnic marl soils (Soil Survey Staff 1999) are classed as Aquents. The marl soils are properly classed in the suborder Aquents and should be classed in the proposed Great Group Limnaquents (Ahrens and Hurt 1999). Limnaquents, as proposed, would include soils comprised of mineral materials composed of marl, coprogenous earth, and diatomaceous earth.

HYDRIC SOILS

Hydric soils are defined as soils that formed under conditions of saturation or inundation (flooding or ponding) for periods long enough during the growing season to develop anaerobic conditions in the upper part of the soil (*Federal Register* 1994). Saturation is characterized by zero or positive pressure in the soil water with most of the soil pores filled with water. Inundation is characterized by a water table above the soil surface (Soil Survey Staff 1996b). Hydric soils have seasonal high saturation and/or inundation for a significant period (more than a few weeks) during the wettest period of the year (Figures 14.4 and 14.5). In the following section on landforms, hydric soils, as they occur on each landform, are identified. These hydric soils have one of the indicators identified in *Field Indicators of Hydric Soils in the United States* (Hurt et al. 1996) and *Soil and Water Relationships of Florida's Ecological Communities* (Hurt and Henderson 1997). Figure 14.6 is a graphic representation of the hydric soils that occur on the wetland landforms. Where the soils are non-hydric, they do not have one of the hydric soil indicators. Figure 14.7 is a graphic representation of the non-hydric soils that occur on the upland landforms.

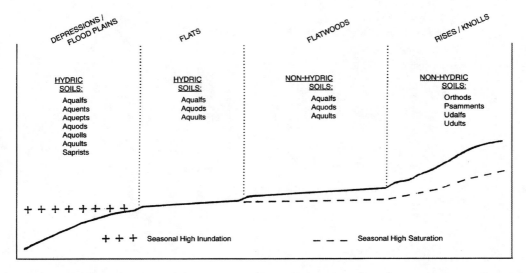

Figure 14.4 A common soil toposequence of flatwoods and associated landforms. Depth of seasonal high inundation and depth to seasonal high saturation are shown. Some landforms may not be present in all landscapes.

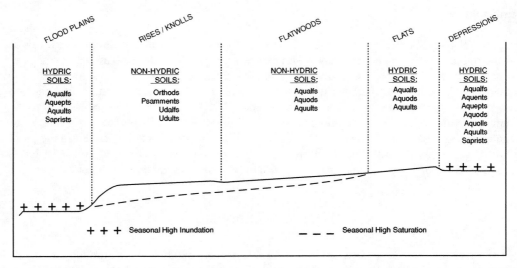

Figure 14.5   Most common soil toposequence of flatwoods and associated landforms. Depth of seasonal high inundation and depth to seasonal high saturation are shown. Some landforms may not be present in all landscapes.

HYDRIC SOILS

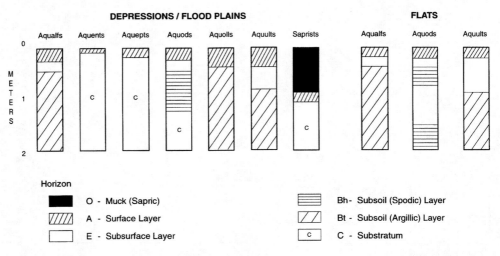

Figure 14.6   Idealized pedons that represent the hydric components of the soils associated with the flatwoods and associated landforms of the Atlantic and Gulf coastal lowlands landscape. These soils occur on flats and flood plains, and in depressions.

NON-HYDRIC SOILS

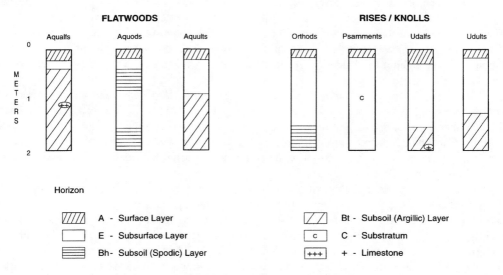

Figure 14.7   Idealized pedons that represent the non-hydric components of the soils associated with the flatwoods and associated landforms of the Atlantic and Gulf coastal lowlands. These soils occur on flatwoods, rises, and knolls.

LANDFORMS

Flatwoods

Flatwoods landforms (Figures 14.2 and 14.3) are typically broad, nearly level, flat areas with slightly convex and concave relief. The convex relief exists where flatwoods abut flats and depressions. The concave relief exists where flatwoods abut rises and knolls. Flatwoods are 8 cm or more higher than adjacent flats, 30 cm or more higher than adjacent depressions and flood plains, and 15 cm or more lower than adjacent rises and knolls. The only higher closely associated landforms in the coastal lowlands are rises and knolls.

Characteristic native vegetation is slash pine (*Pinus elliottii*), longleaf pine (*Pinus palustris*), loblolly pine (*Pinus taeda*), saw palmetto (*Serenoa repens*), gallberry (*Ilex glabra*), wax myrtle (*Myrica cerifera*), chalky bluestem (*Andropogon capillipes*), broomsedge bluestem (*Andropogon virginicus*), and pineland threeawn (*Aristida stricta*). This vegetation normally appears as a shrub-dominated community with an open canopy of pine trees and a sparse herbaceous layer.

The soils that occur on flatwoods characteristically are poorly drained, acid, have a low cation exchange capacity, and a low to medium moisture-holding capacity. Flatwoods soils are predominantly not hydric and have seasonal high saturation at a depth of 15 to 45 cm below the soil surface, although, during periods of high rainfall, they may be episaturated for more than a few days. Aquods, Aquults, and Aqualfs (Figures 14.4 and 14.5) are the most common soils. Aquods are dominant, and often have more than one spodic layer (Figure 14.2) or are underlain by an argillic horizon (Figure 14.3). The nearly level flat to slightly convex relief of flatwoods allows the landform to be readily identified even where not vegetated.

During wet seasons, flatwoods transmit water to adjacent, lower-lying flats and depressions through subsurface flow and overland flow and some water to the underlying aquifer through deep seepage. During dry seasons they receive water from the flats and depressions (Crownover et al. 1995).

Depressions

Depressional landforms (Figures 14.2 and 14.3) are typically sunken, lower parts of the earth's surface; they have concave relief, and do not have natural outlets for surface drainage. They are 30 cm or more lower than adjacent landforms. Depressions commonly occur at the lower elevations of a soil toposequence (Figure 14.4). Most commonly, however, depressions of various sizes are interspersed throughout the landscape (Figure 14.3) as low-lying areas with frequent seasonal high inundation (Figure 14.5).

Characteristic native vegetation is baldcypress (*Taxodium distichum*) and water tolerant hardwoods. Some depressions are treeless expanses of grasses, sedges, rushes, and other herbaceous plants. Other depressions are dominated by shrubs. Locally, depressions have many names: pocosin in North Carolina, Carolina bay in South Carolina, and freshwater marsh and sawgrass marsh in Florida. Cypress dome is also a common name for depressional landforms throughout the coastal lowland landscapes.

The soils in depressions characteristically are very poorly drained, acid, have a low to high cation exchange capacity, a medium to high moisture-holding capacity, and are saturated or inundated much of the year. These soils are predominantly hydric. Surface soil horizons are frequently slightly higher in organic matter content as compared with upland soils of similar suborders. Aqualfs, Aquents, Aquepts, Aquods, Aquolls, Aquults, and Saprists are the most common soils. The strongly concave relief and lack of natural outlets allow this landform to be readily identified even where it is not vegetated.

Groundwater flow generally follows the broad elevational gradients of the surface (Crownover et al. 1995). During wet seasons, depressions store water as soil water but mostly as surface water. They transmit some water to the underlying aquifer through deep seepage. During dry seasons they transmit water back to the surrounding areas of higher-lying flats and flatwoods. This phenomenon is due to the higher evapotranspiration potentials of the flats and flatwoods, and the amount of water is dependent upon the elevational gradients and is also known to occur in other landscapes of the United States where this landform occurs (Lide et al. 1995).

Depressions are important landforms in the landscapes of the south Atlantic and Gulf Coastal Lowlands. They provide habitat for a large diversity of plants and animals; however, because they frequently occur as small areas dispersed throughout the landscape, many have been drained or partially drained prior to being converted to agriculture and timber production. Where undisturbed, depressions filter pollutants from the surrounding higher-lying landforms.

Flood Plains

Flood plains (Figures 14.2 and 14.3) are constructional landforms built from sediments deposited during overflow and lateral migration of drainageways. Similarly to depressions, flood plains are 30 cm or more lower than adjacent landforms. They also occur at the lower elevations of most soil toposequences (Figures 14.4 and 14.5); however, flood plains, unlike depressions, have natural outlets and have nearly level to concave relief with slightly elevated natural levees adjacent to drainageways.

Characteristic native vegetation is a wide and diverse variety of water-tolerant deciduous hardwoods. A few of the flood plains in these landscapes are treeless expanses of grasses, sedges, rushes, and other herbaceous plants. Some are dominated by baldcypress. Casual observers of flood plains that occur on these coastal lowlands landscapes normally consider them "swamps."

The soils on flood plains characteristically are poorly to very poorly drained, acid to neutral, have a medium to high cation exchange capacity, a medium to high moisture-holding capacity, and are saturated or inundated much of the year. Unlike flood plains of many other landscapes, these soils are predominantly hydric. Surface soil horizons are frequently slightly higher in organic matter content as compared to upland soils of similar suborders. Aqualfs, Aquents, Aquepts,

Aquods, Aquolls, Aquults, and Saprists are the most common soils. The nearly level, slightly concave relief and presence of natural outlets allows this landform to be readily identified even where it is not vegetated.

During wet seasons flood plains store water as soil water but mostly as surface water (Figures 14.4 and 14.5). They transmit water to lower elevations on the flood plain, eventually discharging at sea level. Some water is contributed to the underlying aquifer through deep seepage. During dry seasons, like depressions, flood plains transmit water back to the surrounding areas of higher-lying landforms. This is due to the higher evaporatranspiration potentials of the other landforms, and the amount of water is dependent upon the elevational gradients.

Flood plains are mostly undisturbed landforms in the landscapes of the south Atlantic and Gulf Coastal Lowlands. They provide habitat for diverse plants and animals and provide flood protection for the adjacent higher landforms where they are left undisturbed. Flood plains also filter pollutants from the surrounding higher-lying landforms. Where disturbed, increased flooding of adjacent landforms often results, and pollution reduction is lessened.

Flats

Flats landforms are typically smooth, lack any significant curvature or slope, and evidence little change in elevation, with poorly defined outlets. They are 8 to 30 cm lower in elevation than adjacent higher flatwoods landforms. Flats commonly comprise relatively insignificant portions of the coastal lowland landscapes (Figure 14.2). They may, however, in the southernmost range of their extent, dominate the landscape (Figure 14.3). Flats have nearly level, slightly concave to flat relief. Convex relief exists where flats abut flatwoods and other non-hydric landforms.

Characteristic native vegetation is mixed hardwood and pine with a dense understory of shrubs and saplings in its northern range of occurrence and, in its southern range of occurrence, an open canopy of pine and understory of grasses and/or herbs devoid of shrubs. Flats are commonly known as sloughs in south Florida and as swamps, bayheads, and shrub and pitcher plant bogs in other areas.

Soils on flats characteristically are poorly drained, acid to neutral, have a low to medium cation exchange capacity, and a low to medium moisture-holding capacity. They are predominantly hydric and have seasonal high saturation at a depth of less than 15 cm below the soil surface (Figures 14.4 and 14.5). During periods of high rainfall they often have shallow (less than a few cm) inundation for more than a few days. Aquods, Aquults, and Aqualfs are the most common soils. Surface soil horizons are frequently slightly higher in organic carbon content when contrasted to upland soils of similar suborders in the northern range of their occurrence on coastal lowland landscapes. Due to differential biomass production, the converse is true in the southern range of occurrence.

Flats are the most difficult of the flatwoods and associated landforms for untrained observers to identify because the relief differences between the adjacent higher landforms are subtle. Flats are especially difficult to recognize in their northern range of occurrence because vegetation is usually dense. To the trained observer, the nearly level, concave to flat relief and poorly defined outlets are characteristic and observable. Vegetation can also provide a clue for separation in heavily vegetated areas. Flatwoods have the shrub saw palmetto that disappears at the flats landform break to be replaced by other shrubs in its northern range of occupancy and by pineland threeawn grass in its southern range of occupancy.

Groundwater flow generally follows surface elevation gradients. During wet seasons, flats transmit water to adjacent lower-lying depressions via surface and subsurface flow. They transmit water to nonadjacent depressions by lateral flow through the subsurface of the flatwoods soils.

Rises and Knolls

Rises and knolls, frequently called ridges, have convex relief. A rise is an imprecise term for a landform that has a broad summit and gently sloping sides, and a knoll is a landform that occurs as a small, low, rounded, isolated area rising above the lower landforms (Soil Survey Staff 1996a). Rises and knolls are typically 15 cm or more higher in elevation than surrounding wetter landforms.

Characteristic native vegetation is mixed mesic hardwoods and pines forest. This vegetation normally appears as a forest-dominated community with a closed canopy of trees and sparse shrub and herbaceous layers.

The soils that occur on rises characteristically are somewhat poorly drained to moderately well drained, acid, have a low to medium cation exchange capacity, and a low to medium moisture-holding capacity. Soils on rises and knolls are non-hydric and have seasonal high saturation at depths of greater than 45 cm from the soil surface. Orthods, Psamments, Udualfs, and Udults are the most common soils. The nearly level to gently sloping slightly convex relief of rises and knolls allows the landforms to be readily identified even where not vegetated.

Rises and knolls are the most hydrologically isolated of these landforms. They, during wet seasons, contribute water about equally to adjacent lower landforms through lateral flow and to the underlying aquifer through deep seepage. They also contribute some water as overland flow to adjacent lower landforms during high rainfall events. During dry seasons they neither transmit to nor receive from adjacent landforms.

SUMMARY

Soils of the flatwoods and associated landforms of the south Atlantic and Gulf Coastal Lowlands landscape classify into the following orders: Histosols, Spodosols, Ultisols, Mollisols, Alfisols, Inceptisols, and Entisols. Also recognized are 11 suborders: Aqualfs, Aquents, Aquepts, Aquods, Aquolls, Aquults, Orthods, Psamments, Saprists, Udalfs, and Ulults. Soils of the flatwoods are dominantly Aquods. All other associated landforms lack a dominant soil suborder.

The south Atlantic and Gulf Coastal Lowland landscapes commonly have the following landforms: flatwoods, depressions, flood plains, flats, and rises and knolls. Depressions, flood plains, and flats most commonly have hydric soils. Depressions and flood plains primarily function as discharge wetlands, and flats function as flow through/discharge areas to other wetlands. Flatwoods, rises, and knolls characteristically have non-hydric soils. These landforms function as recharge areas to wetlands and to underlying aquifers.

Each of the flatwoods and associated landforms of the Atlantic and Gulf Coastal Lowland landscapes has characteristic shape and relief that make them readily identifiable with or without the presence of vegetation.

REFERENCES

Abrahamson, W.G. and D.C. Hartenett. 1990. Pine flatwoods and dry prairies. pp. 103–149. *In* Myers, R.L. and J.J. Ewel (Eds.). *Ecosystems of Florida.* University of Central Florida Press, Orlando, FL.

Ahrens, R.J. and G.W. Hurt. 1999. Proposed Limnaquents Great Group, USDA, NRCS, Lincoln, NE. *Soil Survey Horizons* 40(1):13–16.

Caldwell, R.E., O.C. Olson, J.B. Cromrite, and R.G. Leighty. 1958. *Soil Survey of Manatee County, Florida.* USDA, Soil Conservation Service, U.S. Govt. Printing Office. Washington, DC.

Crownover, S.H., N.B. Comerford, D.G. Neary, and J. Montgomery. 1995. Horizontal groundwater flow patterns through a cypress swamp — pine flatwoods landscape. *Soil Sci. Soc. Am. J.* 59:1199–1206.

Davis, J.H., Jr. 1967. General map of the natural vegetation of Florida. Inst. Food Agric. Exp. Stn., Circ. S-178, Univ. Florida, Gainesville.

Edmisten, J.E. 1963. *The Ecology of the Florida Pine Flatwoods.* Ph.D. thesis, University of Florida, Gainesville.

Federal Register. 1994. Changes in hydric soils of the United States. Washington, DC, July 13.

Hurt, G.W., P.M. Whited, and R.F. Pringle. 1996. *Field Indicators of Hydric Soils in the United States*. USDA, NRCS, Fort Worth, TX.

Hurt, G.W. and W.G. Henderson. 1997. *Soil and Water Relationships of Florida's Ecological Communities*. USDA, NRCS, Florida Soil Survey Staff, Gainesville, FL.

Lide, R.F., V.G. Meentemeyer, J.E. Pinder III, L.M. Beatty. 1995. Hydrology of a Carolina bay located on the upper coastal plain of western South Carolina. *Wetlands* 15:47–51.

Ober, L.D. 1954. *Plant Communities of the Flatwood Forests in Austin Cary Memorial Forest*. M.S. thesis, University of Florida, Gainesville.

Soil Conservation Service. 1981. *Land Resource Regions and Major Land Resource Areas of the United States*. USDA, SCS, Agriculture 296. U.S. Govt. Printing Office. Washington, DC.

Soil Survey Staff. 1999. *Soil Taxonomy: A Basic System of Soil Classification for Making and Interpreting Soil Surveys*. 2nd ed. USDA Agric. Handbook 436. U.S. Govt. Printing Office. Washington, DC.

Soil Survey Staff. 1996a. *National Soil Survey Handbook*. USDA, NRCS, U.S. Govt. Printing Office. Washington, DC.

Soil Survey Staff. 1996b. *Keys to Soil Taxonomy*. 7th edition. USDA, NRCS, U.S. Govt. Printing Office. Washington, DC.

Tuttle, S.D. 1975. *Landforms and Landscapes*. 1975. Wm.C. Brown Company, Dubuque, IA.

Watts, F.C. and V.W. Carlisle. 1997. Soils of flatwoods and associated landforms of Florida. Soil and Crop Sciences of Florida, *Proceedings*, 57:59–65.

Watts, F.C., T.S. Bowerman, R.A. Casteel, D. Hinz, A. Jenkins, J.C. Remley, T.J. Solem, A. Younk, D. Vyain, P.E. Ayers, and K.C. Bracy. 1996. *Soil Survey of Baker County, Florida*. USDA, NRCS, U.S. Govt. Printing Office. Washington, DC.

PART III

Wetland Soils with Special Conditions

CHAPTER 15

Hydrologically Linked Spodosol Formation in the Southeastern United States

Willie Harris

INTRODUCTION

Coastal flatwoods of the southeastern U.S. are consistently occupied by Spodosols (Soil Survey Staff 1996), soils distinguished by subsoil accumulations of C and associated metals. A significant amount of C is stored in these Spodosols, considering the extent of their occurrence and the thickness of the zones of accumulation (Daniels et al. 1975, Stone et al. 1993). The mechanisms responsible for such large-scale C accumulation are pertinent to global C dynamics but are not well understood. Spodosols occur on excessively to poorly drained landscapes in many regions (Nichols et al. 1990), particularly in cooler climates. However, southern coastal Spodosols are mysteriously found mainly within relatively narrow ranges of hydrological conditions that prevail for flatwoods landforms.

The flatwoods–Spodosol connection apparently relates to the presence of fluctuating water tables. The reason for inferring a hydrological link is that, though they are most abundant on flatwoods landforms, Spodosols can be found wherever hydrological conditions resemble those of flatwoods. Furthermore, diagnostic horizons otherwise definitive of Spodosols are commonly found below the 2-m depth considered for soil classification, and within the zone of groundwater fluctuation.

This chapter summarizes the current state of knowledge regarding hydrology and morphology of flatwoods soils and presents possible explanations for the water table influence. Comprehension of relationships discussed in this chapter requires some familiarity with Spodosol genesis. Therefore, the next section provides a summary of "classical" concepts of Spodosol formation. Some specific characteristics of flatwoods Spodosols will then be presented and followed by a brief depiction of the hydrological setting of flatwoods. Finally, the hydrological scenarios associated with Spodosol formation are highlighted, along with possible mechanisms accounting for the water table influence.

CONCEPTS OF SPODOSOL FORMATION

The term "Spodosol" is an abstraction in the sense of being a class ("Soil Order") defined by detailed criteria set forth in the USDA soil taxonomic system (Soil Survey Staff 1996). The criteria are tailored to capture soils that have strongly expressed features indicative of illuvial noncrystalline organometal accumulations. Generally (with exceptions beyond the scope or intent of this chapter), the requirement is the presence of a diagnostic horizon called "spodic." Spodic horizons are dominated by "spodic materials," which are materials meeting minimal color and chemical requirements characteristic of organometal components. Obviously, not all horizons of organometal accumulations meet the taxonomic requirement of a spodic, and thus there are soils that resemble Spodosols but technically are not. For convenience, I use the term "Spodosol" without qualification, but the processes discussed also apply to "Spodosol-like" soils and to zones below 2-m depth that exhibit Spodosol-like morphology. The latter zones, as revealed in deep borings and excavations, are extensive and quite thick in some areas. They may constitute the bulk of subsurface C stored in the region.

There are some conditions and tendencies common to Spodosols (McKeague et al. 1983). They occur predominantly in sandy- to coarse-loamy textured materials under humid or perhumid climates. Elevated clay content apparently inhibits or precludes the chemical interaction of metals and C (Holzhey et al. 1975). High rainfall promotes (i) biological activity, (ii) generation of reactive organic by-products, (iii) acidification, (iv) release of metals through weathering, and (v) vertical leaching and pedological redistribution of soil components. Vegetation is less definitive, since Spodosols are overlain by a variety of plant species. However, acidifying trees and shrubs (conifers, heath, etc.) are commonly associated with Spodosols (Dalsgaard, 1990).

How does C accumulate within a specific soil zone to form a spodic horizon? A lot of thought and research effort has been devoted to this question. There is still considerable uncertainty regarding the details, and there are probably numerous "variations on the theme" from region to region. The three major components involved in the process are organic C, Al, and Fe. Also, evidence has been presented that inorganic short-range-order alumino-silicates are important components in the formation of some Spodosols (e.g., Farmer et al. 1980, Kodama and Wang 1989, Dahlgren and Ugolini 1991). Not all components are equally represented in spodic horizons. Iron, for example, is commonly very low in poorly drained Spodosols (Holzhey et al. 1975, Sodek et al. 1990). The following paragraph presents a hypothetical scenario of how components may interact in the formation of a spodic horizon. The scenario presumes a central role for C (which is appropriate for flatwoods Spodosols) and draws heavily from the ideas of DeConinck (1980).

Mobile organic substances are released to soil solutions via litter decomposition, root exudation, etc. Initially, these substances are relatively low in molecular weight, anionic, and highly reactive. They strongly attract and complex polyvalent metals such as Al and Fe (Wright and Schnitzer 1963), and hence promote the release and mobilization of these metals from the inorganic solid phases in soils. The organic substances remain mobile as long as they are soluble (or dispersed) and can move downward in the soil with associated metals "in tow." However, continued reaction with metals has two effects which ultimately promote the precipitation of the organometal complexes: (i) accumulating metals lower the negative charge by specifically adsorbing to the organic surfaces, and hence decrease the mutual repulsion of the complexes; and (ii) metals serve as bridges between complexes, linking them together and increasing molecular weight. Precipitation is a consequence of these aggregating effects of metals. Hence, we have a mechanism for the formation of spodic horizon that could be summed up by three concepts: complexation, mobilization, and precipitation.

The mechanism presented above entails a pedological redistribution of components within the soil. Specifically, there is a depletion of metals in one zone and an enrichment of metals-plus-C in a subjacent zone (Franzmeier et al. 1965). This redistribution is dramatically expressed in the morphology of many Spodosols. A white to light-gray E (albic) horizon is the zone of depletion (eluviation), and a darker-colored B (spodic) horizon, the zone of accumulation (illuviation). The spodic B is appended with appropriate subscripts to indicate a prevalence of C (Bh), metals (Bs), or both (Bhs). Flatwoods Spodosols have morphology and components generally consistent with formation via classical concepts. The next section elaborates on their specific characteristics.

FLATWOODS SPODOSOLS

Aquods comprise the dominant suborder of Spodosols in the coastal plain of the southeastern U.S. (Brown et al. 1990). They are distinguished by poor drainage and one or more spodic (usually Bh) horizons. Better-drained Spodosols — Orthods — also occur to a minor extent in the region. The processes and hydrological relations to be discussed for Aquods apply to Orthods as well, because for both suborders, the actual zone of albic/spodic formation is within the influence of a fluctuating water table. Some Orthods are only slightly better drained than Aquods (i.e., they are still imperfectly drained), and some are exceptional in that the water table and Bh horizon are deeper in the soil. Vegetation, as influenced by drainage, is sometimes useful in delineating Aquods and Orthods. However, both may occur under southern pine forest vegetation typical of flatwoods. Common flatwoods species include longleaf and slash pine (*Pinus palustrus* and *elliottii),* saw palmetto (*Seronoa repens),* gallberry (*Ilex gabra),* and various other herbaceous and grass-like plants (Abrahamson and Harnett 1990).

Probably the greatest concentration of Aquods in the world occurs in Florida (Collins 1990), and most of the information presented in the chapter was derived from Florida Aquods (Florida Cooperative Soil Survey Program data–Sodek et al. 1990, the last in a series of data publications). The information may be applicable to Aquods of other regions as well, particularly those of similar coastal landscapes. However, geographic distinctions have been noted. For example, Rourke et al. (1988) reported, from a study of 90 Aquods, that spodic horizons were significantly deeper for southern Aquods. Aquods are distributed worldwide, occurring in Africa, Asia, Australia, Europe, New Zealand, South America, and other regions of North America (Collins 1990).

Flatwoods Spodosols, like most Spodosols worldwide, are sandy. Even spodic horizons are predominantly quartz sand (>90%). However, many Aquod series have a loamy-textured argillic horizon below the spodic. Morphologically, the typical E-over-Bh sequence of flatwoods Aquods and Orthods is visually striking due to the contrasting colors of the two horizons. The boundary between the E and Bh can be abrupt or clear in distinctness. Topography of the boundary can range from nearly flat (smooth) to highly convoluted (irregular or broken).

The spodic materials of flatwoods are dominated by organic C (hence the "h" subscript in "Bh"), as determined by various selective dissolution procedures (Yuan 1966, Holzhey et al. 1975, Lee et al. 1987, 1988, Sodek et al. 1990). These materials also contain Al in amounts roughly 0.1 to 0.2 those of C, but Fe content is generally much lower. Aluminum and C are usually not uniformly distributed within the spodic but show inverse depth trends. The majority of upper spodic horizons (remember, flatwoods Spodosols commonly have a deeper spodic as well) show trends of a decrease in C content and increase in Al content with depth. These trends have also been reported for Spodosols of other regions as well (e.g., Peterson 1976, Higashi et al. 1981, McKeague et al. 1983, Birkeland 1984, Thompson et al. 1996). Exceptions to these trends are rare in shallow spodic horizons and could ultimately be explained by subtle morphological or hydrological anomalies. However, deeper spodic horizons within bisequel flatwoods Spodosols commonly show an increase in C and decrease in Al, with depth. These differences in trends between shallow and deep spodic horizons probably reflect differences in hydrological fluxes.

HYDROLOGIC SETTING OF FLATWOODS

Flatwoods and associated landforms (flood plains, depressions, flats, rises, and knolls) comprise a landscape system that appears essentially flat to the casual observer. However, subtle topographic changes correspond to distinct differences in hydrology, vegetation, and soils. High water tables are a consequence of very low relief in conjunction with a restriction on vertical water movement. The restriction is commonly a sparingly permeable geologic stratum (aquiclude) which perches

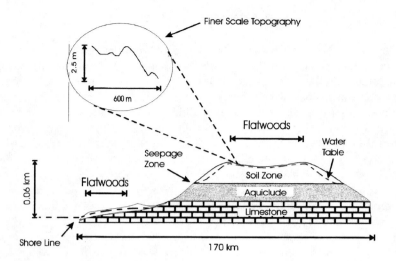

Figure 15.1   Schematic representation of common regional hydrological conditions for coastal flatwoods and associated landforms. Note the extreme vertical scale exaggeration necessary to intelligibly convey topographic, geological, and hydrological relations. The high water table results from the slow permeability of the fine-textured marine sediments that constitute an aquiclude for the inland flatwoods, and from low elevations above sea level for the coastal flatwoods. Flyout depicts finer scale of topographic features, also with high vertical exaggeration.

water and creates endoaquic soil conditions (Figure 15.1). Also, the ocean itself can impose direct base-level limits for landscapes occurring only slightly above sea level. Flatwoods, despite imperfect drainage, are highly leached omnotrophic landforms; in effect, water-transported nutrients come primarily from precipitation. They are the "uplands" in cases (which are common) where the local landscape does not include knolls and rises.

The surficial, unconfined aquifers that affect the soils of these landforms have an appreciable lateral-flow component, with hydraulic gradient and flow rate fluctuating seasonally with rainfall distribution (Crownover et al. 1995). Water tables fluctuate as well, but not uniformly within the landscape. Flatwoods constitute extensive areas where water table fluctuations are generally most extreme within the soils. Thus, in dry periods the water table drops to a greater relative extent in flatwoods than it does in depressions and flats (Figure 15.2; Crownover et al. 1995). There is also

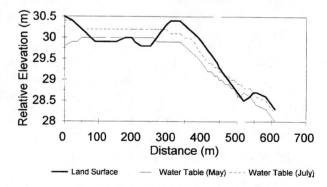

Figure 15.2   Surface and groundwater topography of a research site on a flatwoods-and-depressions landscape. High and low groundwater conditions are depicted. A regional slope is discernible under the vertical exaggeration of this figure, but is not evident while standing on the site itself. Ponds are depressional features resulting from karst features in deep limestone underlying the aquiclude. Note that ponds serve as discharge zones during periods of lower water table. (From data of Crownover, S.H., N.B. Comerford, D.G. Neary, and J. Montgomery. 1995. Horizontal groundwater flow patterns through a cypress swamp in flatwoods landscape. *Soil Sci. Soc. Am. J.* 59:1199–1206. With permission.)

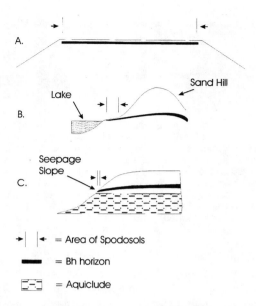

Figure 15.3 Schematic hypothetical representations of ways in which spodic horizons can occur at different depths and landscape settings (A = flatwoods, B = lake margin, and C = slope prone to seepage). Their occurrence generally corresponds with the zone of fluctuating water table, which "outcrops" within 2 m over an extensive area in the case of flatwoods. Note that a Spodosol "transforms" laterally to an Entisol (Quartzipsamment) when the Bh horizon "drops" below 2 m.

evidence that water levels in ponds can be higher than water tables in the surrounding flatwoods soils during dryer periods (Crownover et al. 1995), meaning that ponds seasonally could serve as recharge zones (Figure 15.2). The positive gradient from ponds to surroundings arises from evapotranspiration, a significant mode of water loss from flatwoods ecosystems (Ewel and Smith 1992). Evapotranspiration creates a drawdown in the Spodosol to a greater extent than in the pond (Figure 15.2). During the drawdown, a gravity flow potential is established from the pond to the landscape.

We have arrived at a point where the first "clue" to the flatwoods–Spodosol association "mystery" can be introduced. Flatwoods are landforms where a relatively unusual set of conditions actually "spread out" to constitute areally extensive near-surface domains within the local landscape (Figure 15.3). These conditions are: sandy textures, periods of chemical reduction (Garman et al. 1981, Phillips et al. 1989), and intense leaching. The latter two conditions arise from saturation–leaching hydrological cycles. What does this have to do with Spodosols? Further clues can be found in the relationship between water tables and Spodosol morphology.

HYDROLOGY AS RELATED TO SPODOSOL MORPHOLOGY

The level of certainty in pedology is limited by the complexity of soils. It is therefore appropriate to scrutinize the proposition that hydrology strongly influences the distribution of flatwoods Spodosols. For example, why wouldn't vegetation be an equally important prospective factor? This chapter highlights hydrology as a primary empirically associated variable, with the recognition that biotic (plant, microbial, etc.) factors as influenced by moisture may ultimately be implicated when details of the processes unfold. However, there is "smoking gun" evidence already on hand that implicates hydrological influence: E-over-Bh horizon sequences in sandy materials within zones of water table fluctuation regardless of vegetation, depth, or landscape position (Figure 15.3). For example, deep borings and excavations commonly reveal spodic horizons below excessively drained Quartzipsamments, in proximity to the water table. Spodosols correspond to an "outcropping" of

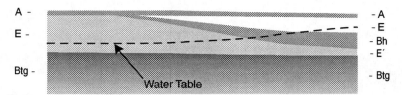

Figure 15.4 Schematic representation of a transition between a Udult (left) and an Aquod (right). The wet season water table is depicted, but was consistently deeper in the Udult even for dryer conditions over the 5-month period of monitoring. Note that the Bh horizon fades in prominence and approaches the surface as the water table slopes downward. This trend is not an anomaly, but has been reported as typical of such a transition by personnel who have mapped flatwoods and associated landforms. Note that the sandy-over-loamy sequence transgresses the transition. Albic-spodic sequences can therefore be deduced to be a hydrologically-linked process superimposed on this or other sequences, though a minimum thickness of the upper sand (about 0.5 to 1.0 m) is essential. (Data from Garman, C.R., V.W. Carlisle, L.W. Zelazny, and B.C. Beville. 1981. Aquiclude-related spodic horizon development. *Soil and Crop Sci. Soc. Florida Proc.* 40:105–110.)

a spodic horizon and associated hydrological conditions within 2 m of the soil surface (a taxonomic constraint). For example, narrow bands of Spodosols occur along the shores of some lakes and within the seep zone on some slopes, in both cases within the 2-m depth range of groundwater fluctuation. Thus a hydrological condition prevalent in flatwoods is also empirically linked with E- and Bh-horizon occurrence.

Further evidence of hydrological linkage to albic–spodic formation is found within transitions between Spodosols and other soils. The bordering soils can be better drained or more poorly drained than the Spodosol, suggesting that there is a hydrological "window" optimal for processes involved. The transition to better-drained soils encompasses a gradual thinning of E and Bh horizons as they "fade upward" to merge with the surface horizon (Garman et al. 1981, personal observations of the author) (Figure 15.4). The color of the Bh also lightens, reducing its distinctiveness within the soil.

The divergence of the Bh horizon and water table shows that the former does not relate to mean values of the latter. The Bh is consistently associated with fluctuating water table on the landscape, yet it does not tend to follow the water table downward within transitions (though Bh horizons do occur at greater depths in deep sandy materials, as previously discussed). Obviously, the position and developmental intensity of the Bh horizon does not directly correspond to water table position, the former being relatively static, and the latter dynamic (Hyde and Ford 1989, Crownover et al. 1995, Comerford et al. 1996). Explaining Bh horizon occurrence requires the consideration of other hydrologically related variables: duration of saturation (residence time), proximity of the water table to soil surface, direction of hydrological flows, and likely other factors as yet unknown. The next section addresses prospective mechanisms accounting for water table influence, along with supporting data.

PROSPECTIVE MECHANISMS OF WATER TABLE INFLUENCE

The formation of an albic–spodic sequence does not require a water table influence in the general case, as evidenced by the fact that Spodosols of many regions are commonly well drained (Nichols et al. 1990). The regional restriction of the process largely to poorly drained conditions, as discussed above, is therefore a special case. There are apparently inhibitions to Spodosol formation in the southeastern U.S. that are reduced below threshold conditions in the presence of a fluctuating water table. Aquods have been studied with respect to properties, composition, and various processes (e.g., Daniels et al. 1975, Holzhey et al. 1975, Harris and Carlisle 1985, Harris et al. 1987 a, b, Skjemstad 1992, Thompson 1992, Mokma 1993, Evans and Mokma 1996, Thompson et al. 1996). Rarely, however, have mechanisms been presented to explain how the water

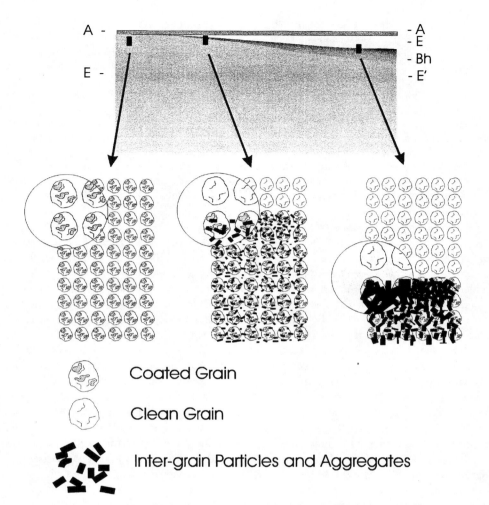

Figure 15.5 Schematic depiction of the "anatomy" of the sandy matrix within the transition between an Aquod and a Udult (left). Sand grains of the sandy eluvial horizons of the Udult are coated with silt and clay, as cemented by metal oxides. Grains of the Aquod E horizon above the Bh are mainly free of inorganic coatings. The Bh horizon has coated grains as well as some intergranular fine material and is enriched in both crystalline clay and noncrystalline organo-Al components. This scenario leads to the inference that the E- over Bh sequence formation entails the mobilization of crystalline grain coatings as well as the accumulation of noncrystalline organo-Al within the Bh matrix.

table could promote their formation (Farmer et al. 1983, Harris et al. 1995). This section reviews some of the data about Aquods of the southeastern U.S. that relate to possible mechanisms for the hydrological linkage.

There are five related details about flatwoods Aquods that may be important clues. They are well-documented observations or reasonably straightforward inferences, and serve as "points of departure" for the subsequent discussion.

i. *The albic/spodic boundary delineates zones of clean and coated grains, both along transitions (where it is slanted; Figure 15.5) and within Aquods in general.*

ii. *The thickness and coloration of the Bh horizon tend to diminish in tandem with the thickness of the overlying E horizon along transitions to better-drained soils.* Some areas of thick white E horizons, deep Bh horizons, and hummocky topography do not show this trend, possibly because better-drained sites are topographic "highs" that formed via redistribution (e.g., eolian) of the incoherent E-horizon sands.

   iii. *The albic horizon is depleted of clay (e.g., depletion of sand-grain coatings is tantamount to depletion of clay and silt for Aquods;* Figures 15.5 and 15.6). It consists of >99% quartz sand grains that are nearly coating-free. Trace amounts of clay are dominantly quartz.

   iv. *The spodic horizon has a greater amount of crystalline clay* (Figures 15.5 and 15.6) *than overlying and subjacent horizons.* Selective dissolution and mineralogical data (as yet unpublished) support this statement. However, it can also be deduced from the fact that only a trivial proportion of the clay in Bh horizons could be accounted for by noncrystalline aluminosilicates, based on particle-size and selective dissolution data (Sodek et al. 1990; Harris and Hollien 1999).

   v. *The spodic horizon clay consists of the same minerals that occur in the sand-grain coatings of subjacent horizons and of surrounding soils of other Orders.* These are typically resistant secondary phyllosilicates, gibbsite, and quartz (Harris and Carlisle 1985, Harris et al. 1987a, 1987b).

The above conditions constitute multiple lines of evidence that some of the crystalline clay in Bh horizons originated from overlying horizons via the destabilization and mobilization of sand-grain coatings. These coatings are common within the sandy soils of the region (even for the brown sands that do not meet the USDA "coated" family criterion), and their absence almost invariably signals the presence of an underlying Bh at some depth (V.W. Carlisle, personal comunication). This is true even for series of Quartzipsamments dominated by clean grains (white to light gray in color); the Bh in this case is simply below the 2-m taxonomic depth cutoff. Thus, particle migration is implicated as an integral part of the mechanism by which Aquods form. The alternative would be that resistant minerals somehow dissolve and reprecipitate, which seems improbable.

The link between hydrological condition and albic/spodic formation might be the water-table induced destabilization of sand-grain coatings. This idea is supported by the results of a study (Harris et al. 1995). The researchers verified that (i) grain coatings are removed by oxalic acid (a common indigenous soil organic acid; Fox and Comerford 1990) more easily from sandy soils located on toeslope positions adjacent to Aquods than from similar soils in summit positions, (ii) the toeslope soils contained significantly less crystalline "free" Fe (citrate–dithionite extractable) and gibbsite [Al(OH)$_3$] than did the summit soils, and (iii) toeslope soils were predisposed to the experimental induction of Aquod-like "E" and "Bh" zones within columns eluted with weak oxalic acid solutions, but *only under conditions of sufficient residence time (22-h water table retention cycles).* Microscopic examination verified a stripping of the coatings from the "E" and an accu-

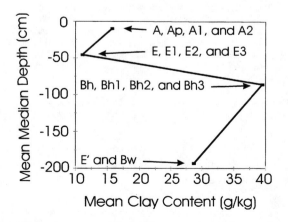

Figure 15.6 Depth trends of mean clay content, by horizon groups, for Alaquod pedons characterized through the Florida Cooperative Soil Survey Program. (n = 244, 313, 369, 37, and 85, respectively, for A, E, Bh, E′, and Bw horizons.) Clay content was determined after organic C removal. (From Sodek, F. III, V.W. Carlisle, M.E. Collins, L.C. Hammond, and W.G. Harris. 1990. Characterization for selected Florida soils. University of Florida, Res. 81-1 Gainesville.)

mulation of fine material as lamellae in the Bh. These lamellae darkened rapidly. Column effluents showed an accumulation of oxalic acid, Al, and Fe in the columns with the 22-h water table (as opposed to the freely drained controls).

Why might grain coating "cement" be weakened by fluctuating water tables? A possible reason is that the solubility of metal oxides is increased by factors such as chemical reduction (e.g., Fe oxides) and/or residence time of complexing organic acids. This suspicion is supported by significantly lower crystalline "free" Fe and gibbsite for toeslope soils adjacent to Aquods, experimental release of coatings with higher residence time (Harris et al. 1995), and extremely low amounts of Fe in most Aquods. The hydrology of flatwoods promotes both reduction and leaching of any Fe initially present. The crystallinity of Fe is also perturbed under cyclic redox conditions, rendering it more soluble in the presence of organic acids. The inorganic soil components are too recalcitrant under well-drained conditions for biological pressures alone to induce sufficient metal release for Spodosol formation.

Grain coatings constitute the only source of metals in the quartzose sandy soils of the region, which are essentially devoid of weatherable minerals (sand fraction generally >99% quartz and resistant heavy minerals). This is true throughout the Aquod and the landscape in regions of deep sands. Therefore, if organo-complexation of metals was the underlying mechanism of albic/spodic formation (as by "classical" theory), then the metal components of coatings must have been involved. Taken a step further, the most likely source of these metal components to undergo complexation are the "free" forms of Al and Fe that serve as the last vestiges of grain coating cement. Their dissolution would release the bulk of the coatings, consisting of crystalline resistant minerals, which would be susceptible to particle transport. The E horizon would thus be formed. Organically complexed metals (at this point, primarily Al) would also be mobilized, ultimately to precipitate in the Bh or be transported out of the soil zone.

The fact that C- and Al-bearing substances have precipitated in the zone of slightly (but significantly) elevated crystalline clay is an important clue. Water tables of Aquods frequently rise above the Bh, to be resident for appreciable periods in the albic horizon (Hyde and Ford 1989). Yet precipitation of organometal components does not occur in this clay-free eluvial zone. Soluble groundwater constituents drain freely from the albic horizon. A possible reason for the sharply defined E-over-Bh (light-over-dark) boundary is that the slightly higher clay content increases moisture retention, and hence the residence time of these soluble constituents, as the groundwater recedes to greater depths during dry periods. The increased residence time, in turn, would favor the attainment of kinetic- and solubility-product thresholds for precipitation of C and Al (together and/or separately). The rapid darkening of experimentally induced lamellae in oxalic acid-leached columns (Harris et al. 1995) supports the idea that slight clay accumulation fosters C precipitation. Some spodosols have a Bh horizon just below a surface horizon, and some have a thick B'h (commonly below 2 meters) that is below another spodic horizon with no intervening albic horizon. These spodic horizons with no overlying albic may form from reactions of organics with *in situ* Al, or with Al in groundwater as proposed by Farmer et al. (1983).

SUMMARY AND CONCLUSIONS

Spodosols occur extensively on coastal flatwoods landforms of the southeastern U.S. A globally significant amount of C is stored in their spodic horizons, but the mechanisms of C accumulation are not well understood. These Spodosols mysteriously are restricted mainly to zones of fluctuating water tables, the prevailing hydrological condition of flatwoods. Hence they are predominantly Aquods. This chapter has addressed the regional hydrological restriction of Spodosol-associated processes.

The albic/spodic diagnostic horizon sequence commonly occurs below the 2-m "soil" zone, underlying shallower spodics or soils of other Orders. The sequence commonly fades upward to disappear at the surface in the transition from Spodosols to better-drained soils. Spodic materials dominantly consist of C, with some Al but very little Fe. Carbon usually decreases with depth in shallow spodics, while Al increases. The albic/spodic boundary delineates zones of clean and coated grains. The albic horizon is depleted of crystalline clay, whereas the spodic has elevated amounts. Spodic horizon clay consists of the same minerals that occur in the sand-grain coatings of subjacent horizons and of surrounding soils. Thus, there are multiple lines of evidence that some crystalline clay in Bh horizons originated from overlying horizons via the destabilization and mobilization of sand-grain coatings.

Metal-oxides cements of sand-grain coatings could be the hydrological link to regional Spodosol formation. These cements are strong under well-drained conditions, but are potentially weakened (partially dissolved, reduced in crystallinity, etc.) by chemical reduction and other conditions induced by a fluctuating water table. Organo-complexation of weakened cementing metals could result in release and migration of coating components (predominantly resistant secondary phyllo-silicates), leaving a coating-free zone (albic) and accounting for elevated crystalline clay in the spodic. The increased moisture retention established by the illuviated particles would extend solution residence time as the groundwater recedes, such that kinetic and thermodynamic thresholds of precipitation for C and Al species could be attained.

EPILOGUE

This chapter does not challenge extant theories of Spodosol formation, but it addresses the special case of a regional hydrological restriction for Spodosol occurrence. I have relied heavily upon simplified graphical models to convey the "big picture" perspective, realizing that the true complexity of soils and hydrology is beyond intelligible illustration. My inferences are based on real data and statistically significant distributional trends, but others could arrive at different conclusions from the same evidence. Science is an iterative process. I hope that the evidence and ideas presented here will stimulate thought, dialogue, and progress toward solutions to the mysteries of Aquods.

ACKNOWLEDGMENTS

My understanding of Aquods was developed in large part from data and samples generated by the Florida Cooperative Soil Survey Program. I am indebted to the many professionals with the U.S. Natural Resources Conservation Service and the University of Florida Soil and Water Science Department who contributed to this program. I have also benefited from field perspectives provided by colleagues within the state, particularly Dr. Vic Carlisle and Mr. Frank Watts.

REFERENCES

Abrahamson, W.G. and D.C. Hartnett. 1990. Pine flatwoods and dry prairies. pp. 143–149. *In* R.L. Myers and J.J. Ewel (Eds.) *Ecosystems of Florida.* Univ. of Central Florida Press, Orlando, FL.

Birkeland, P.W. 1984. *Soils and Geomorphology.* Oxford Univ. Press, New York.

Brown, R.B., E.L. Stone, and V.W. Carlisle. 1990. Soils. pp. 35–69. *In* R.L. Myers and J.J. Ewel (Eds.) *Ecosystems of Florida.* Univ. of Central Florida Press, Orlando, FL.

Collins, M.E. 1990. Aquods. pp. 105–122. *In* J.M. Kimble and R.D. Yeck (Eds.) *Proceedings of the Fifth International Soil Correlation Meeting, Characterization, Classification, and Utilization of Spodosols.* USDA–NRCS, U.S. Govt. Printing Office. Washington, DC.

Comerford, N.B., A. Jerez, A. Freitas, and J. Montgomery. 1996. Soil water table, reducing conditions, and hydrologic regime in a Florida flatwoods landscape. *Soil Sci.* 161:194–199.

Crownover, S.H., N.B. Comerford, D.G. Neary, and J. Montgomery. 1995. Horizontal groundwater flow patterns through a cypress swamp–pine flatwoods landscape. *Soil Sci. Soc. Am. J.* 59:1199–1206.

Dahlgren, R.A. and F.C. Ugolini. 1991. Distribution and characterization of short-range-order minerals in Spodosols from the Washington Cascades. *Geoderma* 48:391–413.

Dalsgaard, K. 1990. Spodosols of Denmark. pp. 133–141. *In* J.M. Kimble and R.D. Yeck (Eds.) *Proceedings of the Fifth International Soil Correlation Meeting, Characterization, Classification, and Utilization of Spodisols.* USDA–NRCS, U. S. Govt. Printing Office. Washington, DC.

Daniels, R.B., E.E. Gamble, and C.S. Holzhey. 1975. Thick Bh horizons in the North Carolina Coastal Plain: morphology and relation to texture and ground water. *Soil Sci. Soc. Am. Proc.* 39:1177–1181.

DeConinck, F. 1980. Major mechanisms in formation of spodic horizons. *Geoderma* 24:101–128.

Ewel, K.C. and J.E. Smith. 1992. Evapotranspiration from Florida pond cypress swamps. *Water Resource Bull.* 28:299–304.

Evans, C.V. and D.L. Mokma. 1996. Sandy wet Spodosols: water tables, chemistry, and pedon partitioning. *Soil Sci. Soc. Am. J.* 60:1495–1501.

Farmer, V.C, J.D. Russell, and M.L. Berrow. 1980. Imogolite and proto-imogolite allophane in spodic horizons: evidence for a mobile aluminum silicate complex in podzol formation. *J. Soil Sci.* 31:673–694.

Farmer, V.C., J.O. Skjemstad, and C.H. Thompson. 1983. Genesis of humus B horizons in hydromorphic humus Podzols. *Nature* 304:342–344.

Fox, T.R. and N.B. Comerford. 1990. Low-molecular-weight organic acids in selected forest soils of the southeastern USA. *Soil Sci. Soc. Am. J.* 54:1139–1144.

Franzmeier, D.P., B.F. Hajek, and C.H. Simonson. 1965. Use of amorphous material to identify spodic horizons. *Soil Sci. Soc. Am. J.* 33:815–816.

Garman, C.R., V.W. Carlisle, L.W. Zelazny, and B.C. Beville. 1981. Aquiclude-related spodic horizon development. *Soil and Crop Sci. Soc. Florida Proc.* 40:105–110.

Harris, W.G. and V.W. Carlisle. 1985. Clay mineralogical relationships in Florida Haplaquods. *Soil Sci. Soc. Am. J.* 51:481–484.

Harris, W.G., V.W. Carlisle, and S.L. Chesser. 1987a. Clay mineralogy as related to morphology of Florida soils with sandy epipedons. *Soil Sci. Soc. Am. J.* 51:1673–1677.

Harris, W.G., V.W. Carlisle, and K.C.J. Van Rees. 1987b. Pedon zonation of hydroxy-interlayered minerals in Ultic Haplaquods. *Soil Sci. Soc. Am. J.* 51:1367–1372.

Harris, W.G., S.H. Crownover, and N.B. Comerford. 1995. Experimental formation of Aquod-like features in sandy coastal plain soils. *Soil Sci. Soc. Am. J.* 59:877–886.

Harris, W.G. and K.A. Hollien. 1999. Changes in quantity and composition of crystalline clay across E-Bh boundaries of Alaquods. *Soil Sci.* 164:602–608.

Higashi, T., F. DeConinck, and F. Gelaude. 1981. Characterization of some spodic horizons of the campine (Belgium) with dithionite-citrate, pyrophosphate, and sodium hydroxide tetraborate. *Geoderma* 25:131–142.

Holzhey, C.S., R.B. Daniels, and E.E. Gamble. 1975. Thick Bh horizons in the North Carolina Coastal Plain: II. Physical and chemical properties and rates of organic additions from surface sources. *Soil Sci. Soc. Am. Proc.* 39:1182–1187.

Hyde, A.G. and R.D. Ford. 1989. Water table fluctuations in representative Immokalee and Zolfo soils of Florida. *Soil Sci. Soc. Am. J.* 53:1475–1478.

Kodama, H. and C. Wang. 1989. Distribution and characterization of noncrystalline inorganic components in Spodosols and Spodosol-like soils. *Soil Sci. Soc. Am. J.* 53:526–532.

Lee, F.Y., T.L. Yuan, and V.W. Carlisle. 1987. Characterization of organic matter in ortstein and non-ortstein horizons of selected Florida Spodosols. *Soil Crop Sci. Soc. Florida Proc.* 47:72–78.

Lee, F.Y., T.L. Yuan, and V.W. Carlisle. 1988. Nature of cementing materials in ortstein horizons of selected Florida Spodosols: II. Soil properties and chemical form(s) of aluminum. *Soil Sci. Soc. Am. J.* 52:1766–1801.

McKeague, J.A., F. DeConinck, and D.P. Franzmeier. 1983. Spodosols. pp. 217–252. *In* L.P. Wilding, N.E. Smeck, and G.F. Hall (Eds.) *Pedogenesis and Soil Taxonomy — II. The Soil Orders.* Elsevier, Amsterdam.

Mokma, D.L. 1983. Color and amorphous materials in Spodosols. *Soil Sci. Soc. Am. J.* 57:125–128.

Nichols, J.D., M.E. Collins, and G.W. Hurt. 1990. Role of water table in Spodosol formation. pp. 238–241. *In* J.M. Kimble and R.D. Yeck (Eds.) *Proceedings of the Fifth International Soil Correlation Meeting, Characterization, Classification, and Utilization of Spodosols.* USDA–NRCS, U.S. Govt. Printing Office. Washington, DC.

Peterson, L. 1976. *Podzols and Podzolization.* DSR Forlag, Copenhagen.

Phillips, L.P., N.B. Comerford, D.G. Neary, and R.S. Mansell. 1989. Simulation of soil water above a water table in a forested Spodosol. *Soil Sci. Soc. Am. J.* 53:1236–1241.

Rourke, R.V., B.R. Brasher, R.D. Yeck, and F.T. Miller. 1988. Characteristic morphology of Spodosols. *Soil Sci. Soc. Am. J.* 52:445–449.

Sodek, F. III, V.W. Carlisle, M.E. Collins, L.C. Hammond, and W.G. Harris. 1990. Characterization for selected Florida soils. University of Florida, Res. 81-1 Gainesville.

Soil Survey Staff. 1996. *Keys to Soil Taxonomy.* 7th ed., USDA–NRCS, U.S. Govt. Printing Office. Washington, DC.

Stone, E.L., W.G. Harris, R.B. Brown, and R.J. Kuehl. 1993. Carbon storage in Florida Spodosols. *Soil Sci. Soc. Am. J.* 57:179–182.

Skjemstad, J.O. 1992. Genesis of podzols on coastal dunes in southern Queensland: III. The role of aluminum-organic complexes in profile development. *Austr. J. Soil Res.* 30:645–665.

Thompson, C.H. 1992. Genesis of Podzols on coastal dunes in southern Queensland: I. Field relationships and profile morphology. *Austr. J. Soil Res.* 30:593–613.

Thompson, C.H., E.M. Bridges, and D.A. Jenkins. 1996. Pans in humus Podzols (Humods and Aquods) in coastal southern Queensland. *Austr. J. Soil Res.* 34:161–182.

Wright, J.R. and Schnitzer, M. 1963. Metallo-organo interactions associated with podzolization. *Soil Sci. Soc. Am. Proc.* 27:171–176.

Yuan, T.L. 1966. Characterization of surface and spodic horizons of some Spodosols. *Soil Crop Sci. Soc. Florida Proc.* 26:163–171.

CHAPTER 16

Soils of Northern Peatlands: Histosols and Gelisols

Scott D. Bridgham, Chein-Lu Ping, J. L. Richardson, and Karen Updegraff

INTRODUCTION

Peatlands are a subset of wetlands that have accumulated significant amounts of soil organic matter. Soils of peatlands are colloquially known as peat, with mucks referring to peats that are decomposed to the point that the original plant remains are altered beyond recognition (Chapter 6, SSSA 1997). Generally, soils with a surface organic layer > 40 cm thick have been classified as Histosols in the U.S. soil classification system (Soil Survey Staff 1998a). Permafrost-affected soils have recently been classified in the new order, Gelisols, and this includes many soils previously classified as Histosols (Soil Survey Staff 1998a).

As discussed in Chapter 6, peatlands have historically been classified based on a number of criteria, such as topography, ontogeny (i.e., landscape developmental sequence), hydrology, soil and/or water chemistry, plant community composition, and degree of soil organic matter decomposition (Moore and Bellamy 1974, Cowardin et al. 1979, Gore 1983, Bridgham et al. 1996, National Wetlands Working Group 1997). Given the confusion in peatland terminology and the emphasis of this chapter on soils, we will discuss here only the dominant ecological paradigm in peatlands — the ombrogenous–minerogenous gradient. Although the fundamental definition of this gradient is based on hydrology, it is often thought to be coincident with (and a primary control over) plant community composition and the biogeochemistry of peatland soils (Bridgham et al. 1996).

Minerogenous peatlands have significant inputs of groundwater and/or overland runoff, generally imparting higher basic cation content and pH to their soils (Heinselman 1963, Moore and Bellamy 1974). These peatlands are generally called fens, whereas treed minerogenous peatlands are often termed swamp forests in North America, although this latter term is also used to describe forested wetlands on mineral soils (National Wetlands Working Group 1997). In contrast, ombrogenous peatlands through deep accumulation of peat have achieved a landscape topographic position where they are isolated from all but atmospheric inputs of water, alkalinity-generating cations, and nutrients. As a result, they have low ash and basic cation content and low pH in their soils, and are commonly termed bogs. Fens exhibit a wide range of minerotrophy due to complicated interactions between hydrology, topographic landscape position, and chemistry of surrounding and/or underlying mineral soils (Bridgham and Richardson 1993, Bridgham et al. 1996, Verry 1997). For example, a region where mineral soils are dominated by sand plains with very low exchangeable cation content can have fens with significant groundwater input but soil chemistry and plant communities more characteristic of bogs.

Fens with more minerogenous characteristics (i.e., higher soil pH and basic cation content) are generally described as "rich," whereas those more similar to bogs in soil chemistry and plant community composition are called "poor." Bridgham et al. (1996) objected to terms such as rich and poor fens, because they essentially describe a gradient of pH and basic cation concentration, while most studies have pointed to nitrogen and/or phosphorus as the limiting nutrients for plant growth in peatlands. They suggested that nutrient availability gradients may not be coincident with the ombrogenous–minerogenous gradient; recent experimental results have demonstrated that nitrogen availability is greater in more minerogenous peatlands, whereas phosphorus availability is higher in more ombrogenous peatlands (Bridgham et al. 1998, Chapin 1998).

The effect of permafrost on peatlands is dramatic, lending support to defining the new soil order Gelisols for permafrost-affected soils. The formation and development of several major peatland types are the direct result of permafrost action (Zoltai and Tarnocai 1971, Moore and Bellamy 1974, National Wetlands Working Group 1988, Botch et al. 1995). Additionally, soil carbon pool sizes, distribution, and bioavailability are strongly affected by (1) cryoturbation, which is the soil-mixing action of freeze/thaw processes, and (2) by the presence of permafrost itself, which has strong controls over soil temperature and moisture and runoff (Michaelson et al. 1996).

The literature on peatlands is vast, and we focus here only on their soils, particularly within the context of the ombrogenous–minerogenous gradient and the effects of permafrost. The objectives of this chapter are to (1) summarize the geographic distribution of the world's peatlands, (2) describe the new USDA soil classifications of Gelisols and compare it to classifications of other countries and organizations, (3) examine the effects of the physical structure and botanical composition of various peats on their hydrologic properties, and (4) compare the physical and chemical characteristics of peats in U.S. wetlands from Florida to Alaska, with an emphasis on the ombrogenous–minerogenous gradient for Histosols and the defining characteristics due to permafrost in Gelisols.

GEOGRAPHIC DISTRIBUTION

Global Peatlands

There have been two recent systematic estimates of global wetland area. Matthews and Fung (1987) estimated wetland area at a 1° latitude by 1° longitude cell resolution based on (1) vegetation classification according to the United Nations Education Scientific and Cultural Organization (UNESCO), (2) soils from Food and Agriculture Organization (FAO) maps, and (3) fractional inundation from navigation charts developed by the U.S. Department of Defense. Global wetland area was estimated to be 5.30×10^6 km^2 (about 4% of the terrestrial surface area), 57% of which was estimated to be forested and nonforested peatlands (2.97×10^6 km^2). Wetland area and type are distinctly nonrandomly distributed across the globe. About half the global wetland area lies between 50 and 70°N latitude, 95% of which is classified as peatlands.

Aselmann and Crutzen (1989) estimated global wetland area to be 5.70×10^6 km^2 based on published regional wetland surveys, and of this amount 33% was classified as bog and 26% as fen (Figure 16.1). Swamps were primarily forested wetlands on mineral soil, so would not be considered peatlands. Thus, they estimated a global peatland area of 3.35×10^6 km^2 (Table 16.1), reasonably similar to the estimate of Matthews and Fung (1987) given the difference in methods. Aselmann and Crutzen also found that the majority of wetlands occur between 50 and 70°N latitude, and this northern wetland peak is comprised primarily of peatlands (Figure 16.1).

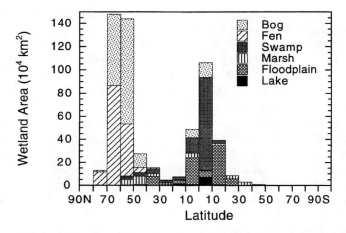

Figure 16.1 Global wetland area in 10° latitudinal belts for various wetland types. (Modified from Aselmann, I. and P. J. Crutzen. 1989. Global distribution of natural freshwater wetlands and rice paddies, their net primary productivity, seasonality and possible methane emissions. *J. Atmospheric Chem.* 8:307–359. With permission from Kluwer Academic Publishers.)

Several other, more limited, estimates of global peatland area have been done (Table 16.1), using methods similar to those of Aselmann and Crutzen (1989). Gorham (1991) estimated that boreal and subarctic peatlands occupy 3.46×10^6 km², whereas Armentano and Menges (1986) estimated that temperate and boreal peatlands occupy 3.49×10^6 km² (Table 16.1). Kivinen and Pakarinen's (1981) often cited early estimates of global peatland area, 4.22×10^6 km², is much higher than most subsequent estimates.

The convergence of recent estimates of global peatland area to 3.0 to 3.5×10^6 km² may suggest greater confidence in these numbers than is warranted. The primary data for these estimates, with the exception of those of Matthews and Fung (1987), are often the same. Problem areas include

Table 16.1   Global Peatland Area (in 10³ km²)

Regions	Reference 1	2	3	4	5	6	7	8
FSU[a]	1448	—	1500	—	1500	1648	1597[f]	—
Canada	1204	—	1190	1223	1500	—	1575[g]	—
Europe	147	—	220[c]	—	270	—	176	—
S.E. Asia	197	—	—	—	287	—	—	—
Alaska	325	—	—	—	494	—	—	130[i]
U.S.[b]	13	—	550[d]	—	102	—	104	102[j]
Other	16	—	—	—	68	—	40	—
Total	3350	2974	3460[e]	—	4221	—	3492[h]	—

[a] Former Soviet Union.
[b] U.S. excluding Alaska, except where noted.
[c] Fennoscandia only.
[d] Includes Alaska.
[e] Includes only boreal and subarctic peatlands.
[f] Includes Finland.
[g] Includes Alaska.
[h] Temperate and boreal peatlands only.
[i] Includes 42.4×10^3 km² of Histosols and 87.5×10^3 km² of Histels (organic Gelisols). From Table 16.2.
[j] From Table 16.2.

Data from Aselmann and Crutzen 1989; Matthews and Fung 1987; Gorham 1991; Tarnocai et al. 1995; Kivinen and Pakarinen 1981; Botch et al. 1995; Armentano and Menges 1986; Soil Survey Staff 1998b.

the former Soviet Union (FSU). A recent, more thorough analysis from that area yielded a higher overall peatland estimate of 1.65×10^6 km^2 (Botch et al. 1995). Considerable recent effort has gone into mapping Canadian peatlands, and the estimate of 1.22×10^6 km^2 by Tarnocai et al. (1995) is an underestimate that will be updated soon (personal communication, I. Kettles, Geologic Survey of Canada). Estimates of Alaskan peatlands are widely divergent (Table 16.1), but the best estimate is probably from the USDA STATSGO (State Soil Geographic) database (Soil Survey Staff 1998b). Alaska's peatlands were divided into 42.4×10^3 km^2 of Histosols and 87.5×10^3 km^2 of organic soils affected by permafrost (former Histosols), which would be classified as Gelisols in the new U.S. soil classification system (Soil Survey Staff 1998a). Using what we feel are the best estimates to date (Botch et al. 1995 for the FSU, Tarnocai et al. 1995 for Canada, STATSGO and MUIR [Map Unit Interpretation Record] databases for the U.S., and Aselmann and Crutzen 1989 for the remainder of the world), we arrive at a global peatland area of 3.46×10^6 km^2. Gorham (1991) estimated the same area for just boreal and subarctic peatlands, but our two estimates vary among specific regions of the world.

Despite these uncertainties, all recent estimates suggest that peatlands comprise half or more of all global wetlands, and they occur primarily in the boreal zone encompassing the former USSR, Canada, Alaska, and Fenno-scandia (Table 16.1). It is interesting that, while northern climates are clearly conducive to peat formation, large areas of tropical peatlands do exist — for example, very deep peat deposits occur on the islands of New Guinea, Borneo, and Sumatra.

Table 16.2 gives the distribution of peatlands within the U.S. There are two related databases maintained by the USDA that provide the best available estimates of peatland area in the U.S. (Soil Survey Staff 1998b). MUIR (Map Unit Interpretation Record) contains digitized soil maps at a scale of 1:12,000 to 1:31,680, but large areas of certain states have not had soil surveys done. This includes states such as Michigan and Minnesota that have large expanses of peatlands. In STATSGO (State Soil Geographic), other sources of information are used to estimate soil information in unmapped areas, but the scale is at 1:250,000, except Alaska, which is at 1:1,000,000. In states that are poorly mapped, STATSGO data are necessary to obtain realistic estimates of peatland area. However, because of the coarse scale, STATSGO fails to recognize many small peatlands. Consequently, it was deemed most accurate to take the highest estimate of STATSGO or MUIR for each state (Soil Survey Staff 1998b). Total peatland area in the U.S. is 231,781 km^2 (Table 16.2), with Alaska alone accounting for 56% of all peatlands. Excluding Alaska, the two regions with the most peatlands are the Midwest and South. In particular, large areas of peatlands occur in Michigan, Minnesota, Wisconsin, Florida, Louisiana, and North Carolina.

The distribution of Alaskan peatlands into Histosols and Gelisols demonstrates that 67% of its peatlands are affected by permafrost. Tarnocai (1998) estimated that 36% of Canadian peatlands had permafrost features (Organic Cryosols). In particular, significant areas of peatlands occur in the zone of discontinuous permafrost (Gorham 1991). Mosses and black spruce tend to enhance permafrost formation in this discontinuous zone (Van Cleve et al. 1991, Camill and Clark 1998).

Global Carbon Storage in Peatlands

Although peatlands only occupy about 2 to 3% of the terrestrial land surface, they represent a globally significant carbon pool because of the deep organic soil deposits that have accumulated over thousands of years. Gorham (1991) recently estimated that boreal and subarctic peatlands comprise a carbon pool of 455 Pg (1 Pg = 10^{15} g), having accumulated at an average rate of 0.096 Pg/yr (23 g C m^{-2} yr^{-1}, or a vertical accumulation of 0.5 mm/yr) during the postglacial period. This is about one third of the global soil carbon pool of 1272 Pg estimated by Post et al. (1985). More recent estimates of soil carbon content in peatlands have been made for the former Soviet Union (214 Pg, Botch et al. 1995) and Canada (153 Pg, Tarnocai 1998). Moreover, the arctic terrestrial soil carbon pool may have been underestimated by a factor of two (Michaelson et al. 1996). Previous

Table 16.2   Area of Organic Soils (km²) in the United States

State	Histosol	Data*	Histel	State	Histosol	Data	Histel
Midwest				**South**			
Illinois	356	M	—	Alabama	809	S	—
Indiana	1490	S	—	Arkansas	—	M	—
Iowa	301	M	—	Florida	15,943	S	—
Kansas	—	M	—	Georgia	1879	S	—
Michigan	16,511	S	—	Kentucky	—	M	—
Minnesota	24,345	S	—	Louisiana	9537	M	—
Missouri	51	M	—	Mississippi	908	S	—
Nebraska	44	M	—	North Carolina	6339	S	—
North Dakota	26	M	—	Puerto Rico	28	M	—
Oklahoma	—	M	—	South Carolina	650	S	—
South Dakota	—	M	—	Tennessee	—	M	—
Wisconsin	13,476	S	—	Texas	52	M	—
Total	56,601			Virginia	549	S	—
				Total	36,693		
Northeast				**West**			
Connecticut	434	S	—	Alaska	42,460	S	87,511
Delaware	356	S	—	Arizona	—	M	—
Maine	3965	S	—	California	617	S	—
Maryland	949	M	—	Colorado	335	S	—
Massachusetts	1364	M	—	Hawaii	1920	M	—
New Hampshire	899	M	—	Idaho	236	S	—
New Jersey	732	M	—	Montana	260	S	—
New York	3131	S	—	Nevada	74	S	—
Ohio	309	S	—	New Mexico	1	M	—
Pennsylvania	163 .	M	—	Oregon	329	S	—
Rhode Island	119	S	—	Utah	28	S	—
Vermont	270	M	—	Washington	790	M	—
West Virginia	—	M	—	Wyoming	30	M	—
Total	12,692			Total	47,079		87,511

Total Histosols = 153,065.

†Total Wetland Histosols = 144,270.

‡Total Peatlands = 231,781.

* S = STATSGO, M = MUIR. The highest Histosol area was taken from either STATSGO (State Soil Geographic database) or MUIR (Map Unit Interpretation Record database). Folist and Histel area were taken from STATSGO.
† Total Histosols – Folists (8795 km²).
‡ Total Histosols – Folists + Histels.

From Soil Survey Staff. 1998b. Query for Histosol soil components in the National MUIR and STATSGO data sets 8/98. Natural Resource Conservation Service, USDA, Lincoln, NE and Statistical Laboratory, Iowa State University, Ames, IA.

studies have failed to consider the significant carbon reserves in the lower soil horizons of tundra soils, which form as a result of cryoturbation.

Peat deposits of the boreal region tend to be deeper than those of the subarctic, and the boreal region has higher long-term net accumulation rates (Ovenden 1990, Gorham 1991, Botch et al. 1995, Ping et al. 1997a). On average, long-term accumulation rates in subarctic and boreal peatlands were estimated to be 7 to 11 and 23 to 41 g m^{-2} yr^{-1}, respectively (Ovenden 1990). Accumulation rates ranged from 12 g m^{-2} yr^{-1} in arctic peatlands to 80 g m^{-2} yr^{-1} in more minerotrophic mires in the boreal and temperate zones of the former Soviet Union, with an average of 30 g m^{-2} yr^{-1} (Botch et al. 1995).

CLASSIFICATION OF ORGANIC SOILS

Gelisols

Histosol soil classification was discussed in Chapter 6. In this section we will elaborate on the classification of Gelisols, which now include some soils formerly considered to be Histosols. The new Gelisol order includes soils with permafrost or Gelic materials in the surface 1 m (Soil Survey Staff 1998a). Organic soils are placed into the suborder Histel. In this hierarchy, more attributes are portrayed in the Gelisol order than in the previous *Soil Taxonomy* (Soil Survey Staff 1996). For the sake of example, we compare classifications of five organic soil pedons from Alaska (discussed further below in "Peat Biogeochemistry") based on both the previous and new *Soil Taxonomy*. Pedons 1 and 2 are intermediate fens, whereas Pedons 3 to 5 are bogs:

Soil Taxonomy (1996)	*Soil Taxonomy* (1998)
Pedon 1. Euic, Pergelic Cryosaprist	Euic, Sapric Glacistel
Pedon 2. Euic, Pergelic Cryohemist	Euic, Terric Hemistel
Pedon 3. Dysic, Fluventic Borosaprist	(not a Gelisol)
Pedon 4. Dysic, Pergelic Cryofibrist	Dysic, Sphagnic Fibristel
Pedon 5. Dysic, Typic Cryofolist	(not a Gelisol)

In the previous system, the Pergelic subgroup represented soils having a pergeic soil temperature regime (i.e., mean annual soil temperature below freezing). The formative element Cry-, meaning cold summer temperatures, was redundant with Pergelic. The properties reflecting the base status or the soil reaction, dysic and euic, enter in the family level. However, in the new Gelisol order, soil temperature and climate are combined into the order level, thus creating more room in the hierarchy for other important properties: fiber content and the presence or absence of ground ice (Glacic) enter in the great group level, and mineral horizons enter in the subgroup level. Family designations remain the same. The new Gelisol order certainly better reflects the cryogenic nature of the permafrost-affected organic soils and the management interpretations.

The new U.S. soil classification with the Gelisol order is now more similar to the Canadian system (Soil Classification Working Group 1998). In the latter system, the Cryosolic order includes soils affected by permafrost, and soils in the Organic Cryosol great group are organic soils with permafrost. Subgroups then are defined by fiber content of the control section or by the depth of peat over mineral soil or ice.

Comparison of Four Classification Schemes

By way of comparison, we examine four alternative methods for classifying organic soils from Florida to Minnesota and two histic beaver meadow horizons (Table 16.4 and see *Peat Biogeochemistry — A Comparative Approach* below). The first method is the USDA protocol (Soil Survey Staff 1998a), as described in Chapter 6. The second is the ASTM protocol (ASTM 1990), with sapric, hemic, and fibric peats having 0 to 32%, 33 to 67%, and > 67% dry-mass unrubbed fiber, respectively. The third method is the Canadian protocol (Soil Classification Working Group 1998), with sapric peat having a rubbed fiber content of < 10% by volume and a pyrophosphate index of ≤ 3, fibric peat having ≥ 75% rubbed fiber content by volume *or* ≥ 40% rubbed fiber by volume and a pyrophosphate index of ≥ 5, and hemic peat failing to meet the requirements of fibric or sapric peat. The fourth method is the von Post scale (Mathur and Farnham 1985, ASTM 1990, Parent and Caron 1993), where sapric, hemic, and fibric peats have von Post ratings of 7 to 10, 4 to 6, and 1 to 3, respectively.

None of the samples had ≥ 75% average rubbed fiber content by volume, but most of the bog and acidic fen soils would be classified as fibric in the USDA and Canadian systems based on their pyrophosphate color. Visually these samples were composed predominantly of moderately to nondecomposed *Sphagnum* fibers. Similar results were obtained with the ASTM classification system. The von Post scale gave a greater variety of classification values for bogs and acidic fens.

The intermediate fens, tamarack swamps, and cedar swamps had hemic peat according to most of the classification systems, whereas the histic layer in the beaver meadows, the ash swamp, and the southern peats had sapric organic matter according to one or more of the classification systems. Correlations between the classification systems ranged from an r^2 of 0.54 (between von Post and ASTM) and 0.88 (between Canadian and ASTM). Thus, quite different classifications can be given by the different systems, even though peats are only divided into three decompositional categories. Overall, the Canadian system tended to give highest values (i.e., the fewest saprists and hemists), and the USDA and von Post systems the lowest values.

Histosols in the U.S. (i.e., excluding Folists) are 15.7% Fibrists, 20.7% Hemists, and 63.6% Saprists (Soil Survey Staff 1998b). In comparison, Canadian Histosols (their Organic order) are 36.8% Fibrists (their Fibrisol), 61.8% Hemists (their Mesisol), and only 1.4% Saprists (their Humisol; Tarnocai 1998). The differences between the two countries probably reflect greater decomposition of peats at lower latitudes (see *Peat Biogeochemistry — A Comparative Approach* below), and the tendency of the Canadian soil classification system to place similar peats into less decomposed categories than the U.S. system, as discussed above.

Malterer et al. (1992) reviewed methods of assessing fiber content and decomposition in northern peats. They compared the von Post method, the centrifugation method of the former Soviet Union (Parent and Caron 1993), the USDA pyrophosphate color test and fiber-volume methods, and the ASTM fiber-weight method. Their analyses indicate that the centrifugation method of the former Soviet Union and the von Post humification field method separate more classes of peat with greater precision than the USDA and the ASTM methods. Stanek and Silc (1977) similarly found the von Post method differentiated more classes of peat than the rubbed and unrubbed fiber volume methods and the pyrophosphate color test of the USDA.

The pyrophosphate method does not differentiate well-humified peats (Stanek and Silc 1977) and is not particularly effective at extracting peat humic substances (Mathur and Farnham 1985). Additionally, the use of pyrophosphate color is limited because it is a qualitative variable, although spectrophotometric alternatives exist (Day et al. 1979). Mathur and Farnham (1985) state, "There is little theoretical basis for assuming that the color intensity of a [pyrophosphate] peat extract should be closely related to the extent of humification or that the extraction would be even semiquantitative in the presence of significant amounts of mineral matter." However, the pyrophosphate color index is reasonably well correlated with other measures of humification in Table 16.5.

HYDROLOGY

Hydrology and Peatland Development

Hydrology is the central factor, by definition, in the formation of all hydric soils, but peatlands are unique in the degree of autogenic (i.e., biotically driven) feedbacks between plant production and community composition, microbial decomposition, soil biogeochemistry, and hydrology (Heinselman 1963, 1970, Moore and Bellamy 1974, Siegel 1992). Under waterlogged conditions, especially in northern climates as noted above, net primary production generally exceeds decomposition, resulting in peat formation. The peat's botanical source, state of decomposition, bulk density, and depth interact to determine its hydraulic conductivity (Boelter 1969, Päivänen 1973, Silins and Rothwell 1998, Weiss et al. 1998). At some point, accumulation of deep, highly decomposed peat may impede vertical groundwater exchange with the surface layers. Additionally, the formation of peat itself increases water retention. As water retention increases, the peatland expands above the regional water table, and often above the surrounding landscape. At this point, an ombrogenous system has developed, with its characteristic soil chemistry and plant communities. Thus, we see a succession over time in many peatlands from fens to bogs, with an increasing state of ombrotrophy as a result of increasing biotic control over hydrology.

There are climatic limitations on this process: fens can occur in any climate because of their dependence on outside sources of water, whereas bogs can only occur in regions where precipitation exceeds evapotranspiration. The preponderance of peatlands in northern climates is at least partially due to lower temperatures limiting evapotranspiration, so that peatland formation is favored in areas of even moderate precipitation. However, substrate permeability, artesian pressure heads, and other groundwater factors can override macroclimate in the formation of large peatland complexes (Heinselman 1970, Siegel and Glaser 1987). In permafrost regions, drainage is further slowed by the seasonal freeze–thaw cycle, underlying permafrost, and low evapotranspiration rates on north-facing slopes (Reiger 1983). Permafrost may also act as a confining bed, creating artesian conditions for groundwater discharge and spring-fed wetlands (Racine and Walters 1994).

Hydrology and Peat Characteristics

As noted above, an important attribute of peats is their ability to hold and retain water. Undecomposed fibric peats are predominantly composed of air or water-filled pore spaces of large diameter (> 600 μm, Boelter 1964, Päivänen 1973, Silins and Rothwell 1998). This, in combination with low-density organic solids, results in a saturated water content often exceeding 1000% of oven-dry mass and 90% of total peat volume (Boelter 1964, 1969, Päivänen 1973, Damman and French 1987) (see Figure 16.2). More decomposed, higher bulk density peats and herbaceous peats have smaller pore spaces and correspondingly lower water-storage capacity under saturated conditions, although they still maintain > 80% saturated water content by volume (Boelter 1964, 1969, Päivänen 1973, Silins and Rothwell 1998) (see Figure 16.2). However, water is held in the large pore spaces of fibric peat primarily by detention storage (i.e., easily drainable porosity), and even moderate soil tensions result in large losses of the stored water (Figure 16.2). Similar to mineral soils, more decomposed, higher bulk density peats, with correspondingly smaller diameter pore spaces, have greater water retention under unsaturated conditions, and this difference increases at higher soil tension (Figure 16.2). The different botanical compositions of peats also have an important effect on water-holding capacity and retention (Boelter 1968, Weiss et al. 1998).

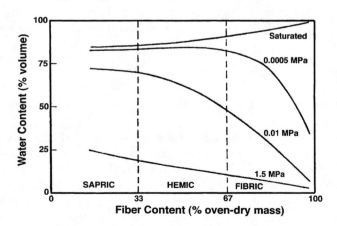

Figure 16.2 Relationship between fiber content of peat, water content, and soil water potential. Note that the definition for sapric, hemic, and fibric peats is somewhat different than used today in the U.S. (Modified from Boelter, D. H. 1969. Physical properties of peats as related to degree of decomposition. *Soil Sci. Soc. Am. Proc.* 33:606–609. With permission.)

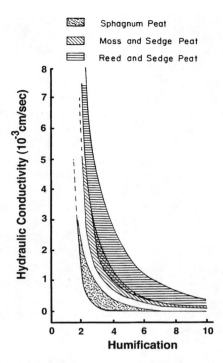

Figure 16.3 Effect of the botanical composition of peat and degree of humification on saturated hydraulic conductivity of peats. Humification is given in the qualitative von Post scale, where 1 is undecomposed and 10 is extremely decomposed. (Modified from Baden, W. and R. Eggelsmann. 1963. Zur Durchlässigkeit der Moorboden. *Z. Zulturtech. Flurbereining* 4:226–254.)

Surface peats have horizontal conductivities that are orders of magnitude greater than downward hydraulic conductivities in deeper peats (Päivänen 1973, Ingram 1982, 1983, Gafni and Brooks 1990). An important cause of this anisotropy is that deeper, more decomposed peat layers tend to have lower saturated hydraulic conductivity (Figure 16.3). In peatland terminology, water flow occurs predominantly in the upper, seasonally aerobic layer of the peat, or acrotelm, with very low flow through the deeper, permanently anaerobic layer, or catotelm (Damman 1986). Interestingly, *unsaturated* hydraulic conductivity is greater in more decomposed peats with smaller-diameter pore spaces (Silins and Rothwell 1998), similar to mineral soils (Brady 1984). Additionally, the plant composition from which the peat was derived has a dramatic effect on saturated hydraulic conductivity, with reed–sedge peat having the highest conductivity, and *Sphagnum* peats the lowest within any particular humification class (Figure 16.3). Undecayed *Sphagnum* moss has very high saturated conductivities, but conductivity decreases rapidly upon humification. Despite the high surface saturated conductivity of peats, horizontal water movement is very slow due to the low slope gradient (Brooks 1992).

These properties of peats have important ecological and economic consequences. The water table is often far below the surface in many peatlands, particularly in bogs, during the growing season (Boelter and Verry 1977, Bridgham and Richardson 1993, Verry 1997), and desiccation is an important constraint on the growth of *Sphagnum* mosses (Titus and Wagner 1984, Rydin 1985). Under drought conditions with a water table far below the surface, more decomposed peats would maintain higher plant-available water and faster transport of water to the roots (Päivänen 1973, Silins and Rothwell 1998).

Water retention and hydraulic conductivity are also important considerations in runoff from peatlands, drainage operations, and in commercial forestry in peatlands (Boelter 1964, Boelter and

Verry 1977, Silins and Rothwell 1998). Drainage of highly decomposed, subsurface peats is quite difficult. Often effective drainage only occurs within 10 m or less of ditches (Bradof 1992a). As an example, failed attempts at draining the large Red Lake peatland complex in northwestern Minnesota from 1907 through the 1930s resulted in virtual bankruptcy of several counties and was only resolved when the state took over large areas of tax-delinquent lands (Bradof 1992b).

We have presented the traditional view of peatland hydrology. However, the work of Siegel and colleagues (Chason and Siegel 1986, Siegel 1988, 1992, Siegel and Glaser 1987) has seriously questioned the assumption that vertical flow is negligible in peatlands, and particularly in bogs because of very low conductivities in deep peat. With both field work and hydrologic modeling studies, they have demonstrated that the hydraulic head in raised bogs is sufficient to drive downward water flowpaths, making bogs recharge zones and adjacent fens discharge zones. Even more interestingly, they have shown some bogs and fens to vary seasonally between being recharge and discharge zones. This dynamic nature of bog–fen hydrology can potentially be explained from the data of Chason and Siegel (1986). They found much higher hydraulic conductivities of deep, decomposed peats than previous studies, which they attribute to discontinuous zones of buried wood, roots, and other structural features in peat that form "pipes" with extremely high conductivities.

Runoff from peatlands outside permafrost areas is low, although it is higher in fens than bogs because of relatively constant groundwater inputs into fens (Boelter and Verry 1977, Verry 1997). However if permafrost is present, the infiltration and surface storage is low, and runoff occurs (Kane and Hinzman 1988). Free water mainly drains laterally above the permafrost following the slope. According to a study conducted in the interior of northeastern Russia, the ratio of water drained laterally to vertically is 8:1 (Alfimov and Ping 1994).

PEAT BIOGEOCHEMISTRY — A COMPARATIVE APPROACH

Conterminous U.S. Peats — the Ombrogenous–Minerogenous Gradient

We examined 39 physical and chemical properties of soils from 20 different wetlands (Tables 16.3 and 16.4), 17 in northern Minnesota, 2 in North Carolina, and 1 in Florida. The Minnesota sites were part of a larger study in carbon and nutrient dynamics in wetlands and were placed along an ombrogenous–minerogenous gradient according to dominant vegetation and soil pH (Bridgham et al. 1998). While this gradient is strictly defined based on hydrology, field data generally show a close correspondence between hydrologic status, vegetation, and soil chemistry (Sjörs 1950, Heinselman 1963, 1970, Glaser 1987, Grootjans et al. 1988, Vitt and Chee 1990, Gorham and Janssens 1992). All sites were classified as Histosols, except for two of the Minnesota sites, Upper and Lower Shoepack, which were beaver meadows in Voyageurs National Park with a surface histic epipedon of from 8 to 21 cm thickness over a mineral layer.

The short pocosin (an ombrotrophic bog dominated by stunted ericaceous shrubs) and gum swamp (minerogenous forested swamp dominated by *Nyssa sylvatica*, *Liquidambar styraciflua*, *Acer rubrum*, and *Taxodium distichum*) sites in the Coastal Plain of North Carolina are described in Bridgham and Richardson (1993). The Florida Everglades site is dominated by sawgrass, *Cladium jamaicensis*. It is part of Water Conservation Area 2A and has not been impacted by agricultural runoff (C. Richardson, Duke University, personal communication). Five replicate cores from 0 to 25 cm depth were taken from hollows in each site, when significant microtopography was present.

We put the 39 variables from all 20 wetlands in Tables 16.3 and 16.4 into a principal component analysis (PCA; Wilkinson et al. 1992). PCA is a multivariate technique that combines the physical and chemical factors into master variables called components that explain the most variation in the data set. The correlation of all 39 variables with the three most important principal components is presented in Figure 16.4. The first principal component had high positive weightings from lignin, the lignin:cellulose ratio, bulk density, and the von Post index. In contrast, variables with high

Table 16.3a Chemical Characteristics of the Average of Five 0 to 25 cm Depth Cores from 20 Sites (All sites are in Minnesota except where noted.)

Site	Type*	Lat. N	Total Org. C %	Total N %	Total P %	C/N	C/P	N/P	Nonpolar Extr.	%AFDM Water Soluble	Acid Soluble
Arlberg	Bog	46° 55'	41.7	1.26	0.044	33.2	971	29.0	7.00	11.60	45.9
Ash River	Bog	48° 24'	44.2	1.13	0.054	39.7	826	20.9	8.41	8.80	53.7
Pine Island	Bog	48° 17'	43.5	1.14	0.055	38.4	849	21.7	8.70	11.42	50.2
Red Lake	Bog	48° 22'	42.7	1.05	0.063	41.2	692	16.8	8.59	10.83	55.4
Toivola	Bog	47° 4'	42.2	1.12	0.039	37.8	1228	31.7	7.01	8.05	55.1
Marcell	Acidic fen	47° 31'	41.6	1.63	0.048	25.6	876	34.4	8.73	10.38	49.5
McGregor	Acidic fen	47° 39'	42.7	1.39	0.077	31.0	583	18.6	8.20	7.14	65.3
Alborn	Int. fen	47 00	38.9	2.61	0.082	14.9	479	32.1	4.75	7.31	48.3
Red Lake	Int. fen	48° 22'	43.3	2.43	0.076	17.9	573	32.0	7.66	7.13	28.8
Ash River	Tamar. sw.	48° 24'	42.9	1.98	0.078	21.8	558	25.6	8.56	6.83	28.3
Meadowlands	Tamar. sw.	47° 4'	42.6	2.57	0.114	16.6	380	22.9	5.32	5.52	39.1
Ash River	Cedar sw.	48° 24'	42.4	1.88	0.077	22.8	791	34.9	7.56	7.42	34.3
Isabella	Cedar sw.	47° 36'	42.6	1.85	0.076	23.3	584	25.0	6.20	8.92	30.5
Meadowlands	Cedar sw.	47° 3'	44.0	2.04	0.079	21.8	574	26.1	7.25	5.64	36.2
Gnesen	Ash sw.	46° 59'	34.7	2.58	0.266	13.5	131	9.7	6.16	8.27	37.0
Voyageurs	Meadow	48° 28'	21.1	1.53	0.127	13.7	166	12.0	6.83	7.22	35.6
Voyageurs	Meadow	48° 28'	23.8	1.59	0.115	15.0	212	13.9	6.57	9.05	30.8
North Carolina	Short pocosin	34° 55'	53.3	1.74	0.029	30.7	1838	59.8	7.25	4.97	25.1
North Carolina	Gum sw.	34° 55'	28.1	1.37	0.126	20.6	232	11.1	7.48	5.49	18.0
Florida	Everglades	26° 30'	45.0	2.99	0.034	15.1	1318	87.7	7.58	5.68	19.6

* Int. = intermediate, Tamar. = tamarack, sw. = swamp, Lat. = latitude, Org. = organic, Extr. = extractable, AFDM = ash-free dry mass, Wat. = water, Sol. = soluble, Carbo. = carbohydrates, Exch. = exchangeable, CEC = cation exchange capacity, Sat. = saturation, Acid-F = acid fluoride.

Table 16.3b Continued.

Site	Type	%AFDM						Mineral Content %	Bulk Density Mg/m³	pH Water
		Wat. Sol. Carbo.	Acid Sol. Carbo.	Soluble Phenolics	Lignin	Lignin/N	Lignin/Cellulose			
Arlberg	Bog	7.67	31.8	0.515	33.3	26.4	0.420	12.1	0.036	3.74
Ash River	Bog	7.89	39.4	0.568	29.8	26.7	0.356	5.2	0.050	3.75
Pine Island	Bog	7.38	35.7	0.649	28.4	24.8	0.361	6.3	0.054	3.80
Red Lake	Bog	7.30	41.7	0.687	24.1	22.9	0.304	6.0	0.052	3.72
Toivola	Bog	7.73	47.8	0.584	28.6	25.6	0.341	8.4	0.030	3.81
Marcell	Acidic fen	4.95	25.0	0.338	29.3	18.1	0.373	11.2	0.020	4.22
McGregor	Acidic fen	4.13	62.6	0.480	18.6	13.3	0.222	5.0	0.018	3.95
Alborn	Int. fen	3.84	14.5	0.203	38.2	14.6	0.441	22.7	0.100	4.84
Red Lake	Int. fen	6.64	24.3	0.215	53.3	22.0	0.650	12.5	0.076	5.57
Ash River	Tamar. sw.	5.66	15.5	0.291	54.3	27.6	0.658	13.4	0.102	5.91
Meadowlands	Tamar. sw.	2.58	17.6	0.247	48.7	19.1	0.555	13.4	0.104	5.63
Ash River	Cedar sw.	9.57	19.1	0.354	48.7	26.2	0.586	13.3	0.095	4.35
Isabella	Cedar sw.	2.15	7.7	0.155	52.2	28.5	0.632	14.7	0.147	6.61
Meadowlands	Cedar sw.	1.53	21.7	0.213	49.3	24.4	0.576	13.0	0.110	5.76
Gnesen	Ash sw.	3.42	8.7	0.096	44.5	17.3	0.546	29.3	0.150	6.13
Voyageurs	Meadow	3.05	5.5	0.070	43.0	28.9	0.548	55.4	0.213	5.75
Voyageurs	Meadow	5.06	4.1	0.070	45.9	28.9	0.598	52.3	0.236	6.16
North Carolina	Short pocosin	3.13	9.0	0.141	62.2	35.8	0.712	3.8	0.087	3.36
North Carolina	Gum sw.	2.43	3.6	0.019	63.2	47.0	0.778	46.1	0.206	3.97
Florida	Everglades	1.69	11.7	0.097	65.1	21.8	0.770	14.8	0.100	6.60

Table 16.3c Continued.

Site	Type	cmoles_c/kg					Extractable bases				Base Sat. %	Acid-F Extr. P. µg/g
		Exch. Acidity	Exch. Bases	CEC pH 7	Na	K	Mg	Ca	Ca/Mg			
Arlberg	Bog	19.9	13.2	33.1	0.382	2.87	3.3	6.7	2.01	39.6	3.26
Ash River	Bog	13.9	11.8	25.7	0.433	1.55	3.5	6.3	1.88	45.1	3.08
Pine Island	Bog	15.7	19.1	34.7	0.544	3.63	4.2	10.7	2.51	56.1	2.68
Red Lake	Bog	17.2	16.5	33.7	0.430	2.57	4.7	8.8	1.89	49.8	3.00
Toivola	Bog	24.0	18.1	42.2	0.465	2.83	3.9	10.9	2.82	43.2	1.12
Marcell	Acidic fen	31.3	21.1	52.4	0.550	1.65	5.0	13.9	2.79	40.8	1.94
McGregor	Acidic fen	30.7	22.0	52.6	0.793	4.42	5.7	11.1	1.97	41.8	36.91
Alborn	Int. fen	11.3	12.3	23.6	0.500	0.90	2.6	8.3	3.29	51.4	1.85
Red Lake	Int. fen	6.4	34.8	41.2	0.465	1.25	8.0	25.1	3.15	85.2	1.45
Ash River	Tamar. sw.	15.4	84.3	99.7	0.344	1.05	21.6	61.3	2.86	84.5	1.32
Meadowlands	Tamar. sw.	14.6	60.2	74.8	0.537	1.01	17.0	41.7	2.45	80.5	13.03
Ash River	Cedar sw.	19.6	31.9	51.5	0.422	1.55	6.8	23.1	3.45	62.8	4.91
Isabella	Cedar sw.	7.8	116.9	124.7	0.524	0.78	24.0	91.6	3.84	93.7	1.58
Meadowlands	Cedar sw.	19.1	67.0	86.1	0.381	0.70	17.4	48.5	2.78	77.8	1.80
Gnesen	Ash sw.	10.6	51.3	61.8	0.679	0.69	10.6	39.3	3.70	82.8	8.49
Voyageurs	Meadow	4.4	26.3	30.7	0.263	1.34	7.5	17.3	2.38	85.5	7.23
Voyageurs	Meadow	3.5	35.2	38.7	0.267	1.13	10.1	23.7	2.45	88.4	3.85
North Carolina	Short pocosin	28.1	7.1	35.2	0.733	0.68	5.3	0.4	0.07	20.1	0.82
North Carolina	Gum sw.	19.3	1.4	20.7	0.294	0.40	0.4	0.3	0.73	6.8	6.56
Florida	Everglades	5.5	139.9	145.4	7.658	1.94	30.3	100.0	3.30	96.2	8.01

Table 16.3d   Continued.

Site	Type	CaCl₂ Ext.		Oxalate Ext.		%AFDM		
		P μg/g	N μg/g	Fe mg/g	Al mg/g	Humin	Fulvic Acids	Humic Acids
Arlberg	Bog	5.66	24.9	2.85	1.88	72.4	20.7	12.0
Ash River	Bog	0.96	0.7	0.90	0.46	72.1	0.6	23.1
Pine Island	Bog	6.16	14.6	0.93	0.57	72.7	23.8	7.6
Red Lake	Bog	11.95	9.6	2.61	1.01	72.3	3.5	21.3
Toivola	Bog	0.43	14.2	3.35	1.08	72.2	6.5	23.4
Marcell	Acidic fen	7.87	72.1	2.95	1.99	75.7	22.9	10.2
McGregor	Acidic fen	36.16	31.6	3.29	0.61	73.8	21.2	7.5
Alborn	Int. fen	0.53	16.1	12.29	3.24	75.9	26.5	23.0
Red Lake	Int. fen	0.91	26.6	6.48	2.08	79.8	14.7	13.8
Ash River	Tamar. sw.	2.38	85.1	2.76	1.63	79.2	3.3	11.5
Meadowlands	Tamar. sw.	1.79	39.0	7.34	1.19	70.6	4.8	11.7
Ash River	Cedar sw.	6.31	41.0	2.98	2.08	80.0	24.2	13.5
Isabella	Cedar sw.	1.02	48.6	6.09	1.09	80.2	0.9	13.4
Meadowlands	Cedar sw.	1.76	54.7	7.88	1.43	79.7	6.3	11.7
Gnesen	Ash sw.	0.20	42.2	10.48	5.61	77.1	13.5	17.2
Voyageurs	Meadow	3.09	19.8	5.70	3.03	73.2	57.0	26.4
Voyageurs	Meadow	0.80	38.9	6.29	2.39	63.5	54.0	23.2
North Carolina	Short pocosin	16.25	5.3	0.62	0.99	72.0	17.6	26.8
North Carolina	Gum sw.	0.41	7.0	1.38	7.64	54.4	42.6	37.1
Florida	Everglades	11.49	35.6	1.71	0.83	77.3	22.5	8.9

negative principal component 1 weightings were pyrophosphate color, rubbed and unrubbed fiber, water and acid soluble components, soluble phenolics, and extractable potassium. These variables suggest that principal component 1 describes a decomposition axis, with peat that has high positive values being highly decomposed.

The second principal component describes an alkalinity/pH axis, with high weightings from extractable Ca and Mg, the Ca:Mg ratio, cation-exchange capacity, total exchangeable bases, % base saturation, and pH (Figure 16.4). Interestingly, % humin, total soil nitrogen, and calcium-chloride extractable N clumped with these alkalinity variables, which suggests a positive relationship between alkalinity/pH, humin formation, and nitrogen pools and fluxes. In contrast, humic acid content had a high negative weighting on this axis, which suggests it has a negative relationship with alkalinity/pH.

The third principal component axis was related to soil carbon and mineral content (Figure 16.4). It had positive weighting from total soil carbon, and high negative weightings from % mineral content, oxalate-extractable Fe and Al, bulk density, and total soil phosphorus. Phosphorus is strongly sorbed by iron and aluminum hydroxyoxides, so it is not surprising that greater mineral content is related to higher total soil phosphorus levels, although this does not necessarily translate into higher available phosphorus (Bridgham et al. 1998). Additionally, more minerogenous peats may receive greater inputs of apatite–phosphorus from weathering.

There is a large cost in labor, time, and expense in doing many of these chemical analyses, and it is heartening that a simple set of physical and chemical variables often measured in peats is closely correlated with many of the more difficult chemical analyses. In particular, mineral content, bulk density, pH, fiber content, and the von Post index are correlated with many other chemical variables (Table 16.5). They are also as effective as the chemical variables in predicting nutrient and carbon mineralization in peats (Lévesque and Mathur 1979, Bridgham et al. 1998).

PCA also allows one to determine "factor scores" for each of the 20 wetlands along these three principal component axes. We used our multivariate data set to discriminate natural groupings of peatlands according to their soil characteristics (Figure 16.5). The first, second, and third

Table 16.4 Physical Characteristics and Various Classification Schemes for the Average of Five 0 to 25 cm Depth Cores from 20 Sites (All sites are in Minnesota except where noted. See text for description of classification schemes.)

Site	Type[a]	Unrubbed Fiber % dry mass	Unrubbed Fiber % volume	Rubbed Fiber % dry mass	Rubbed Fiber % volume	10YR Value Color	10YR Chroma Color	Composite[b] Color	von Post Index	Classifications[c] U.S.	Classifications[c] Canadian	Classifications[c] ASTM	Classifications[c] von Post
Arlberg	Bog	78	73	68	53	7.2	2.2	5	3	2.4	2.6	2.8	2.8
Ash River	Bog	92	82	80	62	8	1.4	6.6	2.6	3	3	3	3
Pine Island	Bog	81	75	68	53	7.8	1.8	6	2.4	2.8	2.8	2.8	3
Red Lake	Bog	86	78	78	60	8	1.6	6.4	3.6	3	3	2.8	2.4
Toivola	Bog	78	73	74	57	7.8	1.8	6	2.8	2.8	2.8	2.8	3
Marcell	Acidic fen	76	72	60	48	8	1.4	6.6	4.6	3	3	3	2
McGregor	Acidic fen	85	78	77	60	8	1	7	2.8	3	3	3	3
Alborn	Int. fen	33	46	26	26	6.2	3.4	2.8	4.6	1.2	2	1.8	2
Red Lake	Int. fen	52	56	41	34	8	2	6	4	2	2	2	2
Ash River	Tamar. sw.	63	63	51	41	7.8	2.2	5.6	4.2	2.4	2.4	2.4	2
Meadowlands	Tamar. sw.	48	53	30	26	7	3.2	3.8	6.8	1.8	2	2	1.2
Ash River	Cedar sw.	53	57	44	36	7.6	2.4	5.2	6.2	2.4	2.4	2	1.6
Isabella	Cedar sw.	47	52	30	26	7	2.8	4.2	8	1.8	2	2	1
Meadowlands	Cedar sw.	58	60	43	35	6.8	2.8	4	6	2	2.2	2.2	1.6
Gnesen	Ash sw.	45	51	11	13	5.2	3	2.2	9.4	1	2	2	1
Voyageurs	Meadow	31	42	20	19	6.8	3.2	3.6	6	1.6	2	1.4	2
Voyageurs	Meadow	35	45	17	17	5.6	3	2.6	6.4	1	2	1.6	1.4
North Carolina	Short pocosin	48	53	16	16	5	3	2	10	1	2	2	1
North Carolina	Gum sw.	28	40	8	10	3	2	1	10	1	1.4	1.2	1
Florida	Everglades	63	63	20	19	5.8	3	2.8	9	1	2	2.2	1

a Int. = intermediate, Tamar. = tamarack, sw. = swamp.
b 10YR Value — 10YR Color (Parent and Caron 1993).
c Average classification for the five cores. 1 = sapric, 2 = hemic, 3 = fibric.

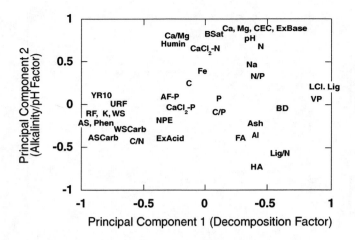

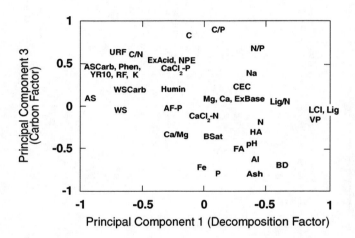

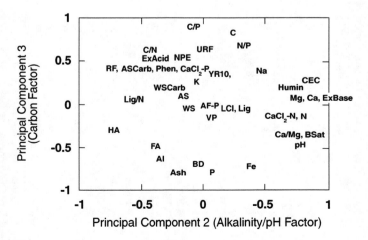

Figure 16.4 The loadings of 39 soil variables from 15 peatlands in northern Minnesota, 2 beaver meadows with histic epipedons in northern Minnesota, 2 peatlands in North Carolina, and 1 peatland in Florida on the first 3 axes of a principal components analysis. The loading is comparable to the correlation coefficient (r) for each variable against each axis. Abbreviations are as in Table 16.5.

Table 16.5a   Pearson Correlations (r) when P < 0.05 for Variables in Tables 16.3 and 16.4

	C	N	P	C/N	C/P	N/P	NPE	WS	AS	WSCarb
C*										
N										
P	-0.57									
C/N	0.48	-0.78	-0.55							
C/P	0.74		-0.74	0.51						
N/P	0.56	0.44	-0.55		0.77					
NPE		-0.52		0.58						
WS		-0.54		0.51						
AS		-0.49		0.63				0.55		
WSCarb		-0.51		0.59				0.59	0.46	
ASCarb		-0.51		0.75			0.47		0.85	0.52
Phen	0.44	-0.62	-0.46	0.88			0.50	0.63	0.80	0.73
Lig		0.58		-0.62				-0.69	-0.97	-0.52
Lig/N									-0.58	
LCI		0.54		-0.64				-0.63	-0.99	-0.49
Ash	-0.95		0.57	-0.64	-0.65					
BD	-0.77		0.58	-0.69	-0.58		-0.44		-0.66	-0.48
pH		0.69		-0.78	-0.45				-0.53	-0.54
ExAcid	0.47			0.55	0.46				0.44	
ExBase		0.60							-0.46	-0.47
CEC		0.55				0.48				-0.46
Na		0.50				0.80				
K		-0.49					0.45	0.50	0.71	0.47
Mg		0.60		-0.44					-0.48	-0.50
Ca		0.60		-0.44					-0.47	-0.47
Ca/Mg		0.51								
BSat		0.57		-0.56						
AF-P										
CaCl$_2$-P										
CaCl$_2$-N										
Fe		0.58	0.59	-0.63	-0.58		-0.77			
Al	-0.62		0.69	-0.49	-0.55					
Humin	0.51									
FA	-0.78									
HA	-0.47									
URF	0.56	-0.48	-0.51	0.82	0.45		0.68	0.52	0.69	0.55
RF		-0.60	-0.51	0.83			0.59	0.57	0.81	0.67
YR10		-0.45	-0.45	0.63			0.62	0.52	0.70	0.65
VP		0.49		-0.59				-0.58	-0.78	-0.68

	ASCarb	Phen	Lig	Lig/N	LCI	Ash	BD	pH	ExAcid	ExBase
C										
N										
P										
C/N										
C/P										
N/P										
NPE										
WS										
AS										
WSCarb										
ASCarb										
Phen	0.87									
Lig	-0.81	-0.80								
Lig/N			0.52							
LCI	-0.84	-0.81	0.99	0.55						
Ash	-0.64	-0.66								
BD	-0.81	-0.78	0.58	0.49	0.64	0.90				

Table 16.5b Pearson Correlations (r) when P < 0.05 for Variables in Tables 16.3 and 16.4 (continued)

	ASCarb	Phen	Lig	Lig/N	LCl	Ash	BD	pH	ExAcid	ExBase
pH	−0.59	−0.65	0.51		0.53		0.57			
ExAcid	0.53					−0.53	−0.63	−0.73		
ExBase			0.49		0.47			0.81	−0.45	
CEC			0.44					0.70		0.98
Na										0.63
K	0.84	0.76	−0.74		−0.73		−0.64	−0.48		
Mg			0.52		0.50			0.81		0.98
Ca			0.50		0.48			0.82	−0.47	1.00
Ca/Mg				−0.58				0.64	−0.45	0.57
BSat								0.90	−0.74	0.76
AF-P										
CaCl$_2$-P	0.53						−0.44		0.54	
CaCl$_2$-N								0.56		0.56
Fe					−0.48			0.55		
Al	−0.53	−0.56				0.68	0.60			
Humin					−0.60	−0.51				0.48
FA		−0.46				0.81	0.61			
HA					0.66	0.54	0.51			−0.49
URF	0.86	0.88	−0.68		−0.70	−0.75	−0.85	−0.51	0.48	
RF	0.91	0.95	−0.80		−0.81	−0.66	−0.81	−0.57	0.47	
YR10	0.81	0.81	−0.71		−0.71	−0.61	−0.75			
VP	−0.79	−0.82	0.80		0.80		0.63			

	CEC	Na	K	Mg	Ca	Ca/Mg	BSat	AF-P	CaCl$_2$-P	CaCl$_2$-N
C										
N										
P										
C/N										
C/P										
N/P										
NPE										
WS										
AS										
WSCarb										
ASCarb										
Phen										
Lig										
Lig/N										
LCl										
Ash										
BD										
pH										
ExAcid										
ExBase										
CEC										
Na	0.62									
K										
Mg	0.96	0.59								
Ca	0.97	0.59		0.97						
Ca/Mg	0.51			0.48	0.60					
BSat	0.64			0.76	0.76	0.74				
AF-P			0.49							
CaCl$_2$-P			0.63					0.75		
CaCl$_2$-N	0.61			0.59	0.57	0.53	0.51			
Fe						0.55	0.46			
Al			−0.49							

Table 16.5c   Pearson Correlations (r) when P < 0.05 for Variables in Tables 16.3 and 16.4 *(continued)*

	CEC	Na	K	Mg	Ca	Ca/Mg	BSat	AF-P	CaCl$_2$-P	CaCl$_2$-N
Humin	0.50			0.45	0.48	0.67	0.53			0.47
FA										
HA	−0.55			−0.47	−0.47	−0.53	−0.45			−0.58
URF			0.74							−0.49
RF			0.75							−0.49
YR10			0.66							
VP			−0.68							

	Fe	Al	Humin	FA	HA	URF	RF	YR10
C								
N								
P								
C/N								
C/P								
N/P								
NPE								
WS								
AS								
WSCarb								
ASCarb								
Phen								
Lig								
Lig/N								
LCI								
Ash								
BD								
pH								
ExAcid								
ExBase								
CEC								
Na								
K								
Mg								
Ca								
Ca/Mg								
BSat								
AF-P								
CaCl$_2$-P								
CaCl$_2$-N								
Fe								
Al								
Humin		−0.49						
FA		0.47	−0.51					
HA		0.57	−0.67					
URF	−0.53	−0.64		−0.54	−0.47			
RF		−0.60		−0.47		0.92		
YR10		−0.63		−0.44	−0.52	0.83	0.92	
VP		0.52				−0.70	−0.88	−0.85

* C = %organic C, N = %total N, P = %total P, NPE = nonpolar extractable, WS = water soluble, AS = acid soluble, WSCarb = water soluble carbohydrates, ASCarb = acid soluble carbohydrates, Phen = soluble phenolics, Lig = lignin, LCI = lignin/cellulose, BD = bulk density, ExAcid = exchangeable acidity, ExBase = exchangeable bases, CEC = cation-exchange capacity, BSat = %base saturation, AF-P = acid fluoride extractable P, CaCl$_2$-N and CaCl$_2$-P = calcium chloride extractable N and P, FA = fulvic acid, HA = humic acid, URF = %unrubbed fiber, RF = %rubbed fiber, YR10 = composite pyrophosphate color (10YR Value - 10YR Color), VP = von Post index.

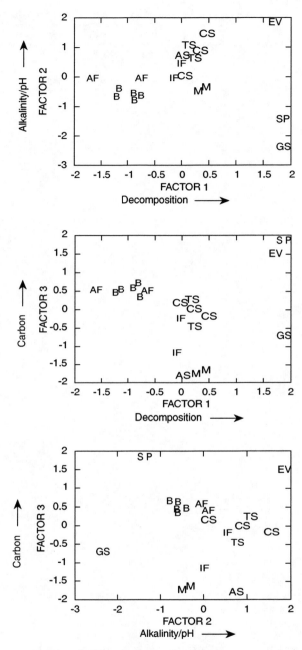

Figure 16.5 Factor scores for the first three axes of a principal components analysis of 39 soil variables from 15 peatlands in northern Minnesota, 2 beaver meadows with histic epipedons in northern Minnesota, 2 peatlands in North Carolina, and 1 peatland in Florida. B = bog, AF = acidic (poor) fen, IF = intermediate fen, TS = tamarack swamp, CS = cedar swamp, AS = black ash swamp, M = beaver meadow, SP = short pocosin (NC), GS = gum swamp (NC), and EV = Everglades (FL). The northern wetlands occur across an ombrogenous–minerogenous gradient in the order listed above, whereas hydrologically SP is an ombrogenous bog, GS is a minerogenous swamp forest, and EV is a "poor" fen.

factors explained 28.8, 22.5, and 21.1%, respectively, of the variance among sites, or 72.4% of the total variance. The first factor (Decomposition Factor) effectively separated three groups of wetlands: acidic fens and bogs, more minerotrophic northern wetlands, and southern peatlands. The second factor (Alkalinity/pH Factor) separated bogs from acidic fens, beaver meadows from minerotrophic northern peatlands, and the alkaline Everglades site from the acidic North Carolina peatlands. The third factor (Carbon Factor) separated minerotrophic northern cedar and tamarack swamps from intermediate fens, the ash swamp site, and the two beaver meadows. The nutrient-deficient short pocosin and Everglades sites were separated from the relatively nutrient-rich North Carolina gum swamp.

The difficulty of applying the ombrogenous–minerogenous gradient to southern peatlands is evident from our data. One sees the expected decrease in rubbed fiber content and increase in mineral ash, lignin, pH, %base saturation, and related variables expressing increasing alkalinity from bogs to ash swamps and beaver meadows in the northern sites, related to increasing minerogenous water inputs and their impact on water chemistry (Table 16.3, Figure 16.6). However, both short pocosins and the Everglades are profoundly phosphorus limited, whereas the gum swamp is relatively fertile (Walbridge 1991, Koch and Reddy 1992, Bridgham and Richardson 1993, Craft and Richardson 1997). Hydrologically, the Everglades site would be considered a "poor" fen, despite its alkaline soil conditions, and the gum swamp is a highly minerogenous "rich" swamp forest, despite its very acidic soil (Table 16.3, Figure 16.6). The sands of the North Carolina Coastal Plain have very low exchangeable basic cation concentrations, so contribute little alkalinity despite being highly minerogenous (Bridgham and Richardson 1993). Additionally, all of the southern peats are highly decomposed hemic or sapric peats with very low fiber and cellulose content, but high lignin content (Tables 16.3 and 16.4; Figures 16.4 through 16.6).

Our data support the traditional concept of an ombrogenous–minerogenous gradient in northern peatlands in terms of alkalinity and degree of decomposition of peats; however, soil nutrient availability is more problematic. We found in these same Minnesota wetlands that more minerogenous wetlands have larger total soil nitrogen and phosphorus pools, but those pools turn over more slowly in minerogenous sites (Bridgham et al. 1998). The large increase in bulk density in more minerogenous sites also has important consequences, because plant roots and microbes exploit a volume and not a mass of soil. The net result of all these factors was that nitrogen availability was higher in more minerogenous Minnesota wetlands, whereas phosphorus availability was higher in more ombrogenous wetlands (Bridgham et al. 1998). Chapin (1998) conducted a detailed fertilization experiment in an intermediate fen and bog in northern Minnesota and found similar results. Interestingly, she found that bog vegetation was not nutrient limited, except for a delayed response in ericaceous shrubs, and *Sphagnum* mosses were actually inhibited at moderate rates of nitrogen addition. The fen vegetation was phosphorus limited. Similar results have been found for both soil nutrient availability (Waughman 1980, Verhoeven et al. 1990, Koerselman et al. 1993, Updegraff et al. 1995) and plant-nutrient response (Clymo 1987, Lee et al. 1987, Boyer and Wheeler 1989, Bridgham et al. 1996) in other northern peatlands.

We suggest that the ombrogenous–minerogenous paradigm is an important and useful concept in northern peatlands, although its relation to a nutrient availability gradient appears to be complicated and worthy of further research. We conclude that the ombrogenous–minerogenous gradient does not appear to be directly translatable into an oligotrophic–eutrophic gradient. Furthermore, traditional concepts of how the ombrogenous-minerogenous gradient affects peat chemistry and physical properties in northern peatlands do not appear to be useful in southern peatlands.

Alaskan Peatlands — Histosols and Gelisols

We also examined a more limited set of soil variables in peats collected in the five pedons from Alaska (Table 16.6). As mentioned above, pedons 1 and 2 are intermediate fens, whereas pedons

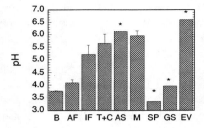

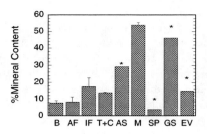

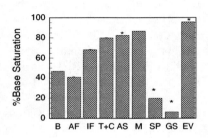

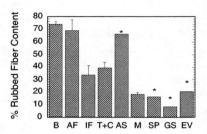

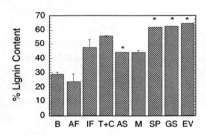

Figure 16.6   Relationship between pH, mineral content, base saturation, rubbed fiber content, and lignin to the ombrogenous–minerogenous gradient in the northern wetlands (going from left to right on the x-axis) and the three southern peatlands (SP, GS, and EV). The northern wetlands occur across an ombrogenous–minerogenous gradient in the order listed above, whereas hydrologically SP is an ombrogenous bog, GS is a minerogenous swamp forest, and EV is a "poor" fen. Average ± 1 standard error, except * indicates N=1 site so standard errors could not be obtained.

3 through 5 are bogs. Pedons 3 and 5 are Histosols, whereas pedons 1, 2, and 4 are Histels within the new order Gelisols.

The bulk density and mineral content of the horizons from the five pedons from Alaska are much higher than those from the Minnesota and southern peats (Table 16.3). Eolian and volcanic deposits (loess and tephra) have been active in many parts of Alaska and northwest Canada since the Late Pleistocene (Péwé 1975, Riehle 1985). Because of this frequent or intermittent input of mineral deposits, the organic soils in these regions have a higher bulk density compared with those developed in the humid maritime zones of southeastern Alaska and British Columbia. The additions of these materials appear in bands and layers in the peat, and thus they can serve as time-stratigraphic markers. In peat developed in bottom lands, mineral layers exist in lamella or bands due to the erosion or washing from surrounding slopes (Pedon 2).

Table 16.6 Characteristics of Selected Histosols from Alaska

Pedon # Lat. N	Horizon	Depth cm	Total Org. C %	C/N	Mineral content %	Bulk Density Mg/m^3	pH $CaCl_2$	Exch. Acidity	cmole/kg CEC	Extractable Bases Na	K	Mg	Ca	Base Sat. %	Fiber Content Unrub. %	Rubbed %	Pyrophosphate Color
1 70° 17'	Oai	0–18	23	35	62	0.39	6.9	17	79	1	tr	4	85	100	52	16	10YR 5/3
	Oa2	18–39	15	13	77	0.49	5.9	13	29	1	tr	1	16	61	26	12	10YR 4/3
	Oa3	39–50	22	14	65	0.38	6.3	18	49	1	tr	4	43	99	58	24	10YR 4/3
	Of	50–100	25	19	61	n.d.	7.1	12	63	2	tr	6	115	100	80	26	10YR 4/3
	Cf	39–80	tr	11	n.d.	1.8	7.7	n.d.	2	0	tr	1	n.d.	100	n.d.	n.d.	n.d.
2 67° 26'	Oi	0–17	51	25	15		7.7	21	168	tr	3	20	183	100	92	64	10YR 7/3
	Oe1	17–35	49	17	20	0.12	6.7	38	197	tr	tr	15	186	100	64	36	10YR 6/3
	Oe2	35–48	52	29	n.d.		5.8	49	160	tr	tr	11	146	98	n.d.	n.d.	n.d.
	C	48–54	24	n.d.	n.d.		5.1	47	89	tr	tr	6	81	99	n.d.	n.d.	n.d.
	Oef1	54–85	42	21	32		5.4	66	139	tr	tr	6	113	86	n.d.	n.d.	n.d.
	Oaf	85–95	29	20	54		n.d.	58	109	tr	tr	5	90	87	52	16	10YR 3/3
	Oef2	95–108	n.d.	n.d.	87		5.6	18	25	tr	tr	1	22	94	20	12	10YR 5/3
3 64° 52'	Oe	0–31	39	19	15	0.13	4.3		136	1	1	15	40	42	70	40	7.5YR7/5
	Oi	31–61	38	31	9	0.1	3.9		116	tr	tr	5	12	16	88	75	10YR 8/2
	Of	61–127	38	26	7	n.d.	4.4		81	tr	tr	5	12	22	90	80	10YR 8/1
4 61° 25'	Oi	0–29	53	68		n.d.	4.1	79	129	1	1	13	63	60			
	Oe	29–47	53	32		0.15	4.3	83	106	1	tr	6	49	52			
	Oa	47–79	49	27		0.4	4.3	82	96	1	tr	4	42	49			
	Oe1	79–97	22	30		0.57	4.5	61	49	1	tr	2	18	40			
	Oe2	97–148	55	28		0.2	4.5	78	106	1	tr	4	48	50			
	O'l	148–165	60	32		n.d.	4.5	85	132	1	tr	6	68	57			
5 56° 30'	Oi	0–3			17		3.3	94	99	1	2	7	13	23	76	56	10YR 8/3
	Oe	3–18			7		3	130	132	1	1	10	13	19	62	42	7.5YR 8/2
	Oa	18–94			54		3.5	72	76	1	1	2	4	9	48	30	5YR 3/4

In northern Alaska, as in the Minnesota sites, vegetation and land cover class show a strong correlation with the base status and pH of the soil (Ping et al. 1998). The pH of Alaskan peatlands decreases from 5.5 to 7.7 in the arctic coast to 4.0 to 4.5 in the boreal forest in the interior, to 3.0 to 3.5 in south central and southeastern Alaska. Most bogs in south central Alaska are extremely acidic and have low base status. Some of the bogs have conductivity less than 10 mS/cm (Clark and Kautz 1997). Péwé (1975) pointed out that there is continuous deposition of carbonate-rich loess in the Arctic Coastal Plain and in interior Alaska if streams are transporting glacial debris. In these soils, extractable Ca and Mg dominate the soluble salts and the exchange sites in the soils (Pedon 1 and 2). Pedon 3 is a raised bog with *Sphagnum* moss as the dominant vegetation. Even though the area also has lower active loess deposition, the soil developed from mosses and is very acidic. The added carbonates are reflected in the Ca-dominance of the exchange sites and the slightly higher base saturation in the surface layer. Although Pedon 4 formed in humid south central Alaska, the base saturation is higher than that of Pedon 3 because it is on a broad flood plain which collects seasonal input of minerals. Pedon 5 is a well-drained Folist in perudic southeastern Alaska. Its soil is strongly acidic (pH at 3.3) and has very low base saturation.

In a recent study, Ping et al. (1997b) found that organic matter in fens of the arctic coast was dominated by cellulose (approximately 50%), whereas the humin fraction was < 20%. Humic acids dominated the soluble fractions, and the C/N ratio ranged from 6 to 17. In comparison, the Minnesota peats had generally < 40% cellulose (i.e., acid-soluble carbohydrates), > 70% humin, a variable humic acid:fulvic acid ratio, and a C/N ratio which ranged from 14 to 41 (Table 16.3). All these data point to a lesser degree of humification of peats as the climate gets colder. This generalization is born out by a similar comparison of the Minnesota peats to those in North Carolina and Florida in Table 16.3.

Peat formed in the zone of continuous permafrost, such as arctic Alaska and northwest Canada, contains cryogenic features such as ice lenses, ice wedges, and other types of ground ice, generally at a depth of 40 to 60 cm (Tarnocai et al. 1993, Ping et al. 1997a, b, 1998). The upper permafrost layer of these soils often contains up to 80% ice by volume. Cryoturbation causes mixing of soil horizons and redistribution of carbon, resulting in significant carbon stores in the permafrost (Michaelson et al. 1996).

Our emphasis in this comparative biogeochemical approach has been on the peatlands of the U.S. A multivariate analysis of numerous soil properties of Canadian bogs was performed by Brown et al. (1990), but their emphasis was not on the ombrogenous–minerogenous gradient, and the study was done within a more limited geographical setting. Additionally, a wealth of information on Canadian peats is found in National Wetlands Working Group (1988). The review by Clymo (1983) emphasizes European peatlands and has long been a classic in this field. Bohlin et al. (1989) examined a wide range of peat properties in a diverse group of Swedish peats and used principal components analysis to examine their results. They found that the peats were differentiated by botanical composition and degree of decomposition, and particularly emphasized the differences between *Sphagnum* (bog) and *Carex* (fen)-derived peats. *Carex* peats were more humified due to microbial decomposition than *Sphagnum* peats. A thorough review of humic substances in peats is provided by Mathur and Farnham (1985).

ACKNOWLEDGMENTS

We would like to thank Curtis Richardson for obtaining the Everglades peat cores, Anastasia Bamford for technical assistance in the laboratory, John Pastor for comments on an earlier version of this manuscript, Inez Kettles for information on the extent of Canadian peatlands, and Sharon Waltman for access to the STATSGO and MUIR NRCS databases. This research was funded by a grant from NASA's Terrestrial Biosphere Program, the National Science Foundation (DEB-

9496305, DEB-9707426), and a Distinguished Global Change Postdoctoral Fellowship from the Department of Energy to Scott Bridgham.

REFERENCES

Alfimov, A. V. and C. L. Ping. 1994. Water regimes of boggy soils on slopes in the Upper Kolyma Basin, NE Russia. *Agronomy Abstr. Agron. Soc. Am.* Madison, WI.

Armentano, T. B. and E. S. Menges. 1986. Patterns of change in the carbon balance of organic soil-wetlands of the temperate zone. *J. Ecol.* 74:755–774.

Aselmann, I. and P. J. Crutzen. 1989. Global distribution of natural freshwater wetlands and rice paddies, their net primary productivity, seasonality and possible methane emissions. *J. Atmospheric Chem.* 8:307–359.

ASTM. 1990. *Annual Book of ASTM Standards.* Volume 04.08. ASTM, Philadelphia, PA.

Baden, W. and R. Eggelsmann. 1963. Zur Durchlässigkeit der Moorboden. *Z. Kulturtech. Flurbereining* 4:226–254.

Boelter, D. H. 1964. Water storage characteristics of several peats *in situ. Soil Sci. Soc. Am. Proc.* 28:433–435.

Boelter, D. H. 1968. Important physical properties of peat materials. pp. 150–156. *In Proc. 3rd Int. Peat Congress.* International Peat Society, Secretariat, Kuokkalantie, 4, FIN-40520, Jyväskylä, Finland.

Boelter, D. H. 1969. Physical properties of peats as related to degree of decomposition. *Soil Sci. Soc. Am. Proc.* 33:606–609.

Boelter, D. H. and E. S. Verry. 1977. *Peatland and Water in the Northern Lake States.* USDA Forest Service General Technical Report NC-31. North Central Forest Experimental Station, Forest Service, USDA, St. Paul, MN.

Bohlin, E., M. Hämäläinen, and T. Sundén. 1989. Botanical and chemical characterization of peat using multivariate methods. *Soil Sci.* 147:252–263.

Botch, M. S., K. I. Kobak, T. S. Vinson, and T. P. Kolchugina. 1995. Carbon pools and accumulation in peatlands of the former Soviet Union. *Global Biogeochemical Cycles* 9:37–46.

Boyer, M. L. H. and B. D. Wheeler. 1989. Vegetation patterns in spring-fed calcareous fens: calcite precipitation and constraints on fertility. *J. Ecol.* 77:597–609.

Bradof, K. L. 1992a. Ditching of the Red Lake Peatland during the homestead era. pp. 263–284. *In* H. E. Wright Jr., B. A. Coffin, and N. E. Aaseng (Eds.) *The Patterned Peatlands of Minnesota.* University of Minnesota Press, Minneapolis, MN.

Bradof, K. L. 1992b. Impact of ditching and road construction on Red Lake Peatland. pp. 173–186. *In* H. E. Wright Jr., B. A. Coffin, and N. E. Aaseng (Eds.) *The Patterned Peatlands of Minnesota.* University of Minnesota Press, Minneapolis, MN.

Brady, N. C. 1984. *The Nature and Properties of Soils,* 9th edition. Macmillan Publishing Company, New York.

Bridgham, S. D., J. Pastor, J. Janssens, C. Chapin, and T. Malterer. 1996. Multiple limiting gradients in peatlands: a call for a new paradigm. *Wetlands* 16:45–65.

Bridgham, S. D. and C. J. Richardson. 1993. Hydrology and nutrient gradients in North Carolina peatlands. *Wetlands* 13:207–218.

Bridgham, S. D., K. Updegraff, and J. Pastor. 1998. Carbon, nitrogen, and phosphorus mineralization in northern wetlands. *Ecology* 79:1545–1561.

Brooks, K. N. 1992. Surface hydrology. pp. 153–162. *In* H. E. Wright Jr., B. A. Coffin, and N. E. Aaseng (Eds.) *The Patterned Peatlands of Minnesota.* University of Minnesota Press, Minneapolis, MN.

Brown, D. A., S. P. Mathur, A. Brown, and K. J. Kushner. 1990. Relationships between some properties of organic soils from the southern Canadian Shield. *Canadian J. Soil Sci.* 90:363–377.

Camill, P. and J. S. Clark. 1998. Climate change disequilibrium of boreal permafrost peatlands caused by local processes. *Am. Naturalist* 151:207–222.

Chapin, C. T. 1998. Plant community response and nutrient dynamics as a result of manipulations of pH and nutrients in a bog and fen in northeastern Minnesota. Ph.D. dissertation, University of Notre Dame, Notre Dame, IN.

Chason, D. B. and D. I. Siegel. 1986. Hydraulic conductivity and related physical properties of peat, Lost River peatland, northern Minnesota. *Soil Sci.* 142:91–99.

Clark, M.H. and D.R. Kautz. 1997. Soil Survey of Matanuska–Susita Valley Area, Alaska. USDA Natural Resources Conservation Service. U.S. Govt. Printing Office. Washington, DC.

Clymo, R. S. 1983. Peat. pp. 159–224. *In* A. J. Gore (Ed.) *Mires: Swamp, Bog, Fen and Moor. Ecosystems of the World, 4A.* Elsevier Scientific Publishing, New York.

Clymo, R. S. 1987. Interactions of *Sphagnum* with water and air. pp. 513–529. *In* T. C. Hutchinson, and K. M. Meema (Eds.) *Effects of Atmospheric Pollutants on Forests, Wetlands and Agricultural Ecosystems.* Springer-Verlag, Berlin, Germany.

Cowardin, L. M., V. Carter, F. C. Golet, and E. T. LaRoe. 1979. *Classification of Wetlands and Deepwater Habitats of the United States.* FWS/OBS-79/31, Fish and Wildlife Service, U.S. Department of the Interior, Washington, DC.

Craft, C. B. and C. J. Richardson. 1997. Relationships between soil nutrients and plant species composition in Everglades peatlands. *J. Environmental Quality* 26:224–232.

Damman, A. W. H. 1986. Hydrology, development, and biogeochemistry of ombrogenous peat bogs with special reference to nutrient relocation in a western Newfoundland bog. *Can. J. Botany* 64:384–394.

Damman, A. W. H. and T. W. French. 1987. *The Ecology of Peat Bogs of the Glaciated Northeastern United States: A Community Profile.* Fish and Wildlife Service Report 85(7.16), U.S. Department of Interior, Washington, DC.

Day, J. H., P. J. Rennie, W. Stanek, and G. P. Raymond. 1979. *Peat Testing Manual.* Technical Memorandum Number 125, Associate Committee on Geotechnical Research, National Research Council of Canada, Ottawa, Canada.

Gafni, A. and K. N. Brooks. 1990. Hydraulic characteristics of four peatlands in Minnesota. *Can. J. Soil Sci.* 70:239–253.

Glaser, P. H. 1987. *The Ecology of Patterned Boreal Peatlands of Northern Minnesota: A Community Profile.* Fish and Wildlife Service Report 85(7.14), U.S. Department of Interior, Washington, DC.

Gore, A. J. P. 1983. Introduction. pp. 1–34. *In* A. J. P. Gore (Ed.) *Mires: Swamp, Bog, Fen and Moor. Ecosystems of the World, 4A.* Elsevier, New York.

Gorham, E. 1991. Northern peatlands: role in the carbon cycle and probable responses to climatic warming. *Ecological Applications* 1:182–195.

Gorham, E. and J. A. Janssens. 1992. Concepts of fen and bog re-examined in relation to bryophyte cover and the acidity of surface waters. *Acta Societatis Botanicorum Poloniae* 61:7–20.

Grootjans, A. P., R. van Diggelen, M. J. Wassen, and W. A. Wiersinga. 1988. The effects of drainage on groundwater quality and plant species distribution in stream valley meadows. *Vegetatio* 75:37–48.

Heinselman, M. L. 1963. Forest sites, bog processes, and peatland types in the Glacial Lake Agassiz Region, Minnesota. *Ecological Monographs* 33:327–374.

Heinselman, M. L. 1970. Landscape evolution, peatland types, and the environment in the Lake Agassiz Peatland Natural Area, Minnesota. *Ecological Monographs* 40:235–261.

Ingram, H. A. P. 1982. Size and shape in raised mire ecosystems: a geophysical model. *Nature* 297:300–303.

Ingram, H. A. P. 1983. Hydrology. pp. 67–158. *In* A. J. P. Gore (Ed.) *Mires: Swamp, Bog, Fen and Moor. A. General Studies.* Elsevier Scientific Publishing Company, Amsterdam, The Netherlands.

Kaila, A. 1956. Determination of the degree of humification of peat samples. *Maatal. Tiet. Aikak.* 28:18–35.

Kane, D. L. and L. D. Hinzman. 1988. Permafrost hydrology of a small arctic watershed. *Proceedings of the 5th International Conference on Permafrost,* Tapir Publishers, Trondheim, Norway.

Kivinen, E. and P. Pakarinen. 1981. Geographical distribution of peat resources and major peatland complex types in the world. *Annales Academiae Scientiarum Fennicae Series A* III. 132:1–28.

Koch, M. S. and K. R. Reddy. 1992. Distribution of soil and plant nutrients along a trophic gradient in the Florida Everglades. *Soil Sci. Soc. Am. J.* 56:1492–1499.

Koerselman, W., M. B. Van Kerkhoven, and J. T. A. Verhoeven. 1993. Release of inorganic N, P and K in peat soils; effect of temperature, water chemistry and water level. *Biogeochemistry* 20:63–81.

Lee, J. A., M. C. Press, S. Woodin, and P. Ferguson. 1987. Responses to acidic deposition in ombrotrophic mires in the U. K. pp. 549–560. *In* T. C. Hutchinson, and K. M. Meema (Eds.) *Effects of Atmospheric Pollutants on Forests, Wetlands and Agricultural Ecosystems.* Springer-Verlag, Berlin, Germany.

Lévesque, M. P. and S. P. Mathur. 1979. A comparison of various means of measuring the degree of decomposition of virgin peat materials in the context of their relative biodegradability. *Can. J. Soil Sci.* 59:397–400.

Malterer, T. J., E. S. Verry, and J. Erjavec. 1992. Fiber content and degree of decomposition in peats: a review of national methods. *Soil Sci. Soc. Am. J.* 56:1200–1211.

Mathur, S. P. and R. S. Farnham. 1985. Geochemistry of humic substances in natural and cultivated peatlands. pp. 53–85. *In* G. R. Aiken, D. M. McKnight, R. L. Wershaw, and P. MacCarthy (Eds.) *Humic Substances in Soil, Sediment, and Water: Geochemistry, Isolation, and Characterization.* John Wiley & Sons, New York.

Matthews, E. and I. Fung. 1987. Methane emission from natural wetlands: Global distribution, area, and environmental characteristics of sources. *Global Biogeochemical Cycles* 1:61–86.

Michaelson, G. J., C. L. Ping, and J. M. Kimble. 1996. Carbon storage and distribution in tundra soils of arctic, Alaska, U.S.A. *Arc. Alp. Res.* 28:414–424.

Moore, P. D. and D. J. Bellamy. 1974. *Peatlands.* Springer-Verlag, New York.

National Wetlands Working Group. 1997. *The Canadian Wetland Classification System.* Wetlands Research Centre, University of Waterloo, Waterloo, Ontario, Canada.

National Wetlands Working Group. 1988. *Wetlands of Canada.* Sustainable Development Branch, Environment Canada, Ontario, and Polyscience Publications, Montreal, Quebec, Canada.

Ovenden, L. 1990. Peat accumulation in northern wetlands. *Quaternary Research* 33:377-386.

Päivänen, J. 1973. Hydraulic conductivity and water retention in peat soils. *Acta For. Fenn.* 129:1–70.

Parent, L. E. and J. Caron. 1993. Physical properties of organic soils. pp. 441–458. *In* M. R. Carter (Ed.) *Soil Sampling and Methods of Analysis.* CRC Press, Boca Raton, FL.

Péwé, T. 1975. *Quaternary Geology of Alaska.* Geological Survey Professional Paper 835, U.S. Geological Survey. U.S. Govt. Printing Office. Washington, DC.

Ping, C. L., J. G. Bockheim, J. M. Kimble, G. J. Michaelson, D. K. Swanson, and D. A. Walker. 1998. Characteristics of cryogenic soils along a latitudinal transect in arctic Alaska. *J. Geophy. Res.* 103:28917–28928.

Ping, C. L., G. J. Michaelson, and J. M. Kimble. 1997a. Carbon storage along a latitudinal transect in Alaska. *Nutrient Cycling in Agroecosystems* 49:235–242.

Ping, C. L., G. J. Michaelson, W. M. Loya, R. J. Candler, and R. L. Malcolm. 1997b. Characteristics of soil organic matter in Arctic ecosystems of Alaska. pp. 157–167. *In* R. Lal, J. M. Kimble and B. A. Stewart (Eds.) *Carbon Sequestration — Advances in Soil Science.* CRC Press, Boca Raton, FL.

Post, W. M., J. Pastor, P. J. Zinke, and A. G. Stangenberger. 1985. Global patterns of soil nitrogen storage. *Nature* 317:613–616.

Racine, C. H. and J. C. Walters. 1994. Groundwater-discharge fens in the Tanana Lowlands, interior Alaska, U.S.A. *Arc. Alp. Res.* 26:418–426.

Rieger, S. 1983. *The Genesis and Classification of Cold Soils.* Academic Press, New York.

Riehle, J. R. 1985. A reconnaissance of the major Holocene tephra deposits in the upper Cook Inlet region, Alaska. *J. Volcano Geothr. Res.* 26:37–74.

Rosendahl, C. O. 1955. *Trees and Shrubs of the Upper Midwest.* University of Minnesota Press, Minneapolis, MN.

Rydin, H. 1985. Effect of water level on desiccation of *Sphagnum* in relation to surrounding *Sphagna. Oikos* 45:374–379.

Siegel, D. I. 1988. Evaluating cumulative effects of disturbance on the hydrologic function of bogs, fens, and mires. *Environmental Management* 12:621–626.

Siegel, D. I. 1992. Groundwater hydrology. pp. 163–172. *In* H. E. Wright Jr., B. A. Coffin, and N. E. Aaseng (Eds.) *The Patterned Peatlands of Minnesota.* University of Minnesota Press, Minneapolis, MN.

Siegel, D. I. and P. H. Glaser. 1987. Groundwater flow in a bog/fen complex, Lost River Peatland, northern Minnesota. *J. Ecol.* 75:743–754.

Silins, U. and R. L. Rothwell. 1998. Forest peatland drainage and subsidence affect soil water retention and transport properties in an Alberta peatland. *Soil Sci. Soc. Am. J.* 62:1048–1056.

Sjörs, H. 1950. On the relation between vegetation and electrolytes in north Swedish mire waters. *Oikos* 2:241–257.

Soil Classification Working Group. 1998. *The Canadian System of Soil Classification.* Agriculture and Agri-Food Canada Publication 1646, NRC Research Press, National Research Council of Canada, Ottawa, Canada.

Soil Survey Staff. 1996. *Keys to Soil Taxonomy.* 7th edition. Natural Resource Conservation Service, USDA, U.S. Govt. Printing Office. Washington, DC.

Soil Survey Staff. 1998a. *Keys to Soil Taxonomy*. 8th edition. Natural Resource Conservation Service, USDA, U.S. Govt. Printing Office. Washington, DC.

Soil Survey Staff. 1998b. Query for Histosol soil components in the National MUIR and STATSGO data sets 8/98. Natural Resource Conservation Service, USDA, Lincoln, NE and Statistical Laboratory, Iowa State University, Ames, IA.

SSSA. 1997. *Glossary of Soil Science Terms*. Soil Science Society of America, Madison, WI.

Stanek, W. and T. Silc. 1977. Comparisons of four methods for determination of degree of peat humification (decomposition) with emphasis on the von Post method. *Can. J. Soil Sci.* 57:109–117.

Tarnocai, C. 1998. The amount of organic carbon in various soil orders and ecological provinces in Canada. pp. 81–92. *In* R. Lal, J. M. Kimble, R. F. Follett, and B. A. Stewart (Eds.) *Soil Processes and the Carbon Cycle*. CRC Press, Boca Raton, FL.

Tarnocai, C., I. M. Kettles, and M. Ballard. 1995. *Peatlands of Canada*. Open File 3152. Geological Survey of Canada, Ottawa, Ontario, Canada.

Tarnocai, C., C. A. S. Smith, and C. A. Fox. 1993. *Guidebook-International Tour of Permafrost-Affected Soils in the Yukon and Northwest Territories of Canada*. Centre for Land and Biological Resources, Research Branch, Agriculture Canada, Ottawa, Canada.

Titus, J. E. and D. J. Wagner. 1984. Carbon balance for two *Sphagnum* mosses: water balance resolves a physiological paradox. *Ecology* 65:1765–1774.

Updegraff, K., J. Pastor, S. D. Bridgham, and C. A. Johnston. 1995. Environmental and substrate controls over carbon and nitrogen mineralization in northern wetlands. *Ecological Applications* 5:151–163.

Van Cleve, K., F. S. Chapin III, C. T. Dyrness, and L. A. Viereck. 1991. Element cycling in taiga forests: state-factor control. *BioScience* 41:78–88.

Verhoeven, J. T. A., E. Maltby, and M. B. Schmitz. 1990. Nitrogen and phosphorus mineralization in fens and bogs. *J. Ecol.* 78:713–726.

Verry, E. S. 1997. Hydrological processes of natural, northern forested wetlands. pp. 163–188. *In* C. C. Trettin, M. F. Jurgensen, D. F. Grigal, M. R. Gale, J. K. Jeglum (Eds.) *Northern Forested Wetlands: Ecology and Management*. CRC Press, Chelsea, MI.

Vitt, D. H. and W.-L. Chee. 1990. The relationships of vegetation to surface water chemistry and peat chemistry in fens of Alberta, Canada. *Vegetatio* 89:87–106.

von Post, L. and E. Granlund. 1926. Södra sveriges torvtillgångar. *Sveriges Geologiska Undersokning Arsbok Series C Avhandlingar och Uppsatser,* No. 335 19:1–127.

Walbridge, M. R. 1991. Phosphorus availability in acid organic soils of the Lower North Carolina Coastal Plain. *Ecology* 72:2083–2109.

Waughman, G. J. 1980. Chemical aspects of the ecology of some south German peatlands. *J. Ecol.* 68:1025–1046.

Weiss, R., J. Alm, R. Laiho, and J. Laine. 1998. Modeling moisture retention in peat soils. *Soil Sci. Soc. Am. J.* 62:305–313.

Wilkinson, L., M. Hill, and E. Vang. 1992. *Systat: Statistics*. Systat, Inc., Evanston, IL.

Zoltai, S. C. and C. Tarnocai. 1971. Properties of a wooded palsa in northern Manitoba. *Arc. Alp. Res.* 3:115–129.

CHAPTER 17

Hydric Soil Indicators in Mollisol Landscapes

James A. Thompson and Jay C. Bell

INTRODUCTION

Mollisols are mineral soils that usually develop under prairie vegetation. They are characterized by relatively thick, dark surface horizons resulting from an increased organic matter content, which can present problems for hydric soil identification due to the lack of visible iron-based redoximorphic features (Chapters 7 and 8) in the upper part of the soil profile. This chapter discusses some of the potential problems encountered when delineating hydric soils in Mollisol landscapes and describes specific hydric soil indicators developed for use in delineating hydric Mollisols.

The thick accumulations of organic matter associated with Mollisols are primarily due to the prairie grass vegetation, which has a dense fibrous root system. The roots, which proliferate in the soil even to depths of 75 cm or greater, have a high rate of annual turnover (Dahlman and Kucera 1965). Because a significant portion of the vegetation biomass is within the soil, root exudates and root death readily contribute substantial organic matter to the upper portions of these soils. When the grassland vegetation is disturbed by grazing or fire, the copious, fibrous roots of prairie grasses and the roots of leguminous forbs create abundant "ligno-protein" molecules of soil organic matter that resist oxidation and solution. The presence of abundant Ca stabilizes organic matter and darkens the soil, creating the characteristic deep black soil characteristic of grassland (Mollisol) soils in temperate regions worldwide. Mollisols, however, can also form under forest vegetation. These soils usually are associated with wetter soil environments or high Ca environments in which the organic matter can be both incorporated and stabilized at rates in excess of decomposition.

A Mollisol must have a mollic epipedon, which, by definition, is a thick (≥ 25 cm), dark (moist Munsell color of value ≤ 3 and chroma ≤ 3), strongly structured surface layer that has high organic carbon content (≥ 6 g kg^{-1}) and has a base saturation $\geq 50\%$ throughout (Soil Survey Staff 1994). Mollisols constitute approximately 22% of the total land area in the U.S. (Brady and Weil 1999) and are commonly found throughout the upper Midwest through the Central Plains. Additional areas of Mollisols in the U.S. are found in the Palouse area of Washington, Oregon, and Idaho (Figure 17.1). Globally, Mollisols are ubiquitous throughout the grasslands of subhumid to semiarid climates, including eastern Europe; Asia, from Turkey and the Ukraine eastward across Russia; the pampas region of South America; and parts of Mexico and Central America.

1-56670-484-7/01/$0.00+$.50

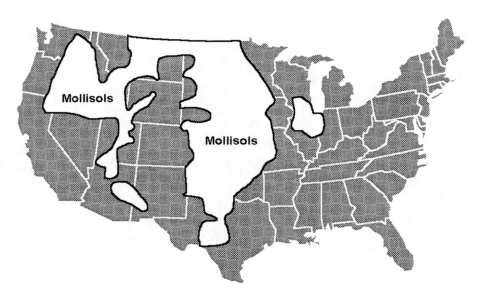

Figure 17.1 Generalized map showing the locations of major land areas of Mollisols in the United States. Moving east to west across these Mollisol regions, soils grade from Udolls to Ustolls to Xerolls in response to decreasing mean annual precipitation from east to west. Smaller areas of Mollisols, including Aquolls and Albolls, are found where vegetation, parent materials, and/or local hydrology favor organic matter accumulation. (Adapted from Brady, N. C. and Weil, R. R. 1999. *The Nature and Properties of Soils.* 12th ed. Prentice-Hall, Inc., New Jersey.

Aquolls and Albolls are Mollisols that formed in seasonally saturated soil conditions. By definition, Aquolls and Albolls are Mollisols with an aquic moisture regime and one or more of several diagnostic soil horizons. These include a histic epipedon above the mollic, redox concentrations within the mollic, or a gleyed subsurface horizon directly below the mollic or within 75 cm of the mineral soil surface (Soil Survey Staff 1994). In the U.S., extensive areas of Aquolls and Albolls are found in broad, flat landscapes with poor natural drainage, such as the Red River Valley of Minnesota and North Dakota, flood plains of the Mississippi, Missouri, Ohio, and Wabash Rivers, and the coastal plains of Louisiana and Texas. These are potentially hydric Mollisols that commonly occur in local depressions where rainfall and slope water can accumulate or in areas of groundwater discharge.

In Mollisol landscapes, delineators of wetlands have had difficulty in separating hydric and non-hydric soils. Common problems are the masking of visible morphologic indicators of hydric soil conditions by the abundant soil organic matter, and the presence of gray-colored carbonates that may mimic accepted hydric soil indicators. Because of these special problems encountered in Mollisols, the following discussion focuses on hydric soil indicators. In particular the focus will be on Aquolls in the humid region of the prairie pothole glaciated area of the U.S.

SEASONALLY SATURATED MOLLISOLS

Soil Formation and Organic Matter Dynamics

There are several factors that favor the development of Mollisols, including semiarid to sub-humid climates, grassland vegetation, and calcium-rich parent materials. These factors control the amount of organic materials added to the soil by favoring (i) increased below-ground biomass production, (ii) deposition of lignin-rich residues, and (iii) development of stabilizing bonds with Ca that slow the rates of organic matter decomposition. The dense, fibrous root systems of prairie

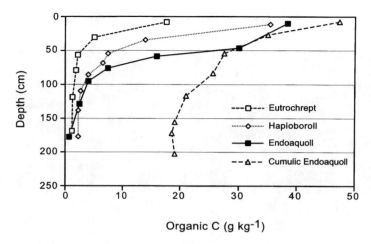

Figure 17.2 Soil organic carbon (C) with depth for an Inceptisol and three Mollisols of varying degree of wetness and mollic development. Differences among the Eutrochrept, the Haploboroll, and the Endoaquoll are greatest in the upper 50 cm; organic C contents are similar below 1 m. The Cumulic Endoaquoll differs in that the organic C content remains high to a depth below 2 m.

grasses with substantial annual root turnover and high lignin contents of the residue promotes high ($\geq$ 6 g kg^{-1}) organic matter levels with depths that are greater in Mollisols than in soils of other orders (Figure 17.2). The extensive root system also favors efficient nutrient cycling within the upper part of the soil, which prevents the loss of organic and mineral materials from below the root zone. A combination of the chemical composition of prairie grasses and the calcareous nature of many prairie soils leads to the formation of stable Ca-organic and clay-organic complexes in the upper part of the soil profile.

Increased organic matter additions and reduced organic matter losses result in higher organic matter levels in wetter Mollisols. For example, along a soil moisture gradient in the northern Great Plains, Munn et al. (1978) found that annual plant productivity increases as soil moisture increases. This results in greater organic matter accumulates in the soil. Also in wetter soil environments, the rates of organic matter decomposition are substantially lowered, mainly due to the lack of oxygen in the soil. Anaerobic decomposition of soil organic matter is slower than aerobic decomposition because anaerobic conditions produce end products that inhibit microbial activity or are toxic to soil microorganisms (Ross 1989). Increased organic matter incorporation coupled with decreases in soil organic matter losses due to wetness results in wet Mollisols that have even thicker and darker surface horizons, such as the soils examined by Richardson and Bigler (1984).

Mollisol Landscapes

These morphological differences in Mollisols that result from increased wetness can most easily be seen along soil moisture gradients found along many hillslopes (Figure 17.3). In general, with increasing soil wetness the A horizon(s) will (i) increase in thickness, (ii) decrease in Munsell value, and (iii) decrease in Munsell chroma. The wettest soils within these landscapes can have 1 to 2 meters of black (N 2/0) soil at the surface. These trends in soil morphology from well-drained to very poorly drained soils in Mollisol hillslopes have been described recently in Minnesota (Bell et al. 1995, 1996, Thompson and Bell 1996, 1998, Bell and Richardson 1997), Iowa (James and Fenton 1993, Khan and Fenton 1994, 1996), and elsewhere in the prairie pothole region (Richardson et al. 1994).

The color of the soil directly below the mollic epipedon may also be a useful indicator of wetness in Mollisol landscapes. In general, a low-chroma ($\leq$ 2) matrix is indicative of prolonged

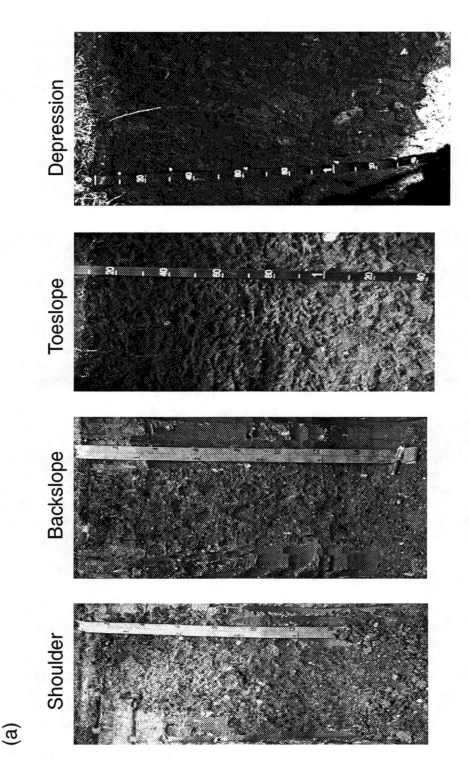

Figure 17.3 Soils along a hillslope transect in a Mollisol landscape (a) in west central Minnesota and (b) southeastern Minnesota. Note the thickening and darkening of the surface horizons. Though not visible here, the subsurface horizons change color as relative wetness of these soils increases.

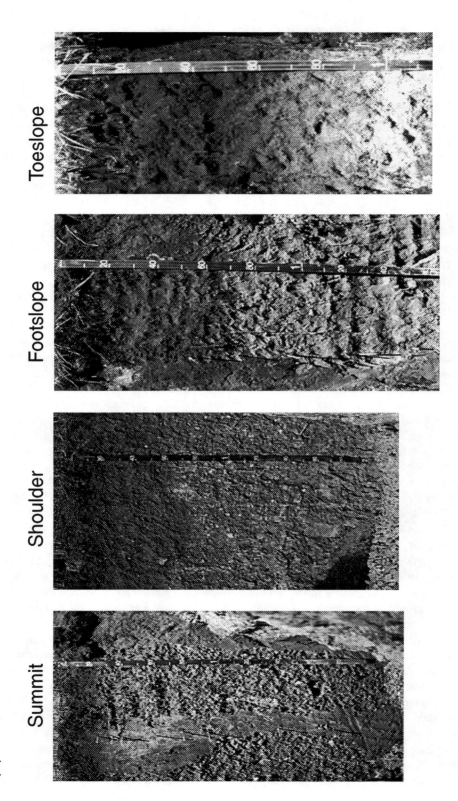

saturated soil conditions (Vepraskas 1994). However, in areas where parent materials are naturally gray, observation of low chroma colors alone does not confirm the presence of seasonally saturated soil conditions. In landscapes underlain by low-chroma parent materials it is advisable to look at relative differences in the soil color from well-drained soils to very poorly drained soils instead of relying on Munsell value and chroma observations from individual points. With increasing soil wetness there is a decrease in the chroma and an increase in the value of the upper B horizon. In Minnesota, changes in subsoil color were found to be only 1 or 2 chroma units (Thompson and Bell 1996, 1998). More pronounced dulling of the subsoil matrix color is seen in data presented by Khan and Fenton (1994, 1996) for a Mollisol hillslope in Iowa. Their data show a decrease in subsoil chroma from 4 to 2, an increase in value from 5 to 6, and a more yellow hue, with a change from 10YR to 5Y.

Redoximorphic features that indicate soil wetness are sometimes observed in or directly below the A horizon(s). In general, the depth to high-chroma mottles is shallower in wetter Mollisols. However, the lack of observable iron-based redoximorphic features does not preclude the occurrence of prolonged soil wetness in the mollic epipedon. The lack of observable redoximorphic features has been attributed to masking by organic coatings on ped and particle surfaces (Parker et al. 1985) that are frequently a direct result of the anaerobic conditions that inhibit organic matter decomposition.

Other Hillslope Processes

Erosion, groundwater discharge, and evaporative discharge on depression edges can also create problems for hydric soil delineation. Because many wetter Mollisols are found in local landscape positions that collect slope water (such as depressions and flood plains), they collect any materials carried by the water, e.g., sediments or dissolved solids. Erosion of A horizon material from upslope positions with redeposition in lower landscape positions can add significant amounts of darker soil materials to the surface of soils in lower landscape positions. Many wet Mollisols found in lower hillslope positions have buried A horizons. However, the more recently deposited surface materials in these wetter Mollisols tend to be lighter in color (chroma 1 or 2) than the underlying buried A horizon, which tends to be black (N 2/0).

Along with accumulation of organic matter, the depth to carbonates in Mollisols can impart information on hillslope hydrology (Chapter 3). The subhumid to semiarid climates in the upper Midwest favor retention of carbonates that are derived from calcareous parent materials. The depth to these carbonates, especially in the wetland areas, is related to wetland hydrology. In groundwater recharge wetlands where water flow is predominantly downward, leaching of carbonates can produce soil profiles that are relatively free of carbonates in the upper 1 to 2 m (King et al. 1983, Knuteson et al. 1989, Mausbach and Richardson 1994, Richardson et al. 1994). In groundwater discharge wetlands, where water flow is predominantly upward, carbonate contents are high throughout the profile, sometimes even accumulating at the soil surface (Arndt and Richardson 1988, Richardson et al. 1994).

Development of a highly calcareous soil horizon at the edge of depressions that trap surface water is a common feature in most young glaciated landscapes (Richardson et al. 1994) extending from humid climates in Iowa (Steinwand and Fenton 1995), northward into subhumid areas of North Dakota (Steinwand and Richardson 1989), and into the semiarid Canadian prairies (Miller et al. 1985) (Figure 17.4). The edges of the wetland have plants and a near-surface water table that combine to evaporate and transpire far more water from the soil than that which moves downward into the soil. The result is a strong reversal of leaching in which dissolved solids are transferred from the landscape and the wetland to the edge and concentrated by evapotranspiration. The concentrations reach levels that allow for formation of calcite and sometimes gypsum. These evaporites are naturally a gray color and may resemble depletions. Their occurrence can create a

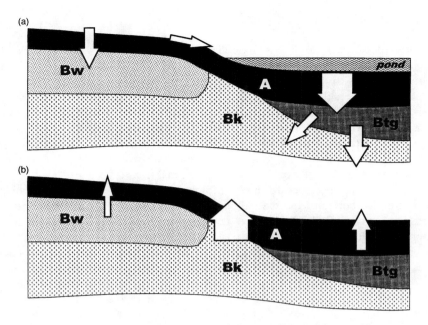

Figure 17.4 Generalized local hydrology of a wetland basin showing seasonal shifts between depression-focused recharge and edge-focused discharge. (a) During the wet season, water flow into the depression and infiltration promotes saturated downward flow. (b) During the dry season evapotranspiration promotes unsaturated upward flow. The edges of the depression have the longest period of time with upward flow and lack downward flow in the wet periods, producing shallow calcic horizons. Arrows are proportional to the amount of water flow. (Figure by J. L. Richardson, personal communication and used by permission).

false indication of redoximorphic features. However, depletions on surfaces can be noted in these gray evaporites. If the iron-depleted surfaces in gray calcareous horizons exceed 5% and occur within 30 cm of the surface, soil scientists in North Dakota use these features as a positive identification of a hydric soil, verifying the presence of a depletion feature (J. L. Richardson, personal communication).

HYDRIC SOIL INDICATORS

Hydric Soil Definition and Criteria

By definition, a hydric soil is a soil formed under saturated, flooded, or ponded conditions long enough during the growing season to develop anaerobic conditions in the upper part (*Federal Register* 1994). In most soils, the result of anaerobic conditions is reflected in the general hydric soil indicators (Environmental Laboratory 1987): organic soil or histic epipedon; sulfidic material; gleyed, low chroma, and low chroma/mottled soils; or iron and manganese concretions. However, some soils have morphologies that are difficult to identify as hydric because of (i) low chroma or red parent materials, (ii) high or low organic matter contents, (iii) high pH (which inhibits iron reduction), or (iv) natural or anthropogenic site disturbance. Hydric Mollisols are particularly difficult to identify because of high organic matter content and natural mixing by soil organisms. Consequently, Mollisols were designated as problem soils that require special consideration for the development of reliable field indicators of hydric soil conditions (Federal Interagency Committee for Wetlands Delineation 1989).

Field Indicators of Hydric Soils for Mollisols

The Natural Resources Conservation Service (NRCS) developed within their *Field Indicators of Hydric Soils in the United States* (Hurt et al. 1996) 11 field indicators targeted for soils with thick, dark A horizons, five of which are test indicators that require additional field verification. All indicators are the result of field-based investigations of actual wetlands and associated soils, and are intended for delineation of the edge of wetlands, not the wetter interiors. The indicators reflect available information derived from hydrology, vegetation, landscape position, and the best professional judgment of wetland scientists. The indicators are now officially accepted by the National Technical Committee for Hydric Soils and have been widely tested. As more testing occurs, new indicators are being developed for local and regional use.

The field indicators developed for use in Mollisol landscapes reflect the thickening and darkening of the surface horizons and the dulling and gleying of the subsurface horizons associated with increasing wetness. Soil morphologies that indicate hydric soil conditions in soils with dark surface horizons include (Hurt et al. 1996): (i) a depleted matrix immediately below the dark surface that does not have observable redoximorphic features, (ii) a dark surface with redoximorphic concentrations or depletions, (iii) yellow soil colors (hue of 2.5Y or yellower) below the dark surface, or (iv) depleted, gleyed, or yellow matrix below a calcic horizon below the dark surface. While the indicators will work in most field situations, there are instances where special care must be taken to properly identify hydric Mollisols. As is noted in the field indicators guide, a soil without an indicator may still be classified as hydric.

An important provision in the field indicators guide is included for soils where anaerobic conditions develop within the upper 30 cm, but short durations of saturated conditions in the upper part are not sufficient to lead to the development of anaerobic conditions that result in low chroma soil colors throughout the upper 30 cm (Hurt et al. 1996):

"Unless otherwise noted, all mineral layers above any of the indicators have dominant chroma 2 or less, or the layer(s) with dominant chroma of more than 2 is less than 15 cm thick."

In many soil landscapes we have observed, this exception can be applied to profiles with surface erosional sediments that presumably were deposited following European settlement after hydric soil morphology developed. The lighter-colored layers or horizons targeted by this provision must be less than 15 cm thick.

We have observed soils from lower landscape positions in steeply sloping agricultural sites that have up to 90 cm of lighter-colored (10YR 3/2 and 2/2) materials deposited above >60 cm of black (N 2/0) soil with distinct (7.5YR 5/8) redox concentrations (unpublished data). These may have formed under hydric soil conditions because of the thick, black buried A horizon and the toeslope landscape position of these soils. However, these soils would not be considered hydric based on the indicators due to the thick overlying accumulation of lighter-colored erosional sediments. While this represents an extreme example, any accumulation of erosional sediments greater than 15 cm would not permit a soil to be classified as hydric based on the current field indicators. In steeply sloping and/or intensively farmed landscapes, the presence of excessively thick erosional sediments is common.

A potentially useful iron-based field indicator of hydric soils in Mollisols is the presence of oxidized root channels (Mendelssohn 1993, Mendelssohn et al. 1995). When present, these features are distinct against the dark matrix colors of the mollic epipedon. We have observed oxidized root channels in hydric Mollisols in the late spring. However, they were not present in the some soils in the late summer of the previous year. Oxidized root channels may be ephemeral features in these soils and, therefore, only indicate recent soil anaerobic conditions. Also, cultivation or other soil mixing can obliterate these features.

Landscape Position

The importance of landscape position is recognized in certain field indicators. We feel that in Mollisol landscapes, landscape position is exceptionally important for hydric soil determinations. In the prairie pothole region, water accumulates in closed or nearly closed depressions. These depressions focus water from the local landscape. Hydric soils may occur in the depressions, and the nearly level toeslopes to concave footslopes surrounding the depression (Mausbach and Richardson 1994, Richardson et al. 1994, Thompson et al. 1997). In areas with open drainage, hydric soils may develop in flood plains, concave and convergent slopes, and areas with nearly level toeslopes to concave footslopes below steep slopes (Mausbach and Richardson 1994, Thompson et al. 1998). In general, converging slopes tend to concentrate water and allow it to accumulate long enough for hydric soils to develop (Mausbach and Richardson 1994). However, because other processes, such as erosion and deposition, contribute to the development of thick, dark surface horizons, some cumulic soils in low sloping landscape position may not be hydric. Thompson and Bell (1996) describe a soil with over 1 m of black (10YR 2/1) surface horizons, but their data indicate that the water table was not within 80 cm of the soil surface during 2 years of monitoring.

Differences in soil hydrology and resultant soil morphology between soils of different landscape positions on a single hillslope are illustrated in Figure 17.5. A very poorly drained soil located in a drainageway of a low-order, intermittent stream (Figure 17.5a) shows high water tables throughout the year. The thick, black (N2/0) surface horizons and high organic matter content reflects the high water table conditions observed in this soil. In a well-drained soil located on the summit position of this same landscape (Figure 17.5b), water tables are lower and fluctuations are greater than in the very poorly drained soil. While this soil is still a Mollisol, the thickness, darkness, and organic matter contents of the upper horizons are considerably less than the very poorly drained soil.

Profile Darkness Index

Thompson and Bell (1996) proposed a soil color index, the Profile Darkness Index (PDI), that quantified the trends of increasing A horizon thickness and darkness in Mollisol catenas. Calculated for each horizon with a Munsell value ≤ 3 and a Munsell chroma ≤ 3, PDI is equal to the sum (over all dark horizons within the profile) of the horizon thickness, divided by the quantity one plus the Munsell value times the Munsell chroma:

$$PDI = \sum_{i=1}^{n} \frac{A \ horizon \ thickness_i}{(V_i C_i) + 1}$$

where thickness is measured in centimeters, V is Munsell value, C is Munsell chroma, and n is the total number of A horizons described. A plot of PDI along a transect from summit to depression in a Mollisol landscape in west central Minnesota (Figure 17.6) illustrates the landscape-scale trends in PDI, which reflect the observed thickening and darkening of the surface horizons as soil wetness increases.

The use of PDI for hydric soil identification and delineation requires setting a threshold value that separates hydric from non-hydric soils. Based on variations in PDI among three study sites (Thompson and Bell 1996, 1998), the threshold value would be specific to soils of similar climate and/or parent materials. These differences among only three sites accentuate the necessity for regionalization of this approach. As with other proposed indicators of hydric soils, the PDI threshold value will change among climates and parent materials. This approach is still considered to be experimental and requires local calibration for detailed delineations of hydric soils.

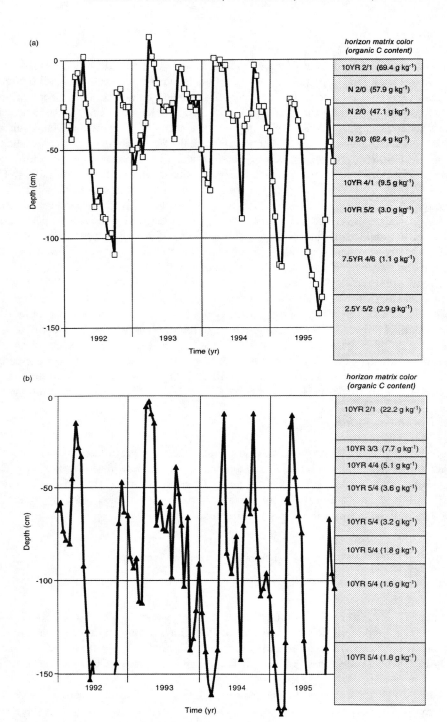

Figure 17.5 Observed water table position and morphological and chemical properties of (a) a very poorly drained soil in a drainageway landscape position, and (b) a well-drained soil in a summit landscape position. The water table is above or near the soil surface in the spring and fall at both landscape positions. However, while the water table falls considerably in the well-drained soil, in the very poorly drained soil it remains within 50 cm of the soil surface for extended periods even during the summer. These differences in hydrology are reflected in the higher organic C content and thicker, darker soil colors of the wetter soil.

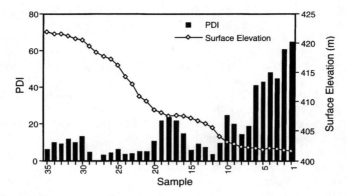

Figure 17.6   Variation in the Profile Darkness Index (PDI) and surface elevation along a hillslope transect in west central Minnesota. The soils of this landscape are shown in Figure 17.3(a). The PDI values are higher where slope gradients are lowest (e.g., from samples 1 to 10, 16 to 19, and 30 to 35), with a general increase across the transect toward the wetter, lower landscape positions (e.g., the toeslope [sample 5] and depression [sample 1]).

CONCLUSIONS

The identification of hydric soils in Mollisol landscapes is problematic because of thick, dark surface accumulations of organic-rich soil materials, particularly in wet Mollisols. While common Fe-based field indicators of hydric soils are not always useful for the identification of hydric Mollisols, other diagnostic soil and landscape features can be used. The accumulation of organic C to form thick, dark surface horizons with the presence of redoximorphic features — or without redoximorphic features but with a reduced or depleted horizon immediately below the dark surface — usually reflects seasonal saturated conditions in Mollisol landscapes. In addition to soil morphology, descriptions of landscape position, which aid in understanding where in the landscape water will tend to accumulate, are often also useful in field identification of hydric soils.

REFERENCES

Arndt, J. L. and J. L. Richardson. 1988. Hydrology, salinity, and hydric soil development in a North Dakota prairie-pothole wetland system. *Wetlands* 8:94–108.

Bell, J. C. and J. L. Richardson. 1997. Aquic conditions and hydric soil indicators for Aquolls and Albolls. pp. 23–40. *In* M. J. Vepraskas and S. Sprecher (Eds.) *Aquic Conditions and Hydric Soils: The Problem Soils.* SSSA Spec. Publ. Soil Sci. Soc. Am., Madison, WI.

Bell, J. C., J. A. Thompson, and C. A. Butler. 1995. Morphological indicators of seasonally saturated soils for a hydrosequence in Southeastern Minnesota. *J. Minn. Aca. Sci.* 59:25–34.

Bell, J. C., C. A. Butler, and J. A. Thompson. 1996. Soil hydrology and morphology of three Mollisol hydrosequences in Minnesota. *In* J. Wakely and S. Sprecher (Eds.) *Preliminary Investigations of Hydric Soil Hydrology and Morphology in Alaska, Indiana, North Dakota, Minnesota, and Oregon.* Wetlands Research Program Technical Report. U. S. Army Corps of Engineers, Waterways Exp. Sta. Vicksburg, MS.

Brady, N. C. and Weil, R. R. 1999. *The Nature and Properties of Soils.* 12th ed. Prentice-Hall, Inc., New Jersey.

Dahlman, R. C. and C. L. Kucera. 1965. Root productivity and turnover in native prairie. *Ecology* 46:84–89.

Environmental Laboratory. 1987. *Corps of Engineers Wetlands Delineation Manual.* Tech. Rep. Y-87-1. U.S. Army Engineer Waterways Exp. Stn., Vicksburg, MS.

Federal Register. 1994. *Definition of Hydric Soils.* USDA–Nat. Res. Cons. Ser., Vo. 59 (133)/Wed. July 13/p. 35681. U.S. Govt. Printing Office. Washington, DC.

Federal Interagency Committee for Wetlands Delineation. 1989. *Federal Manual for Identifying and Delineating Jurisdictional Wetlands.* U.S. Army Corps of Engineers, U.S. Environmental Protection Agency, U.S. Fish and Wildlife Service, and U.S.D.A. Soil Conservation Service, Washington, DC. Cooperative technical publication.

Hurt, G. W., P. M. Whited, and R. F. Pringle (Eds.). 1996. *Field Indicators of Hydric Soils in the United States*. USDA, NRCS, Fort Worth, TX.

James, H. R. and T. E. Fenton. 1993. Water tables in paired artificially drained and undrained soil catenas in Iowa. *Soil Sci. Soc. Am. J.* 57:774–781.

Khan, F. A. and T. E. Fenton. 1994. Saturated zones and soil morphology in a Mollisol catena of central Iowa. *Soil Sci. Soc. Am. J.* 58:1457–1464.

Khan, F. A. and T. E. Fenton. 1996. Secondary iron and manganese distributions and aquic conditions in a Mollisol catena of Central Iowa. *Soil Sci. Soc. Am. J.* 60:546–551.

King, G. J., D. F. Acton, and R. J. St. Arnaud. 1983. Soil–landscape analysis in relation to soil distribution and mapping at a site within the Wyburn Association. *Can. J. Soil Sci.* 63:657–670.

Knuteson, J. A., J. L. Richardson, D. D. Patterson, and L. Prunty. 1989. Pedogenic carbonates in a Calciaquoll associated recharge wetland. *Soil Sci. Soc. Am. J.* 53:495–499.

Lissey, A. 1971. Depression-focused transient groundwater flow patterns in Manitoba. *Spec. Pap. Geol. Assoc. Can.* 9:333–341.

Mausbach, M. J. and J. L. Richardson. 1994. Biogeochemical processes in hydric soil formation. *Current Topics in Wetland Biogeochemistry* 1:68–127.

Mendelssohn, I. A. 1993. *Factors Controlling the Formation of Oxidized Root Channels: A Review and Annotated Bibliography*. Technical report WRP-DE-5, U.S. Army Engineer Waterways Experiment Station, Vicksburg, MS.

Mendelssohn, I. A., B. A. Kleiss, and J. S. Wakeley. 1995. Factors controlling the formation of oxidized root channels: a review. *Wetlands* 15:37–46.

Miller, J. J., D. F. Acton, and R. J. St. Arnaud. 1985. The effect of groundwater on soil formation in a morainal landscape in Saskatchewan. *Can. J. Soil Sci.* 65:293–307.

Munn, L. C., G. A. Nielsen, and W. F. Mueggler. 1978. Relationships of soils to mountain and foothill range habitat types and production in western Montana. *Soil Sci. Soc. Am. J.* 42:135–139.

Parker, W. B., S. P. Faulkner, and W. H. Patrick. 1985. Soil wetness and aeration in selected soils with aquic moisture regimes in the Mississippi and Pearl River deltas. pp. 91–107. In *Wetland Soils: Characterization, Classification, and Utilization*. International Rice Research Institute, Los Baos, Laguna, Philippines.

Richardson, J. L., J. L. Arndt, and J. Freeland. 1994. Wetland soils of the Prairie Potholes. *Adv. Agron.* 52:121–171.

Richardson, J. L. and R. J. Bigler. 1984. Principal component analysis of prairie pothole soils in North Dakota. *Soil Sci. Soc. Am. J.* 48:1350–1355.

Ross, S. 1989. *Soil Processes: A Systematic Approach*. Routledge, New York.

Seelig, B. D. and J. L. Richardson. 1994. Sodic soil toposequence related to focused water flow. *Soil Sci. Soc. Am. J.* 58:156–163.

Soil Survey Staff. 1994. *Keys to Soil Taxonomy*. SCS–USDA, Washington, DC.

Steinwand, A. L. and T. E. Fenton. 1995. Landscape evolution and shallow groundwater hydrology of a till landscape in central Iowa. *Soil Sci. Soc. Am. J.* 59:1370–1377.

Steinwand, A. L. and J. L. Richardson. 1989. Gypsum occurrence in soils on the margin of semipermanent prairie pothole wetlands. *Soil Sci. Soc. Am. J.* 53:836–842.

Thompson, J. A. and J. C. Bell. 1996. Color index for identifying hydric soil conditions in seasonally-saturated Mollisols. *Soil Sci. Soc. Am. J.* 60:1979–1988.

Thompson, J. A. and J. C. Bell. 1998. Hydric conditions and hydromorphic properties within a Mollisol catena in southeastern Minnesota. *Soil Sci. Soc. Am. J.* 62:1116–1125.

Thompson, J. A., J. C. Bell, and C. A. Butler. 1997. Quantitative soil-landscape modeling for estimating the areal extent of hydromorphic soils. *Soil Sci. Soc. Am. J.* 61:971–980.

Thompson, J. A., J. C. Bell, and C. W. Zanner. 1998. Hydrology and hydric soil extent within a Mollisol catena in southeastern Minnesota. *Soil Sci. Soc. Am. J.* 62:1126–1133.

Vepraskas, M. J. 1994. *Redoximorphic Features for Identifying Aquic Conditions*. Tech. Bulletin 301. North Carolina Ag. Research Service, North Carolina State Univ., Raleigh, NC.

CHAPTER 18

Saline and Wet Soils of Wetlands in Dry Climates

Janis L. Boettinger and J. L. Richardson

INTRODUCTION

Wet soils occur in the arid and semiarid climates typical of central and western North America. Wet soils, and wetlands in general, are not prevalent in dry and seasonally dry climates, but their morphology and characteristics are considerably different from those in more humid climates. It is, therefore, important to understand the factors and processes involved in their distribution and formation. These wet soils and wetlands are commonly found at general groundwater discharge sites, such as flood plain edges, lacustrine plains (playas), artesian spring areas, and as a result of irrigated agriculture. Many of the wet soils in dry climates have formed under high evapotranspiration rates coupled with low effective precipitation, which concentrates salts in the soils of these wet areas. Therefore, geomorphology of water concentration, geological sources of the water and sediment, and evapo-concentration of salts are important.

Although these soils are wet, they usually contain carbonates, gypsum, and even more labile salts. Biomass accumulation is limited in many saline wetlands, and soil organic matter, not surprisingly, is limited. Productivity of some saline wetlands such as salt marshes is high, however, and detritus from some of these landscapes feed abundant wildlife. Nonetheless, accumulation of soluble salts and low concentration of organic matter probably hinder typical processes expected in wet soils, such as chemical reduction of Fe^{3+}, which limits the formation of redoximorphic features in saline wet soils.

Boettinger (1997) summarized the geographic distribution, parent material, landform, and vegetation of saline and wet soils mapped as Aquisalids (formerly Salorthids) in soil surveys of the United States; therefore that topic will not be covered here. We add that the soils in many wet areas in dry climates were possibly placed in Miscellaneous Land Classes (MLC) in soil surveys due to the small amount of vegetation cover (<10%). As the importance of saline wetland resources is recognized, more soils series will be established and more soil detail will be mapped (for example, even freshwater wetlands were placed in MLCs through the 1970s).

This chapter summarizes the factors and processes involved in the distribution and formation of saline and wet soils and wetlands of dry climates. These include environmental setting, pedogenesis, and the influence of human activity. The challenges of separating saline and wet soils into hydric and non-hydric soils are also addressed.

ENVIRONMENTAL SETTING

Hydrology

Wet soils of dry climates occur in areas of specific water discharge or accumulation. Open flow systems that have a stream flowing through the area, such as flood plains, have points of discharge (mentioned in Chapters 3, 9, and 12) where water can accumulate. Closed basin systems that lack external drainage, such as playas, serve as sinks for surface runoff and groundwater discharge. Wet soils can occur in the vicinity of springs in either open or closed flow systems. In addition, wet soils can be created in areas irrigated for agriculture, agriculture with drainage ditches where drainage water accumulates, or where water seeps from drainage ditches or irrigation delivery systems.

Climate

Arid and semiarid regions can receive a limited amount of precipitation distributed throughout the year, or can be seasonally dry. In addition, these climates are typified by high evapotranspiration rates, due to high amounts of incoming solar radiation, high albedo, and low humidity. Because annual evaporation exceeds effective precipitation, the soils in these regions frequently accumulate calcium carbonate, gypsum, and salts more soluble than gypsum. Occasionally, saline and wet soils occur in subhumid areas with a pronounced dry season or in a landscape position that has a source of salts and restricted drainage.

Geomorphology and Salt Source

Playas and lake basins represent the most common geomorphic setting for saline and wet soils and wetlands of dry climates and have been studied extensively. For example, playa systems are common throughout the Basin and Range physiographic province of western North America and provide the setting for almost half the soils mapped as Aquisalids in the United States (Boettinger 1997). Playas and lake plains in arid and semiarid regions range from nearly salt-free intermittent ponds to hypersaline salt flats. These landforms occur in basins that lack outlets or do not have an integrated drainage system. Because of the closed nature of the basin and impact of long-term climatic variation, the amount of water in these basins has varied widely over recent geologic time. During pluvial cycles, lakes of significant depth and areal extent exist in these basins. With a climate change to drier and warmer conditions, lack of water causes lake levels to drop or disappear entirely.

Initial salt concentration in closed basin pluvial lake water was probably low. Dissolved products in these lakes were primarily derived from chemical weathering of minerals in the hydrologic and geologic source areas. In the Great Basin, geologic sources are dominantly sedimentary rocks in the east, grading to igneous and metamorphic rocks in the west. Other sources include rain, snow, and aerosol particulates. As lake water evaporated, dissolved products concentrated in lake water and sediments. The characteristics of playa sediments are determined by the balance between clastic sedimentation during pluvial cycles and salt deposition during nonpluvial conditions. Soils formed during dry nonpluvial periods are subject to diurnal and seasonal climatic fluctuations, which can cause fluctuations in the presence and mineralogic composition of salts (e.g., Keller et al. 1986, Eghbal et al. 1989).

On a smaller scale, saline and wet soils occur in recharge–throughflow–discharge wetland systems, typical of glaciated terrain in subhumid climates of the northern Great Plains of central North America (Arndt and Richardson 1989). These landscapes are characterized by little vertical relief (<20 m) on expansive plains of till derived from dolomitic and sulfur-rich marine sedimentary rocks that are underlain by bedrock, which restricts downward movement of water. Arndt and Richardson (1989) found that water is subject to increasing evaporation and concentration as it moves from recharge areas to intermediate throughflow areas, and last to the local sink in the discharge area. In general, soils in recharge wetlands are nonsaline, whereas soils in discharge wetlands are often saline.

Saline wet soils and wetlands also occur near artesian springs in either open or closed flow systems. In closed flow systems, water and salt from artesian springs may not be easily differentiated from overland sources. In contrast, the contribution of saline groundwater from artesian springs is more easily distinguished in open flow systems. For example, the saline water source for about 600 ha of the Cache series (fine, mixed, mesic, semiactive Typic Aquisalid) in Cache Valley, northern Utah, is artesian springs. These springs occur in a north–south trend, apparently at the southern termination of the Dayton fault (S.U. Janecke, Utah State University, Department of Geology, personal communication, 1998). The electrical conductivity (EC) of the spring water is about 5 dS m^{-1} (U.S. Geological Survey 1970). Although the very slowly permeable soils in this saline wetland area are externally drained, salts have accumulated in these soils, evidenced by saturated paste EC up to 47 dS m^{-1} (Erickson and Mortensen 1974).

Saline wetlands can also occur on coastal flats or terraces that are periodically inundated by salt water. Seven of the 24 soil series mapped as Aquisalids in the U.S. occur in these environmental settings (Boettinger 1997). Tidal water and storm surges can quickly provide both water and dissolved salts. If these areas are in warm and/or seasonally dry climates with high evaporation rates, saline wet soils may form.

Some saline and wet soils have been saturated and salinized due to human activity. Seepage from drainage ditches along roads has caused the water table to rise in adjacent areas, and salts are concentrated by evaporation (Skarie et al. 1987). Irrigated agriculture in arid and semiarid regions, however, is probably the major anthropogenic cause of saline and wet soils and artificial wetlands (Boettinger 1997). Seepage from irrigation water canals, such as those in central Utah, constructed in marine shale is related to salinization of adjacent soils. Irrigation may cause ponding in depressional areas and raise the local water tables. These areas of water accumulation may salinize as water is evaporated and salts are concentrated.

PEDOGENESIS

Saline and wet soils in dry climates, such as Aquisalids, may form if the following three conditions are met: (1) there must be a source of water; (2) there must be a source of ions or dissolvable minerals that can be translocated by water; and (3) there must be a process of solution concentration (i.e., evapotranspiration exceeds precipitation).

A major pedogenic process involved in the formation of saline and wet soils in dry climates is salinization, or the accumulation of evaporite minerals. Hardie and Eugster (1970) clearly explained evaporite accumulation and the various mineral assemblages that result (Figure 18.1). Their paper is based on the sequences of mineral accumulation that occur by progressive salt precipitation via evaporation in closed basins. Whittig and Janitzky (1963), Gile and Grossman (1979), Gumuzzio et al. (1982), Last (1984), Keller et al. (1986), Arndt and Richardson (1989), and Eghbal et al. (1989) related evaporite mineral formation and salt accumulation in general to pedogenesis.

The Hardie–Eugster (1970) evaporite sequence is based on "chemical divides," and depends on the initial ionic composition of the solution undergoing evaporation. Salts are ionically bonded, binary compounds that precipitate from solution when the product of the activity of the ions in solution equals or exceeds the solubility product for that particular mineral. A dilute solution becomes more concentrated with ions as water evaporates. The first salt to precipitate from solution is calcite, a mineral of the compound calcium carbonate ($CaCO_3$), which has one mole of Ca^{2+} that

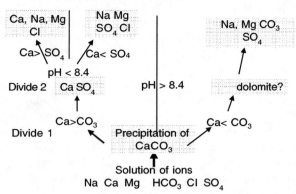

Divide 3 Salt efflorescence or solutions

Figure 18.1   An illustration of a closed basin brine evolution model for soils. Starting with a series of common ions found in soil solutions calcite precipitates followed by gypsum in one pathway. The precipitation of equal molar amounts of ions tends to concentrate the ion with the largest solution concentration and all the unaffected ions in solution. (Modified from Hardie, L. A. and H. P. Eugster. 1970. The evolution of closed-basin brines. *Miner. Soc. Am. Spec. Paper* 3:273–290.)

combines with one mole of CO_3^{2-} to produce one mole of $CaCO_3$. Calcite precipitation creates a "chemical divide." In solution, it is highly unlikely that Ca^{2+} and CO_3^{2-} will occur in exactly equal portions. If Ca^{2+} is dominant, then the available CO_3^{2-} in the concentrating solution is quickly consumed during precipitation of calcite. The resulting solution will be enriched with Ca^{2+} with respect to carbonate. If CO_3^{2-} is dominant, the Ca^{2+} is consumed as calcite is precipitated, causing the resulting solution to be enriched with CO_3^{2-} with respect to calcium. We believe there are two distinct systems of chemical divides or pathways of pedogenesis, "alkaline" and "gypsiferous."

If sulfate (SO_4^{2-}) is present in the solution enriched with Ca^{2+}, the excess Ca^{2+} combines next with SO_4^{2-} and gypsum ($CaSO_4 \cdot 2H_2O$) precipitates. This direction is termed the "gypsiferous" path of pedogenesis. Precipitation of gypsum creates another chemical divide, allowing either Ca^{2+} or SO_4^{2-} to enrich the solution, depending on whether Ca^{2+} or SO_4^{2-} was in higher molar concentration before gypsum formation (Figure 18.1).

The opposite side of the first divide has CO_3^{2-} higher than Ca^{2+}. As CO_3^{2-} concentration increases, the pH increases. This direction is termed the "alkaline" path of pedogenesis. The ratio of bicarbonate to carbonate is 1:1 at a pH of 10.33 (Lindsay 1979). Heat, agitation, or other action depletes the carbon dioxide and bicarbonate from solution. When bicarbonate is depleted, the carbonate increases, as does the pH. Note that carbonate is nearly nonexistent until calcite starts to precipitate but increases logarithmically with a decrease in bicarbonate. A 10× increase in carbonate increases the pH by one unit. Hanson et al. (1990) related this occurrence in a prairie pond. During the winter, pH was <6.0, whereas in summer pH was >9.0. Warm weather and wave agitation caused carbon dioxide to degas from water and increase the carbonate at the expense of bicarbonate, clouding the water with fine precipitates of calcite.

Degassing of carbon dioxide during warm weather, high evaporation, and perhaps some wave action enhances calcite precipitation and increases alkalinity. We expect two distinct pathways of pedogenesis based on climate: (1) sodium carbonate-enriched, highly alkaline systems (pH > 8.5); and (2) calcium carbonate systems with moderately alkaline conditions with pH 7.8 to 8.7 (Arndt and Richardson 1989, Steinwand and Richardson 1989). The geochemistry of these chemical interactions is discussed in some detail in Arndt and Richardson (1992). Alkaline Natraquolls associated with ponds should be expected in warmer climates where evaporation exceeds precipitation. Calciaquolls should occur in cooler climates. The alkaline pathway has consumed its Ca in calcite precipitation. There is no Ca remaining to form gypsum. Further calcite precipitation depletes Ca but has no impact on Mg. Dolomite [$CaMg(CO_3)_2$] should be the next evaporite to form along the alkaline branch of the Hardie–Eugster (1970) chemical divides system. Last and DeDeckker (1992) measured several meters of evaporites in Lake Beeac in Australia and found that the sediments were largely authigenic dolomite and magnesite ($MgCO_3$). Sherman et al. (1962) observed dolomitic Bk horizons in soils of glacial Lake Agassiz. Although there is controversy about whether dolomite can form in soils, the junior author of the present chapter reexamined these soils, and indeed they were dolomite, and the C and O isotopes proved that the materials were deposited *in situ*. Kohut et al. (1995) observed authigenic dolomite in Alberta soils. After dolomite forms, either Mg or carbonate will be in short supply, creating another chemical divide. The low carbonate pathway, contains Mg chloride or sulfate, whereas the high carbonate (low Mg) pathway contains Na carbonates, chlorides, and sulfates.

Further evaporation, whether it be along the alkaline or gypsiferous pathway, results in the precipitation of highly soluble evaporite minerals. The presence of these labile minerals may fluctuate diurnally and seasonally, depending on fluctuations in environmental conditions such as temperature, relative humidity, amount of solar radiation, and amount and intensity of precipitation events. The most important point here is that with evaporation, a sequence of evaporites forms and each of the preceding evaporites affects the next generation of evaporites by removal of material. Initial solution composition, evaporation, and precipitation effectively control the types of evaporites (Hardie and Eugster 1970, Last 1984, Keller et al. 1986, Arndt and Richardson 1992).

SOIL MORPHOLOGY

Aquisalids and other saline and wet soils usually have an A horizon that is enriched in organic matter with respect to the underlying strata. With sparse vegetation, the horizon may be difficult to locate or not exist at all. Boettinger (1997) stated that soils that lack vegetation or are very sparsely vegetated, such as soil in the intermound areas of mound–intermound complexes in saline lake basins in the Great Basin in Nevada and western Utah, may lack an A horizon. Thus, C horizons are probably present at the soil surface. Soluble salt crystals and salt crusts on the soil surface (Figure 18.2), or Az or Cz horizons can disappear quickly during rain events but can reappear during times of high evaporation.

Conventions for describing Bz horizons are in flux. The presence or absence of soil structure is applied inconsistently when naming subsurface horizons. Boettinger (1997) suggested that Bz horizons should be recognized if soluble salt and soil structure can be recognized. We certainly concur.

Additionally, lack of vegetation at the time a "soil" is viewed should not be used as criterion that an area is "not soil." Both the potential for plant growth and the presence of bacteria or other

Figure 18.2   Salt crust formed by evaporation of water that has seeped through marine shale of the Mancos Formation in central Utah, locally known as Emery County "snow."

microbial plants could be at the site and be overlooked by an observer. If the "soil" has structure or other pedogenic features and has formed by processes that can be considered soil-forming processes, the "soil" should be soil without quote marks.

As a general rule, classic redoximorphic features are not present or are poorly expressed in saline and wet soils. Boettinger (1997) reviewed the morphological properties and depths to redox concentrations present in the typical pedons of Aquisalids. Only 10 of the 24 series of Aquisalids had redoximorphic features in the upper 30 cm. Boettinger (1997) listed several possible reasons, but we speculate that the major reason is the lack of carbon as an electron donor and microbial energy source not allowing the Eh to decrease to a level where iron can reduce.

Many soils rich in carbonates, gypsum, and/or soluble salts will change color upon exposure to air. Yellowing of the hue, such as 2.5Y changing to 5Y, or a change in chroma, such as from 3 to 2, upon drying may be due to altered hydration or further precipitation of evaporites (Boettinger 1997). Changes in crystal structures and size upon drying can also change the appearance and perhaps color patterns of soils (Last 1984, Keller et al. 1986).

Some saline wet soils can develop very unusual color patterns. Timpson et al. (1986) noted natrajarosite bright yellow colors and red gypsum crystals associated with sulfatic soils in saline seeps in North Dakota. In these soils, local pH can be low because of sulfide oxidation. The bizarre color association that occurs in sodic–saline seeps with jarosite needs to be considered as exceptional but not rare because Cretaceous shales and Tertiary lignites are very widespread.

INDICATION OF HYDRIC CONDITIONS (CONCLUSIONS)

With no data published on the saturation and redox potential of saline and wet soils, Boettinger (1997) concluded that some highly saline soils might not experience reducing conditions. The major limitations to reducing conditions in soils are the chemical or physical constraints on microbial activity. Very negative osmotic potential, low organic C and N supplies in barren or sparsely vegetated soils, as well as salinity, alkalinity, and induced nutrient deficiencies were all cited as potential limitations to microbially mediated reduction during periods of saturation.

However, new data show that saline and wet soils subject to saturation within 30 cm of the surface for several weeks during the growing season can experience reducing conditions. Sutcliffe (1999) monitored water table depth, pH, EC, temperature, and dissolved oxygen (DO) in four soils in a hillslope catena formed in marine shale of the Mancos Formation in central Utah. These soils, ranging from nonsaline and never saturated to various degrees of salinity (slight to strong) and saturation, were affected by seepage from upslope irrigation canals. One soil in the catena, a Typic Halaquept with an EC of about 20 dS m^{-1}, clearly became anaerobic as the soil warmed and remained anaerobic until the water table dropped to about 30 cm (Table 18.1). This soil develops a fluffy salt crust for part of the year, similar to that shown in Figure 18.2, but does not express any redoximorphic features in the upper part of the soil (Figure 18.3).

In light of these data, we suggest that the presence of salts more soluble than gypsum in the upper 30 cm indicates a hydric soil. The presence of a salt crust should be sufficient as a field indicator of a hydric soil subject to periodic saturation of the upper part of the soil from a saline water table. The morphology of the crust, especially crystals, can be used to identify composition, precipitation history, and the nature of ponding during either annual or diurnal hydrologic activities (Last 1984). Ponded Aquisalids develop brittle salt crusts on the soil surface, whereas Aquisalids with soil saturation due to a capillary fringe develop fluffy, almost snow-like crusts (Boettinger 1997). If the soil color changes upon drying (increase in value coupled with decreasing chroma or yellowing of hue), and carbonates, gypsum, and/or soluble salts are present, redoximorphic features and hydric soil indicators may not have formed or could be masked by the evaporite minerals.

Table 18.1   **Physical, Chemical, and Redox Properties of a Typic Halaquept Affected by Seepage of Irrigation Water through Mancos Shale (Water table depth, pH, EC, temperature, and dissolved oxygen [DO] averaged from measurements in duplicate 30-cm piezometers. Redox potential [Eh] represents average of four redox probes. Anaerobic Eh threshold, the Eh below the soil is considered to be anaerobic, was calculated using pe = 12-pH, where pe = Eh/59. [McBride 1994])**

Date	Depth to Water (cm)	pH	EC (dS m⁻¹)	Temperature (°C)	DO (mg L⁻¹)	Eh (mV)	Anaerobic Eh Threshold (mV)
5/9/98	12	8.1	23.2	11.5	4.9	547	230
5/19/98	9	8.1	21.2	12.6	4.0	261	231
6/3/98	15	7.8	20.4	14.3	2.2	37	250
6/17/98	14	7.7	20.5	13.1	1.9	−23	254
7/2/98	15	7.6	19.7	17.9	2.3	−72	261
8/21/98	17	7.7	25.7	19.9	NM†	−22	252
8/26/98	29	—‡	—	—	—	160	(not calculated)

† NM, not measured due to equipment failure.

‡ —, data not collected due to insufficient amount of solution in the 30-cm piezometers.

Data from Sutcliffe, K.D. 1999. Dynamics of irrigation-induced and saline wet soils, central Utah. M.S. thesis, Utah State University, Logan, UT.

Figure 18.3   Typic Halaquept pedon in a catena affected by seepage from upslope irrigation canals in central Utah. There are no redoximorphic features within 30 cm of the soil surface; the brownish-gray color is inherited from the Mancos Shale parent material. Water level is at 47 cm; pit had to be bailed several times while sampling in November 1996. A Cr horizon occurs at 78 cm. (Photo taken by W.C. Lynn.)

REFERENCES

Arndt, J.L. and J.L. Richardson. 1989. Geochemistry of hydric soil salinity in a recharge-throughflow-discharge prairie-pothole wetland system. *Soil Sci. Soc. Am. J.* 53:849–955.

Arndt, J.L. and J.L. Richardson. 1992. Carbonate and gypsum chemistry in saturated neutral pH soil environments. pp 179–188. *In* Robarts, R.D. and M.L. Bothwell (Eds.) *Aquatic Ecosystems in Semi-arid Regions: Implications for Resource Management.* N.H.R.I. Symposium series 7, Environment Canada, Saskatoon.

Boettinger, J.L. 1997 Aquisalids (Salorthids) and other wet saline and alkaline soils. Problems identifying aquic conditions and hydric soils. pp 79–97. *In* Vepraskas, M.J. and S. Sprecher (Eds.) *Problems of Identifying Hydric Soils.* Soil Sci. Soc. Am. Spec. Publ. No. 50. Madison, WI.

Eghbal, M.K., R.J. Southard, and L.D. Whittig. 1989. Dynamics of evaporate distribution in soils on a fan–playa transect in the Carrizo Plain, California. *Soil Sci. Soc. Am. J.* 53:898–903.

Erickson, A.J. and V.L. Mortensen. 1974. *Soil Survey of Cache Valley Area, Utah: Parts of Cache and Box Elder Counties.* USDA Soil Conservation Service, U.S. Govt. Printing Office. Washington, DC.

Gile, L.H. and R.B. Grossman. 1979. *The Desert Project Soil Monograph.* U.S. Soil Conservation Service, U.S. Govt. Printing Office. Washington, DC.

Gumuzzio, J., J. Batille, and J. Casas. 1992. Mineralogical composition of salt effluorescences in a Typic Salorthid, Spain. *Geoderma* 28:39–51.

Hanson, M.A., M.G. Butler, J.L. Arndt, and J.L. Richardson. 1990. Calcite supersaturation, precipitation and turbidity in a shallow prairie lake. *Freshwater Biology* 24:547–556.

Hardie, L.A. and H.P. Eugster. 1970. The evolution of closed-basin brines. *Miner. Soc. Am. Spec. Paper* 3:273–290.

Keller, L.P., G.J. McCarthy, and J.L. Richardson. 1986. Mineralogy of soil evaporites in North Dakota. *Soil Sci. Soc. Am. J.* 50:1069–1071.

Kohut, C., K. Muehlenbachs, and N.J. Dudas. 1995. Authigenic dolomite in a saline soil in Alberta, Canada. *Soil Sci. Soc. Am. J.* 59:1499–1504.

Last, W.M. 1984. Sedimentology of playa lakes of the northern Great Plains. *Can. J. Earth Sci.* 21:107–125.

Last W.M. and P. DeDeckker. 1992. Paleohydrology and paleochemistry of Lake Beeac, a saline playa in southern Australia. pp 63–74. *In* R.D. Robarts and M.L Bothwell (Eds.) *Aquatic Ecosystems in Semi-Arid Regions: Implications for Resource Management.* N.H.R. I. Symposium Series 7, Environment Canada. Saskatoon.

Lindsay, W.L. 1979. *Chemical Equilibria in Soils.* Wiley-Interscience, NY.

McBride, M.B. 1994. *Environmental Chemistry of Soils.* Oxford University Press. Oxford, U.K.

Sherman, G.D., F. Schultz, and F.J. Alway. 1962. Dolomitization in soils of the Red River Valley, Minnesota. *Soil Sci.* 94:304–313.

Skarie, R.L., J.L. Richardson, G.J. McCarthy, and A. Maianu. 1987. Evaporite mineralogy and groundwater chemistry associated with saline soils in eastern North Dakota. *Soil Sci. Soc. Am. J.* 51:1372–1377.

Steinwand, A.L. and J.L. Richardson. 1989. Gypsum occurrence in soils on the margin of semi-permanent prairie pothole wetlands. *Soil Sci. Soc. Am. J.* 53:836–842.

Sutcliffe, K.D. 1999. Dynamics of irrigation-induced and saline wet soils, central Utah. M.S. thesis, Utah State University, Logan, UT.

Timpson, M.E., J.L. Richardson, L.P. Keller, and G.J. McCarthy. 1986. Evaporite mineralogy associated with saline seeps in Southwestern North Dakota. *Soil Sci. Soc. Am. J.* 50:490–493.

U.S. Geologic Survey. 1970. Selected hydrologic data, Cache Valley, Utah and Idaho. Utah Basic Release Data No. 21, Salt Lake City, UT.

Whittig, L.D. and P.J. Janitzky. 1963. Mechanisms of formation of sodium carbonate in soils. I. Manifestations of biological conversions. *J. Soil Sci.* 14:322–333.

CHAPTER 19

Wetland Soil and Landscape Alteration by Beavers

Carol A. Johnston

INTRODUCTION

Long before humans began constructing wetlands, the beaver (*Castor canadensis* Kuhl) was changing the face of the North American continent, creating and modifying wetlands by building dams. The family Castoridae arose in the Oligocene, and the genus *Castor* arose in the Pliocene (Jenkins and Busher 1979). Beavers have existed throughout the Pleistocene, and giant beavers the size of bears existed as recently as the deglaciation of the North American continent (Erickson 1962).

There were estimated to be 60 to 400 million beavers in North America prior to European settlement (4 to 26 beavers/km^2; Duncan 1984). Beginning in the 15th century, there was a great demand for beaver pelts because the soft underfur was used for felt in the manufacture of men's hats. By the time the steel trap was invented in the mid-1850s, beavers were extirpated from much of their former range. Thus, although beavers had been engineering landscape and soil changes for millennia, their influence was not as evident during the early part of this century.

As beaver-felt hats fell out of fashion, there was a decrease in the price and demand for beaver pelts, and beaver populations have rebounded throughout the country. The current North American beaver population is estimated to be 6 to 12 million, so although their influence is diminished, this influence is once again widespread (Naiman et al. 1988). Studying the contemporary effects of beavers on soils provides a perspective of what their prehistorical influence might have been.

Beavers influence soils and soil-forming factors by dam building, excavation, and foraging. Beavers exhibit these behaviors to a greater or lesser extent in different parts of the country, depending on the available habitat. Some beaver colonies do not construct dams, utilizing existing lakes and rivers instead of creating new water bodies. Some beaver colonies excavate bank dens, whereas others build lodges from woody material. Some beaver colonies cut down large trees, whereas others consume primarily herbaceous vegetation. Despite this range of behavior, beavers are considered a keystone species, affecting landscape structure and dynamics far beyond their immediate requirements for food and space (Naiman et al. 1986).

This chapter is based on published literature and my own research at Voyageurs National Park in northern Minnesota, where beavers and their effects are abundant. Moisture gradients superimposed by beaver dams on different parent materials (glaciolacustrine silty-clays, glaciofluvial sands) have created two distinct hydrosequences having different soil taxa at Voyageurs National Park. The range of soil characteristics that occur in this diverse landscape provides a good basis for examining beaver–soil interactions. However, relatively few soil scientists have conducted research on soil alteration by beavers, so this chapter also draws on research from the disciplines of geography, limnology, and landscape ecology.

BEAVER/LANDSCAPE INTERACTIONS

Landscapes vary in their susceptibility to beaver activities. Beavers are semiaquatic mammals that use water bodies for travel and protection from predators. Glaciated landscapes that have many lakes and streams are particularly conducive to beaver habitation. There can be as many as five colonies per km², each containing an average of six beavers (Aleksiuk 1968, Voigt et al. 1976, Bergerud and Miller 1977, Rebertus 1986, Smith and Peterson 1991, Lizarralde 1993).

Beavers in glaciated landscapes often occupy a variety of water bodies. Within a 100-km² area of terminal moraine in Minnesota, 58% of the 221 active beaver colonies occupied lakes, 36% occupied bogs, and only 6% occupied rivers (Rebertus 1986). Beavers commonly dam the outlets of lakes (Morgan 1868) and meander scroll depressions (Gill 1972), raising the local water levels.

Streams dammed by beaver are small, generally less than 5th order (Naiman et al. 1986, Johnston and Naiman 1990a). Beaver ponds may be built on local drainage divides, such that two or more dams drain to different watersheds (Figure 19.1). A study of 56 beaver ponds in Mississippi showed that 71% were located on intermittent streams or on seepage areas, whereas only 29% were constructed on perennial streams (Arner et al. 1970). The density of beaver colonies is positively correlated with the degree of stream channel bifurcation (Boyce 1980), and negatively correlated with stream gradient (Slough and Sadleir 1977, Allen 1983, Howard and Larson 1985, Beier and Barrett 1987, McComb et al. 1990). Beaver colonies rarely occur on streams with a gradient > 12%, even in mountainous areas (Rutherford 1955, Retzer et al. 1956, Beier and Barrett 1987). Beavers rely on the sound of running water caused by landscape gradients to trigger dam-building behavior (Wilsson 1971, Hartman 1975).

In forested landscapes, beaver ponds create spot disturbance patches in the forest matrix (Remillard et al. 1987). Where before there was only a narrow stream corridor, a beaver dam creates a patch of water with very different properties than the stream and forest which it replaced (Figure 19.2). After beavers abandon a site, the pond drains and the exposed moist soils revegetate to sedge

Figure 19.1 A beaver pond in Voyageurs National Park, Minnesota, located on a drainage divide. Dams on opposite sides of the pond drain to different watersheds.

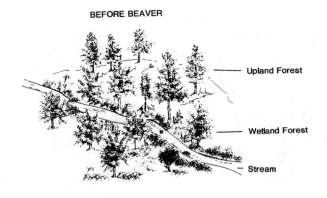

BEFORE BEAVER

Upland Forest

Wetland Forest

Stream

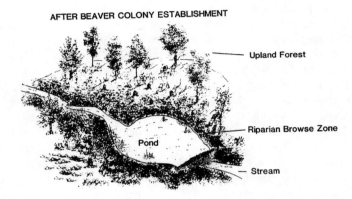

AFTER BEAVER COLONY ESTABLISHMENT

Upland Forest

Riparian Browse Zone

Pond

Stream

Figure 19.2 Landscape patches created by beaver impoundments and foraging, (a) before and (b) after beaver colony establishment. (From Johnston, C.A. and R.J. Naiman. 1987. Boundary dynamics at the aquatic-terrestrial interface: The influence of beaver and geomorphology. *Landscape Ecology.* 1, 47–57. With permission.)

and grass meadows. These "beaver meadows" may persist for a decade or more and are often reflooded when beavers recolonize the site. This cycle of beaver pond flooding, drainage, revegetation, and reflooding is typical in landscapes densely populated by beaver (Remillard et al. 1987, Pastor et al. 1993, Snodgrass 1997).

Studies of beaver pond building at Voyageurs National Park demonstrate that beaver rapidly create new ponds as they spread into new habitats (Johnston and Naiman 1990b). The beaver population of Voyageurs National Park was low during the first half of this century, and beaver ponds covered only about 1% of the landscape as of 1940. With increasing population, the area of beaver impoundments increased. About 100 beaver colonies had impounded 10% of the landscape by 1958, and ~300 colonies had impounded 13% of the landscape by 1986. Between 1940 and 1961, beavers created new ponds at the rate of 0.42% of the landscape per year (i.e., each 10,000-ha area of landscape would have 42 new ha of beaver ponds each year). By comparison, the urbanization of land around the Milwaukee metropolitan area was 0.64% per year (Sharpe et al. 1981), and the abandonment of cropland in a county in Georgia was 0.8% per year (Turner 1987). Therefore, beavers are capable of altering landscapes at rates comparable to those for human activities.

The construction of beaver dams alters streamflow both locally and downstream. Locally, stream impoundment decreases stream velocity and increases surface water storage (Neff 1957, Parker et al. 1985). A single beaver dam near the head of Coal Creek in Wyoming impounded 304,170 cubic meters of water (Grasse 1951). The storage of water within beaver ponds reduces and delays peak

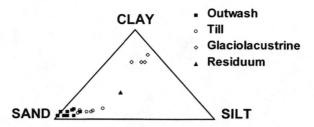

Figure 19.3   Textural triangle showing particle size distribution of parent materials from soils in beaver meadows at Voyageurs National Park.

discharge during summer storm events, so that the outflow from a beaver impounded basin is less flashy than that from a unimpounded basin (Woo and Waddington 1990). However, beaver ponds provide minimal retention during large runoff events such as snow melt (Burns and McDonnell 1998). Beaver impoundments were reported to recharge local groundwater tables in Mississippi (Arner et al. 1970), but the influence of beaver dams on groundwater flow was relatively insignificant in level terrain of the James Bay lowland in northern Ontario due to the lack of hydraulic gradient and low hydraulic conductivity (Woo and Waddington 1990).

Much of the land flooded by beavers was previously wetland (Rebertus 1986, Johnston and Naiman 1987, Johnston 1994, Johnston et al. 1995). When beavers flood uplands, they create new ponds and wetlands; when beavers flood wetlands, they create larger ponds and wetter wetlands. Interpretation of beaver impoundments in 1940 and 1988 aerial photos for Voyageurs National Park showed that 1504 ha of upland were impounded over that time period, creating 668 ha of ponds (i.e., open water) and 836 ha of vegetated wetlands (Johnston 1994). About 2200 ha of wetlands were impounded over the same time period; one third of these flooded wetlands were converted to open water ponds, while two thirds became wetter wetlands. Therefore, beaver impoundments increased the area of open water pond by 1429 ha but had little net effect on the cumulative area of wetland, which was about 2200 ha in both 1940 and 1988 (Johnston 1994). Of the beaver ponds visible on 21 unpublished soil survey field sheets for Voyageurs National Park, all occurred on hydric soils, of which 88% were organic soils and 12% were poorly to very poorly drained mineral soils (Johnston, C.A., unpublished data, 1990).

The glaciated landscapes commonly impounded by beaver contain a diversity of parent materials (i.e., the unconsolidated mineral or organic matter from which soils develop) (Wilde et al. 1950, Johnston et al. 1995). Parent materials collected from mineral soils in beaver meadows at Voyageurs National Park have widely ranging particle size distributions: glacial outwash deposits are sands to loamy sands, till deposits are loamy sands to sandy loams, glaciolacustrine deposits are clays, and residuum from schistose bedrock is a clay loam (Figure 19.3).

ALTERATION OF SOIL FORMING PROCESSES BY BEAVERS

Sedimentation and Erosion

By reducing water velocity, beaver dams reduce the capacity of streams to transport suspended particles, thereby increasing sedimentation rates (Parker et al. 1985, Maret et al. 1987). Knudsen (1962) reported silt accumulations of 1 to 5 cm covering the bottom of 15 beaver ponds in northwestern Wisconsin. Some observers have even attributed the development of special fluvial landforms to valley aggradation resulting from the long-term accumulation of sediment behind beaver dams (Warren 1926, Ruedemann and Schoonmaker 1938, Ives 1942, Rutten 1967, Butler 1991). Unfortunately, there have been few studies actually quantifying the volume of sediment retained or the rate of sediment accumulation in beaver ponds (Naiman et al. 1986, Devito and Dillon 1993, Butler and Malanson 1995), and none of them involved a soil scientist. These studies were done by probing or coring subaqueous beaver pond sediments, determining the mass of accumulated material, and attributing some portion of that mass to retention that has occurred since dam construction. Naiman et al. (1986) attributed all sediments overlying bedrock to the effects of beaver, with volumes ranging from 35 to 6500 m³. Using similar assumptions, Butler and Malanson (1995) reported pond sediment volumes of 11 to 5084 m³ in ponds with sediment depths of 22 to 86 cm. Devito and Dillon (1993) pushed Plexiglas coring tubes approximately 20 cm into beaver pond sediments, and reported sediment accumulations of only 7 to 12 cm, presumably based on stratigraphic evidence. The latter two studies used pond inception dates interpreted from aerial photography to calculate annual accumulation rates. Not surprisingly, annual accumulation rates were reported to be orders of magnitude higher by Butler and Malanson (2 to 28 cm/yr) than by Devito and Dillon (~0.4 cm/yr).

The problem with these methods is the uncertain origin and age of the "sediments." All three of these studies were conducted in glaciated stream channels, so the material underlying any of these ponds could be of glacial, fluvial, or beaver pond origin. Devito and Dillon (1993) apparently tried to distinguish sediment origin, but the other two studies did not, and may have overestimated sediment retention due to beaver activity.

John L. Retzer, a soil scientist with the Rocky Mountain Forest and Range Experiment Station, disputed the allegations by Warren (1926) and Ives (1942) that beaver activity had resulted in unique landforms, saying, "It is believed that glacial deposits played the major role in the origin of those broad valley sections" (Retzer et al. 1956). Given the lack of experimental evidence or reliable dating of beaver pond sediments, the debate is still open about the beaver's role in sedimentation and landform development.

Wave action along the edge of beaver ponds can also deposit mineral matter on the pre-impoundment soil surface. Because this material is moved downslope largely under the influence of gravity, it is less well sorted than sediments carried by flowing water. In the soil described in Table 19.1, 6 cm of sandy colluvium [C] cover the pre-impoundment soil surface [Oa(b)], which overlies glaciolacustrine clay [2Bg(b)]. This soil adjoined a steeply sloping (>20%) upland with sandy till soils. Johnston et al. (1995) also described a Udipsamment soil with a buried pre-impoundment soil surface, which also occurred along the edge of a beaver pond in an area of glaciofluvial parent material. Given that waves have their greatest effect at pond perimeters, the spatial extent of wave-washed soils associated with beaver ponds is probably small.

The catastrophic failure of beaver dams causes flooding and erosion. Extreme precipitation events can cause beaver dams to break suddenly, such that the large volume of water stored behind the dam rushes downstream, scouring out the stream channel and redepositing sediment at some distance downstream (Butler 1989). The spatial extent of such stream channel erosion is usually small, primarily affecting fish habitat (Rupp 1955, Hale 1966, Kondolf et al. 1991, Stock and Schlosser 1991). However, the release of water from a burst beaver dam in eastern British Columbia started a catastrophic landslide that inundated the tracks of the Canadian Pacific Railroad (Dugmore 1914), and a beaver dam outburst flood in Oglethorpe County, Georgia, killed four people (Butler 1989). Based on elevation surveys done before and after a beaver dam failure, Kondolf et al. (1991) reported that the entire streambed of Tinemaha Creek was incised about 0.6 m. The gravels that had been deposited above the beaver dam (median diameter = 7.8 mm) were swept away, leaving a bed of cobbles and boulders.

Pedoturbation

Several beaver activities cause soil mixing, or pedoturbation. Beavers excavate soils for dam and lodge building material, particularly in areas where woody materials are scarce. Although the

Table 19.1   Morphologic and Physical Characteristics for a Pedon with Buried Organic Horizon in a Drained Beaver Pond at Voyageurs National Park

Horizon	Depth (cm)	Matrix Color	Redox Concentration Color	USDA Texture†	Particle-Size Distribution (%)			Structure			Bulk Density (Mg/m³)	pH
					Sand	Silt	Clay	Grade	Size	Shape‡		
Oe	21–5	10 YR 2/2	—	hemic	—	—	—	weak	very fine	gr	0.56	6.6
Oa	5–0	10 YR 2/1	—	sapric	—	—	—	weak	very fine	gr	0.54	6.8
C	0–6	2.5 Y 4/2	—	gsl	76	16	8	single grain	—	—	1.50	7.4
Oa(b)	6–11	N 2/0	—	sapric	—	—	—	moderate	very fine	gr	0.60	6.8
A(b)	11–37	5 Y 3/1	10 YR 3/6	sl	56	30	14	moderate	medium	pl	1.59	6.8
2Bg(b)	37+	5 GY 5/1	—	clay	14	22	64	moderate	fine	sbk	1.39	7.1

† gsl = gravelly sandy loam, sl = sandy loam.
‡ gr = granular, pl = platy, sbk = subangular blocky.

downstream face of a beaver dam is usually covered by sticks, the upstream face is plastered with mud (Warren 1905, Pullen 1975). The volume of earth, sticks, and stones in a 79-m beaver dam was nearly 200 cubic meters (Morgan 1868).

Bank-dwelling beavers excavate burrows that extend several meters into the banks of riparian zones (Gill 1972, Barnes and Dibble 1988, Dieter and McCabe 1989, Butler and Malanson 1994). Morgan (1868) reported that beaver burrows were common along the banks of the Missouri River from the mountains to the mouth of the Big Sioux, and recounted a report that streams in the Cascade Mountains of Oregon and California were lined with beaver burrows. Although the quantity of material excavated is relatively small, a much larger area of soil can be disturbed when such bank dens collapse or are eroded by streams.

Beavers also excavate canals that radiate out from ponds or streams into riparian foraging areas to facilitate transport of woody plant materials (Berry 1923, Townsend 1953). Beavers use their paws to carry excavated soil, removing it from the canal by throwing it out on either side or by carrying it out into the pond. Morgan (1868) mapped several beaver canals constructed in northern Michigan, including one that was 167 m long with a cross-sectional area of about 1m × 1m.

Development of Redoximorphic Features

Soil inundation results in the development of redoximorphic features, such as a reduced matrix and redox concentrations (i.e., zones of apparent accumulation of Fe–Mn oxides; Soil Survey Staff 1994). At Voyageurs National Park, beaver impoundment changed the matrix colors in the B horizons of a glaciofluvial hydrosequence from dark brown (7.5 YR 3/4) in an upland soil to dark grayish-brown (2.5 Y 4/2) and olive gray (5 Y 5/2) in an adjacent soil that had been submerged for at least 14 years and then drained (Table 19.2). Redox concentrations appeared in the formerly ponded soil as strong brown (7.5 YR 4/6) masses in the B and C horizons, but were absent in the upland soil.

In a glaciolacustrine hydrosequence at Voyageurs National Park, redox concentrations consisted of prominent yellowish-brown and dark yellowish-brown (10YR 4/6-5/6) pore linings in both a beaver meadow and an adjacent upland soil (Johnston et al. 1995). The depth to redox concentrations was 30 cm in the upland soil (Eutroboralf; Figure 19.4), but redox concentrations occurred throughout the profile in the beaver meadow soil (Ochraqualf; Figure 19.4).

Eluviation/Illuviation

The related processes of eluviation and illuviation, in which soil material is removed in suspension from one horizon and deposited in another, occurred in soils of the glaciolacustrine hydrosequence at Voyageurs National Park. All three mineral soils in the hydrosequence (Eutroboralf, Ochraqualf, Argiaquoll) exhibited argillic horizons (Bt, Btg), in which clay was translocated downward (Johnston et al. 1995). The thickness of the argillic horizons decreased with increasing wetness (Figure 19.4).

There were also pronounced subsurface peaks (≥ 800 g m^{-3}) of oxalate-extractable Fe in the mineral soils of the glaciolacustrine hydrosequence (Johnston et al. 1995). Depth to these peak concentrations decreased with increasing wetness, occurring in the upper B horizons of the Alfisols and the A2 horizon of the Argiaquoll (Figure 19.5). The source of this Fe was probably the overlying E horizons, which were depleted in both clay and Fe; Fe migrates with clay during interhorizon translocation (Richardson and Hole 1979). There was no comparable illuviation in soils developed in glaciofluvial parent materials, which contained little clay or oxalate-extractable Fe.

The thickness of argillic horizons and the depth to peaks of oxalate-extractable Fe both indicate that downward leaching is greatest in unimpounded soils along the edge of the beaver meadow, and least in wetter soils within the beaver meadow.

Table 19.2   Morphologic and Physical Characteristics for Pedons in a Glaciofluvial Hydrosequence at Voyageurs National Park

Horizon	Depth (cm)	Matrix Color	Redox Concentration Color	USDA Texture†	Particle-Size Distribution (%)			Structure			Bulk Density (Mg/m³)	pH
					Sand	Silt	Clay	Grade	Size	Shape‡		
Upland												
Oe	6–0	7.5 YR 3/2	—	hemic	—	—	—	—	—	—	0.19	5.4
E	0–10	10 YR 5/2	—	lfs	ND	ND	ND	mod	fine	pl	1.13	5.2
Bw1	10–36	7.5 YR 3/4	—	fsl	76	16	8	weak	fine	sbk	1.19	5.1
Bw2	36–46	7.5 YR 4/4	—	fsl	64	24	12	weak	fine	sbk	1.19	5.9
C	46+	10 YR 4/6	—	fsl	64	24	12	single grain	—	—	1.23	6.1
Beaver Meadow												
Oe	10–0	5 YR 3/3	—	hemic	—	—	—	weak	very fine	gr	0.28	4.9
Bg1	0–12	10 YR 4/2	—	sl	64	28	8	weak	fine	gr	1.60	4.9
Bg2	12–27	2.5 Y 4/2	10 YR 4/6	sl	64	28	8	weak	fine	pl	1.52	5.2
Bg3	27–41	5 Y 5/2	7.5 YR 4/6	ls	88	4	8	weak	fine	sbk	1.62	5.3
C1	41–54	2.5 Y 7/2	7.5 YR 4/6	sl	64	24	12	weak	fine	sbk	1.56	5.1
C2	54–72	2.5 Y 5/4	—	cos	ND	ND	ND	single grain	—	—	ND	ND

† cos = coarse sand, fsl = fine sandy loam, lfs = loamy fine sand, ls = loamy sand, sl = sandy loam.
‡ gr = granular, pl = platy, sbk = subangular blocky. ND = not determined.

Data from Johnston, C.A., G. Pinay, C. Arens, and R.J. Naiman. 1995. Influence of soil properties on the biogeochemistry of a beaver meadow hydrosequence. Soil Sci. Am. J. 59:1789–1799; Johnston, C.A. Unpublished data.

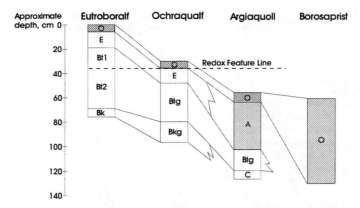

Figure 19.4 Beaver meadow hydrosequence developed in glaciolacustrine parent materials. Redoximorphic feature line denotes the uppermost location of redox concentrations. Other lines connect taxonomic horizons. (From Johnston, C.A., G. Pinay, C. Arens, and R.J. Naiman. 1995. Influence of soil properties on the biogeochemistry of a beaver meadow hydrosequence. *Soil Sci. Soc. Am. J.* 59:1789–1799. With permission.)

A comprehensive solute budget of a beaver pond in New York state showed that the pond retained Al and released Fe^{2+} (Cirmo and Driscoll 1993), which implies chemical redistribution in the sediments. The pond retained aluminum at the rate of 660 moles ha^{-1} yr^{-1} during the summer (retaining 94% of inputs from the inlet stream), and 2100 moles Al ha^{-1} yr^{-1} during October through June (retaining 25% of inputs); retention was attributed to hydrolysis and precipitation of Al from the inlet stream. Outputs of Fe^{2+} exceeded inputs by 1300 equivalents ha^{-1} yr^{-1} in summer and 670 equivalents ha^{-1} yr^{-1} during October through June; this loss was attributed to microbial Fe^{3+} reduction within the anaerobic pond sediments. In the summer, hydrous oxides of Fe and organically bound Fe were much higher in sediments of the stream channel immediately below the pond than in stream sediments above the pond (Table 19.3), and brick-red Fe oxyhydroxide precipitates were noted on the streambed below the pond (Smith et al. 1991). With the exception of exchangeable Al, concentrations of Al and Fe compounds in stream sediments were much higher in September than in June (Table 19.3).

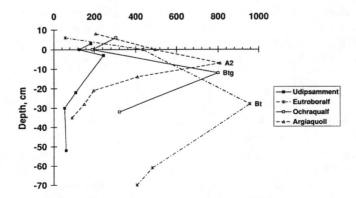

Figure 19.5 Oxylate-extractable Fe in mineral soils of the beaver meadow hydrosequence developed in glaciolacustrine parent materials. (From Johnston, C.A., G. Pinay, C. Arens, and R.J. Naiman. 1995. Influence of soil properties on the biogeochemistry of a beaver meadow hydrosequence. *Soil Sci. Soc. Am. J.* 59:1789–1799. With permission.)

Table 19.3 **Iron and Aluminum Chemistry of Stream Bottom Sediments Collected Upstream and Downstream of a Beaver Pond**

	Upstream of Pond (μmoles/g)	Downstream of Pond (μmoles/g)
June		
Organically bound Al	112	59
Organically bound Fe	50	57
Hydrous oxides of Al	73	46
Hydrous oxides of Fe	67	39
Exchangeable Al	2.3	2.7
September		
Organically bound Al	237	263
Organically bound Fe	126	1066
Hydrous oxides of Al	237	723
Hydrous oxides of Fe	217	786
Exchangeable Al	1.8	2.2

Adapted from Smith, E.E., C.T. Driscoll, B.J. Wyskowski, C.M. Brooks, and C.C. Cosentini. 1991. Modification of stream ecosystem structure and function by beaver (*Castor canadensis*) in the Adirondack Mountains, New York. *Canadian Journal of Zoology* 69:55–61.

Carbonate Accumulation/Dissolution

The Ca content of soils associated with beaver meadows at Voyageurs National Park varied with both parent material and position in the hydrosequence (Johnston et al. 1995). Calcium concentrations were much higher in soil horizons derived from glaciolacustrine parent materials (27.6 to 98.0 mg/kg) than in soil horizons derived from glaciofluvial parent materials (0.6 to 5.1 mg/kg). Within the glaciolacustrine hydrosequence, carbonate accumulation was evident in the upper two soils in the form of visible masses of free carbonates (Bk and Bkg horizons; Figure 19.4). These deposits were interpreted to be secondary carbonates because they occurred as masses, rather than as strata that would be typical of glaciolacustrine parent materials. The accumulation of soil carbonates has been similarly observed at the upland edge of pothole wetlands where there is groundwater recharge and lateral flow toward the wetter sites, flow patterns typical of ponds that have a frequent drawdown from a leached upland (Arndt and Richardson 1988, 1989, Richardson et al. 1994).

Concurrent studies of soil pore water chemistry showed a significant downslope increase in Ca^{+2} and Mg^{+2} (Johnston et al. 1995). Carbonate dissolution occurs when water with low partial pressure of CO_2 (pco_2) is added to soil (Novozamsky and Beek 1976), which could explain the observed increase in Ca^{+2} and Mg^{+2} concentrations with increasing moisture.

Organic Matter Accumulation

Beavers and their ponds have both direct and indirect effects on organic matter accumulation and distribution with the soil profile (Table 19.4). Hodkinson (1975a,b) computed an energy flux balance for an open water beaver pond by measuring various organic matter inputs and outputs. Sources of input to the energy balance were waterborne organic particles from the inlet stream (70%), algal photosynthesis within the pond (22%), and windborne litter from the land surrounding the pond (8%). Of these inputs, 56% were retained within the pond, and 44% passed out of the pond as waterborne organic particles or gaseous endproducts of respiration.

The construction of beaver dams redistributes woody litter, concentrating about 7 Mg of wood (the equivalent aboveground biomass of 24 ten-inch-diameter aspen trees; Pastor and Bockheim, 1981) within the relatively small area of the dam site (Johnston and Naiman 1990c). Rains (1987)

Table 19.4   Effects of Beaver Activities on Organic Matter Accumulation in Soils (Paludification defined by Heinselman 1963; other pedogenic processes defined by Buol et al. 1973)

Beaver Activity	Effect	Pedogenic Process
Construction	Local accumulation of woody plant litter	Littering
Foraging	Accumulation of large woody litter	Littering
Ponding	Accumulation of large woody litter due to tree mortality	Littering
	Submergence of O layer, retention of waterborne and windborne organic matter inputs	Decomposition
	Development of sedge tussocks and floating sedge mats	Paludization
	Bog formation?	Paludification

reported evidence of a mid-Holocene beaver dam buried in alluvium near Edmonton, Alberta, but soil profiles containing relict beaver dams are probably rare due to their small area.

Beavers kill trees by felling or flooding them, greatly increasing the woody biomass deposited on riparian soils and pond bottoms (Figure 19.6). The trees felled can constitute as much as 42% of the total forest biomass within the browse zone around a beaver pond, and the amount cut annually by a beaver colony (8.4 Mg ha^{-1} yr^{-1}: Johnston and Naiman 1990c) is 73% of the net primary productivity of a northern Wisconsin aspen forest (Pastor and Bockheim 1981). The area affected by beaver foraging and flooding may constitute as much as one fourth of the landscape, concentrated in valley bottoms (Naiman et al. 1988), so this influence is much more likely to be detected in soil profiles. There are several published reports of beaver-gnawed wood buried in peat profiles (Sjörs 1959, Kaye 1962, Ferguson and Osborn 1981). Kaye (1962) examined five peat profiles exposed by marine erosion in Massachusetts and Rhode Island, and found that all were rich in beaver-gnawed wood at or near the base of the postglacial organic sediments. A reconnaissance study of 37 peat

Figure 19.6   Woody litter from trees killed by beaver flooding at Voyageurs National Park.

Figure 19.7 Dead sedge tussocks previously submerged by a beaver pond, and then exposed in 1985 by pond drainage. A silty crust of pond sediments coats the litter. The stick visible in the upper right corner has been cut and debarked by beaver.

profiles in the Hearst and Moosonee Districts of northern Ontario revealed that 21 contained a basal layer of beaver-chewed wood (Adams, P.W., unpublished manuscript, 1990).

The ponding of water submerges the soil surface and associated ground cover, plunging it into an anaerobic environment with low decomposition rates. Updegraff and colleagues (1995) used laboratory incubation techniques to compare peat decomposition rates in aerobic and anaerobic environments, emulating submergence by beaver ponds. When a sedge peat collected from the surface of a beaver meadow was incubated at 15°C for 80 weeks, carbon mineralization was significantly higher under aerobic conditions (94.8 mg C/g soil) than under anaerobic conditions (47.0 mg C/g soil), although nitrogen mineralization was comparable between aerobic and anaerobic treatments (24.1 and 22.3 mg N/g soil, respectively). Thus, anaerobic conditions in the bottom of a beaver pond would decrease the rate of carbon mineralization by about half, but would have no effect on the rate of nitrogen mineralization.

The effect of slower subaqueous decomposition is visible when a beaver pond drains, exposing organic matter that had been submerged by inundation. Sedge (*Scirpus cyperinus* and *Carex* spp.) litter and preserved tussocks 75 cm tall were still easily identifiable despite at least 5 years of inundation by a beaver pond at Voyageurs National Park (Figure 19.7). The preserved organic matter in the O layer was still quite thick 2 years after pond drainage, forming a 24-cm histic epipedon (Figure 19.8).

Beaver impoundments indirectly accelerate organic matter accumulation by promoting productive aquatic plants that generate thick litter layers and form peat tussocks. Sedge meadows have moderate primary productivity (Johnston 1988) and relatively slow decomposition rates (Chamie and Richardson 1978, Reader and Stewart 1972), which would optimize organic matter accumulation. At Voyageurs National Park, meadows constituted about 20% of beaver impoundment areas, and their area increased sixfold between 1940 and 1986 with the increase in beaver population (Johnston 1994).

Beaver impoundments also promote the formation of floating peat mats. At Voyageurs National Park, one tenth of beaver impoundment areas had floating mats (Johnston 1994). Although floating peat mats are most commonly associated with bogs (e.g., Lindeman, 1941), they also occur in wetlands subject to fluctuating water levels (Lieffers 1984, Hogg and Wein 1988). This growth

Figure 19.8 Soil profile in area of dead sedge tussocks (Figure 19.7), showing 24-cm histic epipedon that remained 2 years after pond drainage. Top of tape measure is at the interface between the mineral soil and the histic epipedon.

form would promote the accumulation of organic matter because plant productivity in the floating mat will remain stable and high, while decomposition rates will be low due to anaerobic conditions below the mat.

Over millennia, beavers may have played a pivotal role in peatland formation. Researchers may have overlooked this role because of the small amount of material that peat corers excavate (Kaye 1962). Paludification, the process of bog expansion caused by gradual raising of the water table as drainage is impeded (Heinselman 1963, 1970), may have been catalyzed by beaver ponds. Evidence for this hypothesis are multiple reports of beaver-gnawed wood at the contact between mineral substrate and overlying peatland, as well as reports of beaver-gnawed wood alternating with peat and charcoal within peat profiles (Adams, P.W., unpublished manuscript, 1990). Beaver ponds could have served as a nucleus for peat formation, providing sites for the establishment of peat-forming sedges and bryophytes. The coalescence of adjacent beaver impoundments over time as beavers construct new, higher dams that flood old ones (Woo and Waddington 1990) could promote paludification.

Beaver meadows may promote melanization, the darkening of mineral soil by admixture of organic matter. Melanization resulted in the mollic epipedon of the Argiaquoll soil at Voyageurs National Park, consisting of a 39-cm A horizon overlaid by an 8-cm Oa horizon (Figure 19.4). Inspection of historical aerial photographs indicates that the area was a wooded wetland before beaver flooded it in the 1950s, so the melanization may not have been caused by beaver activity, but rather by the accumulation of organic matter in the wetland soil. However, beaver undoubtedly colonized this landscape prior to the arrival of the Europeans who extirpated them, so there may have been historical interaction between wetland soil pedogenesis and beaver activity. Deep incorporation of organic matter is usually associated with deep-rooted herbaceous species, such as the bluejoint grass (*Calamagrostis canadensis*) that pervades drier beaver meadows, so vegetation alteration by beaver may have indirectly promoted melanization.

Beaver meadow soils derived from glaciofluvial parent materials did not exhibit the same degree of organic matter incorporation as did glaciolacustrine soils. The litter layer (Oe horizon) of beaver meadow soils derived from glaciofluvial parent materials was slightly thicker than the litter layer of the adjacent upland soil (Table 19.2), but was not nearly as thick as histic epipedon observed in the glaciolacustrine hydrosequence (Figure 19.8). Organic soils did not occur in the glaciofluvial hydrosequence, and deep organic matter incorporation occurred only when O and A horizons were buried by wave-washed mineral matter (Table 19.1).

WHICH CAME FIRST: THE BEAVER OR THE WETLAND?

Given that beavers have profoundly influenced landscapes for millennia, it is probable that they have also profoundly influenced soil development, but it is difficult to determine with certainty the specific locations and magnitude of influence. Aerial photographs can be reliably used to establish the location and area of beaver ponds over the 60-year period of record (e.g., Rebertus 1986, Remillard et al. 1987, Johnston and Naiman 1990a), and beaver-gnawed wood buried in soil profiles can be radiocarbon dated to establish historical beaver presence (e.g., Kaye 1962, Rains 1987), but either of these techniques must be combined with field studies to establish the effects of beaver activities on soils.

Even when a connection is established between beaver presence and soil characteristics, it is difficult to prove whether the soil characteristics were *caused* by beaver activity, or merely associated with a habitat preferred by beavers. Did beavers create special fluvial landforms by their dam building, or did they merely occupy these fluvial landforms because they were a suitable habitat? Did beavers initiate peatland formation, or did they merely occupy peatlands that were forming in response to other geogenic and pedogenic processes?

At Voyageurs National Park, a relatively undisturbed landscape where beavers and their natural predators still coexist, there are few stream segments or wetlands that are *not* affected by beaver. Assuming that this was the norm elsewhere in the country before Europeans settled here, the effects of beavers on soils were probably pervasive. Beavers and wetlands have co-evolved, and it is fruitless to try to separate the two. With the reintroduction of beaver populations throughout the country, their effects on the hydrology, sedimentation/erosion, pedoturbation, organic matter accumulation, redoximorphic features, eluviation/illuviation, and carbonate accumulation/dissolution of wetland soils will once again be established.

ACKNOWLEDGMENTS

Financial support from the National Science Foundation (DEB-9615326) is gratefully acknowledged. This is Contribution Number 238 of the Center for Water and the Environment.

REFERENCES

Aleksiuk, M. 1968. Scent-mound communication, territoriality, and population regulation in beaver (*Castor canadensis Kuhl*). *J. Mammol.* 49:759–762.

Allen, A.W. 1983. *Habitat Suitability Index Models: Beaver.* FWS/OBS-82/10.30 Revised. U.S. Department of the Interior, Fish and Wildlife Service, Fort Collins, CO.

Arndt, J.L. and J.L. Richardson. 1988. Hydrology, salinity, and hydric soil development in a North Dakota prairie-pothole wetland system. *Wetlands* 8:93–108.

Arndt, J.L. and J.L. Richardson. 1989. Geochemistry of hydric soil salinity in a recharge-throughflow-discharge prairie-pothole wetland system. *Soil. Sci. Soc. Am. J.* 53:848–855.

Arner, D.H., J. Baker, D. Wesley, and B. Herring. 1970. An inventory and study of beaver impounded water in Mississippi. pp. 110–127. in *Proceedings, 23rd Annual Conference, Southeastern Association of Game and Fish Commissioners,* Mobile, AL. Southeastern Association of Game and Fish Commissioners, Columbia, SC.

Barnes, W.J. and E. Dibble. 1988. The effects of beaver in riverbank forest succession. *Can. J. Bot.* 66:40–44.

Beier, P. and R.H. Barrett. 1987. Beaver habitat use and impact in Truckee River Basin, California. *J. Wildl. Manage.* 51:794–799.

Bergerud, A.T. and D.R. Miller. 1977. Population dynamics of Newfoundland beaver. *Can. J. Zool.* 55:1480–1492.

Berry, S.S. 1923. Observations on a Montana beaver canal. *J. Mammol.* 4:92–103.

Boyce, M.S. 1980. Habitat ecology of an unexploited population of beavers in interior Alaska. *Worldwide Furbearer Conference Proceedings,* Frostburg, MD. J.A. Chapman and D. Pursley (Eds.) pp. 155–186.

Buol, S.W., F.D. Hole, and R.J. McCracken. 1973. *Soil Genesis and Classification.* Iowa State University Press, Ames.

Burns, D.A. and J.J. McDonnell. 1998. Effects of a beaver pond on runoff processes: comparison of two headwater catchments. *J. Hydrol.* 205:248–264.

Butler, D.R. 1989. The failure of beaver dams and resulting outburst flooding: a geomorphic hazard of the southeastern Piedmont. *The Geographical Bulletin* 31:29–38.

Butler, D.R. 1991. Beavers as agents of biogeomorphic change: a review and suggestions for teaching exercises. *J. Geogr.* 90:210–217.

Butler, D.R. and G.P. Malanson. 1994. Beaver landforms. *The Canadian Geographer* 38:76–79.

Butler D.R. and G.P. Malanson. 1995. Sedimentation rates and patterns in beaver ponds in a mountain environment. *Geomorphology* 13:255–269.

Cirmo, C.P. and C.T. Driscoll. 1993. Beaver pond biogeochemistry: acid neutralizing capacity generation in a headwater wetland. *Wetlands* 13:277–292.

Chamie, J.P.M. and C.J. Richardson. 1978. Decomposition in northern wetlands. pp. 115–130. *In* R.E. Good, D.R. Whigham, and R.L. Simpson (Eds.) *Freshwater Wetlands: Ecological Processes and Management Potential.* Academic Press, New York.

Devito, K.J. and P.J. Dillon. 1978. Importance of runoff and winter anoxia to the P and N dynamics of a beaver pond. *Can. J. Fish. Aquat. Sci.* 50:2222–2234.

Dieter, C.D. and T.R. McCabe. 1989. Factors influencing beaver lodge-site selection on a prairie river. *American Midland Naturalist* 122:408–411.

Dugmore, A.R. 1914. *The Romance of the Beaver.* J.B. Lippincott Co., Philadelphia, PA.

Duncan, S.L. 1984. Leaving it to beaver. *Environment* 26(3):41–45.

Erickson, B.R. 1962. A description of *Castoroides ohioensis* from Minnesota. *Minnesota Academy of Science Proceedings* 30:7–13.

Ferguson, A. and G. Osborn. 1981. Minimum age of deglaciation of Upper Elk Valley British Columbia. *Can. J. Earth Sci.* 18:1635–1636.

Gill, D. 1972. The evolution of a discrete beaver habitat in the Mackenzie River Delta, Northwest Territories. *The Canadian Field-Naturalist* 86:233–239.

Grasse, J.E. 1951. Beaver ecology and management in the Rockies. *J. Forestry* 49:3–6.

Hale, J.G. 1966. Influence of beaver on some trout streams along the Minnesota North Shore of Lake Superior. pp. 5–29. Minnesota Fisheries Investigations, No. 4. Department of Conservation, Division of Game and Fish, St. Paul, MN.

Hartman, A.M. 1975. Analysis of conditions leading to the regulation of water flow by a beaver. *Psychol. Rec.* 25:427–431.

Heinselman, M.L. 1963. Forest sites, bog processes, and peatland types in the Glacial Lake Agassiz Region, Minnesota. *Ecological Monographs* 33:327–374.

Heinselman, M.L. 1970. Landscape evolution, peatland types, and the environment in the Lake Agassiz Peatlands Natural Area, Minnesota. *Ecological Monographs* 40:236–261.

Hodkinson, I.D. 1975a. Energy flow and organic matter decomposition in an abandoned beaver pond ecosystem. *Oecologia* 21:131–139.

Hodkinson, I.D. 1975b. Dry weight loss and chemical changes in vascular plant litter of terrestrial origin in a beaver pond ecosystem. *J. Ecol.* 63:131–142.

Hogg, E.H. and R.W. Wein. 1988. The contribution of *Typha* components to floating mat buoyancy. *Ecology* 69:1025–1031.

Howard, R.J. and J.S. Larson. 1985. A stream habitat classification system for beaver. *J. Wildl. Manage.* 49:19–25.

Ives, R.L. 1942. The beaver-meadow complex. *J. Geomorph.* 5:191–203.

Jenkins, S.H. and P.E. Busher. 1979. Castor canadensis. *Mammalian Species* 120:1–8.

Johnston, C.A. 1988. *Productivity of Wet Soils: Biomass of Cultivated and Natural Vegetation.* Oak Ridge National Laboratory Report ORNL/Sub/84-18435/1, Oak Ridge, TN.

Johnston, C.A. 1994. Ecological engineering of wetlands by beavers. pp. 379–384. *In* Mitsch, W.J. (Ed.) *Global Wetlands: Old World and New.* Elsevier Science, Amsterdam.

Johnston, C.A. and R.J. Naiman. 1987. Boundary dynamics at the aquatic-terrestrial interface: The influence of beaver and geomorphology. *Landscape Ecology* 1:4–57.

Johnston, C.A. and R.J. Naiman. 1990a. The use of a geographic information system to analyze long-term landscape alteration by beaver. *Landscape Ecology* 4:5–19.

Johnston, C.A. and R.J. Naiman. 1990b. Aquatic patch creation in relation to beaver population trends. *Ecology* 71:1617–1621.

Johnston, C.A. and R.J. Naiman. 1990c. Browse selection by beaver: Effects on riparian forest composition. *Can. J. For. Res.* 20:1036–1043.

Johnston, C.A., J. Pastor, and R.J. Naiman. 1993. Effects of beaver and moose on boreal forest landscapes. pp. 236–254. *In* S.H. Cousins, R. Haines-Young, and D. Green (Eds.) *Landscape Ecology and Geographic Information Systems.* Taylor & Francis, London.

Johnston, C.A., G. Pinay, C. Arens, and R.J. Naiman. 1995. Influence of soil properties on the biogeochemistry of a beaver meadow hydrosequence. *Soil Sci. Soc. Am. J.* 59:1789–1799.

Kaye, C.A. 1962. Early postglacial beavers in Southeastern New England. *Science* 138:906–907.

Knudsen, G.J. 1962. *Relationship of Beaver to Forests, Trout, and Wildlife in Wisconsin.* Wisconsin Conservation Department Tech. Bull. No. 25, Madison, Wisconsin.

Kondolf, G.M., G.F. Cada, M.J. Sale, and T. Felando. 1991. Distribution and stability of potential salmonid spawning gravels in steep boulder-bed streams of the eastern Sierra Nevada. *Transactions, American Fisheries Society* 120:177–186.

Lieffers, V.J. 1984. Emergent plant communities of oxbow lakes in northeastern Alberta: salinity, water-level fluctuation, and succession. *Can. J. Bot.* 62:310–316.

Lindeman, R.L. 1941. The developmental history of Cedar Creek Lake, Minnesota. *American Midland Naturalist* 25:101–112.

Lizarralde, M.S. 1993. Current status of the introduced beaver (*Castor canadensis*) population in Tierra del Fuego, Argentina. *Ambio* 22:351–358.

Maret, T.J., M. Parker, and T.E. Fannin. 1987. The effect of beaver ponds on the nonpoint source water quality of a stream in southwestern Wyoming. *Wat. Res.* 21:263–268.

McComb, W.C., J.C. Sedell, and T.D. Buchholz. 1990. Damsite selection by beavers in an eastern Oregon USA basin. *Great Basin Naturalist* 50:273–282.

Morgan, L.H. 1868. *The American Beaver and His Works.* J.B. Lippincott, Philadelphia, PA, 330 pp.

Naiman, R.J., J.M. Melillo, and J.E. Hobbie. 1986. Ecosystem alteration of boreal forest streams by beaver (*Castor canadensis*). *Ecology* 67:1254–1269.

Naiman, R.J., C.A. Johnston, and J.C. Kelley. 1988. Alteration of North American streams by beaver. *BioScience* 38:753–762.

Neff, D.J. 1957. Ecological effects of beaver habitat abandonment in the Colorado Rockies. *J. Wildl. Manage.* 21:80–84.

Novozamsky, I., and J. Beek. 1976. Common solubility equilibria in soils. pp. 96–125. *In* G.H. Bolt and M.G.M. Bruggenwert (Eds.) *Soil Chemistry, A. Basic Elements.* Elsevier, Amsterdam.

Parker, M., F.J. Wood, B.H. Smith, and R.G. Elder. 1985. Erosional downcutting in lower order riparian ecosystems: have historical changes been caused by removal of beaver? pp. 35–38. *In Riparian Ecosystems and their Management: Reconciling Conflicting Uses.* U.S. Forest Service General Technical Report RM-120.

Pastor, J. and J.G. Bockheim. 1981. Biomass and production of an aspen-mixed hardwood–spodosol ecosystem in northern Wisconsin. *Can. J. For. Res.* 11:132–138.

Pastor, J., J. Bonde, C.A. Johnston, and R.J. Naiman. 1993. A Markovian analysis of the spatially dependent dynamics of beaver ponds. pp. 5–27. *In* Gardner, R.H. (Ed.) *Predicting Spatial Effects in Ecological Systems,* American Mathematical Society, Providence, RI.

Pullen, T.M., Jr. 1975. Observations on construction activities of beaver in west-central Alabama. *J. Alabama Acad. Sci.* 46:14–19.

Rains, B. 1987. Holocene alluvial sediments and a radio-carbon dated relict beaver dam, Whitehead Creek, Edmonton, Alberta. *The Canadian Geographer* 31:272–277.

Reader, R.J. and J.M. Stewart. 1972. The relationship between net primary production and accumulation for a peatland in southeastern Manitoba. *Ecology* 53:1024–1037.

Rebertus, A.J. 1986. Bogs as beaver habitat in north-central Minnesota. *American Midland Naturalist* 116:240–245.

Remillard M.M., G.K. Gruendling, and D.J. Bogucki. 1987. Disturbance by beaver (*Castor canadensis Kuhl*) and increased landscape heterogeneity. pp. 103–122. *In* Turner, M.G. (Ed.) *Landscape Heterogeneity and Disturbance.* Springer-Verlag, New York.

Retzer J.L., H.M. Swope, J.D. Remington, and W.H. Rutherford. 1956. Suitability of physical factors for beaver management in the Rocky Mountains of Colorado. Colorado Dept. Game Fish Tech. Bull. No. 2, 1956.

Richardson, J.L. and F.D. Hole. 1979. Mottling and iron distribution in a Glossoboralf-Haplaquoll hydrosequence in a glacial moraine in northwestern Wisconsin. *Soil Sci. Soc. Am. J.* 43:552–558.

Richardson, J.L., J.L. Arndt, and J. Freeland. 1994. Wetland soils of the prairie potholes. *Adv. Agron.* 52:121–171.

Ruedemann, R. and W.J. Schoonmaker. 1938. Beaver dams as geologic agents. *Science* 88:523–525.

Rupp, R.S. 1955. Beaver-trout relationship in the headwaters of Sunhaze Stream, Maine. *Transactions, American Fisheries Society* 84:75–85.

Rutherford, W.H. 1955. Wildlife and environmental relationships of beavers in Colorado forests. *J. Forestry.* 57:803–806.

Rutten, M.G. 1967. Flat-bottomed glacial valleys, braided rivers and the beaver. *Geologie en Mijnbouw* 46:356–360.

Sharpe, D.M., F.W. Stearns, R.L. Burgess, and W.C. Johnson. 1981. Spatio-temporal patterns of forest ecosystems in man-dominated landscapes of the eastern United States. pp. 109–116. *In* Tjallingii, S.P. and de Veer, A.A. (Eds.) *Perspectives in Landscape Ecology.* Centre for Agricultural Publication and Documentation, Wageningen.

Sjörs, H. 1959. Bogs and fens in the Hudson Bay Lowlands. *Arctic* 12:2–19.

Slough, B.G. and R.M.F.S. Sadleir. 1977. A land capability classification system for beaver (*Castor canadensis*). *Can. J. Zool.* 55:1324–1335.

Smith, D.W. and R.O. Peterson. 1991. Behavior of beaver in lakes with varying water levels in northern Minnesota. *Environ. Management* 15:395–401.

Smith, M.E., C.T. Driscoll, B.J. Wyskowski, C.M. Brooks, and C.C. Cosentini. 1991. Modification of stream ecosystem structure and function by beaver (*Castor canadensis*) in the Adirondack Mountains, New York. *Canadian Journal of Zoology* 69:55–61.

Snodgrass, J.W. 1997. Temporal and spatial dynamics of beaver-created patches as influenced by management practices in a south-eastern North American landscape. *J. Appl. Ecol.* 34:1043–1056.

Soil Survey Staff. 1994. *Keys to Soil Taxonomy,* 6th edition. USDA–SCS, U.S. Govt. Printing Office. Washington, DC.

Stock, J.D. and I.J. Schlosser. 1991. Short-term effects of a catastrophic beaver dam collapse on a stream fish community. *Envir. Biol. Fish.* 31:123–129.

Townsend, J.E. 1953. Beaver ecology in western Montana with special reference to movements. *J. Mammol.* 34:459–479.

Turner, M.G. 1987. Spatial simulation of landscape changes in Georgia: a comparison of 3 transition models. *Landscape Ecol.* 1:29–36.

Updegraff, K., J. Pastor, S.D. Bridgham, and C.A. Johnston. 1995. Environmental and substrate controls over carbon and nitrogen mineralization in northern wetlands. *Ecological Applications* 5:151–163.

Voigt, D.R., G.B. Kolenosky, and D.H. Pimlott. 1976. Changes in summer foods of wolves in central Ontario. *J. Wildl. Man.* 40:663–668.

Warren, E.R. 1905. Some interesting beaver dams in Colorado. *Proceedings of the Washington Academy of Science* 6:429–437.

Warren, E.R. 1926. Notes on the beaver colonies in the Long's Peak Region of Estes Park, Colorado. *Roosevelt Wildlife Annals* 1:192–234.

Wilde, S.A., C.T. Youngberg, and J.H. Hovind. 1950. Changes in composition of ground water, soil fertility, and forest growth produced by the construction and removal of beaver dams. *J. Wildl. Man.* 14:123–128.

Wilsson, L. 1971. Observations and experiments on the ethology of the European beaver (*Castor fiber* L.). *Viltrevy* 8:115–266.

Woo, M-K. and J.M. Waddington. 1990. Effects of beaver dams on subarctic wetland hydrology. *Arctic* 43:223–230.

Index